BRANCHING IN NATURE

Dynamics and Morphogenesis of Branching Structures, from Cell to River Networks

Les Houches School, October 11-15, 1999

Editors

V. FLEURY
J.-F. GOUYET
M. LÉONETTI

Springer

Berlin
Heidelberg
New York
Barcelona
Hong Kong
London
Milan
Paris
Singapore
Tokyo

EDP Sciences

Les Ulis
Paris
Cambridge

Centre de Physique des Houches

Books already published in this series

1 Porous Silicon Science and Technology
J.-C. VIAL and J. DERRIEN, Eds. 1995

2 Nonlinear Excitations in Biomolecules
M. PEYRARD, Ed. 1995

3 Beyond Quasicrystals
F. AXEL and D. GRATIAS, Eds. 1995

4 Quantum Mechanical Simulation Methods for Studying Biological Systems
D. BICOUT and M. FIELD, Eds. 1996

5 New Tools in Turbulence Modelling
O. MÉTAIS and J. FERZIGER, Eds. 1997

6 Catalysis by Metals
A. J. RENOUPREZ and H. JOBIC, Eds. 1997

7 Scale Invariance and Beyond
B. DUBRULLE, F. GRANER and D. SORNETTE, Eds. 1997

8 New Non-Perturbative Methods and Quantization on the Light Cone
P. GRANGÉ, A. NEVEU, H.C. PAULI, S. PINSKY and E. WERNER, Eds. 1998

9 Starbursts Triggers, Nature, and Evolution
B. GUIDERDONI and A. KEMBHAVI, Eds. 1998

10 Dynamical Networks in Physics and Biology
D. BEYSENS and G. FORGACS, Eds. 1998

11 Solid Interstellar Matter: The ISO Revolution
L. d'HENDECOURT, C. JOBLIN and A. JONES, Eds. 1999

12 Optical Solitons: Theoretical Challenges and Industrial Perspectives
V.E. ZAKHAROV and S. WABNITZ, Eds. 1999

13 From Quasicrystals to More Complex Systems
F. AXEL, F. DÉNOYER and J.P. GAZEAU, Eds. 1999

Book series coordinated by Michèle LEDUC

Editors of "Branching in Nature – Dynamics and Morphogenesis of Branching Structures, from Cell to River Networks" (No. 14)
V. Fleury (École Polytechnique, Palaiseau, France)
J.-F. Gouyet (École Polytechnique, Palaiseau, France)
M. Léonetti (Université Paris-Sud, Bât. 510, Orsay, France)

ISSN 1436-6452
ISBN 3-540-41888-1 Springer-Verlag Berlin Heidelberg New York
ISBN 2-86883-527-9 EDP Sciences Les Ulis

Printed in France

List of Sponsors

This school was made possible by the generous financial support of the following agencies:

European Union
Centre National de la Recherche Scientifique
Ministère de l'Éducation Nationale, de la Recherche et de la Technologie
Ministère des Affaires Étrangères
Ministère de la Défense
Institut National de la Recherche Agronomique

Scientific Committee

We thank the following scientists for their kind help:

Herman Cummins (City College of CUNY, New York, U.S.A.), Albert Libchaber (Rockefeller University, New York, U.S.A.), James D. Murray (University of Washington, Seattle, U.S.A.), Yves Pomeau, Alain Prochiantz (École Normale Supérieure, Paris, France), Ewald Weibel (University of Berne, Switzerland)

Organizers

Vincent Fleury, Laboratoire de Physique de la Matière Condensée, École Polytechnique, 91128 Palaiseau Cedex, France

Jean-François Gouyet, Laboratoire de Physique de la Matière Condensée, École Polytechnique, 91128 Palaiseau Cedex, France

Marc Léonetti, Laboratoire de Physique des Solides, bâtiment 510, Université de Paris-Sud Orsay, 91405 Orsay, France

Preface

Branching is probably the most common mode of growth in Nature. From plants to river networks, from lung and kidney to snow-flakes or lightning sparks, branches grow and blossom everywhere, in every realm of Nature. When Galileo Galilei stated that the geometry of Nature was written in terms of planes, cones and spheres, he missed one essential pattern of Nature: the tree. However, the tree has been recognized as a "scientific" objet very early, ever since the classical times. Pliny, Strabo or Theophrastus were well aware of the existence of "dendrites", *i.e.*, stones in the shape of plants or corals, although the existence of such stones was a puzzle to them. In his book *Prodromus Cristallographiae*, in which the very word cristallography appears in print for the first time (1711), Mauricius Capeller, a Swiss naturalist, shows in between facetted crystals, several examples of dendritic crystals. He seemed already to believe in the existence of a general class of branching structures (*arbusculatum in modum*) in Nature. In the same spirit, his friend Jean-Jacques Scheuchzer had demonstrated experimentally that viscous fingering could generate geological dendrites (1699). At the time of Renaissance, Leornardo da Vinci was quite interested in the resemblance between trees and vessels (1508), and later, Nicola Steno did not hesitate in considering branched deposits of silver found in mines as a close relative of snow.

More recently, the french explorer and academician of science Charles de la Condamine was very keen on growing branching deposits, and stated that they formed in a process akin to river formation (1740).

Hence, the idea of a unity behind amazingly different systems has remained latent for two thousand years. However, if we except the work of Scheuchzer, very little true progress was made on the scientific issues linked to branching morphogenesis until very recently. The specialization and segmentation of the scientific fields, which is one characteristic of modern science, has turned each branching pattern into a specific scientific object. In all these fields, geology, botany, physiology, physics, etc., entirely different scientific communities seem to pursue divergent goals. But in each of these fields, branching patterns have been observed and carefully studied: new tools, either intellectual or material have been designed, and independent communities have found different means of investigating the morphogenesis of branches. Despite the existence of new descriptive tools (for example, the fractal geometry), the dynamics of growth of branching structures is far from understood, especially in biology, but in other fields as well (for example, the origin of scaling laws in geology is not really known).

The ambition of the organizers of the School of Physics and Biology which was held in Les Houches in october 1999 and whose title was: "Dynamics and morphogenesis of branching structures, from cell to river networks", was to gather people from entirely different fields, and have them talk to each other. This school of physics and biology was motivated by increasing scientific evidence that branching structures, the issues and concepts which they convey and the technical machinery required to tackle them, might be much closer than expected. There exist universal paradigms.

Now that the possibility of some universality between branching structures has been put forward, it is interesting to gather in a book state-of-the-art scientific information about each of these systems, with a special emphasis on the possible connections or relationships between them. In this spirit, the contributions should help non-specialists to catch up with the information they need, in order to be able to apply or pick-up ideas in other fields. Hence, the chapter about river networks is not so much intended to geologists, but maybe to researchers interested in vascular growth. The chapter about meristem growth is not so much intended to botanists, but rather to physicists or mathematicians attempting to model the growth of living tissue, etc. Many other examples could be invoked, which the reader will discover by him- or herself all along this book. Surprising relationships may be found, which indeed point to a form of universality among different branching patterns, and, also, to some unavoidable differences. We believe that many other examples of straightforward, or maybe only loose relationships between seemingly different systems will be found rapidly, and that they will help in understanding the dynamics of branching patterns. In this context, an interdisciplinary approach will certainly provide rapid progress in fields were progress has been slow.

As an example, we should cite the numerical model of growth known as Diffusion Limited Aggregation, which produces self-organized fractal patterns. This model, which was invented in the field of statistical mechanics by Thomas Witten and Len Sander (1981) now enables geologists or researchers in the field of hemodynamics, to perform rapid and large-scale simulations of branching morphogenesis, apart from providing more than a plausible explanation for the existence of many branching patterns. In the same spirit, the phase field formalism may allow to model growth in fields of morphogens, or in elastic fields. An important methodological issue, here, is that, if many different systems or models fall into the same universality class, then, it is simpler and more rapid to study, among all the possible models, the one which is technically the simplest.

When gathering the chapters for this book, we found it embarrassing to split the topics into separate sections such as "Botany", "Geology" or "Physics", when our very goal is to favour an interdisciplinary approach. This is why we have chosen the systems: Plants, Rivers, Neurons, etc., as titles, and not the scientific fields. However, each system has its emblematic power, and we thought it easier for a reader to start with familiar structures such as plants or rivers. This is why the chapters appear in the following order: Plants, Rivers, Organs, Neurons, Branching Organs, Vessels, Bacteria, Crystals, Fluids. A mathematical contribution appears in a section of its own. An introductory chapter about growth in fields raises beautiful questions linking physics to botany. However, we insist that, eventually, each chapter should be read as a contribution to the field of branching morphogenesis as a whole.

Apart from the main contributions, which correspond to the invited talks which were given at the Les Houches School, a few contributions by participants have been included after the relevant chapters. These contributions may be felt as more specialized and rapid than the main chapters which they follow.

The cover of this book, which shows two alpinists looking into crevasses may be a good metaphor of the current endeavour: to climb as high as possible, in order to have a deeper insight.

We are grateful to all speakers and participants for their efforts and enthousiasm.

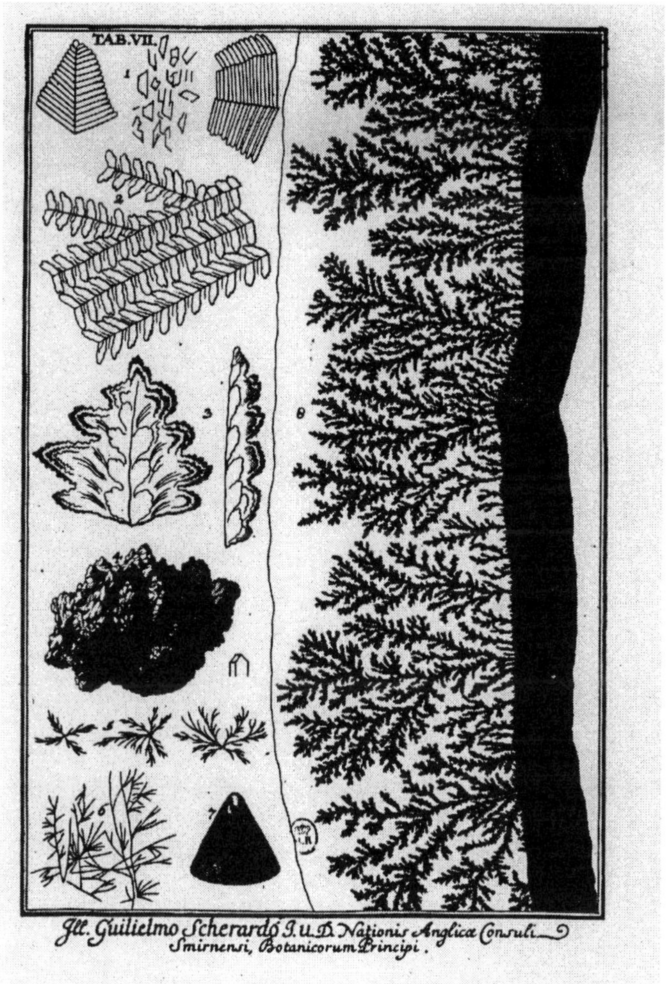

The Swiss naturalist Jean-Jacques Scheuchzer reported a systematic study of viscous fingering to the French academy of sciences in 1699, in order to explain the origin of dendrites.

V. Fleury, J.-F. Gouyet and M. Léonetti

List of Speakers

Only the names of the speakers appear in the table of content, since they were actually present at les Houches to give their talk. The names of the other contributors to the written chapters do not appear in this list.

Savério Bellusci
Institut Curie
UMR 144
26 rue d'Ulm
75248 Paris Cedex 05
France

Eshel Ben-Jacob
School of Physics and Astronomy
Raymond and Beverly Sackler
Faculty of Exact Sciences
Tel-Aviv University
69978 Tel-Aviv
Israel

Yves Couder
Laboratoire de Physique Statistique
École Normale Supérieure
24 rue Lhomond
75005 Paris
France

Francis Hallé
Institut de Botanique
Laboratoire de Botanique Tropicale
UMR 5554
163 rue Auguste Broussonet
34000 Montpellier
France

Alain Karma
Department of Physics
Northeastern University
Boston MA 02115
U.S.A.

Darryl Kropf
Department of Biology
University of Utah
Salt Lake City
Utah 84112
U.S.A.

Patrick Laufs
Department of Cell Biology
John Innes Centre
Colney lane
Norwich
NR4 7UH
U.K.

Ferdinand le Noble
Department of Physiology
Cardiovascular Research Institute
Maastricht
Universiteit Maastricht
P.O. Box 616
6200 MD Maastricht
The Netherlands

Bernard Lévy
Hôpital Lariboisière
Inserm U141
41 boulevard de la Chapelle
75475 Paris Cedex 10
France

Paul Meakin
Department of Physics
University of Oslo
Box 1048
Oslo 0316
Norway

Chaouqi Misbah
Université Joseph Fourier
Laboratoire de Spectrométrie
Physique
BP. 87
38402 Saint-Martin-d'Hères
France

Bernard Sapoval
Laboratoire de Physique de la
Matière Condensée
École Polytechnique
91128 Palaiseau
France

Jean-Pierre Ternaux
Unité de Neurocybernétique
Cellulaire
UPR 9041 INSERM
280 boulevard Sainte-Marguerite
13009 Marseille
France

CONTENTS

CHAPTER II

Inside the Buds: The Meristems

by P. Laufs and J. Traas

CONTRIBUTED PAPER 1

Two Symmetries Linking Biological and Physical Branching Morphogenesis

by V. Fleury

CHAPTER III

Establishing a Growth Axis in Fucoid Algae

by D.L. Kropf, S.R. Bisgrove and W.E. Hable

CONTRIBUTED PAPER 2

Shape Stability during Osmotic Growth

by M. Léonetti

CONTRIBUTED PAPER 3

On Transcellular Ionic Currents

by M. Léonetti and E. Dubois-Violette

RIVERS

CHAPTER IV

Branched Patterns in Geology: Rivers and Other Systems

by P. Meakin, T. Sun and T. Jøssang

NEURONS

CHAPTER V

Neuronal Arborization

by J.-P. Ternaux

CONTRIBUTED PAPER 4

Chemical Waves and Dendrites Navigation during Self-Wiring of Neural Nets

by R. Segev and E. Ben-Jacob

BRANCHING ORGANS

CHAPTER VI

The Mouse Embryonic Lung: A Biological Example of Branching Morphogenesis

by S. Bellusci, A. Mailleux, D. Ndiaye and J.-P. Thiery

CHAPTER VII

Branched Structures, Acinus Morphology and Optimal Design of Mammalian Lungs

by B. Sapoval, M. Filoche and E.R. Weibel

CONTRIBUTED PAPER 5

Quantitative Studies of Branching Morphogenesis in the Developing Kidney

by I.S. Harper, L.A. Cullen and J.F. Bertram

CONTRIBUTED PAPER 6

Morphogenic Responses of Mammary Epithelial Cells Grown in Biological Semi-Solid Substrates

by V. Fafeur, A. Delannoy-Courdent, J.C. Bout and B. Vandenbunder

VESSELS

CHAPTER VIII

Vascular Development: Design Principles and Morphometric Analysis of a Branched Vascular Tree

by F. le Noble, W. Hacking, D. Slaaf and H. Struijker-Boudier

CHAPTER IX

Mechanics of the Large Artery Vascular Wall

by B. Levy and A. Tedgui

CONTRIBUTED PAPER 7

A Link Between Dendritic Growth and Remodeling of Blood Vessels

by V. Fleury and L. Schwartz

BACTERIA

CHAPTER X

Adaptive Branching During Colonial Development of Lubricating Bacteria

by I. Golding, I. Cohen, I.G. Ron and E. Ben-Jacob

CONTRIBUTED PAPER 8

Pattern Formation Modeling of Bacterial Colonies

by I.R. Cantalapiedra, A.M. Lacasta, C.E. Auguet, A. Peñaranda and L. Ramírez-Piscina

---

CRYSTALS

CHAPTER XI

Dendritic Growth

by A. Karma

CONTRIBUTED PAPER 9

Sidebranching in Solutal Dendritic Growth

by R. González-Cinca, L. Ramírez-Piscina, J. Casademunt and A. Hernández-Machado

CONTRIBUTED PAPER 10

Experimental Study of Sidebranching in Directional Solidification

by M. Georgelin and A. Pocheau

FLUIDS

CHAPTER XII

Branching during Dewetting and Wetting

by C. Misbah

CONTRIBUTED PAPER 11

Viscous Fingering in a Gel

by A. Lindner, P. Coussot and D. Bonn

CONTRIBUTED PAPER 12

Branching Transition in Viscous Fingering with a Liquid Crystal

by R. Folch, J. Casademunt and A. Hernández-Machado

MATHEMATICS

CONTRIBUTED PAPER 13

Mathematical Meristems: The Singularities of Laplacian Growth

by M.A. Peterson

INTRODUCTION

Patterns with Open Branches or Closed Networks: Growth in Scalar or Tensorial Fields

Y. Couder

In a variety of physical systems in which there is growth, patterns are spontaneously generated. Since it is also during their growth that living entities acquire their shapes, it is tempting to consider possible relations between the two types of morphogenesis. This direction of research was initiated by D'Arcy Thompson (1917) in his classical book "On Growth and Form". To the present day it is still an important and open problem. The fact that biological growth is under the control of genetic factors does not abolish the fact that it also obeys the laws of physics. Many situations of biological morphogenesis can thus involve an interplay between genetic determination and physical self-organisation. This is, of course, well beyond the field of the present article. We limit ourselves here to the description of some recent results concerning branches and meshworks in physical systems and discuss a few biological systems, mostly in botany, where these results may be relevant.

We consider two-dimensional situations in which two media coexist in a plane. When one medium grows at the expense of the other the interface separating them is in motion. Physicists have investigated the patterns formed when this moving front is unstable. The main result obtained in the last decades is that several of these processes, in spite of being different in their nature, can be related to each other both experimentally and theoretically. They are referred to as free boundary problems and have some general common properties due to an underlying similar mathematical structure. We will discuss two distinct families of phenomena.

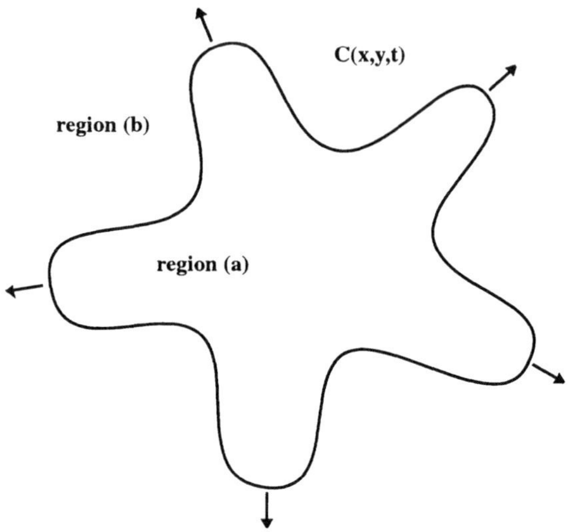

Fig. 1. — Scheme of an interface separating two media a) and b) and moving in a diffusive or Laplacian field $C(x, y, t)$.

1. GROWTH IN A DIFFUSIVE SCALAR FIELD

1.1. The Basic Phenomenon

A first family of processes includes viscous fingering, crystal growth, electrodeposition, growth of bacterial colonies, diffusion limited aggregation, corrosion of solids... Reviews can be found in *e.g.* Kessler *et al.* (1988), Pomeau and Ben Amar (1992), Couder (2000). In all these phenomena the interface grows at a velocity proportional to the gradient of a scalar quantity (concentration, temperature, pressure, etc.) and this scalar field satisfies either a diffusion law or a Laplacian law. As the interface moves it destabilises and forms protrusions. Later, when the protrusions become large they branch repeatedly and form tree-like structures. Two types of branching processes are observed, depending on whether the system is isotropic or anisotropic.

If we forget the specific physical quantities involved in each system and limit ourselves to two-dimensional situations, the generic problem of growth in a diffusive field can be schematised as follows. An interface (Fig. 1) separates two regions of a plane respectively labelled (a) and (b). In this plane exists a scalar field $C(x, y, t)$. In the simplest cases C is constant in region (a) and varies in region (b) where it is a diffusive field governed by:

$$\frac{\partial C}{\partial t} = D_{\mathrm{C}} \nabla^2 C, \tag{1}$$

where D_{C} is a diffusion constant.

In some morphogenetic problems (such as viscous fingering) the field is governed by Laplace's equation:

$$\nabla^2 C = 0, \tag{1'}$$

which also corresponds to the limit obtained when the phenomena controlled by equation (1) are very slow.

We consider here situations where the system is out of equilibrium so that medium (a) expands. The interface moves in a local gradient of C with a normal velocity:

$$V_n \propto \vec{n} \cdot \vec{\nabla} C. \tag{2}$$

In both cases defined by (1) or (1′) the determination, at a given time, of the field $C(x, y)$ depends on the boundary conditions at the interface between regions (a) and (b) and at infinity. The displacement of the interface, because it changes the location of boundaries, modifies the field C and thus, in turn, the interface velocity. This process leads to the instability which generates the pattern. The whole process is non-local, due to the long range interactions introduced by the diffusion law.

Examples: Crystal Growth and the Growth of a Bacterial Colony

One of the archetypes of growth controlled by equations (1) and (2) is crystal growth in a solution. When a solution of a salt as *e.g.* ammonium bromide is cooled solidification occurs, and the salt may form dendrites. The solid keeps growing by condensing the salt of the solution. The initial concentration C_0 of the salt in the solution is depleted in the vicinity of the growing crystal and forms a diffusive field governed by equation (1).

Crystallisation has an exact equivalent in the growth of colonies of bacteria (Fujikawa and Matsushita 1991 and Ben Jacob *et al.* 1992 and this volume). In the simplest experiment a few bacteria are deposited in the centre of a Petri dish on the surface of a gel containing a small concentration of nutrient. The bacteria proliferate when they are well fed but cease to do so when the food becomes scarce. The growth of the bacterial colony spontaneously generates a branched pattern in which most of the growth occurs at the tip of the longest protrusions. The reason for this type of growth can be summarised in the following way.

The initial concentration of nutrient is homogeneous and equal to C_0. However the feeding of the bacteria depletes this concentration in their vicinity. The concentration of nutrient in the gel becomes uneven and two regions can be distinguished. Region (a) is occupied by a dense concentration of bacteria and the growth has stopped by lack of nutrient. The outer region (b) has not yet been occupied by bacteria and the nutrient field $C(x, y)$ is diffusive, following equation (1) where D_C is the diffusion coefficient of the nutrient in the gel. As for the velocity of growth of the colony it occurs at the boundary and depends on the local flux of nutrient so that the interface moves with a normal velocity proportional to the local gradient of C (relation 2).

Saffman Taylor Viscous Fingering

One of the main tools for the investigation of fingering and branching has been a fluid dynamics instability called Saffman Taylor viscous fingering. It is performed in a Hele Shaw cell formed of two parallel plates separated by a narrow gap of thickness b. The instability occurs at the interface between two fluids of very different viscosities μ_i when the fluid of low viscosity displaces the fluid of high viscosity. The velocity of the mean flow of each fluid, driven by pressure and limited by the viscous friction on the wall satisfies:

$$\vec{u}_i = -\frac{b^2}{12\mu_i}\vec{\nabla}p_i. \tag{3}$$

The fluids being incompressible the pressure distribution on the two sides of the interface satisfies Laplace's law

$$\nabla^2 p_i = 0. \tag{4}$$

The interface separating the two fluids is unstable when the less viscous fluid forces the other to recede. When the viscosity of the less viscous fluid can be neglected the situation is exactly that described by equations (1′) and (2) where the scalar quantity is the pressure.

1.2. The Length Scales of the Patterns

The Intrinsic Instability Length Scale

The analogy of equations (1′) and (2) with the laws of electrostatics provides a simple interpretation of the front instability. If we consider a smooth front with a small protrusion, the gradient of C will be increased in front of the protrusion by the point effect so that the front velocity will be locally increased and the protrusion will grow. In principle the sharper the protrusion, the larger the velocity.

However in all known processes governed by equations (1) and (2) there are additional effects at the interface which stabilise it at small scales. In viscous fingering and in crystal growth, surface tension is the stabilising factor. To take this effect into account it is necessary, for each problem, to perform a linear stability analysis of a straight interface. This analysis provides the length scale l_{inst} at which the front is most unstable. Such a stability analysis was performed by Chuoke *et al.* (1959) for viscous fingering and by Mullins and Sekerka (1964) for solidification.

The Other Length Scales of the Problems

In a diffusive field the thickness of the disturbed region near the interface moving at velocity V, is bound by the diffusion length $l_D = D/V$. This length gives an upper bound to the range of the non local interactions of the structures. For very slow diffusive growth it becomes large. When the field is Laplacian this length becomes infinite. Other large scales then become important, usually those imposed by the boundaries of the system (*e.g.* the width W of a channel).

1.3. The Branching Process

Let us now consider a situation where the instability has already grown and formed a pattern. Its foremost part consists of growing protrusions which are called fingers, needles, etc. depending on the system. As they keep growing these structures themselves become unstable and give rise to branches. This branching process differs in nature, depending on whether the system is isotropic or not. We will first consider the isotropic situation.

1.3.1. Branching in an Isotropic Situation: The Tip-Splitting Instability of Viscous Finger

We can take as an example the viscous fingering obtained by injection of air in the center of a circular Hele Shaw cell (Bataille 1968) filled with oil. The first fingers to form appear near the central injection hole and have a width of the order of $l_{\rm inst}$. As these fingers grow radially they widen and become unstable. Repeatedly each finger splits into two equal or unequal parts which will grow and compete. Usually, after such a splitting one of the fingers of new generation stops, while the other keeps growing and eventually becomes unstable again. These events generate the growing complexity of the pattern which exhibits a fractal structure when it has grown to be large enough (Fig. 2).

This "tip-splitting instability" can also be observed and studied in a simpler geometry. We consider a Saffman-Taylor finger growing at a constant velocity in a divergent wedge of angle θ_0 (Fig. 3) a geometry introduced by Thomé *et al.* (1988). The advantage of this system is that its two length scales are well defined. The smaller one is the instability length scale $l_{\rm inst}$ which is constant if the front velocity is kept constant. The larger one is the local width of the channel which grows larger as the structure develops. As air is injected at the apex of the cell, a single stable finger forms in the region of the cell where the cell's width W is of the order of $l_{\rm inst}$. As the finger keeps growing W increases and the finger is observed to grow approximately self similarly, thus broadening with W. This is a characteristic of isotropic growth where the finger width is not selected by $l_{\rm inst}$ but by the larger scale (here W). The shape, the selection and the stability of these self similar fingers growing in wedges are due to the effect of isotropic surface tension and are now well known analytically and numerically (Ben Amar 1991 and Ben Amar *et al.* 1991).

As it keeps growing the finger tip becomes so large that it grows wider than approximately 2 $l_{\rm inst}$ and becomes unstable. This instability, because it affects the extremity of the finger, is often called tip-splitting. In a recent work (Lajeunesse and Couder 2000) this process was investigated specifically. The first sign of the destabilisation of the finger is the formation of a flat in the finger profile. This flat rapidly becomes a dip in which the interface stops moving (Fig. 3). This local inversion of curvature marks the beginning of the

Fig. 2. — A large radial pattern resulting from the Saffman Taylor instability.

splitting of the initial finger into two branches. They are separated by a region, which we will call a fjord, in which the oil remains blocked (Figs. 2 and 3).

Investigating the later evolution of this splitting instability, we observed that this process can lead to two different kinds of long term behaviours.

- In the first case, observed for small angles ($\theta_0 < 70°$), (including the linear channel), the instability fails to create two independent long lived structures. One of the branches, the secondary branch, is rejected on the finger side and stops growing (Fig. 3). The other branch, the main branch, widens and eventually becomes unstable again;

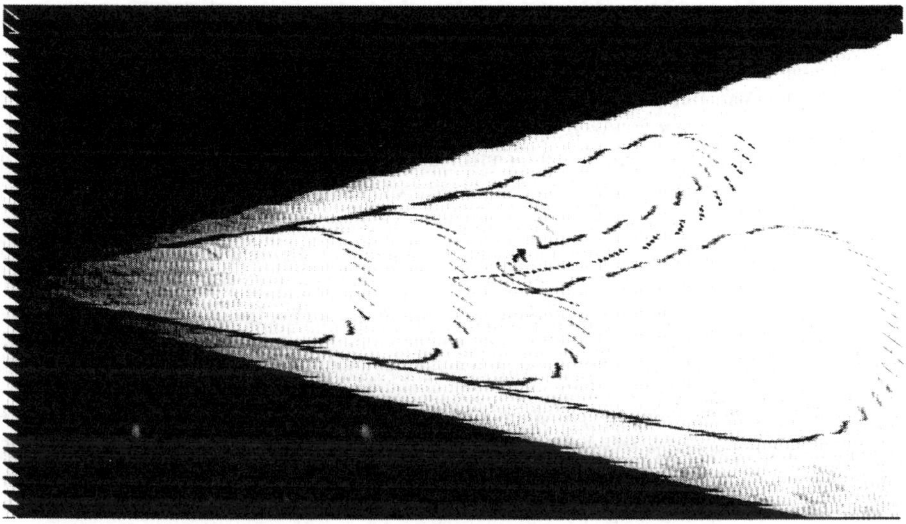

Fig. 3. — A tip splitting instability as observed on a superposition of photographs of a Saffman Taylor finger growing in a wedge of angle 30°. The dotted line is a curve normal to self similar stable profiles and fits the middle of the fjord separating the two branches.

- In the second case, observed in cells of large angles ($\theta_0 > 90°$), both branches can keep growing independently and form long lived structures.

The transition between these two kinds of behaviour occurs in the range $70° < \theta_0 < 90°$, where both perfect and imperfect tip splitting can be observed. To put these observations on a more quantitative ground it is appropriate to study the shape of the fjords instead of that of the secondary branches. Perfect tip splitting always corresponds to radial fjords, on the contrary, in the imperfect tip splitting case, the observed fjords are curved. Their initial axis is always somewhat divergent from the cell's axis (Fig. 3) and with time this axis rotates progressively further away from the cell's axis until the side finger stops growing.

In order to understand this evolution, we propose (Lajeunesse and Couder 2000) an interpretation based on the two following hypotheses:

i) The destabilisation of the Saffman Taylor front occurs by the amplification of a motionless disturbance, *i.e.* having no intrinsic propagation. Its only motion is due to its advection along the curved profile, a kinematic effect pointed out by Zel'dovich (1980) (see also Pelcé 1988);

ii) The formation of a dip in the interface, being very local, affects the nearest of these isobars only. The further ones remain undisturbed and tend to stabilise an envelope of the structure similar in shape to the stable finger.

In other terms the unstable pattern remains embedded in a field which retains a memory of the previous stable structure.

With these two hypotheses, the evolution of the fjords separating the two branches can be predicted from the evolution of the finger had it remained stable. The shape of the fjord is similar to the trajectory of a point of the interface that would be on a stable Saffman-Taylor finger. This trajectory is particularly easy to determine as it is given by the curves normal to the successive finger profiles which are known analytically and numerically. If we assume a random distribution of initial disturbances on the whole width of the finger it is possible to predict the distribution of sizes of the lateral side branches.

1.3.2. Side Branching in Tip Dominated or Anisotropic Situations

There exists a second family of growth profiles characterised by the stability of the growing extremities. Such growth shapes are observed when specific properties allow for the tip of a finger to become a singular region. This is obtained in two situations:

- a local disturbance is applied to the finger's tip. This was observed in locally disturbed viscous fingering (Couder *et al.* 1987; Rabaud *et al.* 1988);
- the system is globally anisotropic as in *e.g.* the crystalline dendrites.

In both these situations the observed structures have a needle shape with a parabolic tip (Fig. 4). As demonstrated by Ben Amar (1990) for dendrites, the radius of curvature of this parabola ρ is proportional to the instability scale: $\rho = \alpha\ l_{\text{inst}}$. The coefficient of proportionality α is a decreasing function of the strength of the tip disturbance or of the anisotropy.

The Nature of the Side Branching Instability

If the disturbance or the anisotropy is strong enough, ρ is of the order of l_{inst}. Therefore the needle tip is not wide enough for an instability to grow. The tip of the parabola is thus stable but not its sides. The weak disturbances due to noise are amplified as they are advected away from the tip and result in a lateral destabilisation of the profile which causes the formation of side branches (Fig. 4a). As shown in Rabaud *et al.* (1988) this instability is usually convective. This means that in the tip's frame of reference it is advected with a large velocity so that there is no feedback from downstream to upstream. In other terms the instability has a spatial growth instead of a temporal one. All convective instabilities behave as selective amplifiers of the upstream noise so that they do not form strictly periodic structures. In the present cases the side branches are not regularly spaced (Fig. 4a). However, a characteristic of convective instabilities is that it is possible to replace the natural noise by a temporal forcing so that the lateral branching can become strictly periodic (Fig. 4b) (Rabaud *et al.* 1988; Bouissou *et al.* 1990; Quian and Cummins 1990).

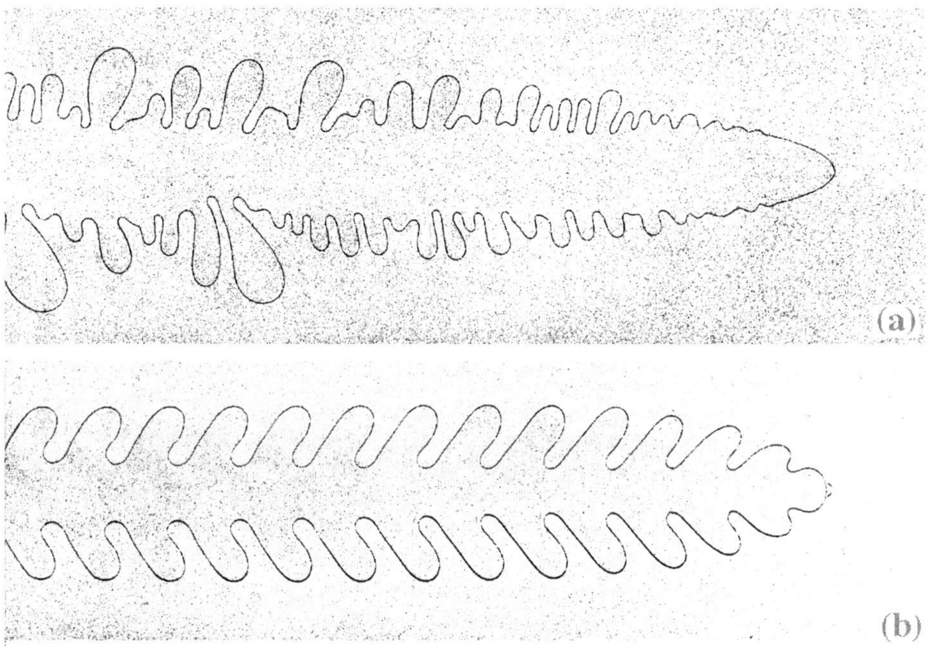

Fig. 4. — a) The parabolic tip of a disturbed Saffman-Taylor finger and its irregular side branching. b) The periodic side branching obtained by a forcing of the tip.

It can be noted that if the tip curvature was weaker, the advection velocity would be smaller and there could be a feedback from downstream to upstream. In the usual terminology (Huerre 2000) this would mean that there would be a region of absolute instability near the tip. In other terms, near the tip the instability growth would be temporal instead of spatial. In this situation there would be, in this region, a spontaneous selection of a well defined periodicity. There is some indication of observation of this phenomenon in dendrites.

The Patterns Generated by the Side Branching Instability

As the side branches keep growing they start competing. This competition can be directly observed (Fig. 5) (Maurer and Couder unpublished data). It starts near the tip where the initial destabilisation has given rise to initially uneven protrusions. When one of these protrusions is larger than its neighbours it will grow faster and screen them off so that their growth will be stopped. As it keeps growing the protrusion will compete with neighbours located further and further along the front. Typically the interaction between two branches occurs when their distance along the front is of the order of their length. Again the competition will be biased and the larger branch will win. It thus appears that the initial destabilisation is dominant; the branch which finally dominates a

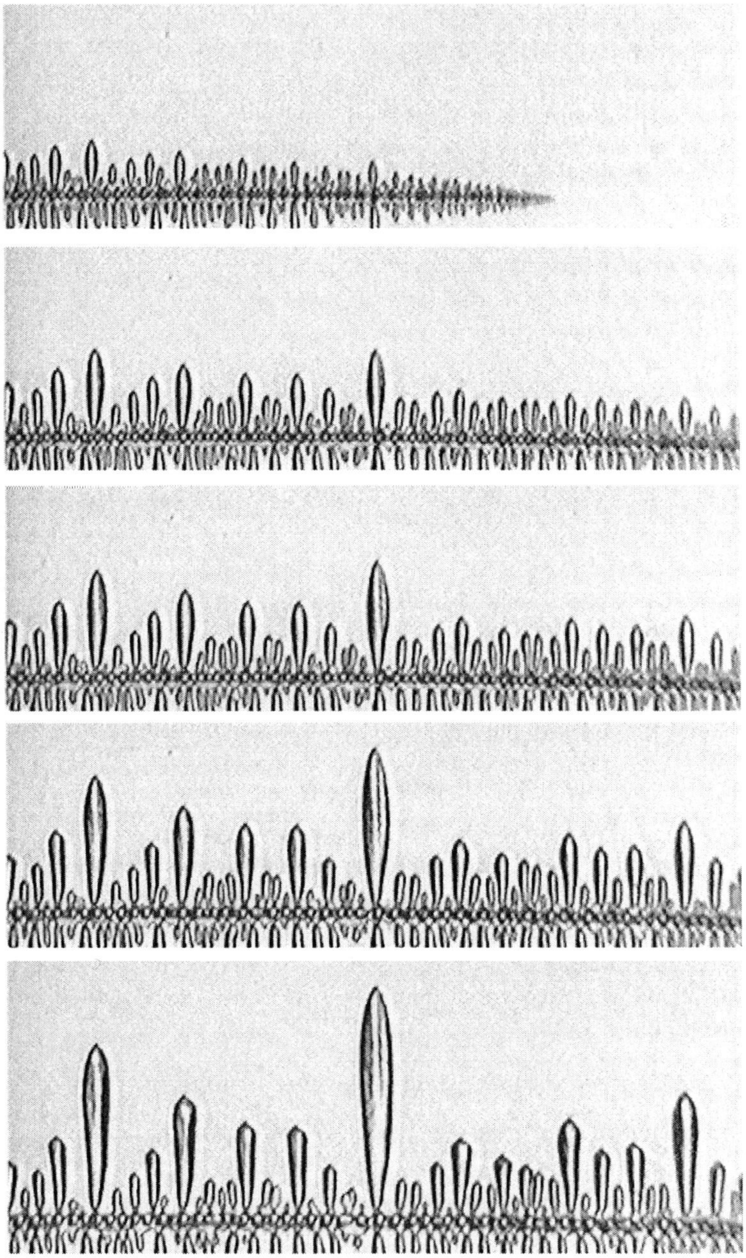

Fig. 5. — The formation of the large scale structure of a slow growing dendrite (with a large diffusion length scale $l_D = D/V$). Successive photographs show the coarsening resulting from the competition of side branches.

large region of the front was originally the largest of this region at its formation. In the world of lateral branches, as in others, the hope of a branch of becoming large and dominant is determined by its birth. If the diffusion length is large (so that the field is quasi Laplacian) the branches keep competing and form a hierarchical structure. This structure is a self affine fractal with a fractal dimension of the order of 1.5 (Couder *et al.* 1990). In contrast, if the diffusion length scale $l_D = D/V$ is finite, the competition between branches stops when the distance between neighbouring branches becomes of the order of l_D. As a result, when separated by a distance larger than l_D the parallel branches do not compete and grow at the same velocity (Fig. 6).

1.4. Speculations on Isotropic or Tip Dominated Growth in Plants

The distinction between isotropic or tip dominated growth and the resulting instabilities is found to be identical in various phenomena such as crystal growth, viscous fingering, the growth of bacterial colonies, all controlled by equations (1) or (1′) and (2).

It can be remarked that this distinction between the isotropic and the tip-dominated growth appears to hold for other types of quasi two dimensionnal growth which are not modelled by these equations. Figure 7 shows the shapes of a brown algae *Fucus* (a) and a fern leaf (b). In *Fucus* there are tip splittings leading to irregular dichotomous partitions. In the fern the leaf the tip never splits and there is formation of periodic lateral protrusions very similar to the side branching of *e.g.* crystalline dendrites. In plants the growth occurs "from inside" with, as far as we know, no direct relation with a field in the surrounding medium. However there is a similarity with the physical systems in that the growth occurs mainly at the periphery of the blades in regions having what the botanists call meristematic activity. It is interesting to note that the distribution of these regions is different in the two chosen examples. In *Fucus* the meristematic activity occurs along the whole periphery of the blade. In contrast, in ferns, there is a singular localised apical meristem at the tip of the leaf. These two types of growth can thus be considered as isotropic and singular at the tip respectively. It is striking that the resulting instabilities and structures are similar to those observed in the above described physical systems.

A three-dimensional tip dominated growth is also observed at the apex of stems of superior plants. Their tip is stable with the shape of a paraboloid which grows steadily. This structure destabilises laterally to give rise to little protrusions, the primordia, which will grow into various organs such as leaves or florets. These protrusions are strongly organised along vertical lines or in spirals. A large number of works have been devoted in botany to the possible morphological role of biochemical or mechanical processes. For instance some models for phyllotaxis rely on purely diffusive processes (Mitchison 1977), others emphasize the role of the mechanics of growing tissues. In the latter model proposed by Green (1992) a differential rate of growth of the inner and outer

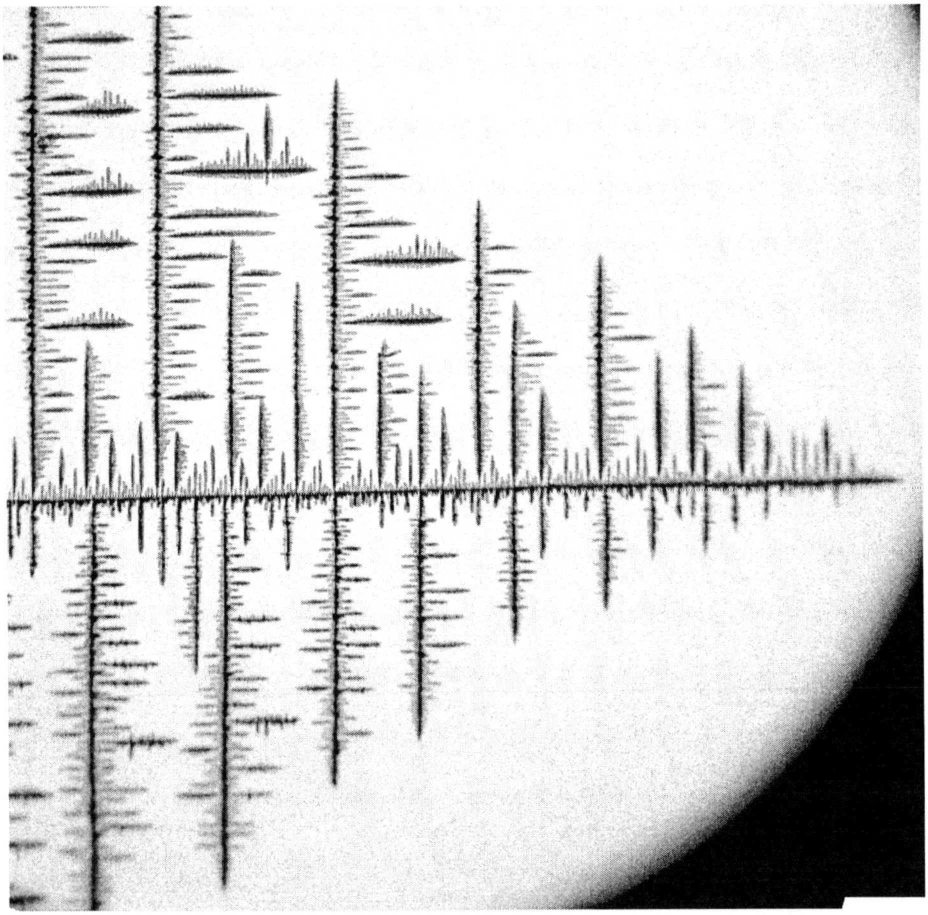

Fig. 6. — The large scale structure of a fast growing dendrite. The diffusion length scale being small, the branches grow independently when they are at a distance exceeding the diffusion length. Note however that the branches interact at short distances so that they never come into contact with each other.

tissues of the apex is responsible for the formation of stresses which will generate a buckling instability. As pointed out by Douady and Couder (1996) the phyllotactic organisations can be obtained by self organisation in both types models so that it is difficult here to discriminate them on the basis of the structure they generate. We will return to this point below.

As a final note we can remark that in both the 2D fern leaf and the 3D phyllotaxis, the observed periodicity of the lateral branching is an indication that the instability is absolute at the tip (see Sect. 1.3.2).

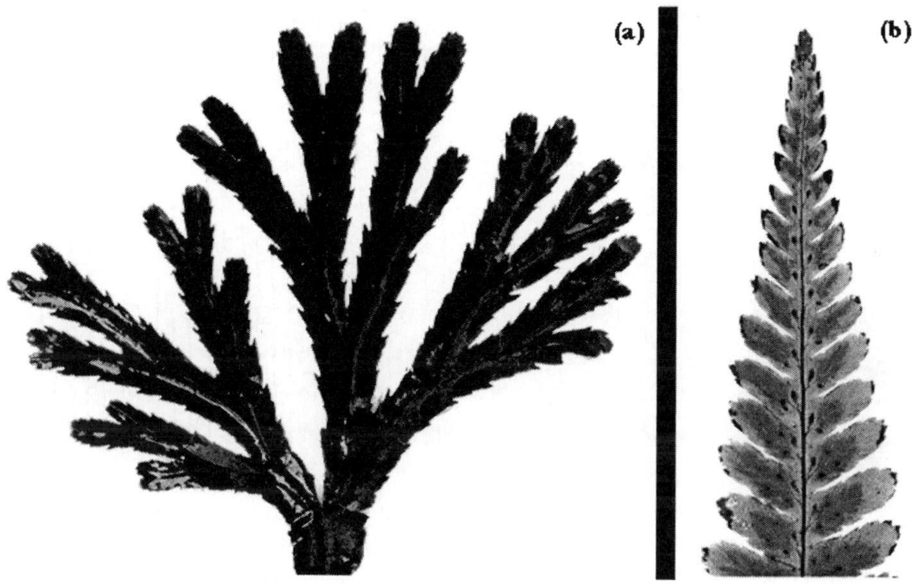

Fig. 7. — a) The pattern resulting from successive tip splitting instabilities as observed in Fucus. b) The side branching instability observed in a fern leaf.

2. GROWTH IN A TENSORIAL FIELD

Returning to physical systems there are other free boundary problems which form very different patterns and belong to another universality class. The origin of their specificity is that the field in which they grow is tensorial instead of being scalar. The archetype of such structures are fractures. A discussion of fracture propagation is beyond the scope of the present work; we limit ourselves here to the main geometric characteristic of the resulting patterns in two dimensions. In collaboration with Pauchard, Adda Bedia, Allain and Douady we revisited these patterns with the aim of comparing them to the vasculature of plant leaves. The result we obtained are discussed below.

2.1. The Formation of a Network of Fractures in a 2D Stressed Material

A rather common pattern forming situation is the fracture due to the stress generated by a different contraction of two media. These are illustrated by the cracks in drying mud or in the glaze of chinaware. In both cases a superficial layer retracts more than its substrate. In the first case this is due to evaporation, in the second to cooling. This differential retraction, together with the adhesion between layers, generates approximately uniform and homogeneous

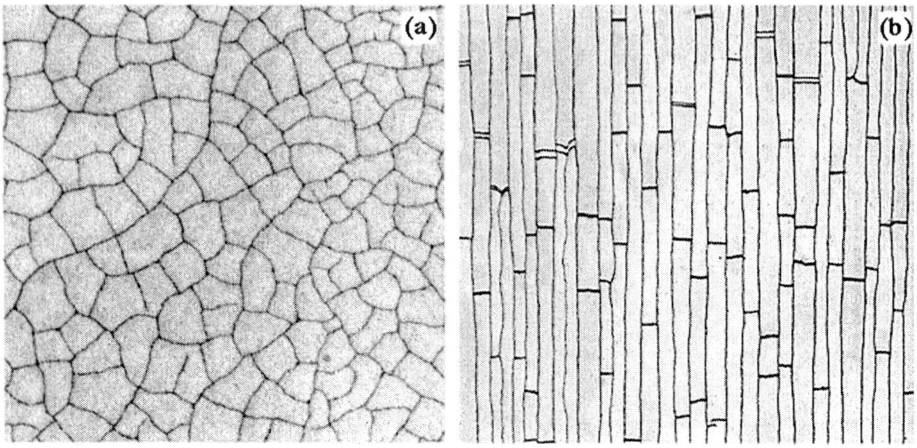

Fig. 8. — a) A pattern of cracks obtained during the drying of a thin isotropic and homogeneous layer of gel. b) An anisotropic pattern of cracks grown in a sample with a weak drying gradient.

stresses. When they reach a threshold the material breaks and a network of fractures forms.

All the patterns shown here were obtained in variants of experiments (Allain and Limat 1995; Pauchard *et al.* 1999) performed with concentrated colloidal suspensions of latex particles of diameter 0.1 μm in water (with volume fraction 0.50 and 0.25). These suspensions are deposited on a glass plate and left to desiccate by evaporation, a process limited by diffusion of water into air. The suspension first becomes a gel which tends to shrink as it dries, the adhesive forces on the substrate oppose this shrinking. This growing mismatch results in stresses which are at the origin of the crack formation.

2.1.1. Isotropic Case

The morphology of these cracks depends on the shape of the sample and on the homogeneity of the drying. When a large homogeneous layer of the suspension is deposited on a glass substrate (Fig. 8a), the resulting pattern is isotropic. It forms a mesh-like network, the fractures being connected to each other at points where they meet at right angles. Their typical distance is scaled on the thickness of the superficial layer (Groisman and Kaplan 1994). This is simply related to the fact that regions of the sample separated by a distance of a few times the layer's thickness behave independently due to the friction on the substrate.

The main characteristic, the reconnection of the fractures, is directly linked with the tensorial character of the stress field. It can be intuitively understood by the following argument. We consider a given region of the sample which we

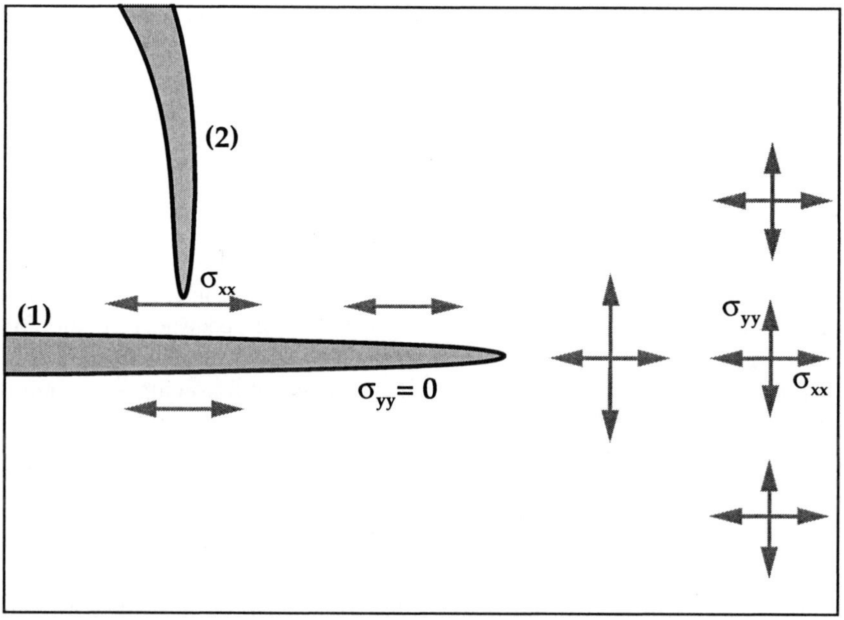

Fig. 9. — Sketch of the propagation of two successive fractures in an initially homogeneously stretched medium. Compression breakdown patterns would correspond to this sketch with all the stresses having the reverse signs.

assume approximately homogeneously stressed. The first fracture will tend to propagate through the material so as to release the main local dilating stress. If we assume that this first fracture has propagated along the Ox axis (Fig. 9) it has released the tensions σ_{yy} in the perpendicular direction. However the material remains stretched along Ox. If a second fracture, moves to the vicinity of the first fracture it can still propagate using this remaining energy. Because of the principle of local symmetry, the second fracture will tend to orientate so has to be submitted to pure opening stresses. It will thus rotate to release the remaining tension σ_{xx}. Ultimately the second fracture will thus collide perpendicularly with the first one.

2.1.2. Anisotropic Case

It is possible, in the previous experiment, to give a preferred direction of growth to the fractures by imposing a gradient to the drying process. If the sample has a thickness gradient, the thinner regions will dry first so that a gradient of water concentration will form. The cracks will start forming in the driest region and then form a parallel array which propagates in the direction of the gradient. This array has a periodicity proportional to the local thickness of the layer (Fig. 8b). Note that when one of the parallel cracks stop growing it connects perpendicularly, first to one of its neighbours, then to the other. This produces a Y shape anastomosis. Later the remaining stresses (in the direction parallel to the cracks) are released by the formation of transverse cracks forming a stepladder so that the closed network necessary to the release of all stresses is also obtained here.

2.2. The Network of the Plant Leaves Venation

The vasculature of plant leaves form very striking patterns which have been classified by Hickey (1973). At first sight they appear to be tree-like structures, but closer examination shows that the veins are all reconnected and form a network structure.

In spite of their diversity the veins' arrays have the following set of common properties (Esau 1953; Nelson and Dengler 1997).

(a) There is a hierarchy of veins of different sizes which have formed successively during the leaf's growth;

(b) The vascular network has a spatially uniform density in the lamina at all stages of the development of the leaf;

(c) There is a common trend of the veins to reconnect (anastomose) and to form a reticulum;

(d) These reconnections occur, mostly at right angles, between veins formed at a given time and veins formed earlier.

2.2.1. Isotropic and Anisotropic Venation

At a late stage of their formation the growth of the dicotyledon leaves becomes isotropic and homogeneous (Avery 1933). The smallest veins (of high order of formation) which appear at this time, form a lattice shown in Figure 10a. They have a hierarchy which results from their successive order of formation. The younger veins are observed to be connected with the older ones at approximately right angles. These patterns are thus geometrically similar to the cracks obtained in a thin homogeneous layer shown in Figure 8a.

In contrast the venation of most monocotyledons appear very anisotropic. Figure 10b shows a detail of the venation in Lily of the valley. The pattern is basically formed of veins approximately parallel to each other. However it is again a network since transverse veinlets connect these veins. As can be seen in Figure 10b, when one of the parallel veins stops growing, it reconnects perpendicularly to its two nearest neighbours in a characteristic Y shaped anastomosis described in the case of bread wheat by Blackman (1971). the resulting shape is identical to the reconnected cracks of Figure 8b.

2.2.2. Growth in a Scalar or a Tensorial Field?

As already discussed in Section 1.4 the dominant theories about botanical growth rely mostly on biochemical or on mechanical processes. For the venation problem the only hypotheses that were considered relied on diffusion or

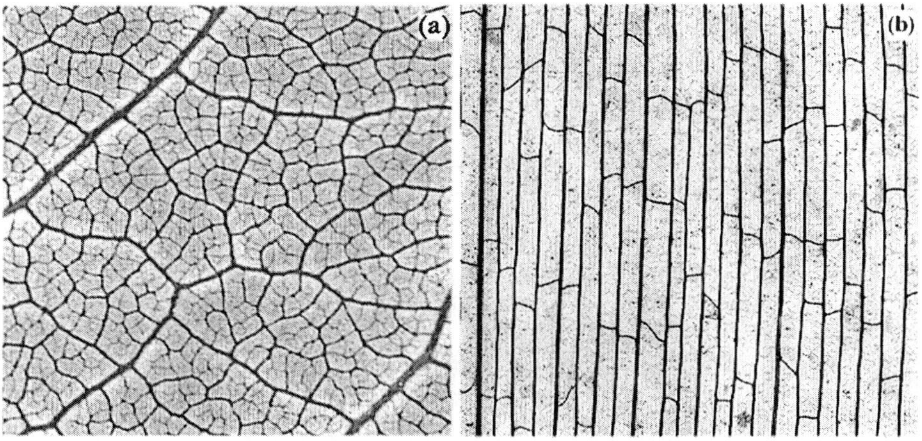

Fig. 10. — a) An isotropic venation pattern: the high order venation of a leaf of *Polygonum Polystachium*. b) An anisotropic venation pattern: a detail of the venation in Lily of the valley (*Convallaria Maialis*). Note the Y shaped anastomosis of a vein with its neighbour.

on reaction diffusion processes. In these models (Mitchison 1980; Meinhardt 1984; Sachs 1991) a growth hormone diffusing in the tissues is canalised by the formation of preferential channels by which it is evacuated.

However, as shown in Section 1, the patterns grown in diffusive scalar fields are characterised by a tree-like organisation with a hierarchy of opened branches and no reconnections. For this reason it is difficult, in these models, to account for the veins anastomosis. In contrast it appears that the patterns (Fig. 10) have the same topology as the crack patterns (Fig. 8a) suggesting that the growth of veins occurs in a tensorial field rather than a scalar one.

2.3. How Can the Venation Patterns Be Similar to the Crack Patterns?

In a recent work (Couder *et al.* 1999) we propose a new hypothesis in which the initial stage of the vein formation happens under the influence of the stress field present in the inner tissues of the leaf. Before presenting it, let us recall a few facts about veins and about stresses in botanical tissues

When it is mature, a vein associates two types of tissue forming distinct ducts. The xylem brings the water up from the roots while the phloem distributes the chemical products synthesised in the leaf, to the rest of the plant. However we are interested here in the initial genesis of the venation. What will develop into the veins appears at an early stage of the leaf formation when its inner structure is still relatively simple. Transverse cuts (Fig. 11a) show that the blade is formed of a small number of cell layers with two epidermis

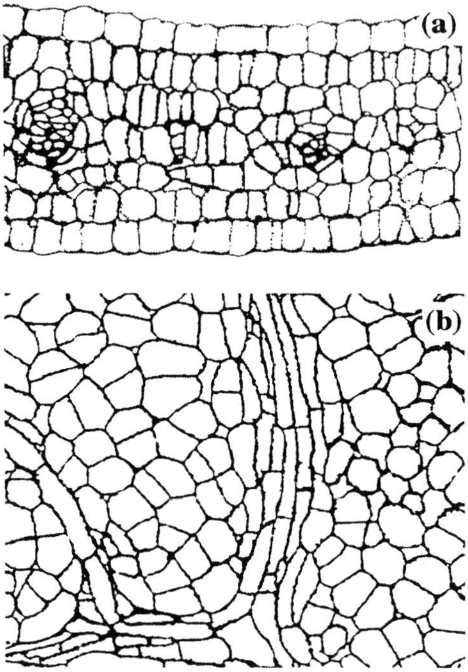

Fig. 11. — a) A transverse cut through a young leaf (redrawn from a figure in Nelson and Dengler 1997). b) A cut parallel to the surface of a young leaf in the mesophyll layer.

separated by a tissue called the mesophyll. The precursor of a vein is observed to form in the middle of this mesophyll by differentiation of strands of cells. These cells undergo specific divisions so that they become narrower and more elongated than their neighbours (Fig. 11b). This new tissue is called the procambium and it is only at a later stage that it will differentiate into xylem and phloem and acquire transport functions. As the procambium forms the draft of the vascular system, the problem is that of its genesis.

Since the analogy in the patterns geometry suggests a growth in a tensorial field it is natural to consider the possibility that the mechanical stresses contribute to the differentiation of procambium. In a different context many botanical works have been devoted to the investigation of the role of mechanical stresses in the growth of plant tissues. Reviews on this subject can be found in *e.g.* Kutschera (1989), Hejnowicz and Sievers (1995) and Peters and Tomos (1996). In growing organs such as stems or hypocotyls, the parallel growth of the epidermis and of the inner tissues generates, in these tissues, stresses of opposite sign. The stress is tensile in the epidermis and compressive in the inner tissues. The situation is similar in leaves so that the mesophyll is submitted to a compressive stress which increases as the leaf grows in size.

2.3.1. *A Hypothesis for the Procambium Formation*

Our main hypothesis is that when a mesophyll cell is submitted to a stress exceeding a threshold this cell reacts by having a specific division oriented in a shear free direction. This division will release the stress so that a neighbouring cell will be submitted to a larger stress and undergo the same evolution. This process will result in the formation of strands of procambial cells. This hypothesis is indirectly supported by the results of experiments (Lintilhac and Vesecky 1981) showing that, in botanical tissues, oriented cell divisions can be forced by an externally applied compressive stress. In these situations the new cell walls form perpendicularly to one of the main directions of stress, a shear free direction.

2.3.2. *Simulation of the Main Veins of Dicotyledon Leaves*

If the formation of procambium results from a breakdown under compression perpendicular to the main compression in a shear free direction and if it releases the stress, the sketch given in Figure 9 would describe the formation of the procambial strand instead of the cracks provided the directions of the arrows are reversed so that the stresses are compressive instead of dilating. For this reason we tried to push further the analogic simulation of venation patterns using cracks.

We are now interested in the two most common organisations of the main veins of dicotyledon leafs. In both of them there is a central main vein and pinately organised secondary veins. In the organisation, called "brochidodromous" by Hickey (1973) and shown in Figure 12a each secondary veins bends as it comes near the leaf margin and connects with its neighbour forming characteristic loops. In the organisation called "craspedodromous" the secondary veins go straight to the border.

The primary and secondary veins form at an early stage when the leaf is still very small and narrow. For this reason we investigated the formation of cracks when the influence of boundaries is dominant. We deposited on the glass plate very narrow, wedge shaped, strips of suspension. The meniscus bordering the deposited fluid is strongly pinned onto the substrate so that the drying results in a narrow sample of gel with pinned borders. The first fracture to appear is central, in the middle of the strip. Then transverse cracks form one by one which can have two different behaviours depending on the structure of the bordering meniscus. If the contact angle is small, each crack, as it comes into the vicinity of the border, rotates away from it. The result is a pattern with a series of anastomosing loops shown in Figure 12b and similar to the brochidodromous organisation. If the contact angle is large the transverse cracks go straight to the border thus forming a craspedodromous pattern. These experiments show that it is possible to simulate the large scale venation as well as the small scale one.

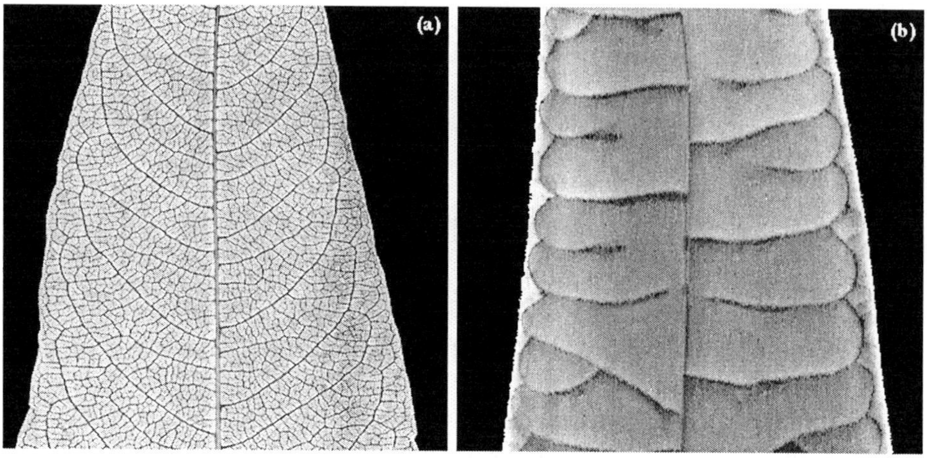

Fig. 12. — a) The brochidodromous organisation of the main veins of *Polygonum Polystachium.* b) The loops formed by cracks near the meniscus bordering a narrow band shaped sample.

2.4. Discussion

In general terms and as discussed in Section 1.4 several hypotheses have associated biochemical or to mechanical processes to botanical morphogenesis. However it is usually difficult to discrimate between the two. The origin of this difficulty lies in the complex interplay which exists in reality between the mechanics and the biochemistry in living tissues.

The various experiments on botanical growth which have shown the importance of biochemical processes have not ruled out a role of mechanical stresses. However it has been hitherto difficult to find discriminant experiments which demonstrate a dominant role for the mechanics of the tissues in the morphological processes. This is the context which gives its meaning to our result. We have shown that the vasculature of leaves have a topology typical of growth in a tensorial field. We have thus obtained a strong element of proof that the mechanical stresses in the growing tissues have a dominant role in, at least, one morphogenetic process: the formation of procambial strands. The hypotheses that we have put forward to interpret this effect are new but they are already supported by previous results. More work is needed for a complete demonstration of their physiological relevance.

3. CONCLUSION

The main point discussed in this lecture is the similarity of patterns obtained in different systems provided they have similar mathematical structure.

In reverse, different types of patterns correspond to different families of problems. Particular emphasis was given to the topological difference between structures growing in scalar or tensorial fields respectively. We have shown that this distinction can be used and how it could help understand the problem of the genesis of leaf venation.

Acknowledgements

Many of the results presented here were obtained over the years during collaborations with M. Adda-Bedia, C. Allain, A. Arnéodo, F. Argoul, M. Ben Amar, S. Douady, N. Gerard, V. Hakim, E. Lajeunesse, J. Maurer, M. Mashaal, L. Pauchard, M. Rabaud and H. Thomé to whom the author expresses his gratitude.

REFERENCES

C. Allain and L. Limat, *Phys. Rev. Lett.* **74** (1995) 2981-2984.

G.S. Avery, *Amer. J. Bot.* **20** (1933) 513-564.

J. Bataille, *Rev. Inst. Pétrole* **23** (1968) 1349-1364.

M. Ben Amar, *Phys. Rev. A.* **41** (1990) 2080-2092.

M. Ben Amar, *Phys. Rev. A* **44** (1991) 3673-3685.

M. Ben Amar, V. Hakim, M. Mashaal and Y. Couder, *Phys. Fluids A* **3** (1991) 2039-2042.

Ben Jacob, H. Schmueli, O. Shochet and A. Tenenbaum, *Physica A* **187** (1992) 378-424.

E. Blackman, *Ann. Bot.* **35** (1971) 653-665.

P. Bouissou, A. Chiffaudel, B. Perrin and P. Tabeling, *Europhys. Lett.* **13** (1990) 89-94.

R.L. Chuoke, P. Van Meurs and C. Van der Pol, *AIME* **216** (1959) 188-194.

Y. Couder, *Perspectives in Fluid Dynamics* (Cambridge University Press, 2000).

Y. Couder, F. Argoul, A. Arnéodo, J. Maurer and M. Rabaud, *Phys. Rev. A* **42** (1990) 3499-3503.

Y. Couder, L. Pauchard, M. Adda-Bedia, C. Allain and S. Douady, *On the self-organisation of leaf venation* (1999) Preprint.

W. D'Arcy Thompson, *On growth and Form*, Vols. 1 and 2 (Cambridge University Press, 1917), Edition of 1952.

S. Douady and Y. Couder, *J. Theor. Biol.* **178** (1996) 255-312.

K. Esau, *Plant Anatomy* (John Wiley, New York, 1953).

H. Fujikawa and M. Matsushita, *J. Phys. Soc. Jpn.* **60** (1991) 88-94.

L.J. Gibson and M.F. Ashby, *Proc. Roy. Soc. London A* **383** (1982) 43-59.

P.B. Green, *Int. J. Plant Sci.* **153** (1992) S59-S75.

A. Groisman and E. Kaplan, *Europhys. Lett.* **25** (1994) 415-420.

Z. Hejnowicz and A. Sievers, *J. Exp. Bot.* **46** (1995) 1045-1053.

L.J. Hickey, *Amer. J. Bot.* **60** (1973) 17-33.

P. Huerre, *Perspectives in Fluid Dynamics* (Cambridge University Press, 2000).

D. Kessler, J. Koplik and H. Levine, *Adv. Phys.* **37** (1988) 255-339.

U. Kutschera, *Physiologia Plantarum* **77** (1989) 157-163.

E. Lajeunesse and Y. Couder, *J. Fluid Mech.* **419** (2000) 125-149.

P.M. Lintilhac and T.B. Vesecky, *Amer. J. Bot.* **68** (1981) 1222-1230.

G.H. Mitchison, *Sci* **196** (1977) 270-275.

G.J. Mitchison, *Proc. Roy. Soc. London B* **207** (1980) 79-109.

H. Meinhardt, Models of pattern formation and their application to plant development, in *Positional controls in plant development*, edited by W. Barlow and D.J. Carr (Cambridge University Press, 1984).

W.W. Mullins and R.F. Sekerka, *J. Appl. Phys.* **35** (1964) 444-451.

T. Nelson and N. Dengler, *The Plant Cell* **9** (1997) 1121-1135.

L. Pauchard, F. Parisse and C. Allain, *Phys. Rev. E* **59** (1999) 3737-3740.

P. Pelcé, *Dynamics of Curved Fronts* (Academic Press, San Diego, 1988).

W.S. Peters and A.D. Tomos, *Ann. Bot.* **77** (1996) 657-665.

Y. Pomeau and M. Ben Amar, Dendritic growth and related topics, in *Solids far from equilibrium*, edited by C. Godrèche (Cambridge University Press, 1992).

X.W. Quian and H.Z. Cummins, *Phys. Rev. Lett.* **64** (1990) 3038-3041.

M. Rabaud, Y. Couder and N. Gerard, *Phys. Rev. A* **37** (1988) 935-947.

T. Sachs, *Pattern formation in plant tissues* (Cambridge University Press, Cambridge, 1991).

H. Thomé, M. Rabaud, V. Hakim and Y. Couder, *Phys. Fluids A* **1** (1989) 224-240.

T. Witten and L.M. Sander, *Phys. Rev. B* **27** (1983) 5686-5697.

Ya.B. Zel'dovich, A.G. Istratov, N.I. Kidin and V.B. Librovich, *Combust. Sci. Technol.* **24** (1980) 1-13.

CHAPTER I

Branching in Plants

F. Hallé

INTRODUCTION

Branching in botany has been much studied for a long time; as a consequence plants, and especially trees, play an important part in the knowledge we have of branching. Depending on which plant species is considered, branching may be nihil, apical or lateral, exogenous (case of stems) or endogenous (case of roots), continuous or rhythmic, immediate (tropical plants) or delayed (plants of temperate climate), spontaneous or traumatic. Trees can be unitarian or, through a special mechanism of branching called reiteration, they can become colonies. A few architectural models - which are basically branching models - shall be presented. The many roles of branching will be discussed: space occupation, increase of energy catching, rapid increase of surface in still a small volume. Comparison with another group of sessile (= fixed) living beings, the corals, shows that branching is not restricted to plants, but that it is a characteristic of sessile life; mobility and branching are mutually exclusive. No plant, either fossil or extant, is completely devoid of branching (= ramification) and this is the reason why the term "branching" in itself evokes the plant kingdom. As a matter of fact, botany is the scientific field in which branches and branching can best be studied. There is a long history of interest in plant form - especially in plant branching, as a plant's overall form is defined by its branching. The first scholar to become fascinated by the subject was, as far as I know, the Greek philosopher Theophrastus (circa 400 BC). Then came Leonardo da Vinci with his treatise on tree ramification [1], for the artist's attention. The German poet Goethe (18th century) then Goebel [2], Velenovsky [3] and Arber [4] have laid the foundations of our modern plant morphology - which is basically the science of plant ramification (or branching). Hallé *et al.* [5, 6], Bell [7] and Kaplan [8] are contemporary heirs of this tradition. In the following pages many concepts are borrowed form Bell and Kaplan. Branching in plants cannot be separated from growth, as these two phenomenons are the two faces of the same coin. Hence, I shall begin with some reflections concerning the relationship between growth and branching, in the general context of the biology of multicellular organisms (plants and animals).

1. GROWTH AND BRANCHING

Plants and animals both begin as a single cell, the fertilized egg. Cell divisions follow and lead to growth; therefore growth is a common feature of plants and animals: to get from small to big, growth is the only way!

Branching, however is not a common feature of the two multicellular groups of living beings. In plants, branching is not only frequent, it is a constant, whereas in most animals it is not involved in the growth and building-phase of the external form. Even limbs, fingers or appendages, which look superficially like branches, are actually created through programmed cell death (or apoptosis, Ameisen [9]). A cat has definitely no branches.

If we, as biologists, are to understand the meaning of branching, we have to ask the question: if growth is common to plants and animals, why is there branching in plants and not in animals?

Answering this question is not simple. The answer has to do with the sessile condition of plants as opposed to animals being mobile organisms; it also depends on whether the energy used for growth gets into the multicellular body through and internal or an external surface, and this point deserves a more detailed discussion.

In animals, cats for instance, food is swallowed. Then, the biochemical energy contained in it enters the body through the surfaces of the digestive tube, hence through an internal surface. Furthermore, this digestive surface is very large (Fig. 1).

A big external surface would hinder mobility, therefore a cat has to keep its external surface small. This is why an animal is essentially a body, transversed by a tube (the digestive tube) having a large internal surface area, and which is made up of an external surface that is as small as possible.

Branching, a mechanism that creates large surfaces, is not involved in the embryological development of external animal form; conversely it is a typical feature of plants, regularly associated with their growth (Fig. 1).

Plants – at least green plants – catch solar energy on their external surface; this is how this energy gets into the body. As compared with biochemical energy of the animal food, solar energy flow is rather weak: less than one kilowatt per square meter per day. As a consequence, over the life of a plant, its external surface, also called the photosynthetic surface, must become very large.

The digestive surface area inside the animal finds its functional counterpart in the photosynthetic external surface of the plant. Take a cat and turn it

Fig. 1. — The internal digestive surface of an animal (bottom) *versus* the external photosynthetic surface of a plant (top) [10].

inside out: with the "internal" surface of the bowels worn outside instead of fur, you get a (sort of) plant!

2. WHY ARE THERE MERISTEMS IN PLANTS?

Another basic difference between these two groups of multicellular organisms is the presence of meristems in plants and not in animals. This also requires explanation. First, I will explain what a meristem is. At the apex of every shoot and root, there is a small group of embryonic cells, where active cell division and enlargement occurs. Growth takes place in these very localized regions which are termed meristems [7].

Why are there meristems in plants and not in animals?

In order to catch as much solar energy (for leafy shoots) and water (for roots) as possible, plants have to have large external surfaces. A more or less

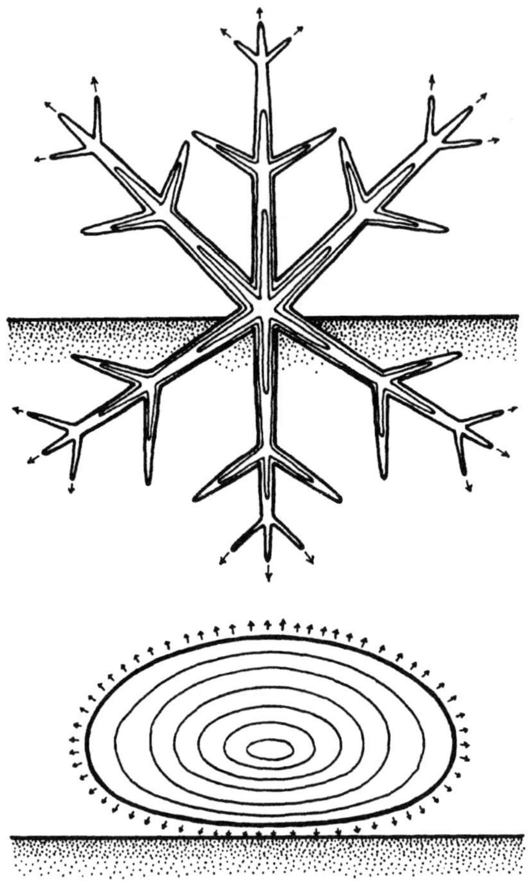

Fig. 2. — With its diffuse growth, a more or less spheric volume is prone to mobility (bottom, the animal) whereas meristematic growth creates a branching system which is sessile and spreads in two contrasted environment (top, the plant).

spherical form would not work, as the surface of a sphere is small in relation to its volume; the greater plant-form diverges from the sphere, the better it will be for plant growth and function.

How to arrive at the largest possible surface? As already stated, branching is part of the answer; the acquisition of a meristem is the other.

During the embryonic phase in animals, growth involves the whole surface, whereas in plants it does not (Fig. 2).

Instead, plant growth involves stems, twigs and roots, each of them having an apical meristem and therefore growing at its tip (Laufs, this volume).

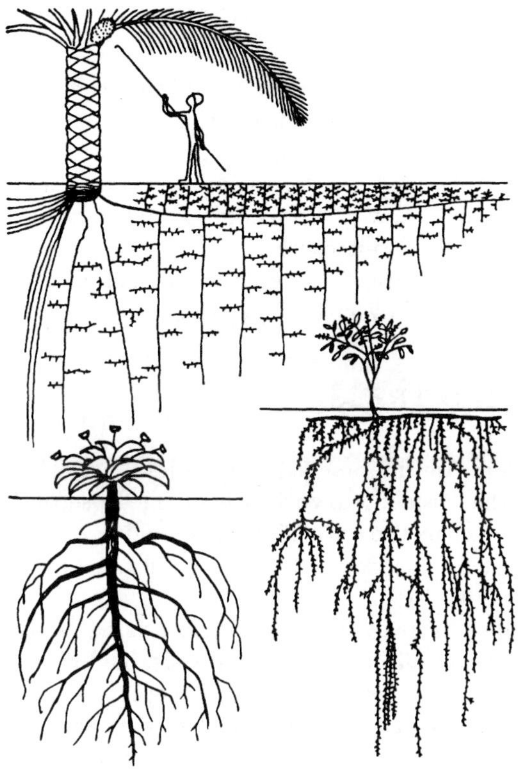

Fig. 3. — In plants, root systems are always profusely branched; the underground surface is much bigger than the aerial one [10].

The end result is a plant area of astonishing magnitude; a tree of normal height (±30 m) has an estimated total surface area of 200 hectares, which can be visualized by a rectangle of 2 kilometers by one [10]. Moreover, the bulk of this huge surface is underground (Fig. 3), root systems being always profusely branched.

Such a large surface makes it difficult for trees to move from one place to another and this is consistent with the fact that they are fixed in place – or sessile. This especially makes sense if one considers that they thrive in two environments of contrasted densities, that is, air and soil; the nascent plant occupies the interface between these two environments, then spreads in both directions.

Solar energy being much the same everywhere, trees have no need to run after their food as do many animals.

3. BRANCHING AND ARCHITECTURE

Thom [11] challenged the botanists when he said that the complexity of branching in trees is beyond any analysis. During the same period of time, Oldeman and I were analyzing tropical tree branching as a part of what is now called "tree architecture". Eventually I expressed my disagreement with Thom [12]. Here I want to summarize how it is possible to analyze the branching of a tree in order to understand its architecture.

First, not all trees have branches. A tree can remain unbranched – that is to stay unbranched in its aboveground parts, as roots are always branched.

4. UNBRANCHED TREES

A palm or a tree-fern are unbranched trees (Fig. 4). Such trees have usually very large leaves, the largest known leaves being those of *Raphia regalis*, an unbranched African palm: they are 25 meters long!

Is it correct or not to consider such trees as unbranched? It depends on the status leaves are endowed with.

A leaf is, in itself, a flattened branch system arising at right angle from the bearing stem, highly specialized in photosynthesis, having an active metabolism and a short life-span. However, leaves are so deeply specialized and differentiated that we, botanists, consider unbranched a tree bearing only leaves.

Although this unbranched condition does exist in ferns (*Cyathea*, Fig. 4) in Cycads (*Encephalartos*), Gymnosperms (*Dicranophyllum hallei*, Fig. 4 [13]), Monocotyledons (the coconut tree) and Dicotyledons (the pawpaw tree), it is not frequent in trees as most of them do produce leaves plus branches: this is branching, within which we have to distinguish two major types, apical an lateral branching.

5. APICAL vs. LATERAL BRANCHING

If the apical meristem of the shoot – or the root – undergoes division into two branch meristems, and if the parent shoot apex gives up its existence in the formation of the two branches, then we have an apical branching (Fig. 5). This occurs in Pteridophytes (*Lepidodendron*, a fossil tree; *Lycopodium* or *Psilotum*, two tropical herbs), *Cycas*, the Monocotyledons (*Hyphaene*, *Nypa*; *Flagellaria*, a vine) and Dicotyledons, where the only example found so far is *Mammilaria*, an herbaceous cactus. However, despite this large distribution, apical branching is even more unfrequent than is the unbranched condition. Much more usual is the lateral branching.

In the vast majority of Gymnosperms and Angiosperms, there are meristems in the axil of any subtending leaf and therefore lateral branching is usually termed "axillary branching".

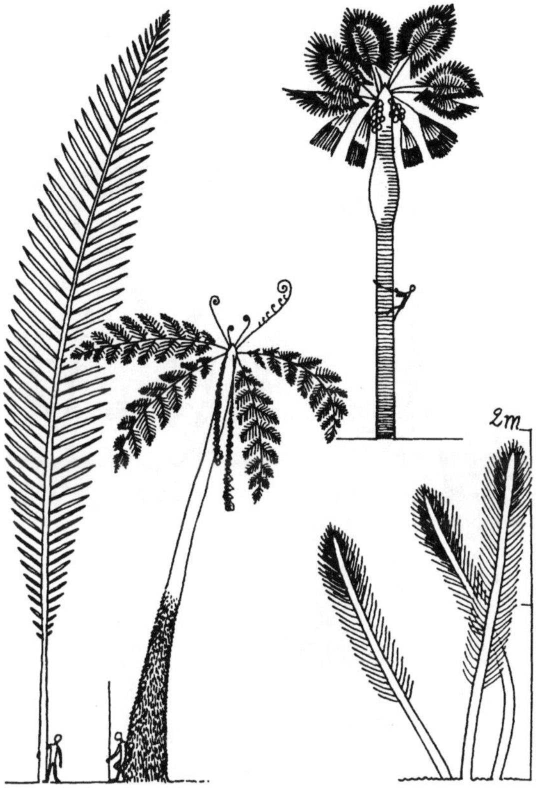

Fig. 4. — Unbranched trees. Top right: the *Palmyra* palm. Bottom left: the largest known leaf, 25 m long, is the one of *Raphia* palm, Congo. Bottom center: a tree-fern. Bottom right: *Dicranophyllum hallei*, a fossil Gymnosperm [13].

Many examples are provided by familiar trees, either European – oak, pine, poplar, beech – or tropical – mango-tree, coffee-tree, rubber-tree or cocoa-tree (Fig. 6). In roots, where there are no subtending leaves, branching is not axillary but still is lateral; roots have a special sort of branching, termed "endogenous".

6. EXOGENOUS vs. ENDOGENOUS BRANCHING

In the aerial parts of the tree, buds are on the surface of the stems; branching, then, is superficial or exogenous. Branching in roots is somewhat different.

In the root, due to the risk of meristem erosion by growing through the soil, buds are replaced by root primordia which are groups of meristematic cells originating below the surface of the parent root. The lateral roots push their

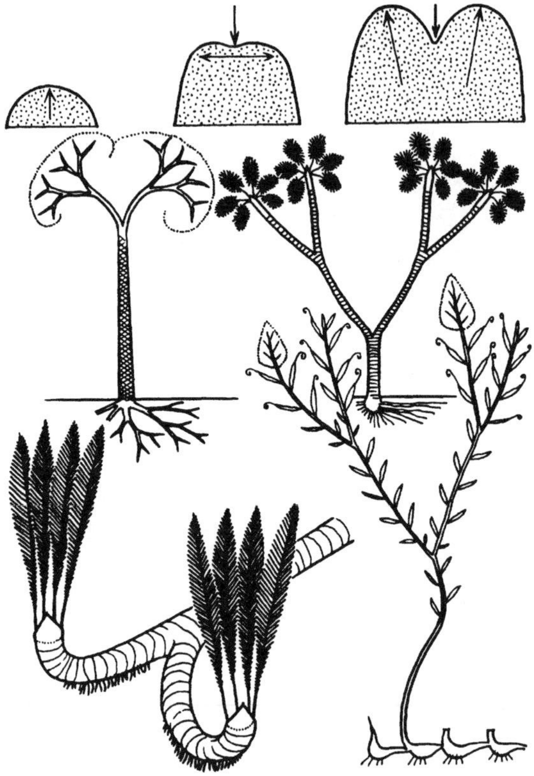

Fig. 5. — Apical branching. Top: schematic view of apical branching [8]. Middle left: a fossil *Lepidodendron* displaying apical branching in its branches and roots. Middle right: the Doum palm of the African Sahel (*Hyphaene thebaica*). Bottom left: the *Nypa* palm is creeping, with apical branching. Bottom right: the tropical vine *Flagellaria* [6].

way out through the parent root cortex. Branching proceeds from inside and is called endogenous (Fig. 7).

Within lateral branching two types have to be distinguished, continuous and rhythmic branching.

7. CONTINUOUS vs. RHYTHMIC BRANCHING

Figure 6 gives an example of continuous branching in the Coffee-tree. The continuous growth of the vertical axis results in the regular disposition of leaves along this axis. As every leaf produces an axillary branch, branching is continuous. Continuous branching tends to be restricted to tropical trees, submitted

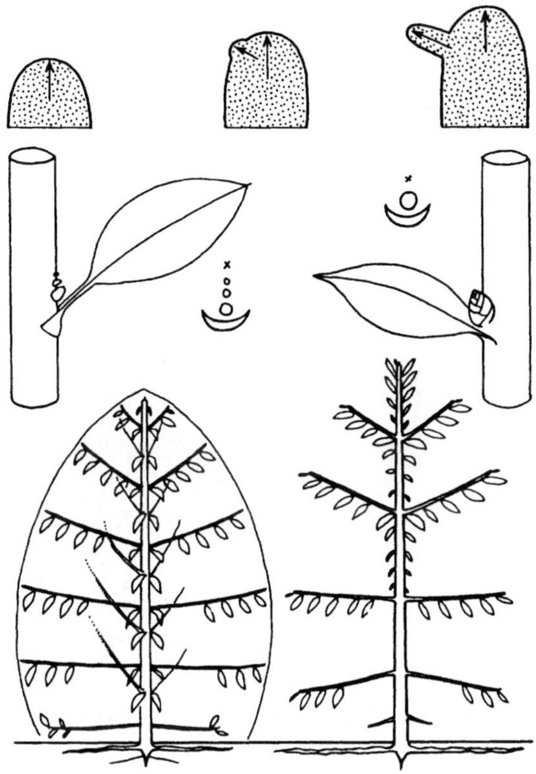

Fig. 6. — Lateral branching. Top: Schematic view of lateral branching [8]. Middle; two examples of axillary buds, with their schematic representations. Bottom left: continuous lateral branching in the Coffee-tree. Bottom right: rhythmic lateral branching in the Nutmeg-tree.

to a climate in which hot weather and high humidity are more or less constant. Nevertheless, even in the humid tropics, many trees exhibit rhythmic branching (ebony, kapok-tree, nutmeg-tree, *Araucaria*).

Rhythmic branching becomes the rule in higher latitudes, Europe for instance, where the regular return of winter, by stopping the growth, imposes a rhythmic growth, a rhythmic position of leaves on stems and eventually a rhythmic branching of annual wave-length (Fig. 6).

Latitude and climate impose another constraint on branching, which can be delayed or immediate.

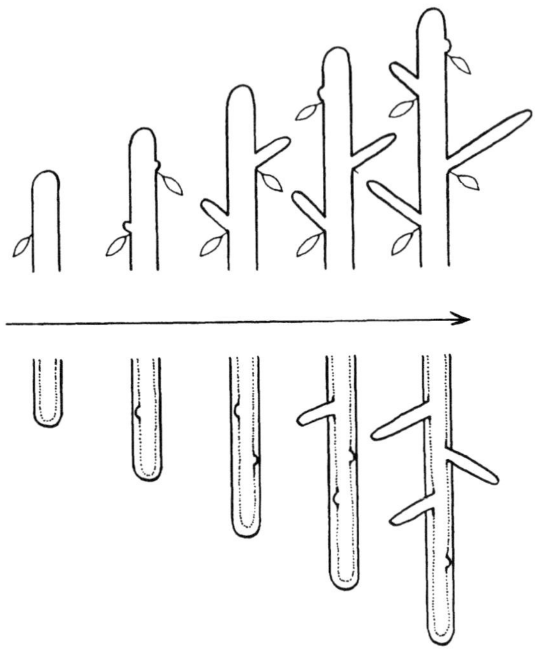

Fig. 7. — Exogenous branching of stems (top) *versus* endogenous branching of roots (bottom).

8. IMMEDIATE vs. DELAYED BRANCHING

In the humid tropics, where climatic conditions are permanently the best possible for plant growth, the axillary meristem may grow and develop without any delay and thus extends simultaneously with the apical meristem of the parent-axis (Fig. 8): this is immediate branching, so frequent in tropical trees.

On the contrary, in seasonal climates like the one we have in Europe, summer is the season where the axillary meristem becomes "organized into a temporarily protected resting structure, a bud" [7]. The bud rests during winter and branching is postponed until the next spring: this is why it is said to be delayed (Fig. 8).

Branching and branching again, the tree soon turns into a huge collection of branches. Depending on the tree species, two contrasting situations may occur; branching can be homogeneous, or differentiated.

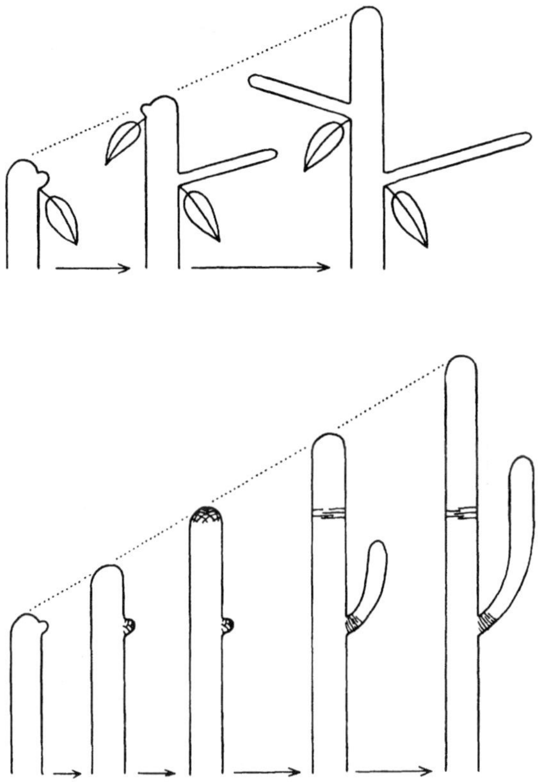

Fig. 8. — Immediate branching (top) *versus* delayed branching (bottom).

9. HOMOGENEOUS vs. DIFFERENTIATED BRANCHING

In a big oak, for instance, all the twigs forming themselves into the crown's canopy tend to grow vertically, they all have a rhythmic growth and they all bear flowers and fruits. The collection of branches is then homogeneous, and this is correlated with small branching angles (Fig. 9). A first possible situation is properly described as homogeneous branching.

If the branching angles are wider, a differentiation is usually observed between the many axis forming the tree crown. Axis which are more or less vertical have a big primary diameter, a rapid growth, a long lifespan and they are specialized in the exploration of new spaces – whereas those which are more or less horizontal have a small diameter, a limited growth, a short life span, and they are specialized in the exploitation of the incident energy of light.

The wider the branching angle, the deeper the differentiation, the smaller the branch diameter and the shorter the branch lifespan.

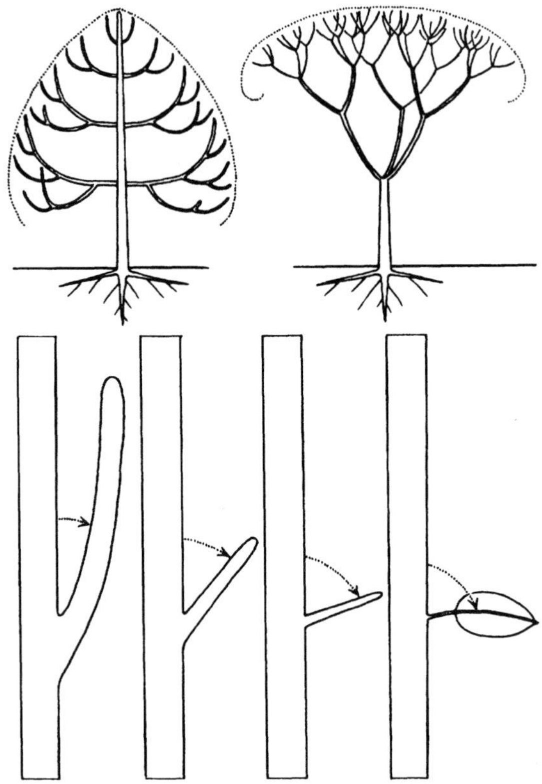

Fig. 9. — Homogeneous branching (top) in Oak (left) and cassava (right). Bottom: the wider the branching angle, the deeper the branching differentiation.

With a right angle of branching, what is laterally produced is either a branch resembling a leaf (a phyllomorphic branch) or the leaf itself, which is an extreme in branch differentiation to catch the light energy; a leaf has a thin petiole and a short life span.

Road systems show equivalent structures. A highway divides with a small angle, and the two ways keep approximately the same big diameter whereas a small track may arise at right angle on the main road. Flows are operating in both branch systems and road systems.

10. SPONTANEOUS vs. TRAUMATIC BRANCHING

So far, in all the examples quoted above, branching was spontaneous; merely linked to growth, it allowed the tree to catch more space, more light and more water.

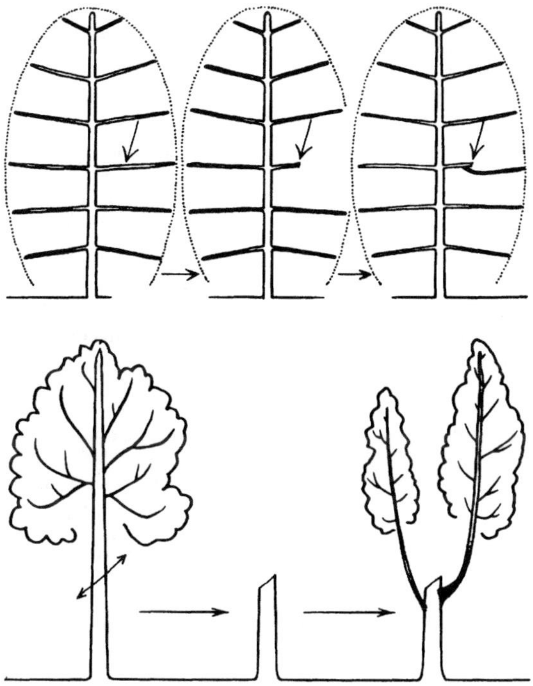

Fig. 10. — Spontaneous *versus* traumatic branching. Without a traumatism (arrows) no branching would have occurred. The traumatism induces the regeneration of the damaged axis, which could be either a trunk (bottom, a young oak) or a branch (top, an Araucaria).

But branching of a given stem can also be triggered by a damage on this stem. Figure 10 shows two cases in which, without the stem having been chopped, no branching would have occurred; this is what tree architects call traumatic branching. In these two cases (Fig. 10), the traumatism induces a regeneration of the damaged axis.

Leaving the scale of the leafy shoot, let's have a look at the tree as a whole, in order to introduce the concept of architectural model.

11. THE ARCHITECTURAL MODELS

Observations of the sequence of events that takes place during the lifespan of different trees indicate that each species has a recognizable "blue print" to which the young tree conforms itself. This "blue print" – or architectural unit – represents a particular developmental sequence of branching with a finite number of branch categories [14]. The architectural units can be grouped into

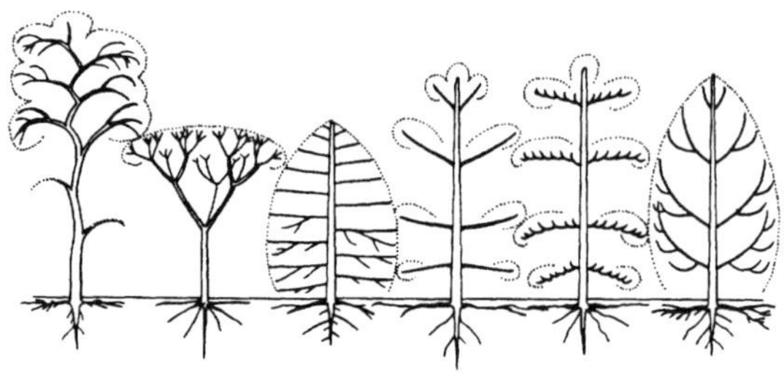

Fig. 11. — Six architectural models. From left to right: Troll's, Leeuwenberg's, Roux's, Massart's, Aubréville's and Rauh's. Thirty architectural models are already described in plants [5, 6].

branching models – or architectural models [5, 6]. Around thirty architectural models are already described in plants, either tropical or temperate, trees or herbs, fossil or extant. For the sake of simplicity, only six of them are shown (Fig. 11).

The same architectural model occurs again and again in plants which are botanically very remote. Pines, oaks and the rubber-tree (*Hevea brasiliensis*, Euphorbiaceae) belong to the same model of Rauh.

Architectural models are usually easy to observe in young trees, growing in good environmental conditions. Full grown big trees can either remain unitarian or become colonies.

12. UNITARIAN TREES vs. COLONIES

Some trees are unitarian. This means that they grow according to one model; the old tree is therefore isomorphic with the young one. Araucaria is a good example (Fig. 12).

Most trees, either temperate or tropical, are not unitarian; they are instead colonies of architectural units, growing on top of each other (Fig. 12).

We call reiteration the process through which the tree duplicates its architecture, producing new copies of its architectural unit [15]. Reiteration is a special sort of branching giving birth, not to a branch, but to a "new tree" growing on top of the former one.

This "new tree" is usually complete, having trunk, branches, leaves, flowers, fruits and even sometimes a root system (Fig. 13). The reiterated architectural unit can also be pauperized, being then deprived of roots, of trunk or even of leaves.

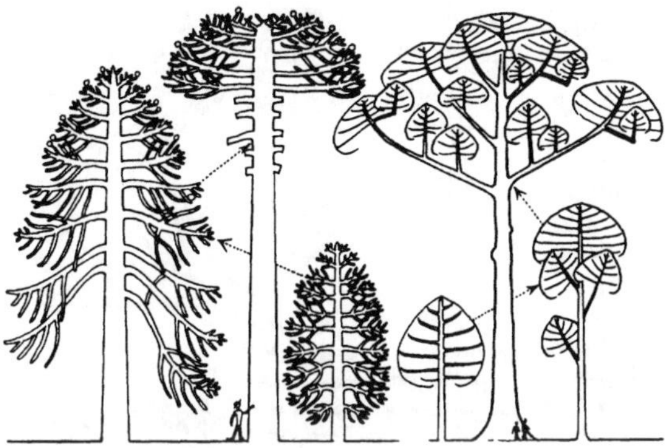

Fig. 12. — Unitarian trees *versus* colonies. left: three stages in the growth of an unitarian tree, *Araucaria araucana*. Right: three stages in the growth and reiteration process of a Dipterocarp. The adult is clearly a colony of architectural units [14].

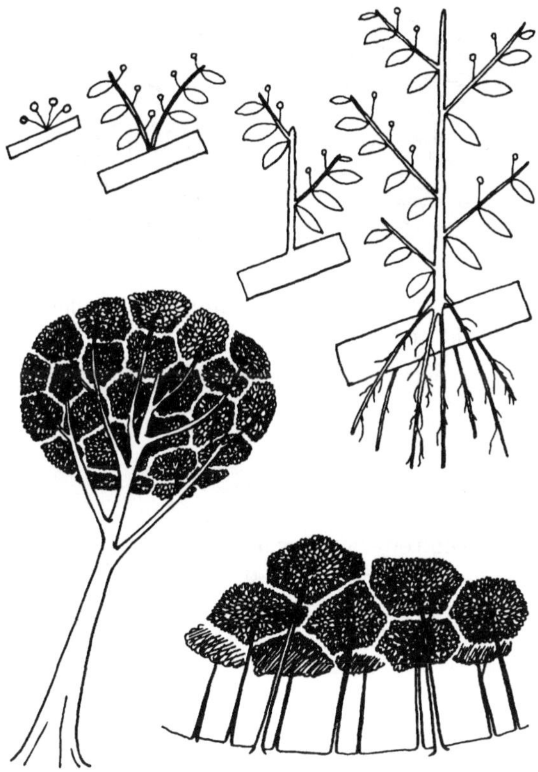

Fig. 13. — Pauperization of the reiterated architectural unit (top) and "crown-shyness" (bottom).

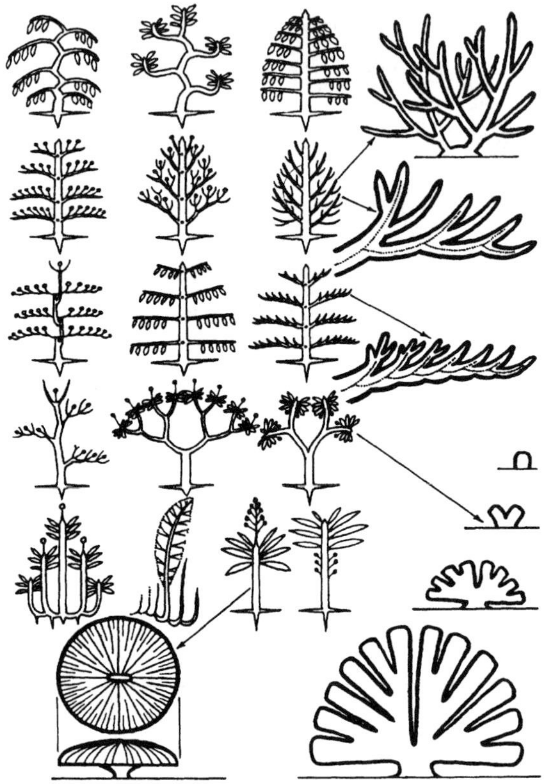

Fig. 14. — Trees and corals [10]. Several architectural models apply to trees (left) and to corals as well (right and bottom).

13. CROWN-SHYNESS

For unknown reasons, the reiterated units, in some tree species, avoid to touch each other. This has been called "crown-shyness", and the term is now classical, although it is clearly anthropomorphic. Most of the very tall trees – pines, oaks, Dipterocarps, and so on – display some sort of crown-shyness (Fig. 13).

14. BRANCHING IN TREES AND CORALS

In biology, it frequently occurs that several characteristics are linked together, in such a regular way that some degree of prediction becomes possible.

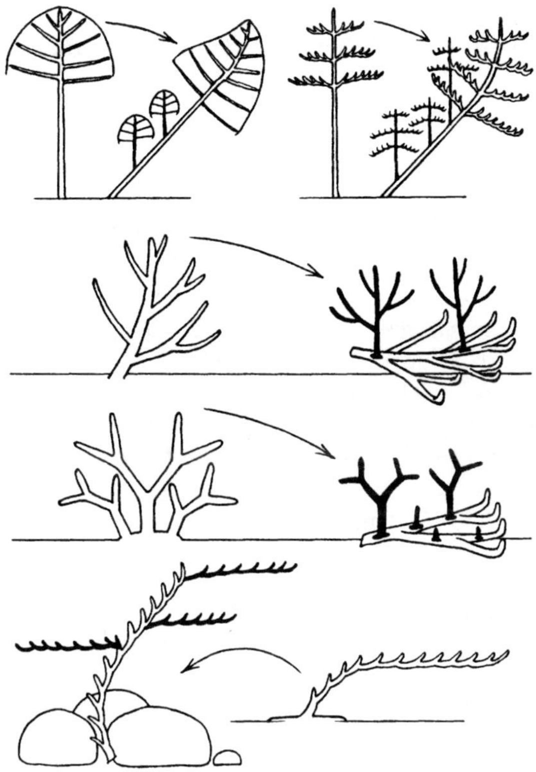

Fig. 15. — Traumatic reiteration in trees (top) and in corals (middle and bottom) [10].

For instance, if the energy enters the body through an external surface, and if the organism is of macroscopic size, then it can be predicted that:

- This organism is fixed and it has branches;
- Its external surface is very large;
- To keep it large, this organism has meristems at the tip of its stems;
- It is likely to be a colony – instead of an individual – when fully grown.

All this applies to plants in general, and especially to trees. Interestingly, it also applies to at least one zoological group, the one of corals (Figs. 14 and 15). A thorough comparison between trees and corals would certainly be outside the scope of this contribution. At this point, it is enough to indicate that corals display meristems, branching (apical *vs.* lateral, spontaneous *vs.* traumatic, etc.), architectural models, reiteration and shyness [10].

As a concluding remark, I want to state that branching in biology is by no means restricted to plants; instead it should be considered as a regular feature of the fixed (or sessile) life. The antagonism between branching and mobility comes first, and the distinction between plants and animals takes the second place.

REFERENCES

[1] L. da Vinci, *Traité de la peinture. Sixième partie: les arbres et les verdures* (Paris, 1651).
[2] K. Goebel, *Organography of plants. Part I. General organography. Part II. Special organography*, Authorized English edition by I.B. Balfour (Clarendon Press, Oxford, 1900 and 1905).
[3] J. Valenovsky, *Vergleichende morphologie der pflanzen* (Verlagsbuchhandlung von F. Rivnac. Druck von Eduard Leschinger, Prag, 1907).
[4] A. Arber, *The natural philosophy of plant form* (Cambridge University Press, 1950).
[5] F. Hallé and R.A.A. Oldeman, *Essai sur l'architecture et la dynamique de croissance des arbres tropicaux* (Masson, Paris, 1970).
[6] F. Hallé, R.A.A. Oldeman and P.B. Tomlinson, *Tropical trees and forests: An architectural analysis* (Springer, Berlin, 1978).
[7] A.D. Bell, *Plant form. An illustrated guide to Flowering Plant Morphology* (Oxford University Press, 1991).
[8] D.R. Kaplan, *Principles of Plant Morphology* (Odin Readers, Berkeley, 1997).
[9] J.C. Ameisen, *La sculpture du vivant. Le suicide cellulaire ou la mort créatrice*, Collection Science Ouverte (Le Seuil, Paris, 1999).
[10] F. Hallé, *Éloge de la plante. Pour une nouvelle biologie*, Collection Science Ouverte (Le Seuil, Paris, 1999).
[11] R. Thom, *Une théorie dynamique de la morphogénèse*, in: Towards a theoretical biology (Waddington, C.H. Edinburgh University Press, 1968).
[12] F. Hallé, *Les modèles architecturaux chez les arbres tropicaux ; une approche graphique*, in: Elaboration et justification des modèles ; applications en biologie (Delattre and Thellier, Maloine, Paris, 1979).
[13] M. Barthel, E. Bettag and D.R. Noll, *Veröffentlichungen des Museums für Naturkunde Chemnitz* **21** (1998) 5-20.
[14] D. Barthélémy, C. Edelin and F. Hallé, *Architectural concepts for tropical trees*, in: Tropical forests. Botanical dynamics, speciation and diversity, edited by Holm-Nielsen *et al.* (Academic Press, London, 1989) pp. 89-100.
[15] R.A.A. Oldeman, *L'architecture de la forêt guyanaise* (Orstom, Paris, Mémoire 73, 1974).

CHAPTER II

Inside the Buds: The Meristems

P. Laufs and J. Traas

1. INTRODUCTION

In animals most organogenesis takes place during embryogenesis and further development consists mostly of growth and maturation of the embryonic organs. This is in contrast to plant embryo development, where only a very basic structure is set up [1]. Mature plant embryos are composed of a single shoot-root axis. The primary shoot consists of a small stem, and, depending on the species, one or several cotyledons (specialized embryonic leaf-like structures) and a limited number of leaves can be formed. Almost all the organs found in an adult plant are formed after germination by small groups of dividing cells called meristems. This implies that multiple ramifications that characterize the architecture of many adult plants take place during postembryonic plant development. The continuous mode of development in plants makes them very flexible, as they can adapt their growth and architecture (development) to environmental conditions. This adaptability counterbalances their inability to move and to flee local unpropitious conditions.

Several meristems exist. The root apical meristem, lying in the embryo at the lower end of the shoot-root axis, will initiate the root system. The shoot apical meristem, located on the upper end of the shoot-root axis, will ultimately produce all the above ground organs. The flowers are produced by particular meristems, the floral meristems.

Here, we will not review the current knowledge of all aspects of branching in plants, but will focus on examples taken from the development of the aerial part: branching of stems and leaves. First, we will discuss general aspects of branching in plants. This will highlight the central roles of the meristems during plant development. We will then discuss the basis of meristem organization and function and how this can be related to branching.

This chapter will be illustrated by examples taken from two model-species, tomato (*Lycopersicon esculentum*) and a small Crucifer *Arabidopsis thaliana*, sometimes called wall cress. These two species are classical models in plant development studies and a lot of morphological, cellular, genetic and molecular data are available. Furthermore, they have a very different architecture and physiology. Some complementary information may however be taken from other examples, when relevant.

2. THE INTERNODE: THE BASIC UNIT OF PLANT ARCHITECTURE

After germination, the shoot apical meristem (SAM) produces the shoot, which can be divided into repetitive modules called phytomers. These are formed by a portion of the stem, called the internode, and the node, composed of one or several lateral leaves, each bearing an axillary meristem in its axil (Figs. 1A and 1B). The general architecture of the plant is determined by relatively few parameters associated with each phytomer: these include the length of the internode, number and position of the leaves and their associated axillary meristem, outgrowth and identity of the axillary meristems. These parameters vary between different plant species, giving rise to the great diversity observed in plant architecture.

Each axillary meristem can turn into a new SAM, producing part of the vegetative organs of the plant or part of an inflorescence, or alternatively an axillary meristem may turn into a flower. In addition depending on the species, the SAMs themselves can be transformed into flowers. Therefore, flowers can have a different position relative to the stem. For instance, they can be in a terminal position, as it is for instance the case in tulips, and result from the transformation of the SAM into a floral meristem. Flowers may also be in a lateral position to the main stem as for example in *A. thaliana* (Fig. 2).

Flowers can also be divided into phytomers, however these differ from shoot phytomers in several ways (Figs. 1C and 1D). First, in the flower, lateral organs develop into floral organs (sepals, petals, stamens and carpels, Fig. 1C) instead of leaves. Secondly, little internode elongation takes place during development of most flowers (Fig. 1D) whereas shoots often show internode elongation (as it the case in tomato, Fig. 2). Finally, flowers are determinate structures whereas shoots are often indeterminate (as in *A. thaliana* (Fig. 2)) *i.e.* the SAMs which produce the shoot, can function over a long period of time and produce many organs. On the other hand, floral meristems will form only a limited and predetermined number of organs (the different organs found in a mature flower, Fig. 1C) before stopping their growth (apical determinacy) and no axillary meristems are formed in the axils of floral organs (lateral determinacy).

These differences between flowers and shoots are controlled by only a few regulatory genes. Mutations in these genes will lead to a partial or a full conversion of a flower into a shoot. For instance, mutations such as *floricaula* in

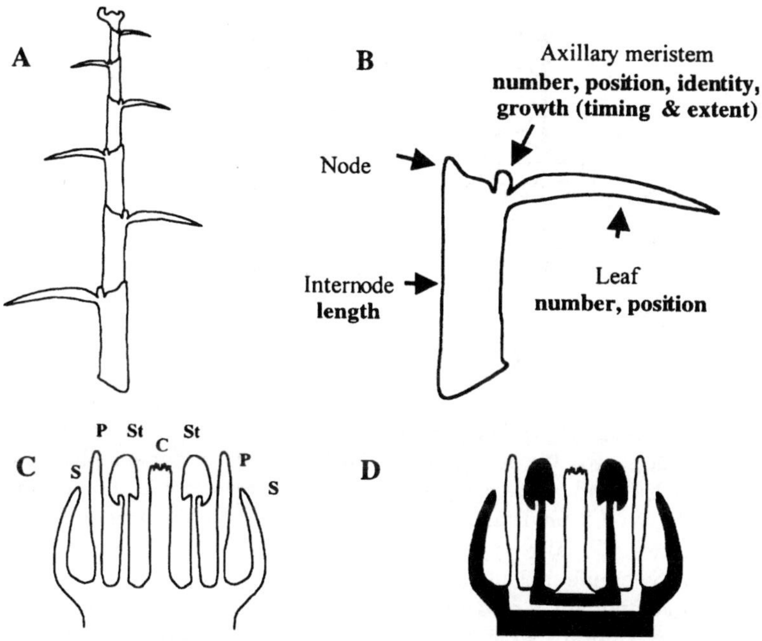

Fig. 1. — The aerial part of a plant is formed by repetitive units, the PHYTOMERS. A) The vegetative part of a plant can be divided into units called phytomers which are produced by the shoot apical meristem. B) Each phytomer is formed by a portion of the stem called the internode and a node where one or several leaves are inserted. In the axil of each leaf an axillary meristem is found. Several parameters can be defined for each phytomer such as the length of the internode, the numbers of leaves and axillary meristems. These parameters determine the general architecture of the plant. C) Median section through a typical flower. Four types of organs are found in four concentric whorls. From the periphery to the centre are found successively sepals (S), petals (P), stamens (St) producing pollen and carpels (C) where the seeds will finally develop. D) The flower can be divided into compressed phytomers (represented in black or white).

Antirrhinum majus (snapdragon) or *leafy* in *A. thaliana* result in floral meristems being replaced by SAMs [2, 3]. The specialized floral organs are transformed into leaf-like organs when the genes required for determining their floral organ identity are not functional [4]. Some mutations disrupt the determinacy of floral meristems: lateral flowers are observed in the axils of *apetala-1* sepals [5] and loss-of-function in the *agamous* gene leads to apically indeterminate flowers [6]. The *fimbriata* mutation breaks down both the lateral and apical indeterminacy of *A. majus* flowers [7].

In addition, several mutations affecting meristem size or maintenance have similar effects on shoot apical and floral meristems. This indicates that although differences exist, flower meristems and SAMs are related structures.

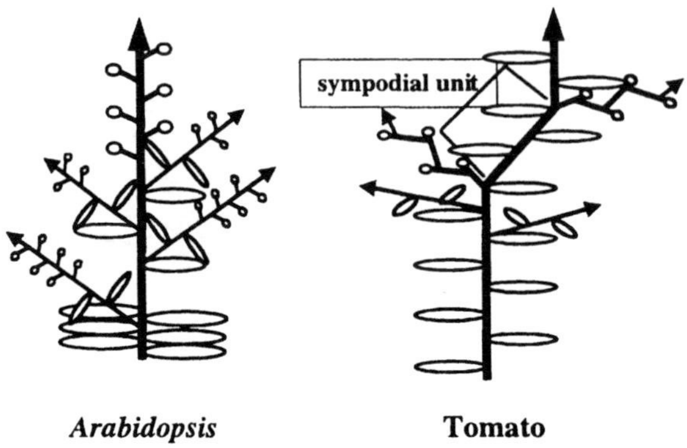

Fig. 2. — Architecture of *Arabidopsis thaliana* and tomato. In *Arabidopsis thaliana*, after germination, a rosette is produced which consists of type 1 phytomers (short internode, one leaf and an axillary meristem). In response to environmental and internal signals, type 2 phytomers are produced (elongated internode, one leaf and axillary meristem). Finally, type 3 phytomers are produced (elongated internode, lateral flower). The main axis is produced by the SAM originating from the embryo, and is hence monopodial. Growth of *A. thaliana* is indeterminate because the SAM never turns into a floral meristem and has therefore the potentiality of producing additional phytomers. Axillary meristems of type 1 and type 2 phytomers will reproduce the same pattern as the main shoot: producing a few type 2 phytomers before switching to type 3 structures. In tomato, the shoot apical meristem forms 6 – 10 vegetative type phytomers consisting of an elongated internode, a leaf and an axillary meristem. The apical meristem will then produce the first inflorescence. Growth will resume from the uppermost axillary meristem, which is called the sympodial meristem. Three more leafy phytomers are formed before the meristem produces the second inflorescence, giving rise to a sympodial unit. Although each sympodial unit is determinate, the whole shoot has an indeterminate growth because the number of units that can potentially be formed is not limited. The growth of the axillary meristem will push the inflorescence to an apparent lateral position. The outgrowth of other axillary meristems gives rise to a few vegetative phytomers before switching to the inflorescence stage. The arrows represent shoot apical meristems, the ellipses leaves and the circles flowers.

They share common regulatory mechanisms and the flower can be seen as a modified shoot. Such a concept was already proposed by Goethe in 1790 and de Candolle in 1813 (cited in [8]).

3. MERISTEM IDENTITY AND BRANCHING PATTERNS

3.1. Modified Meristem Identity Can Affect Branching Patterns

Modifications in phytomer parameters are observed during the development of each plant and two main phases are usually distinguished: a vegetative phase

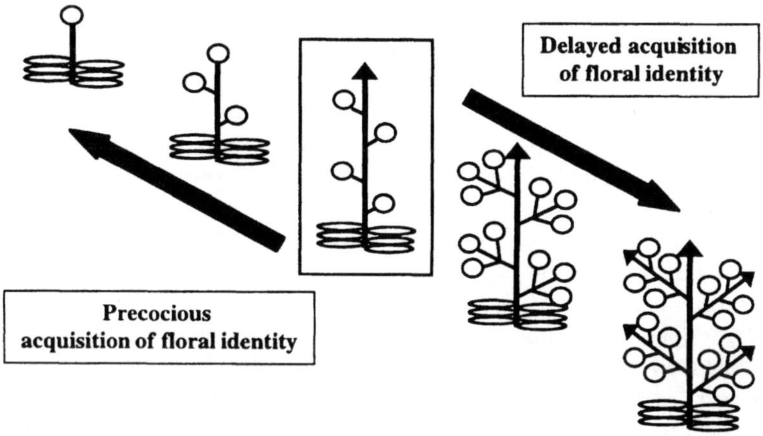

Fig. 3. — Meristem identity and branching pattern. The effect of different meristem identities on the architecture of a hypothetical plant is represented. In the normal situation (centre, boxed) the SAM produces a few leaves and then forms an indeterminate inflorescence axis, which initiates flower meristems in lateral position. If the acquisition of floral identity in the lateral meristems is delayed (*i.e.* they have a temporary inflorescence identity), a more complex branching pattern will be observed (from centre to right). Each lateral meristem will form a short secondary inflorescence axis, before producing an apical flower. An indeterminate secondary inflorescence axis can be formed if the axillary meristem remains with its inflorescence identity. On the other hand, if the primary inflorescence meristem switches to the floral identity a shorter inflorescence axis with a terminal flower is formed (from centre to left). In the most extreme case only a single terminal flower can be formed.

during which leaves and side branches are produced, and a reproductive phase during which the inflorescence or the flowers are formed. The regulation of the identity of the meristems is an important parameter for the branching pattern of the plant as illustrated in Figure 3.

An illustration of the effect of a change in meristem identity on plant architecture is shown in the *leafy* mutant of *A. thaliana* previously cited, in which flowers are replaced by shoot-like structures subtended by a leaf-like organ [3]. Therefore in the *leafy* mutant, lateral meristems have an inflorescence identity instead of a floral identity and this conversion from an apically determined to apically undetermined lateral meristem leads to an increased branching.

Another classical example of modified plant architecture which is due to differences in meristem identity is observed in the cabbage family. Wild cabbage, *Brassica oleracea* var. *oleracea*, has a typical Crucifer structure close to the one described for *A. thaliana*. During domestication, varieties as different as kohlrabi, Brussels sprouts, cabbage, broccoli or cauliflower have been selected [9]. The formation of curds of cauliflower or "romanesco" variety and of broccoli spears is due to a perturbed meristem identity. In the wild variety, as in *A. thaliana* (Fig. 2) lateral meristems produced by the inflorescence meristem will have a floral identity and produce a flower. In the three varieties, cauliflower, romanesco and broccoli, lateral meristems produced by the inflorescence do not adopt a floral meristem identity as in the wild variety but an inflorescence meristem identity. This second generation of inflorescence meristems will also produce new modified lateral meristems with an inflorescence identity. The iteration of this pattern combined with a reduced internode elongation will lead to the headed phenotype of these three cabbage varieties. The iterative production of inflorescence meristems is brief in broccoli and the mature spear will consist of immature flower buds. On the other hand, both inflorescence meristems and very early arrested floral meristems form the mature curd of cauliflower [10]. Flower development will resume only after a vernalisation period. The differences between cauliflower and "romanesco" curd morphologies have been explained by the difference in onset and the rate of new meristem production on the flanks of the older meristem [11].

These two examples show that the overall branching pattern and therefore the architecture of the plant depends partly on the regulation of the identity of the meristem produced. The most extreme case is when a determinate meristem (flower) replaces or is replaced by an indeterminate meristem (shoot). Early modifications of the identity of the meristem (affecting for instance the meristem of primary axis) will result in more severe modifications of the branching pattern than later modifications affecting secondary or tertiary meristems.

3.2. Regulation of Meristem Identity

Changes in meristem identity are associated with phase changes during the life cycle of the plant. Light has been shed on the mechanism of phase transition in plants by the analyses of *A. thaliana* plants mutated in the *TERMINAL FLOWER1* (*TFL1*), gene and of plants overexpressing *TFL1* [12, 13]. Phytomer numbers in *TERMINAL FLOWER1* mutants are reduced during each phase and the inflorescence produces a terminal apical flower. Alternatively, *A. thaliana* plants overexpressing *TFL1* show an increased duration of each phase and even the formation of a new phase in which lateral secondary shoots are not subtended by a leaf. This suggests the existence of a common mechanism regulating phase transition during the whole life cycle of the plant, in which *TFL1* is involved. Such a mechanism seems to be at least partially conserved in different plant species since it has been also found in tomato which has a different architecture from *A. thaliana* (Fig. 2). In the case of tomato, the *SELF-PRUNING* gene, the functional homologue of *TFL1*, is involved in the regulation of the alternance of leafy phytomers and inflorescences during the formation of the sympodial units (Fig. 2 and [14]). Loss-of-function of the *SELF-PRUNING* gene leads to a progressive reduction of the leafy phytomer number in each sympodial unit, until the production of two successive inflorescences and the termination of the shoot. This can be related to the termination of shoot growth in the *A. thaliana TFL1* mutant by the production of a terminal flower.

The development of the shoot of the plant is achieved by growth of the meristems. These meristems can be formed during embryogenesis, but most of them are initiated during later development. In the next sections, we will discuss the organization and development of the shoot apical meristem and floral meristem. Based on these meristem properties and by using some examples, we will discuss how meristems can be formed and how their formation and outgrowth can be regulated.

4. THE SHOOT APICAL MERISTEM -SAM-: A MAJOR ACTOR IN THE ESTABLISHMENT OF ARCHITECTURE

The SAM is located at the growing tip of the stem, hidden and protected by the young developing leaves. It can have different morphologies depending on the species and developmental stage, ranging from flat to dome-shaped or finger-like structures. Most of the meristems are a few hundreds of micrometers wide but they can range from 50 μm to more then 1 mm.

4.1. Organization of the SAM: Cells and Domains

The word "meristem" originates from the Greek word "meristos" meaning partitioned. In flowering plants, the SAM is a small group of dividing cells (usually around a few hundred). However, divisions also occur outside the meristem (in young developing leaves, for instance). Consequently, although all cells from the plant do ultimately derive from the meristem only a small proportion are directly produced by the meristem. Meristematic cells have some particularities that make them recognizable from their neighboring cells. They are small, dense, with a high nucleus cytoplasm ratio and are rather isodiametric. Classical cytological analyses and chimera analyses have shown that different domains can be defined in the meristem [1].

In a first model, different concentric zones have been proposed on the basis of specific cytological features (Fig. 4A). The central zone (CZ) lays at the summit of the apex, the peripheral zone (PZ) forms a ring surrounding the CZ and the underlying rib zone (RZ). Cells in the CZ are usually larger, with bigger nuclei and more vacuoles than the surrounding PZ cells. They are less deeply stained by classical cytological stains indicating that they could have a lower metabolic activity (for instance, see [15,16]. The RZ is formed by vertical cell lines, in which the cells become progressively enlarged. Cells in the PZ divide $2-4$ times more quickly than in the CZ. The mean doubling time of the cells in a meristem varies from around 24 h to 10 days from species to species [17].

In a second complementary model, cell layers are defined within the meristem (Fig. 4B). These cell layers are maintained due to the restriction of the orientation of the cell divisions. Two main compartments are defined, the tunica,

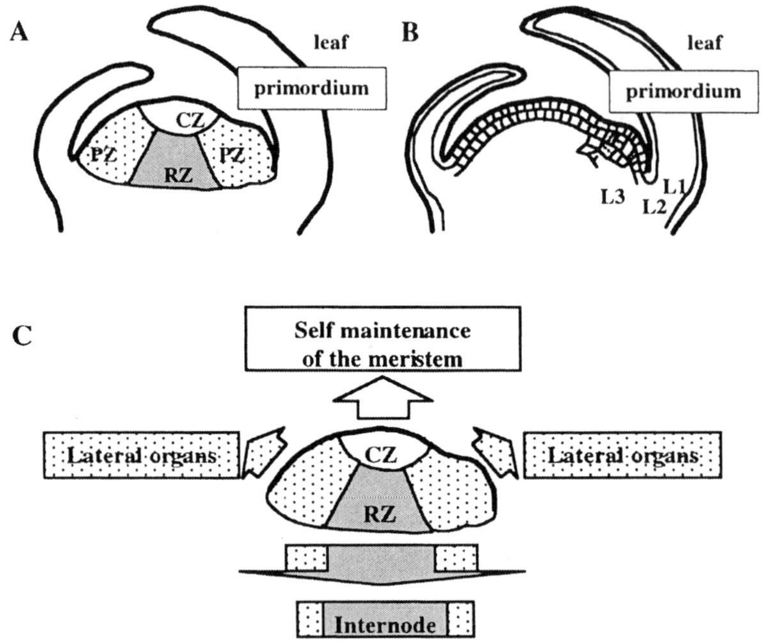

Fig. 4. — Meristem organization (longitudinal sections through a generalized shoot tip). A) Concentric zones model. The meristem is subdivided into different zones based on cytological characteristics. CZ, Central Zone; PZ, Peripheral Zone; RZ, Rib Zone. B) Cell layers model (dicots). In dicots species 3 cell layers called L1, L2 and L3 can be defined. They are characterized by the orientation of the cell division planes and their contribution to the mature organs of the plant. C) Meristem zonation and functions. The CZ is involved in the self maintenance of the meristem throughout the life of the plant whereas the PZ and RZ are involved in tissues and organ formation.

which is the outermost and where the cells divide preferentially in the anticlinal plane (the new cell wall is perpendicular to the meristem surface), and the inner corpus, where cells show no preferential division plane orientation. One layer of cells in Monocots and two layers in Dicots compose the tunica. In Dicots the two tunica layers are called L1, L2 and the corpus L3. The layers will contribute to different parts of the adult plant [18, 19]: cells of the L1 will give rise preferentially to the epidermis of the shoot whereas the L2, together with the L3, will form the inner tissues of leaves and stem. The gametes are produced by L2-derived cells.

4.2. Distribution of Meristem Functions Between the Zones

The SAM has two main functions. First, it has to produce organs and tissues (stem, leaves, and axillary meristems). Second, it has to maintain itself during a long period which in monopodial structures can reach the whole life of the plant. These two main functions have been assigned to different zones of the meristem (Fig. 4C). Initiation of lateral organs takes place in the PZ. The stem tissues are formed partly by the PZ and partly by the RZ. The PZ and RZ are therefore involved in organ and tissues production. The CZ is required for self-perpetuation of the meristem and cells within the CZ have the function of "stem" cells as found in animals [20]. A dynamic view of meristem function would be the following: cells "are produced" within the meristem by cell division and "leave" the meristem when they are incorporated in newly formed organs or tissues. As a consequence, there is a general cell flow from the centre of the meristem to the periphery. The cell flow does not result from movement of the cells (plant cells are unable to move) but is the consequence of the meristem growth. Proper regulation of the meristem development will require a tight regulation of this flow to ensure the equilibrium between the different zones.

These zonation models have been established for SAMs. During floral development, the outer floral organs are first initiated on the meristem periphery and then the inner organs are successively initiated towards the centre. The zonation model can therefore be applied to young floral meristems where organogenesis, as in SAMs, is restricted to the periphery but cannot be extended to more developed floral meristems where reproductive organs develop in the centre of the meristem.

4.3. Self-Organization of the Meristem

4.3.1. Evidence for a Self-Organization Property of the Meristem

The meristem is a very dynamic structure, composed of several functional domains whose integration is required for proper development. One major characteristic of the meristem is its capacity to regenerate after removal of a part of it, or in the more extreme case, to self-organize from only a few undifferentiated cells. This is illustrated by some classical observations.

Regeneration of two meristems was observed after longitudinal splitting of the apex of several species [21,22]. Sussex [23] excised a large part of the SAM of potato, and observed meristem regeneration when he left as few as twelve cells in the L1 layer, which corresponds to one twentieth of the original meristem. This demonstrated the regeneration capacity of the meristem. Further evidence of this meristem property was obtained with the development of *in vitro* culture methods. Excised meristems were able to grow and initiate leaves when cultivated *in vitro*. When fragments of meristems were cultivated *in vitro*, leaf initiation was preceded by a regeneration step of the meristem [24,25].

Meristems can also be formed from non-meristematic cells. This property is used commonly in plant transformation protocols (see *e.g.* [26]). An explant (leaf or root for instance), cultivated *in vitro* in the presence of nutrients (inorganic salts plus carbon source) and phytohormones, can be induced to form a callus (a group of unorganized, undifferentiated dividing cells). By changing the hormonal balance (basically by increasing the cytokinins/auxins ratio) shoots derived from meristematic cells can be induced.

4.3.2. Involvement of the Self-Organization Property of the Meristem for the Meristem Branching Mechanisms

The observations presented above pinpoint two possible mechanisms by which new meristems can be formed. Firstly, new meristems can be formed starting from a few meristematic cells from any part of the meristem. This potential must be regulated during meristem development and it has been proposed that the CZ could be involved in the repression of additional meristem formation. This is based on microsurgery experiments where the destruction of the apex centre is followed by the regeneration of new meristems on the remaining flanks [15, 22, 27]. The mechanism of this repression is not known but it could be due either to nutritional competition or to hormonal signaling. One potential mechanism of meristem branching would be the transient release of this repression, allowing the formation of additional meristems. The second potential mechanism of meristem formation is the initiation of a new meristem by one or several non-meristematic cells of adult tissues as this is observed in explant cultures.

4.4. The Meristem as a Functional Unit

As illustrated in the previous sections, meristem development requires the integration of the behavior of individual cells into a larger multicellular complex. This occurs in a horizontal direction (between CZ and PZ) but also in a vertical direction (for instance, the initiation of a leaf requires a coordinated development of the different cell layers). This integration is dramatically illustrated by numerous cases of periclinal chimeras, in which a meristem is composed of cell layers of different genotypes. In most cases, a coordinated development between the layers of different genotypes is observed, with one or several layers imposing their developmental pattern on the others (see *e.g.* [28–31]. One possible mechanism leading to this integration is through plasmodesmata. Plasmodesmata are pores connecting adjacent plants cells. The regulation of the pores opening and of the transport of proteins and RNAs through these channels is complex and is not yet fully understood (see [32–34] for reviews, and [35, 36] for some examples). It has been shown that some proteins that are important for the regulation of meristem development may travel from cell to cell, possibly via plasmodesmata [37, 38]. This would allow the formation of fields, formed by interconnected cells that share a common developmental fate [36].

4.5. Genetic Regulation of Meristem Development and Molecular Markers

In recent years, important progress has been made in the understanding of the genetic regulation of meristem development. This has been based on several approaches, with possibly the most informative being the isolation of mutants perturbed in meristem development. The subsequent analysis of these mutants and the identification of the corresponding genes give new insights to the genetic and molecular regulation of meristem development. The expression patterns of some of these meristem genes overlap with the domains defined through cytological and functional criteria [39]. We will not go into details of the genetic regulation of meristem function; several recent reviews are available [20, 39–43]. Rather, we will present two classes of mutants and the corresponding genes that are especially relevant for the understanding of meristem branching.

4.5.1. SHOOT MERISTEMLESS: A Marker for Meristematic Cells

The expression pattern of the *SHOOT MERISTEMLESS* (*STM*) gene and the phenotype of *STM* mutants suggest that *STM* expression may be a molecular marker of meristematic cells in the SAM of *A. thaliana*. *STM* is expressed in the meristem (CZ and PZ) and is absent from PZ domains corresponding to the incipient leaf primordium [44]. It must be noted that *STM*-expressing and non-expressing cells have similar cytological features. The down-regulation of *STM* is one of the earliest events associated with leaf formation by the meristem. Loss-of-function mutants of *STM* show a defect in meristem formation during embryogenesis or meristem maintenance during post-embryonic development [45, 46]. *STM* is a member of a multigene family found in Monocots and Dicots assigned as the class I *KNOX* genes after the name of the founding member, the *Knotted* homoeobox gene from maize [47, 48]. This class of proteins contains a conserved motif called the homeodomain and they act as transcription factors, regulating the transcription of other genes. All class I *KNOX* genes in plants have been proposed to be involved in some aspect of meristem development. Tobacco or *A. thaliana* plants overexpressing *KNOX* genes initiate ectopic meristems on their leaves, suggesting that these genes are able to trigger the formation of meristems [49, 50]. Interestingly, *KNOX* genes have been associated with hormone signaling. Transgenic plants overexpressing the *ipt* (isopentenyl transferase) gene involved in the biosynthesis of cytokinins also initiate ectopic meristems on their leaf blade [51]. In addition, increased endogenous levels of cytokinins were found when *KNOX* genes were overexpressed in the whole plant [52], but it was not possible to distinguish a direct effect from an indirect one resulting from overall plant morphological changes. The reverse effect of cytokinins on *KNOX* genes expression is clearer. Using an inducible *ipt* expression system it has been shown that cytokinins can increase the expression levels of *KNOX* genes [53].

4.5.2. *CLAVATA3*: A Marker for CZ Cells

Mutations in any of the *CLAVATA*1, 2 or 3 genes leads to an increase in meristem size and organ number [54–57]. It has been shown that these defects are due to an increased size of the CZ and that therefore, the *CLAVATA* genes could be required for the regulation of cell transition from the CZ to the PZ [58]. The three *CLAVATA* genes have been isolated and the corresponding proteins are involved in signal transduction pathways [59–61]. Interestingly, *CLV3* is expressed in a small group of cells within the meristem which corresponds to the putative CZ, suggesting that *CLV3* expression is a marker for this zone [60].

4.6. Forming a Primordium, What Makes a Meristem Bulge?

As we will see later, some axillary meristems are formed directly by the main SAM. This transformation of one meristematic dome into two meristematic units requires the formation of a bulge on the meristem. The mechanism of this bulging is not understood but, from a morphological point of view, it can resemble the formation of a leaf primordium by the SAM. The first sign of leaf formation is a down regulation of *STM* in a zone of the meristem where the next primordium will form. This is visible one plastochrone (the time between two successive leaf initiation events) before the effective primordium becomes morphologically visible [43]. The leaf primordium grows out as a small symmetrical bulge. It will then acquire a visible dorso-ventral polarity (the dorsal or adaxial side is oriented towards the meristem and will be the upper face of the mature leave, the ventral or abaxial side is the opposite side). In the case of dicots leaves, the blade (expanded part of the leaf) and the petiole will become distinct (for review see *e.g.* [62, 63]).

One way to induce a bulge on a meristem is to have spatial differences in the growth rate between two parts of the meristem. This can be achieved by either increasing growth rate in zones where primordia will be formed, by inhibiting growth between primordia or by a combination of both. Genes involved in these events are expected to be differentially expressed between primordium and non-primordium domains. In addition, when these genes are not functional, one would expect to see perturbations in some aspects of primordium outgrowth (size, number, or separation). The observation of such perturbations in the mutant would also allow us to identify genes actually involved in primordium growth from genes which could be associated with the primordium identity for instance.

Several genes have their expression limited to sites between primordia. Mutations in these genes lead to partial organ fusion. Genes of the *NAC* family have been described in petunia (*NAM* gene) and *A. thaliana* (*CUC2* gene) and mutations in them lead to various levels of fusion between cotyledons or floral organs [64, 65]. During *A. thaliana* embryo development, *STM* in addition to being expressed in the future meristem is also expressed between developing cotyledons. *STM* loss-of-function mutants show a partial fusion of the cotyledons indicating that in some conditions *STM* can be required for growth inhibition [66]. The mechanism by which *NAC* genes and *STM* may inhibit growth is not known. *FIM* in *A. majus* and *UFO* in *A. thaliana* are required for proper flower development including position and separation of floral organs [67,68]. They are expressed between some developing floral organs. Based on the sequence of the protein, it has been suggested that they may be involved in cell cycle regulation and could decrease the proliferation rate between organs [7].

Several genes are expressed in primordia but for only a few there are indications that they could regulate growth. One candidate is the *CYCLIN D3a* gene, which is expressed in *A. majus* leaf primordia [69]. Another candidate is an *EXPANSIN* gene whose expression is increased in incipient tomato leaf primordia [70]. Although no genetic evidence yet implicates these genes in the actual growth of the primordium, the nature of the proteins suggests that they could. *CYCLIN D3a* is a potential regulator of the cell cycle progression and may therefore regulate mitotic activity. EXPANSINS are proteins that can increase the plant cell wall extensibility and may therefore promote cell growth. Local application of EXPANSIN protein to a tomato meristem can induce the development of leaf primordia [71]. These observations suggest that increasing both cell division and growth rates may regulate primordium outgrowth.

4.7. Conclusion

Meristems are formed by a relatively small number of dividing cells. They have two main functions: producing organs and tissues and perpetuating themselves. These functions have been assigned to different zones within the meristem. These zones differ in their cytological features and their gene expression patterns. Although meristems may be long lasting structures and seem to be unchanged when observed at the macroscopic level, they appear very dynamic when observed at the cellular level.

5. MECHANISMS AND REGULATION OF BRANCH FORMATION

5.1. Hypothetical Mechanisms of Branch Formation

Branching requires the formation of a new meristem. As we have seen, meristems can either be directly formed from an existing meristem, or, alternatively, from non-meristematic cells. In the light of the meristem organization, of its characteristics, and of some basic elements of plant architecture, several hypothetical mechanisms of meristem branching can be proposed (Fig. 5).

To give a more precise description of these mechanisms, the expression domains of two molecular markers will be used, *STM* and *CLV3* for meristematic cells and the central zone respectively. A meristem is then defined as a group of *STM*-expressing cells with an apical subgroup of cells expressing both *STM*

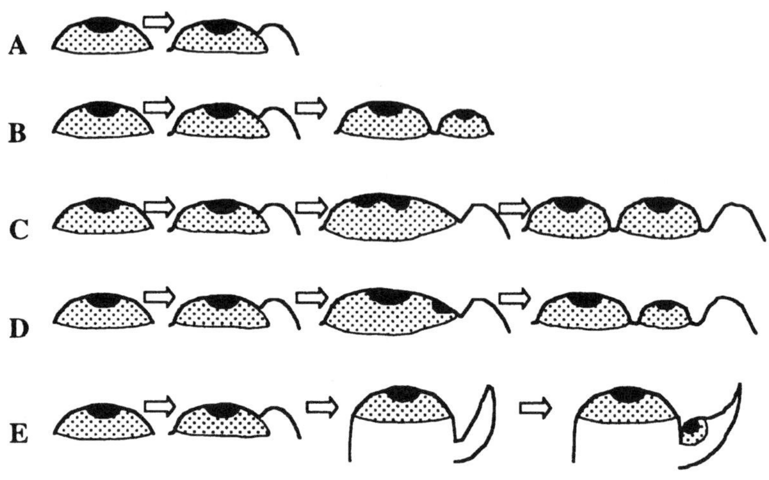

Fig. 5. — Hypothetical mechanisms of meristem branching. A) Leaf development. B) Mechanism I: transformation of a primordium into a meristem. C) Mechanism II: fragmentation of the CZ. D) Mechanism III: formation of a new CZ. E) Mechanism IV: meristem neoformation. The expression domain of *CLV3* and *STM* is represented in black and marks the CZ. The zone expressing only *STM* is represented by points and corresponds to the meristem. The bulge on the right represents a leaf primordium. See text for further details.

and *CLV3*. As soon as the cells leave the meristem, they will no longer express *STM*. The formation of a new meristem will require the formation of a new centre, marked by the formation of a new *CLV3*-expressing CZ domain.

5.1.1. Transformation of a Primordium into a Meristem (Mechanism I, Fig. 5B)

In this case, the formation of the new meristem begins with the formation of a leaf primordium. A *STM* non-expressing domain of the meristem will form a bulge. This bulge, instead of developing into a leaf, will acquire again a meristematic state, expressing *STM* and *CLV3*.

5.1.2. Fragmentation of the CZ (Mechanism II, Fig. 5C)

The branching of the meristem starts with an increase in the size of the meristem in one direction, also including an increase in the size of the CZ. The CZ will then split into two units, generating two individual CZs. The cells located between the two CZs will change from their initial CZ identity to a PZ identity. Therefore, this mode of branching requires a switch of some apical cells from CZ identity to PZ identity.

5.1.3. Formation of a New CZ (Mechanism III, Fig. 5D)

The meristem enlarges in one direction. A new CZ is formed ectopically in the PZ. In this mechanism some PZ cells loose their PZ identity to acquire a CZ identity. One could predict that during the early phases of this mechanism the ratio between the size of the PZ and CZ would increase.

5.1.4. Meristem Neoformation (Mechanism IV, Fig. 5E)

In this mechanism, cells from the new meristem do not directly derive from the main meristem. Meristems are formed in the axils of well-developed leaves which can be considered to be a meristem neoformation. It is likely that this mechanism involves a dedifferentiation of differentiated leaf or stem cells.

In summary, four different branching mechanisms of the meristem can be proposed. Two of them (mechanisms II and III) involve branching of the meristem itself whereas the two others (mechanisms I and IV) need a formation of a meristem *de novo* either near to the pre-existing one or further away. Mechanisms I, III and IV are unequal with respect to the origins and sizes of the meristems formed whereas mechanism II is equal.

Note that some intermediary mechanisms could exist. For instance, an intermediary figure between III and IV can be hypothesized in which just a few meristematic cells related directly to the apical meristem stay in the axils of developing leaf primordium and acquire their proper organization in a meristem only later. As we will see below, the classification of some precise branching processes is hindered by the lack of detailed morphological and molecular analyses.

5.2. Some Examples of Branching Mechanisms

5.2.1. Initiation of Floral Meristems in A. thaliana

Floral meristems are initiated on the flanks of the apical inflorescence meristem in a regular pattern (Fig. 6). This enables the position of the next primordium to be predicted. The formation of the floral meristem is clearly unequal and has been well described from a morphological and molecular point of view. As in the formation of leaves, *STM* is down regulated at the position of the next incipient floral meristem. The young bulge does not express *STM* or *CLV3* (Figs. 6D and 6E [44, 60]). However, these two markers are expressed again in the older floral meristems before initiation of the first floral organs. This suggests that the *A. thaliana* floral meristem formation follows mechanism I.

Two mutants, *pin-formed* and *pinoid* show defects in floral meristem formation and are affected in genes involved in the transport or signal transduction

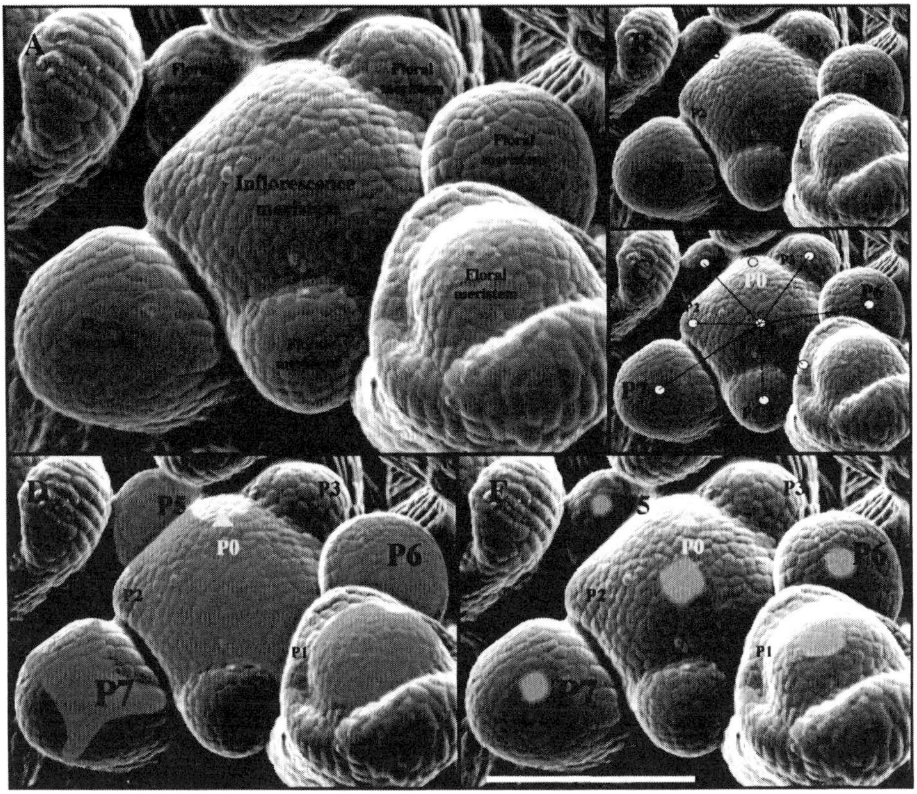

Fig. 6. — Formation of floral meristems in *A. thaliana.* A) Floral meristems arise on the flank of the inflorescence meristem. The outer surface of the L1 cells is visible. B) The floral meristems can be numbered from the youngest meristem visible (P1, here hidden by an older floral meristem) to the oldest (here P7). C) Floral meristems are produced in a regular pattern. This allows prediction of the position of the next floral meristem to arise called P0. D) The expression pattern of *STM* is represented in light-grey. *STM* is expressed throughout the inflorescence meristem except where the primordium P0 is expected. No expression is visible in young floral meristems (P1 to P4) but *STM* expression is turned on again in older floral meristems (P5 onwards). E) The expression pattern of *CLV3* is represented by a light-grey spot. *CLV3* is expressed in the centre of the inflorescence meristem and older floral meristem. It is not expressed in young floral meristems (P1 to P4). Bar = 100 μm. Adapted with permission from Laufs, Grandjean, Jonak, Kiêu and Traas, Cellular parameters of the shoot apical meristem in *Arabidopsis*, *Plant Cell* **10** (1998) 1375-1390.

of a class of phytohormones, the auxins [72, 73]. Therefore, floral meristem formation in *A. thaliana* is associated with hormone signaling.

5.2.2. Initiation of Floral Meristems in Tomato

After 6 – 10 vegetative phytomers the vegetative apical meristem of tomato is converted into an inflorescence meristem. This conversion is marked by a flattening and elongation in one horizontal direction of the meristem [74]. The elongated meristem will then divide into two apparently equal parts. One will develop as a flower, the other remains as an inflorescence meristem. Later floral meristems will be initiated in a similar manner at approximately 90 degrees from the previous one.

Although the formation of floral meristems in tomato has not been investigated at the molecular level, the equal branching of the meristem appears to be similar to mechanism II.

A similar mechanism of flower initiation has been reported for petunia although the architecture of the petunia and tomato are different. At each node of the petunia inflorescence, two leaf-like lateral organs are formed, and the inflorescence meristem will form a lateral flower. Morphological observations have shown that in petunia, as in tomato, the inflorescence meristem bifurcates in two parts, yielding a floral meristem and an inflorescence meristem [75]. In the *extra petals* mutant of petunia the branching of the meristem does not occur and a single terminal flower is produced instead of the inflorescence [75]. The way in which the *EXTRA PETALS* gene induces meristem branching is not known. The petunia *NAM* gene, which belongs to the *NAC* family, could also be involved in meristem branching. It is expressed in a strip of cells corresponding to the groove of the branching meristem [64]. No defect in the initiation of the floral meristem is observed in the *NAM* mutant suggesting that there may be a redundant role of several *NAC* genes during this process.

5.2.3. Leaf Axillary Meristem Formation in Tomato

Axillary meristems are formed in the axils of the leaves in tomato. Anatomical studies have shown that they are derived directly from the shoot apex [76]. In the axil of the fifth youngest leaf, cells still appear meristematic. These cells start to divide and in the axil of the sixth leaf a protruding axillary meristem is clearly visible. The cells between the main apex and the axillary meristem will then start to differentiate, breaking the meristematic continuity between these two meristems. Such a mechanism of axillary meristem formation has been described for numerous plant species. The small group of undifferentiated cells that is maintained in the leaf axil and will give rise to the axillary meristem has been called a detached meristem [1] or a residual meristem [77]. So far it is not clear to what extent it shows a typical meristem organization. For instance it is not known if it expresses *STM* and if it is organized in *CLV3* expressing and non-expressing domains. Therefore the classification of this mode of axillary meristem formation is difficult. It could for instance be related to mechanism III, if the detached meristem showed an early typical meristem organization. On the other hand it could be related to mechanism I, if the cells composing the detached meristem were at first non-distinguishable from leaf primordium cells and were organized into a typical meristem, only later, when a morphological distinct meristem had appeared.

Several genes involved in the formation of these meristems have been identified. Two tomato mutants, *torosa-2* and *lateral suppressor*, have no meristems in the leaf axils [76, 78]. In both cases a relation with hormone signaling was proposed. Although, in *torosa-2* a lower content of cytokinins was found, application of cytokinins could not restore axillary meristem formation in the mutants [78, 79]. Treatments of *torosa-2* with gibberellins increased the number of axillary meristems and their outgrowth [78]. The *LATERAL SUPPRESOR* gene has been cloned and found to encode for a protein potentially involved in the gibberellin signal transduction pathway [80].

5.2.4. Leaf Axillary Meristem Formation in A. thaliana

Formation of axillary meristems in *A. thaliana* was found to occur by a different mechanism. No meristematic-like cells are anatomically apparent in the leaf-axils of *A. thaliana* [81]. Therefore, axillary meristem formation in *A. thaliana* may follow mechanism IV, involving dedifferentiation of differentiated cells and meristem neoformation.

In *A. thaliana*, like in most plant species, axillary meristems are located on the upper, adaxial side of the leaf axil. Ectopic meristems formed in transgenic plants overexpressing *KNOX* genes or the *ipt* gene are also always found on the adaxial side of the leaf, not the lower, abaxial side. This suggests a tight relationship between adaxial cell fate and meristem initiation. This view was supported by the isolation of the *phabulosa* mutant in *A. thaliana* [82]. In this mutant, there is no leaf blade expansion and the whole leaf exhibits adaxial characteristics. In the mutant, axillary meristems are no longer restricted to the adaxial side of the axil but are also found in the lower part. This suggests that the apical meristem may release a meristem inductive signal, to which only the adaxial side of the leaf can respond. In the *phabulosa* mutant, the lower part of the leaf, that now has an adaxial identity, would also be able to form axillary meristems. Alternatively, the adaxial side of the leaf may not only be competent for meristem formation but could itself promote it.

5.2.5. Branching of Fasciated Stems in A. thaliana

5.2.5.1. What is Fasciation? Fasciation, *i.e.* the transformation of a cylindrical stem into a flat, enlarged, band-shaped structure has puzzled scientist for centuries (see *e.g.* [83,84]). This modification of the stem morphology has been traced back to a perturbed organization of the SAM, in which growth occurs preferentially in one direction leading to a ridge-shaped structure. Fasciations always induce perturbed position of the leaves along the stem (perturbed phyllotaxy). Abnormal branching events, in which the stem seems to split into two or several branches, are observed in fasciated stems. This type of branching leads to a decrease in the degree of fasciation and can be interpreted as "correcting events". Fasciations can occur in response to various stresses [15,85,86] or are genetically determined [54–57, 87–92]. The analysis of abnormal meristem development as observed during fasciations and the study of the mechanism leading to such modifications can give us some hint about normal regulation of meristem development. In this respect, the study of genetically determined fasciations can be especially useful. Genetically determined fasciations have been reported in pea [87], chickpea [88], soybean [89] and *A. thaliana*. In the later case mutations in several different loci lead to fasciations that have been reported [54–57, 90–92]. In the next section we will discuss the branching associated with fasciation in two classes of *A. thaliana* mutants, *mgoun* and *CLAVATA*, and focus our attention to the perturbations that occur at the meristem level.

5.2.5.2. Fasciation and Branching in the *mgoun* Mutants of *A. thaliana*. In both classes of mutants, *mgoun* and *CLAVATA*, a fasciated stem is observed. However, they differ in their degree of abnormal branching. In the case of *mgoun*, abnormal branching is usually observed along the fasciated stem. This leads to narrower stems, which can have a normal morphology or still be fasciated. In the case of the *CLAVATA* mutants, the branching points exhibit the same characteristics as in *mgoun* but are far less frequent [54]. The differences in the degree of branching can be traced back to differences in meristem morphology (Fig. 7). In the case of *CLAVATA*, fasciation of the meristem leads to a single, enlarged ridge-shaped structure. On the other hand, in *mgoun* mutants, as the meristem enlarges, several distinct meristematic units are formed. How does the process of meristem fragmentation occur in the *mgoun2* mutant? Why is this fragmentation not observed in the meristem of *CLAVATA*?

The fragmentation process in *mgoun* was followed using *STM* expression as a marker. In enlarged meristems, a wider expression domain of *STM* was observed. When several morphological units were visible, they corresponded to distinct *STM*-expressing domains, confirming the branching of the meristem. In some cases, an apex forming a single morphological unit already had two distinct *STM*-expressing domains. This suggests that the formation of two *STM*-expressing domains is an early event and precedes the morphological fragmentation of the meristem [91].

The mechanism of meristem branching in *mgoun* could correspond to mechanism II or III, involving a fragmentation of the CZ or a CZ neoformation within the meristem, respectively. The two hypotheses can be distinguished using a genetic approach. It has been shown that the increase in meristem size in *CLAVATA3* is due to an enlarged CZ whereas in the *mgoun2* mutant a larger PZ is found [58]. Furthermore, the effects of the mutations are additive at the morphological level. It can therefore be hypothesized that they also have an additive effect at the meristem level, and that in the double mutant *clavata3 mgoun2*, both the CZ and the PZ would be enlarged. If the meristem fragmentation relied on a fragmentation of the CZ, an increased stem branching would be expected in the double mutant. This was not observed although the double mutants show a dramatic increase in stem fasciation. This observation

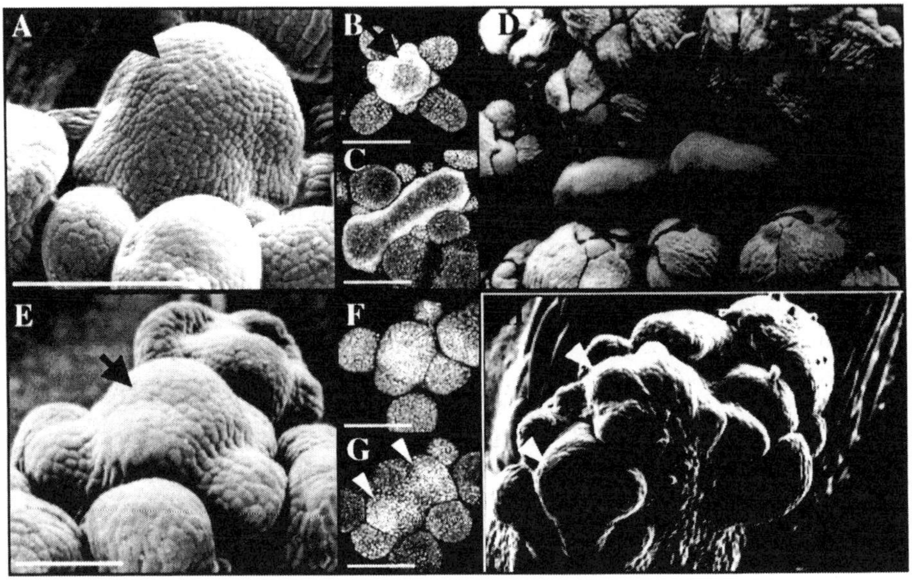

Fig. 7. — Branching of fasciated meristems in *clv3* and *mgo2* mutants in *A. thaliana.* A–D) Meristems from *clv3* mutants. E–H) Meristems from *mgo2* mutants. Young meristems from *clv3* A, B) or *mgo2* mutants E, F) form an enlarged dome (black arrows). Older *clv3* meristems form an enlarged ridge-shaped structure C, D). In *mgo2* mutants, the apices are formed by the juxtaposition of several meristematic units (arrowheads, G, H). Bars = 100 μm. Adapted with permission from Laufs, Grandjean, Jonak, Kiêu and Traas, Cellular parameters of the shoot apical meristem in *Arabidopsis, Plant Cell* **10** (1998) 1375-1390, and Laufs, Dockx, Kronenberger and Traas, *MGOUN1* and *MGOUN2*: two genes required for primordium initiation at the shoot apical and floral meristems in *Arabidopsis thaliana, Development* **125** (1998) 1253-1260.

suggests that meristem fragmentation is not based on the fragmentation of the CZ.

5.2.5.3. Hypothetical Mechanism of Meristem Fragmentation in *mgoun2* and the Role of the CZ in the Control of Meristem Integrity. The observations above suggest that abnormal meristem branching in the *mgoun2* mutant occurs through the formation of a new CZ. Such a mechanism has been reported in the case of ferns [93]. The experiments of meristem regeneration have shown that, in higher plants too, the periphery of the meristem has the potential to form a new meristem. This potential is inhibited during normal growth possibly by the CZ. In the case of *mgoun2* meristems, the inhibition by the CZ would not be strong enough to suppress the neoformation of a CZ. This could be due to the enlargement of the meristem, allowing some meristematic cells to escape the CZ inhibition or due to the *mgoun2* mutation itself which could lead to a less efficient inhibiting signal. In the *CLAVATA* apex only rare branching events would be observed because the PZ is not enlarged and in the double mutant *CLAVATA3 mgoun2* the enlargement of the CZ would reinforce its inhibition on the meristem periphery.

5.3. Conclusions

The formation of a new meristem requires two steps: an increase in meristematic cell number and the formation of a new growth centre, marked by the CZ. These two steps can occur simultaneously or sequentially, in either order. In a given species, different mechanisms can be involved in formation of different meristems types. Reciprocally, different mechanisms can be involved in formation of a given meristem type in different species.

Two major components may have a role in axillary meristem formation: the content of phytohormones (*e.g.* cytokinins, gibberellins and auxins) and the *KNOX* genes. How these two major components do interact to regulate meristem formation is not yet known, although a reciprocal effect between cytokinins and *KNOX* genes has been suggested. Although most of the meristems are formed in the axils of the leaves, new meristems may also arise directly from the leaf blade or leaf margins or from the petiole in various wild-type species. This phenomenon is called epiphylly [8, 94]. In most examples of epiphylly, the shoots form on the adaxial side of the leaf, but little is known about the regulation of the formation of these shoots.

6. REGULATION OF MERISTEM OUTGROWTH

As we have seen in the previous sections, the formation of meristems is tightly regulated in time and space. In addition, once the meristems are formed they may or may not grow further. "Apical dominance" is the inhibition imposed on the outgrowth of the axillary meristems by the main apical meristem and/or its associated young leaves and internodes. When the main apical meristem is pruned off (experimentally or naturally by herbivores, for instance), apical dominance is broken and one or several lateral meristems will develop. Usually the uppermost axillary meristem will develop most, and impose a new apical dominance on the lower axillary meristems. The intensity of apical dominance varies from species to species: strong apical dominance leads to a clear main stem, whereas a more bushy habit is found in species with a weaker apical dominance [95].

Apical dominance is under hormonal control. A simplified view is that the outgrowth of axillary buds is controlled by two antagonistic factors (a more detailed analysis can be found in [96]. Auxins inhibit outgrowth of the axillary meristems whereas cytokinins promote it. This is based on numerous observations such as the following examples:

- in decapitated plants, apical dominance is maintained when auxins are added to the cut stem [97];
- *A. thaliana* mutants showing an increased auxin response have an enhanced apical dominance [98], whereas reduction of the endogenous auxin content in tobacco leads to reduced apical dominance [99];
- a general reduced apical dominance is observed in plants having an endogenous increase of cytokinins (mutants or transgenic plants overexpressing the *ipt* gene [100, 101];
- axillary meristem outgrowth can be locally promoted by applications of cytokinins or by locally endogenous production [102, 103];
- a decreased apical dominance is observed in plants overexpressing *KNOX* genes, this could be mediated by a modification of cytokinins level or response [49];
- although a general role of the auxins and cytokinins in apical dominance is widely accepted, others factors such as the regulation of nutrient levels or other hormones such ethylene or abscisic acid, have also been proposed to have a role in apical dominance [104].

7. BRANCHING IN LEAVES

7.1. Different Leaf Morphologies

A great diversity in leaf morphology can be observed and can be related to the various functions they have. However leaves can be subdivided into two main groups: simple leaves in which a unique blade is found and compound leaves formed by several leaflets. The compound leaves can be subdivided into palmately compound leaves, when all the leaflets are attached to the same point or pinnately compound when they are attached to an axis. The leaf can show one level of branching when all the leaflets are attached at the same axis, or alternatively, several levels of branching if the main axis branches into several generations of axis before bearing leaflets. In addition, each leaf or leaflet can have different shapes (reniform, lanceolated or oblong for instance) and the margins can be entire or lobed or dentate. The present knowledge of leaf development can be found in recent reviews [1, 63, 105, 106].

7.2. The Compound Tomato Leaf

7.2.1. Architecture and Development of the Compound Tomato Leaf

Leaves in tomato show a pinnately compound architecture. They are formed by an axis called a rachis that bears $3-4$ pairs of lateral leaflets and

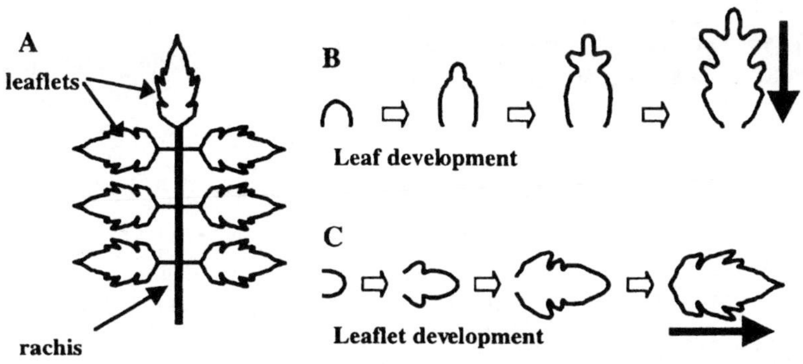

Fig. 8. — Development of the tomato leaf. A) Tomato leaves are formed by an axis called a rachis that bears 3 – 4 pairs of lateral leaflets and 1 terminal leaflet. B) During development of the complex tomato, the terminal leaflet is the first to be formed. The other leaflets are formed following a basipetal order, from the distal part to the basal part of the leaf. C) Each leaflet is lobed and has an acropetal maturation gradient. The lobes first appear on the proximal part of the leaflet.

1 terminal leaflet (Fig. 8A). Thus, the wild type tomato leaf shows one degree of ramification [107]. Each leaflet is lobed.

The leaf follows a complex development pattern. The leaflets have a basipetal morphological gradient: the first developing leaflets are located on the distal part (tip) of the leaf primordium (Fig. 8B). Each leaflet itself has an acropetal maturation gradient: the lobes first appear on the proximal part of the leaflet ([107], Fig. 8C).

7.2.2. *Compound Tomato Leaf and KNOX Genes*

7.2.2.1. Overexpression of *KNOX* Genes Yields an Ultra-Compound Tomato Leaf. Overexpression of *KNOX* genes dramatically changes the tomato leaf architecture as transgenic leaves are transformed in ultra-compound leaves (Fig. 9). These leaves show an increased ramification reaching 6 levels of ramification instead of 1 and a leaflet number reaching in some cases 2000 instead of 7 – 9. It must be noted that the overall size of the leaf is only slightly increased, indicating that the new leaf phenotype results from a repeated ramification process of the wild-type structure [107, 108].

When *KNOX* genes were overexpressed in *A. thaliana* or tobacco no compound leaves were observed although the leaf morphology was modified [49, 50]. In addition to the ectopic meristems, the leaves were crinkled and showed increased lobing. In some instances, ectopic stipules were observed on the base of the lobes. Stipules are usually found on the base of leaf primordia and their ectopic presence could therefore indicate that the lobes might originate from cells that have some meristematic characteristics.

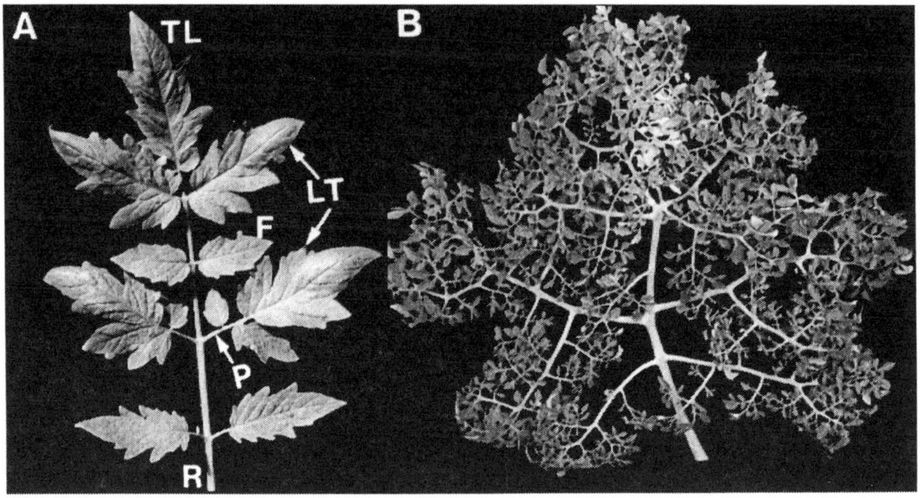

Fig. 9. — Tomato leaf structure and *KNOX* genes. A) The wild-type tomato leaf is formed by an axis called a rachis (R) that bears 3 – 4 pairs of lateral leaflets (LT) and 1 terminal leaflet (TL). B) When the *KNOX* gene *Knotted-1* from maize is overexpressed, an ultra-compound leaf is observed. Reprinted with permission from the authors and Cell Press from Hareven, Gutfinger, Parnis, Eshed and Lifschitz, The making of a compound leaf: genetic manipulation of leaf architecture in tomato, *Cell* **84** (1996) 735-744.

Because both *A. thaliana* and tobacco have simple leaves it was suggested that the different effects of the overexpression of *KNOX* genes in these two species from those in tomato was due to a fundamental difference in the development of simple and compound leaves [108]. This hypothesis was reinforced by the effect of *KNOX* gene overexpression in different tomato mutants showing abnormal leaf morphologies. *KNOX* gene overexpression could not induce any compoundness in the simple leaf of the lanceolate mutant of tomato. Instead, lobing of the mutant leaf was observed, as it is the case in *A. thaliana* and tobacco. Leaves in the tomato mutant called potato-leaf, which have usually only two lateral leaflets, show an increased branching when *KNOX* genes are overexpressed, but this was limited to the part of the leaf where the two leaflets were expected and the wild-type leaf morphology was not restored. This was interpreted as the capacity of ectopic *KNOX*-gene expression to increase the ramification of a pre-existing compound pattern but its inability to create *de novo* compoundness [108].

It has been observed that in the ultracompound tomato leaf, the extra leaflets are produced in an acropetal gradient, just as the leaflet lobes would be produced. Therefore, it was suggested that leaflets and lobes are equivalent structures and that the expression of *KNOX* genes may transform the lobes into leaflets [107].

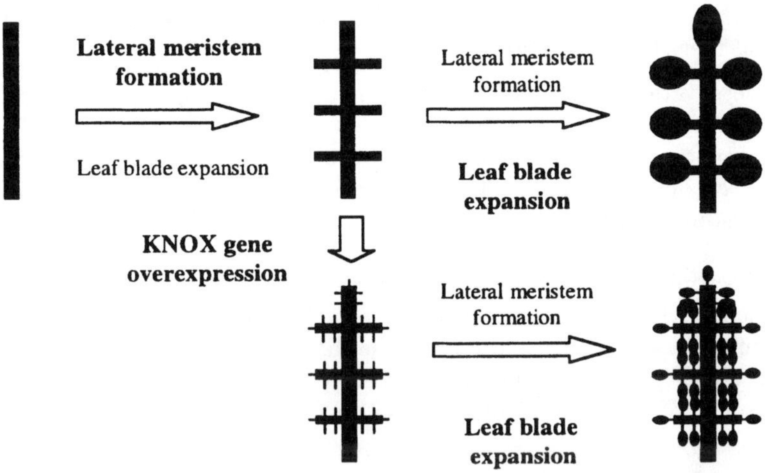

Fig. 10. — A model for branching in the tomato leaf. The formation of the complex leaf is based on the equilibrium between two antagonistic events: formation of new lateral meristems and leaf blade expansion. In the case of normal development, a unique round of lateral meristem formation followed by blade expansion leads to the wild-type tomato leaf which has only one degree of compoundness. When *KNOX* genes are overexpressed, extra rounds of lateral meristems formation are induced (here one extra round is represented) and leaf blade expansion is at the same time inhibited. This leads to an increased ramification of the transgenic leaf (here, two levels of ramification).

7.2.2.2. Tomato *KNOX* Genes Are Expressed in Leaf Primordia. The expression pattern of the *KNOX* genes in tomato differs from the expression pattern found in species like *A. thaliana* and tobacco. In these two species, *KNOX* genes are restricted to the meristem: they are not expressed in the incipient or in the developing leaf primordium. Contrastingly, in tomato, two *KNOX* genes, *Tkn1* and *Tkn2* (also called *LeT6*) are expressed in the meristem and in the zone where a primordium will form [107, 108]. They are also expressed in the developing lobes of the leaf. It has been proposed that the broader expression pattern of *KNOX* genes could account for the larger morphogenetic potential of tomato compound leaves.

7.2.3. A Model for Leaf Development in Tomato

Hareven *et al.* proposed that the establishment of the basic branching pattern of the compound leaf in tomato is achieved by the formation of lateral meristems along the rachis and the concomitant inhibition of blade outgrowth ([108] and Fig. 10). During a second step, each lateral meristem will give rise to a leaflet. During this process the expansion of the blade is activated and the formation of new lateral meristems inhibited. Ectopic expression of *KNOX* genes induced a delay in this activation of blade expansion and one or several additional rounds of lateral meristem formation occur.

The observation that in plants overexpressing *KNOX* genes, additional leaflets result from transformed lobes suggests that the lobe regions have intermediary morphogenetic potential between the lateral meristems and other parts of the leaflet. The development of the leaf therefore involves different identities (lateral meristems, lobes and non-lobe regions) during which the morphogenetic potential is progressively lost and the degree of differentiation increased. The effect of ectopic *KNOX* genes can be interpreted as a delay in these transitions, in particular the lobe region acting as new lateral meristems and originating new ramifications. If this is the case, the difference in effect of the overexpression of *KNOX* genes on compound and simple leaves can then be explained by the presence of lateral meristems in the developing compound leaves.

8. FINAL REMARKS

The plant apex is composed of multicellular domains. If a cell is displaced (by growth), from one zone to the other, it will adopt a new identity and behavior, according to its new position. Exactly how this is achieved is not known, but it is likely that elaborate signaling mechanisms, giving positional information, are involved. Thus, the meristem proper can be subdivided into the CZ, PZ and RZ. In the case of compound leaves, the lobe regions and inter lobe regions can also be distinguished. Going from the CZ to the RZ into the primordia the differentiation degree is increased and at the same time the morphogenetic potential decreases (Fig. 11). Cells are also displaced along the proximo-distal axis (the CZ can be viewed as the proximal region and the fully expanded leaf blade as the most distal part). Although some domains, such as those in the primary SAM may be stable for years or even centuries, others are not maintained or are newly created. If this is so, branching may either result from the creation of new meristematic domain, or from an extension of existing domains (*e.g.* enlargement of the CZ or PZ) (see Fig. 11 for examples).

KNOX genes appear to be important elements in the regulation of branching, as they seem to be involved in the regulation of the transitions between the different stages [109]. When they are overexpressed, they will delay the transitions and this can eventually result in an increased branching depending on the biology of the species. For instance overexpression of *KNOX* genes in Monocots such as maize induce a proximalisation in the leaf without an increase of the branching [110, 111]. Overexpression in Dicots with simple leaves such as tobacco or *A. thaliana*, induce the ectopic formation of meristem on the leaf surface [49,50]. Finally, overexpression in Dicots with compound leaves increases the ramification degree of the leaf [107, 108].

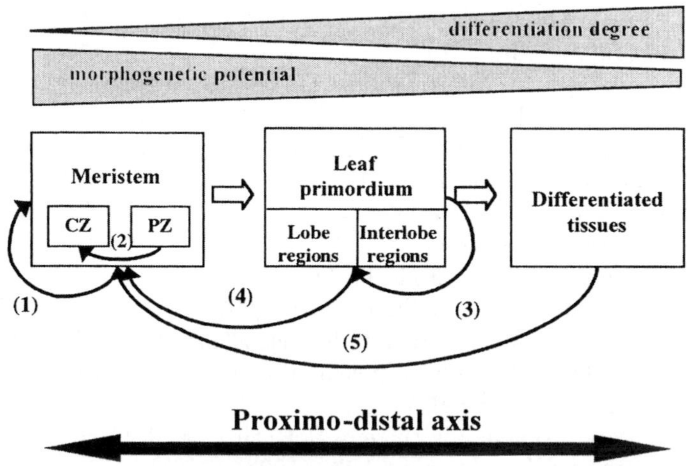

Fig. 11. — A general model for branching of the aerial parts in plants. During development cells go through different stages possibly from a meristematic to leaf primordium identity and ultimately differentiate. As this occurs, their differentiation degree progressively increases and their morphogenetic potential decreases. At the same time they are displaced along the proximo-distal axis. Branching can result from delayed transitions or from "backwards steps". Meristems can split into two units (1, as for example during the development of the inflorescence meristem in tomato). Cells from the peripheral zone can adopt a new central zone cell identity (2, as it is suggested for the splitting of the fasciated meristem in *mgoun* mutants). Leaf primordia can develop new leaf primordia (3, as it is the case during tomato leaf development). Groups of cells which are initiated in the same way as leaf primordia can adopt again a meristem identity (4, as for the formation of floral meristems in *Arabidopsis*). Finally new meristems can be formed from fully differentiated tissues (5, formation of axillary meristems in *Arabidopsis*).

Acknowledgements

We thank P. Lunness, M. Towers and Drs. D. Bradley, J. Doonan and M. LeBris for comments on the manuscript. We thank ASPP, cell Press and Dr. Lifschitz for the authorization to reproduce figures.

REFERENCES

[1] T.A. Steeves and I.A. Sussex, *Patterns in Plant Development*, 2nd Ed. (Cambridge University Press, New York, 1989).

[2] E.S. Coen, J.M. Romero, S. Doyle, R. Elliot, G. Murphy and R. Carpenter, *Cell* **63** (1990) 1311.

[3] D. Weigel, J. Alvarez, D.R. Smyth, M.F. Yanofsky and E.M. Meyerowitz, *Cell* **69** (1992) 843.
[4] D. Weigel and E.M. Meyerowitz, *Cell* **78** (1994) 203.
[5] V.F. Irish and I.M. Sussex, *Plant Cell* **2** (1990) 741.
[6] J.L. Bowman, D.R. Smyth and E.M. Meyerowitz, *Plant Cell* **1** (1989) 37.
[7] G.C. Ingram, S. Doyle, R. Carpenter, E.A. Schultz, R. Simon and E.S. Coen, *EMBO J.* **16** (1997) 6521.
[8] T.A. Dickinson, *Bot. Rev.* **44** (1976) 181.
[9] D.R. Smith, *Curr. Biol.* **5** (1995) 361.
[10] S.M. Carr and V.F. Irish, *Planta* **201** (1997) 179.
[11] M. Kieffer, M.P. Fuller and A.J. Jelling, *Planta* **206** (1998) 34.
[12] D. Bradley, O. Ratcliffe, C. Vincent, R. Carpenter and E. Coen, *Sci* **275** (1997) 80.
[13] O. Ratcliffe, I. Amaya, C. Vincent, S. Rothstein, R. Carpenter, E.S. Coen and D. Bradley, *Development* **125** (1998) 1609.
[14] L. Pnueli, L. Carmel-Goren, D. Hareven, T. Gutfinger, J. Alvarez, M. Ganal, D. Zamir and E.E. Lifschitz, *Development* **125** (1998) 1979.
[15] J.E. Loiseau, *Ann. Sci. Bot. Ser.* **11** (1959) 1.
[16] T.A. Steeves, M.A. Hicks, J.M. Naylor and P. Rennie, *Can. J. Bot.* **47** (1969) 1367.
[17] D. Francis, *Cell size and organ development in higher plants*, edited by D. Francis, D. Dudits and D. Inz, Plant Cell Division (Portand Press, London, 1998) p. 187.
[18] S. Satina, A.F. Blakeslee and A.G. Avery, *Am. J. Bot.* **27** (1940) 895.
[19] E.J. Szymkowiak and I.M. Sussex, *Ann. Rev. Plant Physiol. Plant Mol. Biol.* **47** (1996) 351.
[20] M. Lenhard and T. Laux, *Curr. Opin. Plant Biol.* **2** (1999) 44.
[21] E. Ball, *Am. J. Bot.* **42** (1955) 509.
[22] M. Pilckington, *New Phytol.* **28** (1929) 37.
[23] I.M. Sussex, *Brookhaven Symp. Bio.* **16** (1964) 1.
[24] R.H. Smith and T. Murashige, *Am. J. Bot.* **57** (1970) 562.
[25] E.A. Ball, *Ann. Bot.* **45** (1980) 103.
[26] I. Heidmann, N. Efremova, H. Saedler and S. Schwarz-Sommer, *Plant J.* **13** (1998) 723.
[27] I.M. Sussex, *Nat* **170** (1952) 755.
[28] S.S. Hantke, R. Carpenter and E.S. Coen, *Development* **121** (1995) 27.
[29] R. Carpenter and E.S. Coen, *Development* **121** (1995) 19.
[30] L.E. Sieburth, G.N. Drews and E.M. Meyerowitz, *Development* **125** (1998) 4303.
[31] N.A. McHale and M. Marcotrigiano, *Development* **125** (1998) 4235.
[32] B.G. McLean, F.D. Hempel and P.C. Zambryski, *Plant Cell* **9** (1997) 1043.
[33] K.M. Crawford and P.C. Zambryski, *Curr. Opin. Plant Biol.* **2** (1999) 382.
[34] W.J. Lucas, *Curr. Opin. Cell Biol.* **7** (1995) 673.

[35] K.J. Oparka, A.G. Roberts, P. Boevink, S. Santa Cruz, I. Roberts, K.S. Pradel, A. Imlau, G. Kotlizky, N. Sauer and B. Epel, *Cell* **97** (1999) 743.
[36] P.L.H. Rinne and C. Van der Schoot, *Development* **125** (1998) 1477.
[37] M.C. Perbal, G. Haughn, H. Saedler and Z. Schwarzsommer, *Development* **122** (1996) 3433.
[38] W.J. Lucas, S. Bouch-Pillon, D.P. Jackson, L. Nguyen, L. Baker, B. Ding and S. Hake, *Sci* **270** (1995) 1980.
[39] T. Vernoux, D. Autran and J. Traas, *Plant Mol. Biol.* (in press).
[40] S.E. Clark, *Plant Cell* **9** (1997) 1067.
[41] D. Weigel and S.E. Clark, *Plant Physiol.* **112** (1996) 5.
[42] S.J. Liljegren and M.F. Yanofsky, *Curr. Opin. Cell Biol.* **8** (1996) 865.
[43] M.M.S. Evans and M.K. Barton, *Ann. Rev. Plant Physiol. Plant Mol. Biol.* **48** (1997) 673.
[44] J.A. Long, E.I. Moan, J.I. Medford and M.K. Barton, *Nat* **379** (1996) 66.
[45] K. Endrizzi, B. Moussian, A. Haecker, J.Z. Levin and T. Laux, *Plant J.* **10** (1996) 967.
[46] M.K. Barton and R.S. Poethig, *Development* **119** (1993) 823.
[47] R. Kerstetter, E. Vollbrecht, B. Lowe, B. Veit, J. Yamaguchi and S. Hake, *Plant Cell* **6** (1994) 1877.
[48] S. Hake, E. Vollbrecht and M. Freeling, *EMBO J.* **8** (1989) 15.
[49] N.R. Sinha, R.E. Williams and S. Hake, *Genes Dev.* **7** (1993) 787.
[50] G. Chuck, C. Lincoln and S. Hake, *Plant Cell* **8** (1996) 1277.
[51] J.J. Estruch, E. Prinsen, H. Van Onckelen, J. Schell and A. Spena, *Sci* **251** (1991) 1364.
[52] M. Tamaoki, S. Kusaba, Y. Kano-Murakami and M. Matsuoka, *Plant Cell Physiol.* **38** (1997) 917.
[53] H.-S. Rupp, M. Frank, T. Werner, M. Strnad and T. Schmülling, *Plant J.* **18** (1997) 557.
[54] H.M.O. Leyser and I.J. Furner, *Development* **116** (1992) 397.
[55] S.E. Clark, M.P. Running and E.M. Meyerowitz, *Development* **119** (1993) 397.
[56] S.E. Clark, M.P. Running and E.M. Meyerowitz, *Development* **121** (1995) 2057.
[57] J.M. Kayes and S.E. Clark, *Development* **125** (1998) 3843.
[58] P. Laufs *et al.*, *Plant Cell* **10** (1998) 1375.
[59] S.E. Clark, R.W. Williams and E.M. Meyerowitz, *Cell* **89** (1997) 575.
[60] J.C. Fletcher, U. Brand, M.P. Running, R. Simon and E.M. Meyerowitz, *Sci* **283** (1999) 1911.
[61] S. Jeong, A.E. Trotochaud and S.E. Clark, *Plant Cell* **11** (1999) 1925.
[62] R.S. Poethig, *Plant Cell* **9** (1997) 1077.
[63] L.G. Smith and S. Hake, *Plant Cell* **4** (1992) 1017.
[64] E. Souer, A. Van Houwelingen, D. Kloos, J. Mol and R. Koes, *Cell* **85** (1996) 159.
[65] M. Aida, T. Ishida, H. Fukaki, H. Fujisawa and M. Tasaka, *Plant Cell* **9** (1997) 841.

[66] J.A. Long and M.K. Barton, *Development* **125** (1998) 3027.
[67] R. Simon, R. Carpenter, S. Doyle and E. Coen, *Cell* **78** (1994) 99.
[68] G.C. Ingram, J. Goodrich, M.D. Wilkinson, R. Simon, G.W. Haughn and E.S. Coen, *Plant Cell* **7** (1995) 1501.
[69] V. Gaudin, P. Lunness, P. Fobert, M. Towers, C. Riou-Khamlichi, J. Murray, E. Coen and J.H. Doonan, *Plant Physiol.* (2000) (in press).
[70] D. Reinhardt, F. Wittwer, T. Mandel and C. Kuhlemeier, *Plant Cell* **10** (1998) 1427.
[71] A.J. Fleming, S. McQueen-Mason, T. Mandel and C. Kuhlemeier, *Sci* **276** (1997) 1415.
[72] L. Gälweiler, C. Guan, A. Müller, E. Wisman, K. Mendgen, A. Yephremov and K. Palme, *Sci* **282** (1998) 2226.
[73] S.K. Christensen, N. Dagenais, J. Chory and D. Weigel, *Cell* **100** (1999) 469.
[74] K.D. Allen and I.M. Sussex, *Planta* **200** (1996) 254.
[75] E. Souer, A. van der Krol, D. Kloos, C. Spelt, M. Bliek, J. Mol and R. Koes, *Development* **125** (1998) 733.
[76] J.C. Malayer and A.T. Guard, *Am. J. Bot.* **51** (1964) 140.
[77] R. Garrison, *Am. J. Bot.* **42** (1955) 257.
[78] S. Mapelli and J.M. Kinet, *Plant Growth Regul.* **11** (1992) 385.
[79] S. Mapelli and L. Lombardi, *Plant Cell Physiol.* **23** (1982) 751.
[80] K. Schumacher, T. Schmitt, M. Rossberg, G. Schimtz and K. Theres, *Proc. Natl. Acad. Sci. USA* **96** (1999) 290.
[81] V. Grbic and A.B. Bleecker, *Development* **122** (1996) 2395.
[82] J.R. McConnell and M.K. Barton, *Development* **125** (1998) 2935.
[83] J. Gorter, *Origin of fasciation*, edited by W. Ruhland, Encyclopedia of Plant Physiology (Springer-Verlag, New-York, 1965) p. 330.
[84] W.C. Worsdell, *New Phytol.* **4** (1905) 55.
[85] M. Crespi, E. Messens, A.B. Caplan, M. Van Montagau and J. Desomer, *EMBO J.* **11** (1992) 795.
[86] O. Nilson, T. Moritz, B. Sundberg, G. Sandberg and O. Olsson, *Plant Physiol.* **112** (1996) 493.
[87] W. Gottschalk, *J. Nucl. Agr. Biol.* **6** (1977) 27.
[88] E.J. Knights, *Euphytica* **69** (1993) 163.
[89] M.C. Albertsen, T.M. Curry, R.G. Plamer and C.E. Lamotte, *Bot. Gaz* **144** (1983) 263.
[90] G. Felix, T. Altmann, U. Uwer, A. Jessop, L. Willmitzer and P.-C. Morris, *J. Exp. Bot.* **47** (1996) 1007.
[91] P. Laufs, J. Dockx, J. Kronenberger and J. Traas, *Development* **125** (1998) 1253.
[92] B. Moussian, H. Schoof, A. Haecker, G. Jürgens and T. Laux, *EMBO J.* **17** (1998) 1799.
[93] J.E. Gottlieb and T.A. Steeves, *Phytomorphology* **11** (1961) 230.
[94] R.A. Kerstetter and S. Hake, *Plant Cell* **9** (1997) 1001.
[95] M.G. Cline, *Am. J. Bot.* **84** (1997) 1064.

[96] C.A. Napoli, C.A. Beveridge and K.C. Snowden, *Curr. Top. Dev. Biol.* **44** (1999) 127.
[97] K.V. Thimann, F. Skoog and W.G. Kerckhoff, *Proc. Royal Societ. Bot.* **114** (1934) 317.
[98] H.M. Leyser, F.B. Pickett, S. Dharmasiri and M. Estelle, *Plant J.* **10** (1996) 403.
[99] C.P. Romano, M.B. Hein and H.J. Klee, *Genes Dev.* **5** (1991) 438.
[100] A.M. Chaudhury, S. Letham, S. Craig and E.S. Dennis, *Plant J.* **4** (1993) 907.
[101] J.I. Medford, R. Horgan, Z. El-Sawi and H.J. Klee, *Plant Cell* **1** (1989) 403.
[102] M. Faiss, J. Zalubilova, M. Strnad and T. Schmulling, *Plant J.* **12** (1997) 401.
[103] I. Pillay and I.D. Railton, *Plant Physiol.* **71** (1983) 972.
[104] M.G. Cline, *Physiol. Plantarum.* **90** (1994) 230.
[105] N. Sinha, *Annu. Rev. Plant Physiol. Plant. Mol. Biol.* **50** (1999) 419.
[106] D. Jackson, *Curr. Biol.* **6** (1996) 917.
[107] B.-J. Janssen, L. Lund and N. Sinha, *Plant Physiol.* **117** (1998) 771.
[108] D. Hareven, T. Gutfinger, A. Parnis, Y. Eshed and E. Lifschitz, *Cell* **84** (1996) 735.
[109] A. Hudson, *Curr. Opin. Plant Biol.* **2** (1999) 56.
[110] R.G. Schneeberger, P.W. Becraft, S. Hake and M. Freeling, *Genes Dev.* **9** (1995) 2292.
[111] M. Tsiantis, R. Schneeberger, J.F. Golz, M. Freeling and J.A. Langdale, *Sci* **284** (1999) 154.

CONTRIBUTED PAPER 1

Two Symmetries Linking Biological and Physical Branching Morphogenesis

V. Fleury

1. INTRODUCTION

As the different contributions in this book show, there exist a lot of branching patterns, both in physics and biology, so it is expected that general principles might govern the ubiquity of this class of patterns. However, it is not so easy to find straightforward relationships between the physical and the biological processes, although they are dearly desired. In particular, the very complex biochemistry involved in any living tissue, which eventually supports any morphogenic event, makes it difficult to sort out what is truly relevant for morphogenesis and what provides only the general support and satisfaction of metabolic demands. For example, it is clear that a growing tissue which adopts a branching mode of growth must in the first place be able to grow. Therefore, it is not at all obvious that growth factors play any role in the morphogenesis *per se*, except for the mere fact that they allow the tissue to get larger. Also, it should be added in such an introduction, that it is not very easy either to find relationships between biological branching processes. For example, Leonardo da Vinci thought that vessel growth and plant growth were similar, but was not able to provide a link between their growth mechanisms, except for a purely structural or metaphorical similarity.

Several lectures explain how to "map" one problem onto another, *i.e.*, how to find the mathematical path that relates one problem to another. Generally, the first step in this mapping consists in finding, on a given problem, the set of equations that describes the growth process. Once this set of equations is found, it is often simple to show that these equations are identical, or similar to the equations of the growth in another system. Sometimes, this similarity is found only in one limiting case. This kind of effort is examplified by Misbah's contribution on dewetting. We show here how symmetry concepts may also help in finding a relationship between different branching patterns. These symmetry concepts will shed some light on two branching morphogenesis processes in biology.

2. DENDRITIC GROWTH, BOTANY AND IN/OUT SYMMETRY

2.1. General Principles of Vegetal Growth

The contributions of Hallé and Laufs in this book describe the growth and branching in plants. It is clear that the main morphogenic events occur in a small mass of only slightly differentiated cells, called the meristem. The tissues, there, are not very specialized, and it is only afterwards that stomata, epidermis, venation, bark, etc. will form. Henceforth, lobes and invaginations are generated on the meristem at stages where the cells are very similar, despite the fact that differences in gene expression can already be observed (see Lauf's contribution). However, even if differences in gene expression are observed, it is not so clear whether these differences induce the morphogenic events, or *vice versa.*

Meristems may have rather different shapes [1], ranging from quite flat and compact shapes, to rather conical and rocket-like structures. These differences will ultimately lead to important structural differences in the leaf shape, and also, in part, in the plant shape. These meristems have several remarkable properties: they may "propagate" for ever, growing and emitting side-branches eternally. They have self-organized properties, as well explained by Patrick Laufs, and they exhibit specific symmetry-broken patterns, with Fibonacci series in the numbers of clockwise and anticlockwise spiral arrangements (this is known as the phyllotactic arrangement of leaves, or phyllotaxis).

After learning these properties of meristems, it is difficult at first glance to find any connection between plant growth and the physical processes of growth, as examplified by the crystalline dendrites (typically: the snow-flake).

However, in many instances, dendrites and meristems share several of these remarkable properties. Let us look at Figure 1. We see a meristem (left), and a dendrite (right).

They share the following characteristics: one observes a more or less conic, stable structure, which grows forward eternally. This growth cone, as we shall call it, "emits" lateral instabilities which grow as the tip moves forward. These instabilities may form leaves, or new meristems, which might themselves emit and form new meristems etc. This is close to what is observed in crystals. Indeed, let us turn to the image of the dendrite: the "growth cone" of the dendrite emits side branches, which will themselves behave again as dendrites etc. These features are remarkable enough to prompt an analogy between crystal

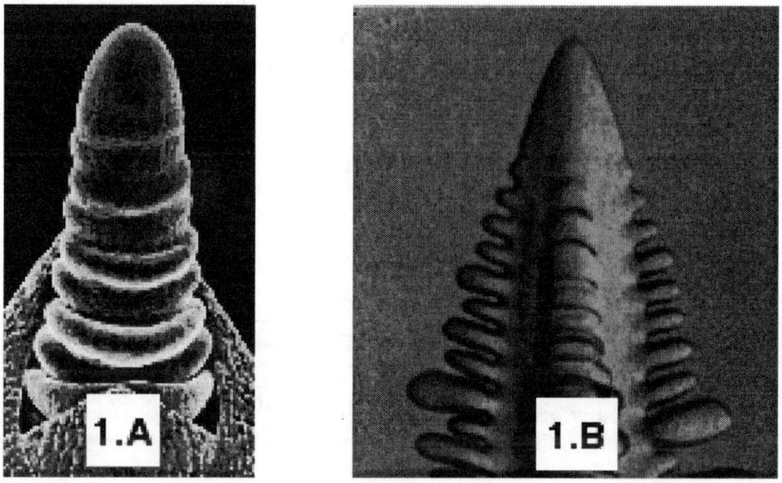

Fig. 1. — A meristem A). A dendrite B). Both look like a stable tip that grows forward eternally, and that emits "waves" which increase in size to become leaves, or new entities like independent meristems or independent dendrites.

growth and meristem growth. Also, the other mode of branching: dichotomous growth, is observed both in dendritic growth ("tip-splitting") and in plants.

However, if one wishes to map the crystal growth onto the meristem growth, one faces several problems. The principal one is simply that dendrites grow by accretion of atoms which come from the *outside* of the dendrite. The inside of the dendrite is supposed to be a perfect solid, with no diffusion of the atoms whatsoever. Conversely, whatever the field that drives the growth of meristems, it will be found *inside* the vegetal and not outside. Especially, if the gene expression fields identified by biologists are indeed the sources of the morphogenic events, it is obvious that these fields must be inside the plant.

2.2. A Possible Relationship Between Vegetal and Crystal Morphogenesis

These apparently trivial remarks may help in connecting crystal growth to plant growth, because if one wishes to map one problem onto the other, one needs to consider internal fields in botany, and external fields in crystal growth. Let us then take the physics of crystal growth, and consider heuristically a case in which the outer field is uniform, and in which the "crystal" grows by internal fields, and not by external fields, as classical in crystallogenesis.

The origin of the instabilities in the dendritic growth process, as explained by several authors in this book, is the Mullins-Sekerka instability: tips tend to grow, because when a protrusion grows ahead, it sees a sharper gradient

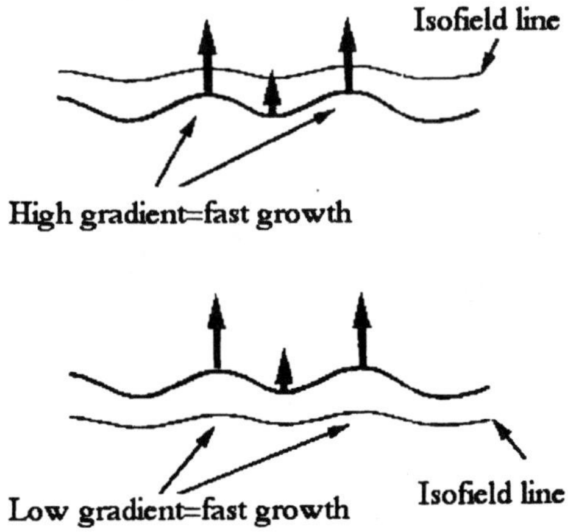

Fig. 2. — Scheme of an interface which grows in a field. If the growth is by external fields (top), gradients become sharper as the interface bulges forward. If the growth is by internal fields (bottom), gradients become lower in the same instance.

of the field (pressure field, concentration field, thermal field, etc.), and so it grows even further ahead. If, now, we assume that a structure, say a vegetal structure, grows by internal fields, we see that a tip growing forward will see lower and lower gradients as it moves (see Fig. 2), or rather, as it grows.

Therefore, we may expect that a structure whose surface grows faster in the direction of higher gradients will become actually flatter and flatter. But instead, we demand that tips emerge and self-organize, as in the "external" dendrites.

As a consequence, on theoretical grounds, we must not consider that the interface grows faster in the directions of higher gradients, but, conversely, that the interface grows *faster* in the directions of *lower* gradients. This means that, instead of considering the internal field as a *promoter* of growth, we must consider the internal field as an *inhibitor* of growth.

In the end, we can internalize the physics of crystal growth if we invert the effect of the field. We then get an internal mechanism of growth, which, in principle, is also able to generate dendritic structures with lobes and invaginations [2]. However the symmetry inside/outside that we have done, requires a symmetry promoter/inhibitor.

This means that the origin of invaginations and lobes ("foliation" in botany), is to be found in inhibitors of growth. Surprisingly, this has been hypothesized, if not proven, for some time in botany [3].

2.3. A Numerical Simulation Showing the Growth Patterns

There exist several ways of implementing numerically or mathematically a model of growth in which growth is faster in the direction of either low or high gradients. We used a set of differential equations first put forward by Meinhardt [4, 5], and which we have studied recently [6].

The set of equations is (note that there are only 3 equations):

$$\frac{\partial A}{\partial t} = cA^2(S_0 \pm S^\alpha) - \mu A + D_A \Delta A \tag{1}$$

$$\frac{\partial S}{\partial t} = co - cA^2 S - \gamma S \pm \varepsilon Y S + D_s \Delta S \tag{2}$$

$$\frac{\partial Y}{\partial t} = dA - eY + Y^2/(1 + fY^2). \tag{3}$$

The justification of such a set of equations is not so complex, and it has, to some extent, a biological support. In a medium of finite concentration S (either a promoter or inhibitor of growth), a small excess concentration of A triggers autocatalytically a peak of A, which propagates in the shape of a traveling wave. This wave corresponds to a "peal" of cells currently undergoing differentiation. Y is a two states dynamical system, Y falls from the "zero" to the "1" state when A reaches a large enough value. This jump of Y (the change in cell state) is irreversible. The A wave depends on S only, especially, its speed depends on the prefactor of the term A^2 in equation (1). The symbol $\pm$ in equations (1) and (2) means that we have to put $+1$ for a classical external dendritic growth, and -1 for a less classical internal dendritic growth. Only the front part of the A wave is important, and A can be made to return to zero in the back of the wave in different manners (this corresponds to cells reaching rest in their differentiated state), by adding Y-dependent terms in equation (1). For example a sink of A may be added wherever Y is not zero.

2.3.1. *External Dendrites Growing Under the Influence of Growth Promoters*

When this set of equations is used to describe the evolution of an interface in an external diffusible field, it is perfectly able to produce "dendrites", as we have shown recently. Figure 3 shows a (+1) dendrite for a typical set of parameters.

The initial condition is a small square of Y and A, in the center, and a constant field of S outside the small square. The pattern grows in the diffusion field of S, as the A wave tends to *rush* towards higher values of S, and the Mullins-Sekerka instability (see chapter by Couder), *i.e.* the "tip-effect", generates side-branches.

2.3.2. *Internal Dendrites Growing Under the Influence of Growth Inhibitors*

By changing very little in the parameters of these equations, we may create a model of growth by internal fields, with higher growth speed in the directions

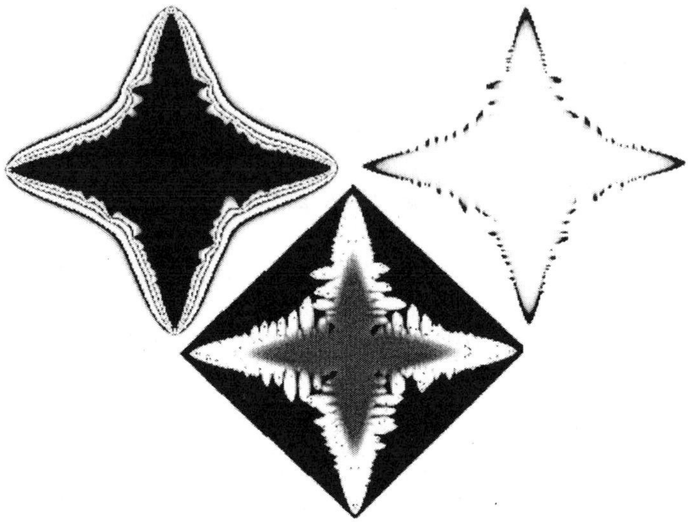

Fig. 3. — A dendrite obtained in a model of growth with external fields using the three equations above. The upper left image shows the field of growth promoter (field S); the gradient is sharper at tips. The upper right image shows the regions undergoing a change in cell state (A field). The lower image shows the pattern which is laid down as the A wave moves forward (Y field).

of low gradients. The internal model requires to consider a term of the form $S_0 - S$, in the first equation.

In this picture, the "standard" state of the tissue is a state in which the vegetal tissue grows at a constant speed. Then, in the presence of inhibition, the growth speed of the tissue is that constant, minus the flux of a given quantity which is an inhibitor of growth. This inhibition will modulate the growth, and it will spontaneously generate undulations and invaginations on the growth cone. This eventually leads to morphogenesis by internal fields.

We have again implemented a numerical solution of these equations. This time, the fields are set to zero outside the pattern, and the pattern grows from the inside, as the A wave tends to *escape* from S. (The diffusion constant of S is also set to zero outside the pattern.) The wave speed is larger when the gradient is small, so the growth speed is maximal at the tips, as demanded. On the sides, a symmetrical "tip-effet" in the field of S is able to generate side-branches. In this instance "a tip" is the dip of a "fjord" in the structure, which mirrors the tip of the branch of the other growth mode. It comes as a surprise that the shape selected by the anisotropy (Fig. 4) is qualitatively the same: a sharp growth cone ("internal dendrite") with side branches (at 45° of the "external dendrite").

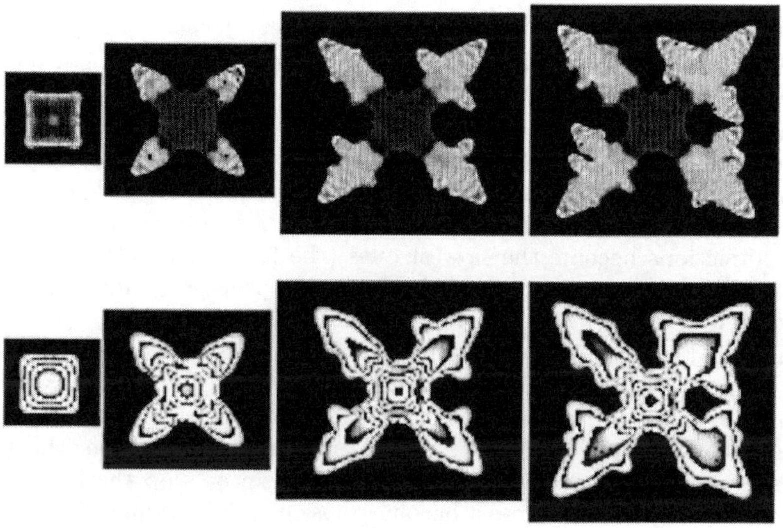

Fig. 4. — A "meristem" obtained in a model of growth with internal fields using the three equations above. A stable tip emits side branches which will themselves behave as independent entities. This is the simplest model which is able to form a growth cone which has spontaneous self-organized properties, and which "reiterates".

The fact that the shape is also cone-like when diffusing from the inside is not so surprising. Indeed, the shape of the growth cone is due to the combination of a forward speed proportional to t and a lateral diffusion proportional to $t^{1/2}$. Apparently, this crude argument is valid in the external mode as well as in the internal mode. The actual dimensions of the meristem will of course depend on the values of the parameters in the equations, which are rooted in properties of biochemicals.

Apart from the actual size of the problem, and depending on these parameters, this model generates by itself, very simply, complex biological shapes with lobes and foliations. Although, we do agree, the patterns are more "squarish" than true roots or leaves, because of a numerical artefact (namely, the use of a square grid for the numerical solution).

At this stage it should be said that the model is much more general than could be thought from the equations (1–3). Although we use here a set of equations coming from reaction-diffusion mathematics *à la* Turing [11], we may say that the internal mode of growth is much more general, as is the external mode of growth (which can be recovered with all sorts of models which can be mapped onto each other). The important issue, here, is that internal and external dendrites are linked to each other by a mirror symmetry. The idiosyncratic implementation with reaction-diffusion equations is not essential, this is why we are not so much concerned, at this stage, by a specific correspondence between the quantities A, S, and Y and biological fields.

2.3.3. Two Interesting Results

Two interesting, and unexpected findings of this model are the following. First, in dendritic growth, the conic shapes grow in a specific direction selected by the anisotropy (see lectures by Karma and Ben Jacob in this book). There exist directions of more rapid growth which, in Figure 3, are simply vertical and horizontal. For the same reason, snow-flakes exhibit six-fold symmetry. Now, when we internalize the growth dynamics, as in Figure 4, we see that the fastest directions become the slowest ones (the favored directions are rotated by 45°). Hence, the symmetry inside/outside + promoter/inhibitor is accompanied by a symmetry fast directions/slow directions.

The second interesting finding is the sequence of shapes which is observed, as a function of parameters. Indeed, if the gradient sensitivity is always too low, the mass of cells grows at a constant pace, and a uniform "leaf", or vegetal is produced (in the numerical model, a regular square). However, as the gradient is increased, it becomes high enough to slow down or stop the interface, and more interesting morphogenesis becomes possible. The sequence of shapes is from a square with small indentation to root-like or filamentary-like structures, with sporadic spontaneous reiteration (as in Fig. 4). This points to the differences in leaf shape, from simple to compound, which are observed in botany, and which can also be artificially obtained with mutants (see for example the tomato leaf in Lauf's chapter). The mathematical calculations suggest that the transition from a simple to a compound leaf might in fact be the consequence of a very simple modification at the molecular level which modifies slightly the magnitude of gradients.

2.3.4. About the Actual Morphogenic Fields

The principle difficulty, in the sort of ideas presented here, is to identify an internal field generating such an instability, among all the biochemical or biophysical fields which are involved in plant growth. We speculate that the main field is just the mechanical strain field in the vegetal tissue. In support of this view we may put forward several facts: first that mechanical action on plants is well known to have morphogenic effects; second, that when one looks at meristem pictures, one is struck by the fact that not only does the meristem have an overall curvature, but individual cells have a curvature which matches the global curvature of the meristem. Therefore, individual cells are bent, which is a clue of a deformation exerted on each of them, which adds up to form the global shape of the nascent leaf or branch while individual cells strive to relief the stress. Another piece of circumstantial evidence is provided by the origin of vascular systems (see Fleury and Schwartz in this book). Indeed, vascular systems in animals form under the influence of mechanical fields, and apparently, so does the vegetal vasculature (see Couder's contribution).

This implies that the morphogenesis of the tissue making up the venation is driven by mechanical forces. It would not be very surprising, then, that the

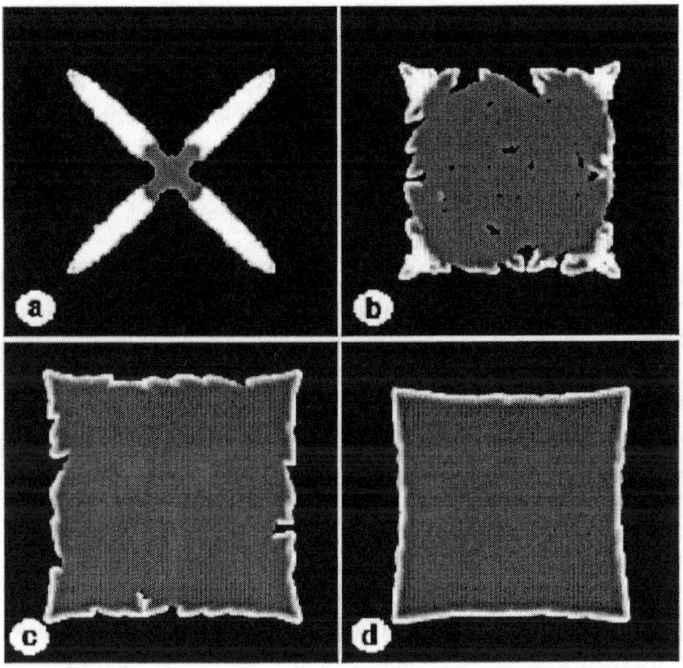

Fig. 5. — Sequence of shapes, from simple patterns to compound and filamentary, as obtained with different parameters (here, we changed the value of D_A, which amounts to change the mitotic rate).

morphogenesis of the rest of the tissue were also under control of mechanical forces.

Finally, it must be insisted that differential rate of cell division cannot by itself be the sole cause of morphogenesis, because cell division implies mechanical displacement and push of surrounding cells to find the final arrangement of new cells in the meristem and eventually in the body of the plant. *A contrario*, cell division is certainly modulated by the strain distribution.

Hence, we may construct a very simple model of foliation based on mechanical strain in the following way. When a tissue (say a vegetal tissue) is under tangential stress, it is quite favorable for it to buckle. Indeed, a flat shape generates a uniform stress field and strain field, in which the stored elastic energy is not minimal. It is more favorable for it to adopt an undulated shape, because the grooves and bumps generate a distribution of stress and strain corresponding to a lower energetical configuration (integral of the product of stress by strain). Now, in such a situation, the tissue will strive to relieve the internal strain. This process is classical in physics and leads to growth of instabilities and wavy structures in the context of the Grinfeld instability [8]. In the Grinfeld instability, mass flow of atoms relieves stress by increasing the bumps, and digging the grooves, whose net effect is to reduce the stored elastic energy. However, the relief of stress has a totally different origin in living tissues. Relief of stress occurs mainly by cell division. Growth of the tissue may be viewed as a sequential process. Metabolites and water uptake induces high internal stresses of cells which is relieved by cell division (indeed, cell division consists in reducing the size of the cell, and hence in storing more mechanical energy inside the tension of the membrane). This stress relief is akin to a "fragmentation" of the tissue into smaller entities, which reduces the stored energy. This is loosely similar to mud cracking in dry ponds, and an analogous concept is used by Yves Couder in this book to explain the origin of leaf venation. In the case presented here, the "fragmentation", *i.e.*, the generation of new cellular walls by cell division, appears at a much smaller scale, the scale of individual cells.

So, the tissue grows by generating (under the action of membrane channels) an internal mechanical stress. The mechanical stress induces a deformation of the entire tissue, by action of all the individual forces. In the deformed state the stress is relieved by mitosis, and the deformed shape finds itself unstressed. The process repeats itself while growth continues, and, at every time, a tissue with similar mechanical properties is generated. At every step, the tissue contains very little residual stress, although it grows under its action. The net result is a forward growth, which is unstable. The mass flow is globally from the grooves towards the bumps, and the bumps actually move forward, while the grooves lay behind or even recede, just as predicted from the symmetry arguments explained above. Regions of high stress will be found naturally in the axils. Depending on the actual genetic feedback which generates cell division under strain, cells will tend to deform or grow in different manners before splitting, but the general principle remains the same. Of course, too high a stress stops mitosis, since cells cannot get arbitrarily small. In this picture, the axillary region is interesting in that it may be a region of both very high mitotic rates, and of dormant cells, when the cell size reaches the lower cut-off.

On fundamental grounds, another reason why mechanical fields are more satisfactory as morphogenic fields is simply that mechanical fields are always present. It is obviously simpler for nature to do morphogenesis by playing with existing and unavoidable mechanical fields, than to create diffusion fields with the correct fine tuning of parameters which will generate branching plants as we know them. (In practice: the only long-ranged equation in the set $1-3$, the one dealing with S, would be provided by mechanics, and the others would imply processing of local information only.)

Needless to say, meristem growth looks certainly more complex than what has been shown here. However, a simple model of the sort presented here is a good start before implementation of genetic "controls". In retrospect, this sort of mechanism might also have been a good start to further evolutionary improvement.

3. VISCOUS FINGERS, ORGAN FUNCTION AND TIME REVERSAL

3.1. Reversibility of the Dendritic Growth Process

In the previous section we changed space by a mirror symmetry across the interface. We shall now invoke another kind of symmetry: time reversal. The equations of the dendritic growth process, and the growth process itself have an interesting feature to them: upon reversal of the growth field, the equations are stable, and the branched patterns retract gently. We mean here that if the sign is changed in the gradients, then the branched pattern retracts, as if the movie was played backwards. Inverting the initial conditions, amounts exactly to a time reversal. This is at variance with other "patterning" processes in modern physics. If one considers for example the equations of deterministic chaos, the equations can be reversed, in theory, but not in practice, because they are chaotic in both directions. Suppose that a set of particles shows a complex, self-organized trajectory, and suppose we change the sign of all velocities, then a chaotic trajectory is still obtained, because it is impossible to restart the system in exactly the same state. Reverting the sign of all relevant quantities is not identical to a time reversal in non-linear dynamics, because the system shows sensitivity to initial conditions in both directions.

This is not the case with branching patterns: if branching patterns are generated by a dendritic growth mechanism from an initially smooth and regular "seed", they will gently retract and return to a smooth shape if the force is reversed, because the instability that generates them is *stabilizing* in the reverse mode.

3.2. Transposition to Organ Morphogenesis and Function

The general principle of reversibility of the branched formation points to the mechanism of morphogenesis of branched organs, and to the way these organs function. Indeed, it has been observed recently that there exists a very important mechanical component to the growth of glands, lungs and kidney [7,9,10].

The morphogenesis of these organs appears to be driven, to a large extent, by the difference between the pressure inside the lumen, and the pressure in the mesenchyme. The growth of the tubes is akin to a viscous fingering process, which is driven by the pressure gradient on the mesenchyme side.

It is known in neo-natal medicine, that increasing (resp. decreasing) the internal pressure of the lumen greatly increases (resp. decreases) the distal ramifications of the pulmonar tree. If the trachea is just clamped distal branches are more numerous, or, conversely, if it is left open, the distal branches are much less numerous.

Now, let us consider how these branched organs function. The liquids (milk, sperm, saliva etc.) may be excreted either spontaneously (passively) under the pressure of the epithelium which secretes the fluid, or it may be expelled by action of a specific muscle (this is the case of the prostate, the peritony, and of specific mammal muscles of the cetacean), or it may be expelled by an external action, such as negative pressure (suction in terrestrial mammals), or even by direct mechanical pressure on the entire gland.

It seems that few people have seriously questioned how all of this complex hydrodynamics is mechanically possible, and why so different mechanical actions may produce the same effect: the easy excretion of the fluid. In other words, why is it that contractions, suctions, global pressure are effective in pushing the fluid out? Why is it that the peritoneum motion does in fact expel the air from the lung, and bring it back in? The efficiency of the process is apparently admitted without further question. However hard it may be to believe, there is no reason *a priori* that a pressure exerted on the abdomen, or a pressure exerted inside a gland will expel the fluid. If the tree were full of a very viscous fluid, like wax, honey or a thick oil, the tubes forming the branches of the tree would undergo a Mullins-Sekerka instability that would lead to closure of the tubes and malfunction of the system. If air was very viscous, the animals would smother: large pockets of air would remain trapped in the lung. That is, it is not merely the case that, in such instances, a branching organ will function less favorably because of higher viscous drag. A lung would work much worse under water, but it would work somewhat (expelling the water would be more demanding). But the lung would not work *at all* in a thick gel, because the tubes would undergo an instability that would strangle them with pressure, instead of gently pushing the air out. It is only because the fluid (air, milk, saliva, tears, sperm etc.) is much less viscous than the mesenchyme that the distal branches gently retract while the fluid is pushed towards the output.

In summary, the fact that the epithelium produces a fluid (which, by its very definition, is less viscous than the mesenchyme) implies that a growing epithelium will grow in a tubular fashion, because this direction is destabilizing. This implies in the same time, that it will function and expel the fluid gently under pressure, because that direction is stabilizing. In some sense, the viscous fingering instability provides a perfect natural piston. Under any mechanical pull in one direction, the tube elongates and "stores" liquid, under any mechanical push in the other direction, the tube shortens and gives back, or excretes, the stored liquid. This is possible only because the deformation of an interface is stable upon reversal when the less viscous fluid is inside: *i.e.* changing the sign of the dynamics is identical to a time-reversal. The very first "liquid storage", in the embryo, provides also the morphogenesis.

Now, one may argue that organ growth and back-and-forth function, is not comparable to a viscous fingering process, because there are visco-elastic effects rather than viscosity alone. However, we speculate that the simplest viscous fingering process may have existed in very primitive animals to form and make function the first rudimentary glands. Moreover, viscous fingering also works well in non-Newtonian fluids such as clay [12, 13] or even in organic gels, as clearly shown in this book by Lindner.

Also, the fact that the growth process, as observed today, is certainly more "active" than actual viscous fingering does not represent a true difficulty for the general principle that I have just presented. Indeed, if a complex mechanical pull is able to form a finger, then reversing the pulling action – which amounts to reversing the direction of time – will retract the distal branches. This will happen even if these branches are not true viscous fingers and the pull is not just a passive "pressure gradient" pull. It does not matter so much if the acinal regions, bronchioles, nephrons or the like are not just viscous fingers, if the mechanics that generated them functions also in reverse mode.

Finally, it must be added that the distal regions, during morphogenesis, are very soft, close to liquids (indeed, these region secrete a lot of matrix degradation enzymes), and in mature organs, the distal branches are much softer than the first generations of branches.

In conclusion, in/out symmetry may be a clue of how plant morphogenesis relates to dendritic growth, and stability under time-reversal may be a clue of how viscous fingering relates to morphogenesis and function of branched organs. In the end, the simplicity of these general principles may explain why these structures: dendrites, viscous fingers, plants and branched organs, have so much in common, and how they were selected by nature.

REFERENCES

[1] J. Troughton and L.A. Donaldson, *Probing Plant Structure* (McGraw Hill, New York, 1972).

[2] V. Fleury, *C. R. Acad. Sci. Ser. B (Paris)* **322** (1999) 725-734.

[3] Schwabe and A.G. Clewer, *J. Theor. Biol.* **109** (1984) 595-619.

[4] H. Meinhardt, *Diff.* **6** (1976) 117-123.

[5] A.J. Koch and H. Meinhardt, *Int. J. Bif. Chaos* **7** (1997) 1-26.

[6] V. Fleury, *Phys. Rev. E.* **61** (2000) 4156-4160.

[7] M.E. Probyn, M.J. Wallace and S.B. Hooper, *Pediatric Res.* **47** (2000) 806-812.

[8] J. Müller, *Study of stress induced morphological instabilites* (Mc Gill University, Montréal, Québec, 1998).

[9] V. Fleury, *Arbres de Pierre* (Flammarion, Paris, 1998).

[10] V. Fleury and L. Schwartz, *Mechanical forces as the primary agent of branching morphogenesis* (submitted).

[11] A.M. Turing, *Phil. Trans. Roy. Soc. B* **237** (1952) 37.

[12] G. D'Accord, J. Nittman and H.E. Stanley, *Phys. Rev. Lett.* **56** (1986) 336-339.

[13] H. Van Damme, *Flow and interfacial instabilities in Newtonian and colloidal fluids*, Fractal approach to heterogeneous chemistry, edited by D. Avnir (John Wiley & Sons, 1989) pp. 199-227.

CHAPTER III

Establishing a Growth Axis in Fucoid Algae*

D.L. Kropf, S.R. Bisgrove and W.E. Hable

1. EMBRYONIC MORPHOGENESIS

Plant morphogenesis is governed by intricate, coordinate regulation of the orientations of cell expansion and cell division. The polarity of expansion and division are influenced significantly by extrinsic cues, including local, systemic and environmental signals. Understanding how these cues are perceived by cells and transduced into molecular, structural and physiological asymmetries that govern morphogenesis is a major focus of plant developmental biology. Although much has been learned about morphogenesis in later stages of development, embryonic morphogenesis largely remains a mystery because the embryo is enclosed within the ovule and is inaccessible to experimental manipulation.

Fucoid algae populate the coastline of much of the northern hemisphere where they grow attached to rocks in the intertidal zone, and eggs and zygotes of these marine brown algae provide an excellent opportunity for investigating embryonic morphogenesis. Large numbers of gametes are released by sexually mature plants into the seawater and thousands of zygotes developing synchronously, free of surrounding tissue, can be easily obtained in less than an hour. As in higher plants, extrinsic signals orient cellular growth and division, and thereby the morphogenetic pattern of the young embryo. The fucoid zygote monitors environmental cues locally in the seawater, integrates that information and establishes a developmental axis that orients both rhizoid growth and the first embryonic division. Importantly, the polarity of an entire population of zygotes can be oriented in unison by application of external vectors. For these reasons, the fucoid algae (including *Pelvetia compressa* (proposed to be renamed

* This paper was already published in *Trends in Plant Sci.* **4** (1999), pp. 490-494, with permission of Elsevier Science.

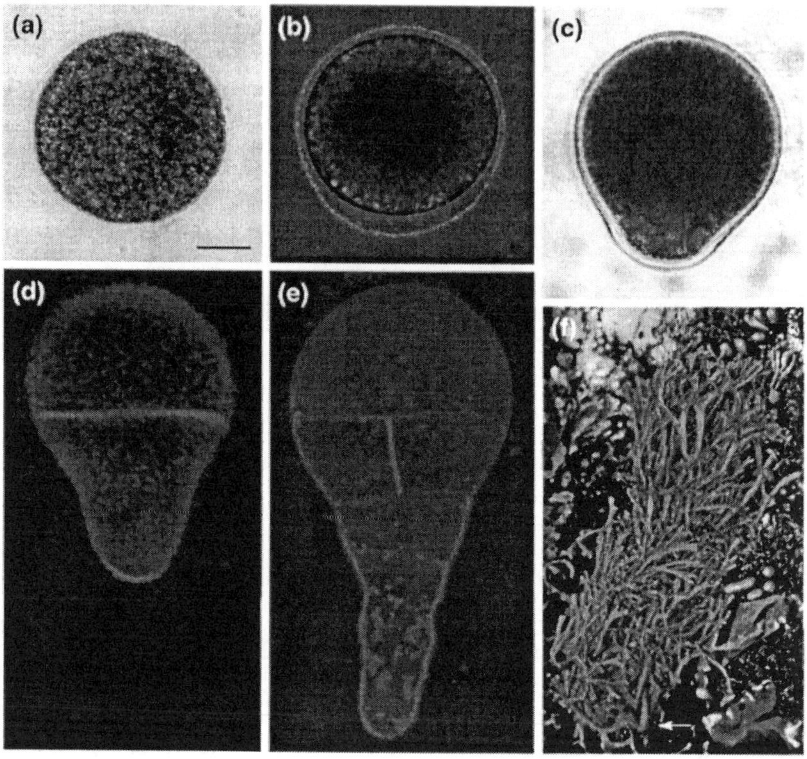

Fig. 1. — Early development in *P. compressa*. (a) Bright field image of an unfertilized egg. (b) Confocal image of a young zygote bearing polar adhesive (outlined with fluorescent microspheres). The region of thickest adhesive (bottom of the zygote) indicates the rhizoid pole. (c) Bright field image of a germinated zygote with emerging rhizoid. Zygote has been plasmolyzed to visualize the cell wall (d, e) Confocal images of embryos in which plasma membranes have been labeled with FM 4-64. (d) The first division is transverse to the growth axis and partitions the zygote asymmetrically. (e) Division pattern in an older embryo is nearly identical to that in higher plants. (f) The mature alga is approximately 0.5 m in length. Arrow denotes the holdfast, derived from the rhizoid, that attaches the mature plant to the rock; the leafy fronds derive largely from the thallus. Scale bar = 25 μm for (a-e).

Silvetia compressa [1]) and several species of *Fucus*) have long served as model organisms for experimental investigation of embryonic pattern formation.

The fucoid egg is a sphere, 60 μm (*Fucus* sp.) to 100 μm (*P. compressa*) in diameter, and by all accounts is radially symmetric (Fig. 1a). Sperm entry activates the egg and the zygote then establishes a rhizoid-thallus axis. Although zygotes can sense a wide array of extrinsic vectors when establishing polarity, light is probably the most important signal in nature. The rhizoid pole forms on the shaded hemisphere of zygotes experiencing unidirectional light, a process termed photopolarization. This polarity is first visually apparent as a local secretion of adhesive at the rhizoid pole (Fig. 1b). Increased secretion at this pole later results in outgrowth of a tip-growing rhizoid, giving the zygote a pear-shaped appearance (Fig. 1c). The first division occurs approximately 1 day after fertilization (AF) and is an invariant, asymmetric division oriented transverse to the growth axis (Fig. 1d). This division creates morphologically distinct rhizoid and thallus cells. The rhizoid cell continues to elongate by tip growth and to divide transverse to the growth axis, creating a file of cells (Fig. 1e). Rhizoids are negatively phototropic and so grow into cracks and crevices in the rocky substratum, eventually forming a holdfast for the plant. In contrast, the thallus cell expands by diffuse growth and undergoes proliferative divisions (each division is transverse to the previous one) giving rise to much of the stipe and fronds of the plant (Figs. 1e and f). Interestingly, the morphology of the young fucoid embryo (Fig. 1e) is very similar to that of a typical higher plant embryo.

In the past few years there have been exciting advances in our understanding of the mechanisms that control morphogenesis in fucoid embryos. Below, we will present our current perspective on the early stages of morphogenesis in which the symmetrical egg (Fig. 1a) is transformed into a polar zygote exhibiting local secretion (Fig. 1b). For simplicity this transformation will be discussed in two phases, axis selection and axis amplification, but in reality it is a continuous process without defined stages. For additional discussions on fucoid morphogenesis, the reader is directed to other excellent articles concerning fertilization [2], tip growth [3,4], division plane orientation [3,5,6] and cell fate specification [7].

2. AXIS SELECTION

Axis selection is the process in which the orientation of the growth axis is specified and marked, and recent evidence indicates that an initial axis is selected at fertilization.

2.1. Sperm Entry

When fucoid zygotes first acquire polarity has long been an open question [5]. In 1931, Knapp reported that the sperm entry site becomes the rhizoid pole of an axis in the distantly related brown alga, *Cystosira barbata* [8]. Consistent with this finding, fucoid zygotes do not need to detect extracellular signals to polarize, and therefore may have an inherent default axis. We have recently investigated this issue in *P. compressa* zygotes and found that fertilization does indeed establish polarity and, like *C. barbata*, the rhizoid pole forms at the sperm entry site. In the absence of environmental cues, polar adhesive deposition and germination occur at the sperm entry site [9].

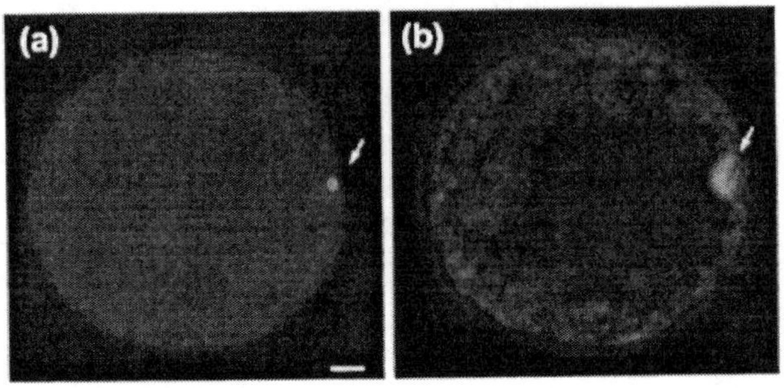

Fig. 2. — An F-actin patch forms at the site of sperm entry. (a) The sperm pronucleus (arrow) is labeled with Hoechst in a zygote 30 minutes AF. (b) In the same cell, an F-actin patch, labeled with rhodamine phalloidin (arrow), colocalizes with the sperm pronucleus. Scale bar = 10 μm.

This sperm-induced axis apparently provides a default axis that can be used if the zygote subsequently fails to detect useful environmental vectors.

What marks the rhizoid pole at the sperm entry site? Egg activation begins with a depolarizing fertilization potential carried by Na^+ and Ca^{2+} influx [2,10]. This depolarization provides a fast block to polyspermy and induces a small rise in intracellular Ca^{2+} that stimulates exocytosis of vesicles containing cell wall components. However, unlike many animal cells, there is no apparent Ca^{2+} wave beginning at the site of sperm entry and there is no evidence for any local increase in Ca^{2+} at the sperm entry site [11]. Although transcellular ion currents are detectable in recently fertilized eggs, they are unstable and shift positions for several hours [12]. Thus, there is no evidence that the ionic fluxes that activate the egg mark an axis.

Instead, sperm entry induces cortical F-actin rearrangements that mark the sperm entry site. Cortical F-actin, which is uniformly distributed in eggs, localizes in a patch at the sperm entry site within minutes of plasmogamy (Figs. 2, 3a). What causes the F-actin patch to assemble at the sperm entry site is completely unknown. Presumably, other proteins, possibly actin binding proteins or rho GTPases, recruit F-actin to this site [13]. These putative proteins may be introduced into the egg cell cortex from the sperm cytosol, similar to the cytosolic sperm factor that is postulated to activate mammalian eggs [14]. In contrast, microtubules, which emanate from the nuclear envelope (Fig. 3), play no role in axis selection or amplification, but function later in alignment of the asymmetric division [15].

2.2. Adhesion and Vector Perception

Following egg activation and axis selection at fertilization, the zygote synthesizes a cell wall but remains unattached to the substratum for approximately 2 hours. (All times are for *P. compressa*; *Fucus* sp. develop somewhat more

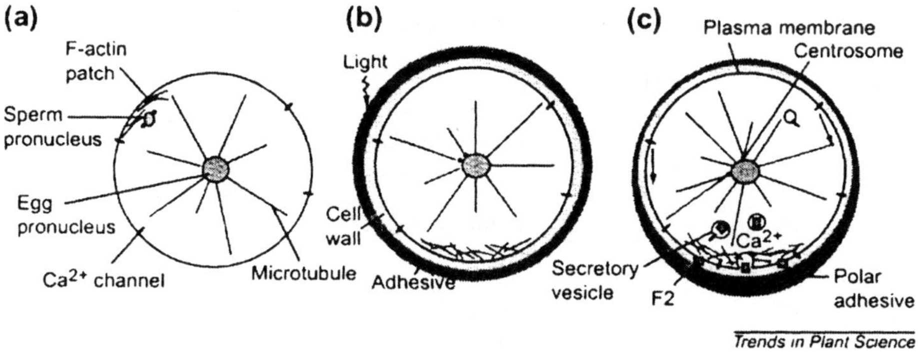

Fig. 3. — Spatial aspects of axis selection and amplification. (a) At fertilization, a cortical F-actin patch forms at the sperm entry site, Ca^{2+} channels are evenly distributed in the plasma membrane and microtubules emanate uniformly from the surface of the egg pronucleus. (b) By 3 hours AF, zygotes have synthesized a cell wall. At this time they begin to secrete adhesive uniformly over their surface and become receptive to environmental cues. Photopolarization of the young zygote causes reassembly of an F-actin patch at the new rhizoid pole (shaded hemisphere). (c) Beginning about 4 hours AF, the axis is amplified as molecules and processes, including Ca^{2+} channels, cytosolic Ca^{2+}, and adhesive become localized to the rhizoid pole. Later in amplification, the sulfated fucan F2 localizes to the rhizoid wall.

slowly.) Because gametes are released from mature receptacles when the sea is calm [16], young zygotes are not washed out to sea in mass, but instead settle rapidly onto the rocky substratum. Monitoring their environment for spatial cues during this period would likely prove useless because their orientation is constantly changing as they roll and tumble over the uneven surface. In fact, young zygotes are incapable of sensing light vectors, even if artificially held in place, until they attach. Attachment (2 – 3 hours AF) is mediated by a strong adhesive secreted uniformly over the surface of the zygote [17] (Fig. 3b) and is composed of polyphenolics and acidic carbohydrates, crosslinked by bromoperoxidase activity [18]. Once attached, zygotes become sensitive to directional light and other environmental cues; if spatial information is perceived, the zygote selects a new axis in accordance with the perceived vector. Rather than rotating the existing axis, the zygote abandons the sperm-induced axis and generates a new axis *de novo* [19]. The F-actin patch at the sperm entry site is disassembled and a new patch is assembled at the nascent rhizoid pole (Fig. 3b) [20].

Because many diverse extrinsic signals are perceived by the zygote, multiple signal transduction pathways are likely used, but very little is known about them. Photopolarization is the best studied pathway (Fig. 4), and light in the UV and blue wavelengths are most effective. The recent discovery that *P. compressa* zygotes contain large quantities of retinal suggests that the

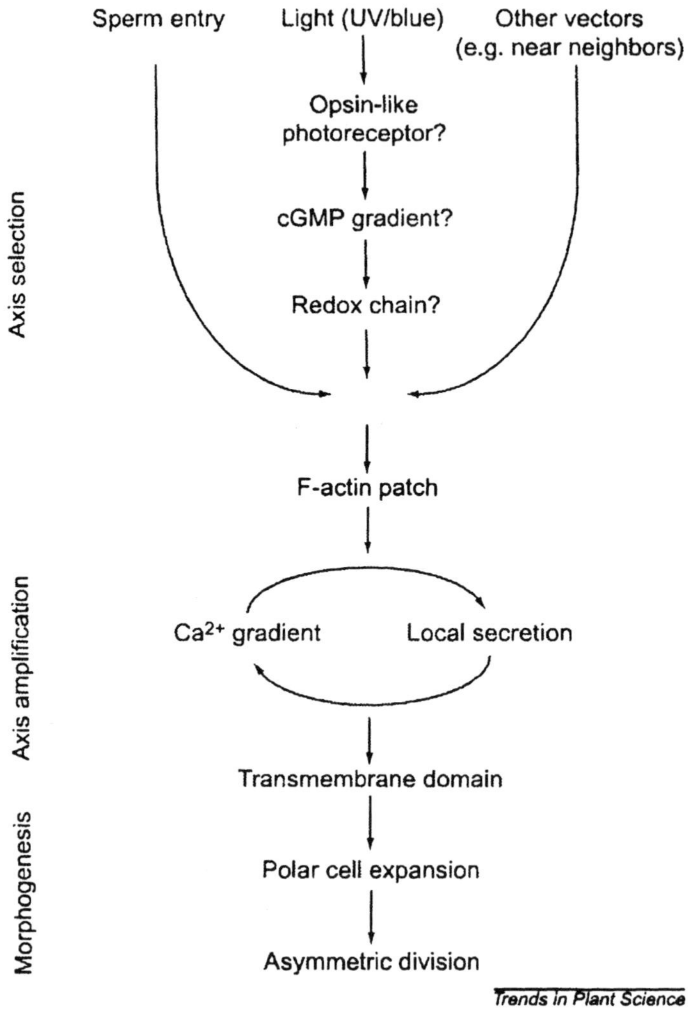

Fig. 4. — From signal transduction to morphogenesis. Multiple signal transduction pathways converge to a common polarization pathway. F-actin localization is an early step in the common pathway, providing a target site for subsequent amplification of the axis which creates a specialized domain that controls morphogenesis. F-actin is colored black because it is important for both axis selection and amplification.

photoreceptor may be an opsin-like molecule, similar to the photoreceptors of the vertebrate visual system [21, 22]. Phototransduction in the vertebrate visual system involves changes in cGMP, and it has recently been shown that cGMP levels rise, albeit somewhat slowly, in *P. compressa* zygotes following irradiation with blue light [23]. One enigma in phototransduction is that the rhizoid pole forms where the zygote is least irradiated. Robinson and colleagues have suggested that light reception on the lighted hemisphere may result in a local increase in cGMP, and that the rhizoid pole forms where the cGMP concentration is lowest [22]. There is also evidence that phototransduction of blue light involves a plasma membrane redox chain. A blue light stimulated redox activity is associated with the plasma membrane, and short-circuiting the redox chain by treatment with an electron acceptor renders zygotes blind to directional blue light [24]. At present, the relation between cGMP and the redox chain during phototransduction is unclear.

The signal transduction pathways must converge into a common polarity establishment pathway, and several recent findings suggest that localization of an F-actin patch may be early in the common pathway (Fig. 4). First, F-actin localizes to the rhizoid pole in response to different polarizing cues; in addition to sperm entry, unilateral light or the presence of a near neighbor (group effect) induce F-actin assembly at the presumptive rhizoid [20]. Second, the local F-actin assembly occurs rapidly following fertilization or photopolarization. Although previous studies indicated that F-actin does not localize until just before germination [5,25], it is now clear that localization occurs within minutes of axis selection (Figs. 3a, b). When the time courses of F-actin localization and photopolarization were measured on a single population of zygotes, it was not possible to separate the two processes temporally [20]. Finally, treatment with inhibitors of F-actin function (latrunculin B or cytochalasins) prevents localization of components of the common amplification loop (see below), indicating that amplification is downstream of F-actin localization.

3. AXIS AMPLIFICATION

Following F-actin localization, the nascent axis is amplified by a positive feedback loop at the rhizoid pole involving local increases in cytosolic Ca^{2+} and secretion (Fig. 4). The recent finding that, like F-actin, Ca^{2+} localizes to the rhizoid pole early in development has significantly advanced our understanding of amplification. In earlier studies, an accumulation of Ca^{2+} at the rhizoid pole was not detected until just prior to germination, approximately 10 hours AF. For this reason, amplification was thought to be a slow process of overlapping stages [5]. However, improved imaging techniques have recently permitted detection of elevated Ca^{2+} at the rhizoid pole by 6 hours AF, within 1 hour of exposure to unilateral light [26]. Polar secretion at the rhizoid pole is also detectable within 1 hour of photopolarization [17,27]. Taken together, these findings suggest that localization of secretion and cytosolic Ca^{2+} occur simultaneously, and that amplification begins earlier and progresses more rapidly than previously thought.

Axis amplification likely proceeds according to the following scenario. The F-actin patch provides a target site for amplification; beginning about 4 hours AF, Ca^{2+} channels start to accumulate in the plasma membrane overlying the cortical patch and simultaneously vesicle exocytosis becomes slightly preferential at this site (Fig. 3c). Two processes probably contribute to channel localization; movement of existing channels through the plane of the membrane [25] and local insertion of channel-containing vesicles [5]. Redistribution of Ca^{2+} channels is supported by the finding that fluorescent dihydropyridine, which binds to L-type Ca^{2+} channels, shows a gradual accumulation at the rhizoid pole in photopolarizing zygotes [28]. (Note, however, that the nature of the dihydropyridine receptor in fucoid algae is controversial [22].) Channel asymmetry causes greater influx at the rhizoid pole [29], resulting in a slight elevation of Ca^{2+} activity beneath the rhizoid plasma membrane. Because exocytosis in plants is Ca^{2+} dependent, vesicle fusion at the rhizoid site increases, more channels are inserted, the cortical Ca^{2+} elevates further, stimulating more vesicle fusion, and hence a positive amplification loop is established. By about 6 hr AF, a measurable Ca^{2+} gradient is present and secreted adhesive is clearly more abundant at the rhizoid pole.

Interestingly, axis amplification apparently begins at the same time developmentally (approximately 4 hours AF), regardless of the nature or timing of the polarizing vector. This premise is based on comparison of dark-grown zygotes bearing a sperm-induced axis to light-grown zygotes bearing a photo-induced axis. The axis in dark-grown zygotes is older than that in light-grown zygotes because it has been present since fertilization. Yet, dark-grown zygotes do not amplify their axes precociously; dihydropyridine receptors [28], adhesive [17], and cytosolic Ca^{2+} [26] remain uniformly distributed in dark-grown zygotes for several hours following fertilization. A direct comparison of the timing of localization in light- and dark-grown zygotes has been made for adhesive; adhesive becomes more abundant at the rhizoid pole at the same time in the two treatments [17]. It may be that the zygotes do not amplify their axes until they have had time to monitor their local environment for spatial cues.

Amplification presumably continues until germination, approximately 10 hours AF, and, surprisingly, the axis remains labile throughout most of this amplification period. If a new vector is perceived, the zygote selects and amplifies a new rhizoid site. The original axis is not rotated, rather a new axis is assembled [19]. The F-actin at the abandoned rhizoid site is disassembled [20], but remnants of the original axis, such as polar adhesive [5], remain. Interestingly, the new axis is assembled and amplified rapidly, within 1 to 2 hours of vector perception, perhaps because these older zygotes have already acquired competence to amplify an axis. Late in amplification, at approximately the time when the axis becomes unresponsive to new vectorial cues, a sulfated fucan (F2) is secreted locally into the rhizoid wall (Fig. 3c) [6].

Axis amplification in fucoid zygotes appears to be similar to polarity establishment in the budding yeast *Saccharomyces cerevisiae*, especially with regard to the role of F-actin. During asexual budding and sexual mating in yeast, F-actin patches assemble at the selected growth site [30] and provide a target site for assembly of a cortical complex that localizes secretion. Several protein components of the cortical complex require F-actin for their localization [31,32].

Likewise in fucoid zygotes, several components of axis amplification have been shown to be dependent on F-actin. Treatment with cytochalasins or latrunculins eliminates the F-actin patch [20] and inhibits localization of dihydropyridine receptors [28] and inward current [25] in young zygotes, suggesting that the F-actin patch may capture and anchor ion channels at the rhizoid pole. Polar secretion of adhesive is also F-actin dependent; in low concentrations of these inhibitors, abundant adhesive is secreted but it is deposited uniformly over the cell surface [17], perhaps because the F-actin target site is missing. However, the cortical F-actin patch may not be the sole F-actin array participating in amplification. For example, an actomyosin system likely translocates secretory vesicles to the rhizoid pole [33] and may also move ion channels through the plasma membrane.

4. TRANSMEMBRANE DOMAINS AND MORPHOGENESIS

Axis amplification generates a specialized domain that spans the cortex, plasma membrane and cell wall at the rhizoid site. The cortex uniquely contains an F-actin patch and elevated Ca^{2+}, the plasma membrane has an accumulation of Ca^{2+} channels and the wall contains F2 (Fig. 3c). Due to the asymmetry in channel localization (and perhaps function), a steady electrical current flows into the zygote at the rhizoid pole [12]. The targeting of secretory vesicles to this domain also has several consequences. The rhizoid cortex becomes optically clear (cortical clearing [12]), probably due to the accumulation of vesicles, and recently-inserted, particle-free membrane patches become abundant in the rhizoid plasma membrane [34]. Interestingly, the zygote preferentially lyses at the rhizoid pole in hypotonic seawater [17], indicating that the rhizoid cell wall is weakened during amplification, perhaps by secretion of wall loosening or degrading enzymes. This unique transmembrane domain becomes consolidated into physical linkages between the plasma membrane and the cell wall around the time of germination. These membrane-wall adhesions can be visualized by plasmolyzing germinated zygotes using hypotonic solutions; strands of cytoplasm remain attached to the cell wall specifically at the tip of the elongating rhizoid [35].

The transmembrane domain is postulated to direct many aspects of subsequent embryogenesis including polar cell expansion, spindle orientation, asymmetric cell division, and cell fate specification (Fig. 4) [5–7]. With regards to morphogenesis, regulation of cell expansion and division are most relevant. Cell expansion occurs primarily by tip growth at the rhizoid pole, and the transmembrane domain defines the growth site. In one sense, the domain is an assemblage of tip growth machinery; F-actin, a Ca^{2+} gradient and targeted vesicle transport and secretion are associated with, and required for, tip growth [4]. But the transmembrane domain clearly has other functions, including orienting the invariant, asymmetric division [3]. This division is oriented by a rotational alignment of the nucleus just prior to mitosis. Microtubules extending from the perinuclear centrosomes anchor in the cortex of the specialized domain. The transmembrane domain apparently provides positional information regarding the location of the rhizoid and also serves as an anchor for microtubule-based force generation that rotates the nucleus until the two centrosomes align parallel with the growth axis [15]. This rotation ensures that the spindle forms axially and the first division, which bisects the spindle, is transverse to the growth axis.

Similar domains have been postulated to control asymmetric divisions in higher plants [13], and may serve other, as yet unknown, functions. We anticipate that the recent advances in our understanding of fucoid morphogenesis will provide a useful framework for investigating this and other aspects of pattern formation during plant embryogenesis.

Acknowledgements

We wish to thank Diana Steller for collecting algae and for providing Figure 1f, and Dave Henderson, Stephen Ruth and Chris Plaisier for comments on the manuscript. Work from our lab has been supported by NSF award IBN 9807811 to DLK and NIH postdoctoral research award 1 F32 HD08470-01 to WEH.

REFERENCES

[1] E.A. Serrao, L.A. Alice and S.H. Brawley, *Phycology* **35** (1999) 382-394.
[2] C. Brownlee, *New Phytol.* **127** (1994) 399-423.
[3] J.E. Fowler and R.S. Quatrano, Cell polarity, asymmetric division, and cell fate determination in brown algal zygotes, in *Seminars in Developmental Biology: Simple Systems for the Analysis of Important Developmental Problems*, edited by D. Kirk, Vol. 6 (Academic Press, 1995) pp. 347-358.
[4] D.L. Kropf, *Microbiol. Rev.* **56** (1992) 316-339.
[5] D.L. Kropf, *Plant Cell* **9** (1997) 1011-1020.
[6] R.S. Quatrano and S.L. Shaw, *Trends Plant Sci.* **2** (1997) 15-21.
[7] C. Brownlee and F. Berger, *Trends Genetics* **11** (1995) 344-348.
[8] E. Knapp, *Planta* **14** (1931) 731-751.
[9] W.E. Hable and D.L. Kropf, *Development* **127** (2000) 493-501.
[10] S.H. Brawley, *Dev. Biol.* **124** (1987) 390-397.
[11] S.K. Roberts, I. Gillot and C. Brownlee, *Development* **120** (1994) 155-163.
[12] R. Nuccitelli, *Dev. Biol.* **62** (1978) 13-33.
[13] J.E. Fowler and R.S. Quatrano, *Annu. Rev. Cell Dev. Biol.* **13** (1997) 697-743.
[14] K. Swann and F.A. Lai, *BioEssays* **19** (1997) 371-378.
[15] S.R. Bisgrove and D.L. Kropf, *Dev. Biol.* **194** (1998) 246-256.
[16] E.A. Serrao *et al.*, *Proc. Natl. Acad. Sci. USA* **93** (1996) 5286-5290.
[17] W.E. Hable and D.L. Kropf, *Dev. Biol.* **198** (1998) 45-56.
[18] V. Vreeland, J.H. Waite and L. Epstein, *J. Phycol.* **34** (1998) 1-8.

[19] L.F. Jaffe, *Exp. Cell Res.* **15** (1958) 282-299.
[20] L. Alessa and D.L. Kropf, *Development* **126** (1999) 201-209.
[21] K.R. Robinson *et al.*, *Biophys. Res. Comm.* **243** (1998) 776-778.
[22] K.R. Robinson *et al.*, *Curr. Top. Dev. Biol.* **44** (1999) 101-125.
[23] K.R. Robinson and B.J. Miller, *Dev. Biol.* **187** (1997) 125-130.
[24] F. Berger and C. Brownlee, *Plant Physiol.* **105** (1994) 519-527.
[25] S.H. Brawley and K.R. Robinson, *J. Cell Biol.* **100** (1985) 1173-1184.
[26] R. Pu and K.R. Robinson, *J. Cell Sci.* **111** (1998) 3197-3207.
[27] K. Schroter, *Planta* **140** (1978) 69-73.
[28] S.L. Shaw and R.S. Quatrano, *J. Cell Sci.* **109** (1996) 335-342.
[29] K.R. Robinson and L.F. Jaffe, *Sci* **187** (1975) 70-72.
[30] A.E.M. Adams and J.R. Pringle, *J. Cell Biol.* **98** (1984) 934-945.
[31] K.R. Ayscough *et al.*, *J. Cell Biol.* **137** (1997) 399-416.
[32] K.R. Ayscough and D.G. Drubin, *Curr. Biol.* **8** (1998) 927-930.
[33] S.H. Brawley and R.S. Quatrano, *Dev. Biol.* **73** (1979) 193-205.
[34] H.B. Peng and L.F. Jaffe, *Planta* **133** (1976) 57-71.
[35] C. Henry, J.R. Jordan and D.L. Kropf, *Protoplasma* **190** (1996) 39-52.

CONTRIBUTED PAPER 2

Shape Stability during Osmotic Growth

M. Léonetti

1. INTRODUCTION

Plant and microbial physiologists generally agree that turgor pressure provides the driving force for the expansion of walled cells [1–3]. For example, the role of turgor pressure in the expansion of fungal cells was first discussed by Reinhardt as early as 1892. This hypothesis is indeed supported by a wealth of biophysical data which are consistent with the standard growth equations developed by Lockhart in 1965 [4] (see a critical review of turgor [5]). What is really turgor pressure? If the intracellular medium is more concentrated than the extracellular medium, there is an entering flux of water which increases the internal pressure (external pressure is assumed to be equal to zero). This pressure is called the turgor pressure. In some algae, turgor can reach a few atmospheres. To maintain the cellular integrity, the plant cell is bound by an extracellular matrix, the plant cell wall, which can sustain such a pressure. This is not so for animal cells which would explode. The turgor pressure is in dynamic balance between wall yielding and plant wall expansion. Of course, at the same time, new material is included in the cell wall. This allows the wall to extend. Lockhart has formulated a quantitative analysis of steady-state growth of single isolated cells. Firstly, he considers the rate of volume increase as a function of water uptake by the cell:

$$(\mathrm{d}V/\mathrm{d}t) = AL_{\mathrm{p}}\,(\Pi - P) \tag{1}$$

where A is the cellular surface, Π the difference between internal and external osmotic pressure, P the intracellular hydrostatic pressure or the turgor pressure and L_{p} the hydraulic conductance.

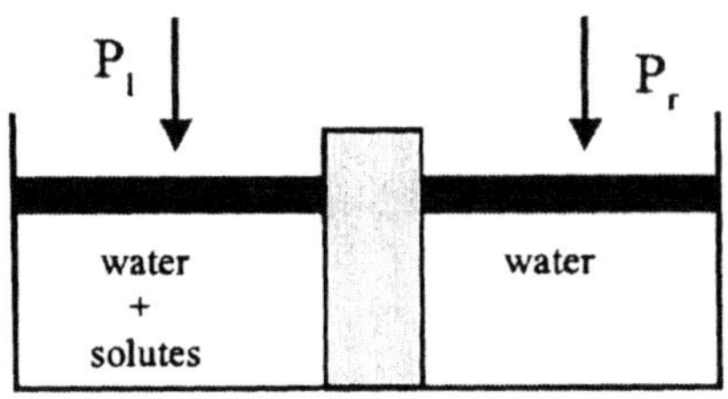

Fig. 1. — Pressures P_r and P_l are applied to maintain constant the volume on each side. At equilibrium, $P_l = P_r + \Pi$ where Π is the osmotic pressure.

Secondly, assuming that cell expansion occurs primarily by elongation, the equation for the rate of wall yielding is:

$$(\mathrm{d}V/\mathrm{d}t) = m(P - P_c) \tag{2}$$

where m is the longitudinal wall extensibility and P_c a critical value of the turgor pressure. In steady-state, the expression of the growth rate is:

$$(\mathrm{d}V/\mathrm{d}t) = (AL_p m/AL_p + m)\,(\Pi - P_c) \tag{3}$$

In elongating *Nitella* internodal cells, $AL_p \gg m$. Then, water transport does not limit the growth. This equation is the standard model of plant cell growth. The aim, here, is not to discuss the relevance of the mechanism (see biological reviews). In the following, there is no elongational growth in the sense of Lockhart but, rather, a competition between the turgor pressure which tends to expand the membrane and a surface tension which tends to limit this expansion.

2. OSMOTIC GROWTH

2.1. Osmosis

Osmosis is the passage of a pure solvent (water in our case) through a semipermeable membrane from one solution to another. Semipermeable means that only the solvent can flow through it and not the solutes.

Consider, for example, two solutions separated by such a membrane (Fig. 1). Each volume of the solution is kept constant by an applied pressure on either side (arrows). The right solution is only a pure solvent at pressure P_r while the left solution at pressure P_l contains some solute at a concentration C.

At $t < 0$, the membrane is impermeable and the pressures are equal $P_r = P_l$. At $t = 0$, we assume that the membrane becomes semipermeable. The solvent chemical potentials read in the limit of low concentrations (Raoult's law):

$$\mu_{\text{right}} = \mu_w^0(P_r) \tag{4}$$

$$\mu_{\text{left}} = \mu_{\text{w}}^{0}(P_{\text{l}}) + k_{\text{B}}TLnX \tag{5}$$

where μ_{w}^{0} is the standard water chemical potential and X the molar fraction of water. k_{B} is the Boltzman constant.

Initially, $\mu_{\text{left}} < \mu_{\text{right}}$ and water flows from the right to the left. This increases the pressure P_{l}. At equilibrium (no flow of water), the pressure P_{l} is equal to $P_{\text{r}} + \Pi$ where Π is the osmotic pressure. Using classical thermodynamics, one has, at constant temperature, $(\partial G/\partial P)_{\text{T}} = V$ where G is the Gibbs energy, and it is quite easy to show that:

$$\mu_{\text{w}}^{0}(P_{\text{l}} = P_{\text{r}} + \Pi) = \mu_{\text{w}}^{0}(P_{\text{r}}) + \int_{P_{\text{r}}}^{P_{\text{r}}+\Pi} V_{\text{m}}\,\mathrm{d}p \tag{6}$$

where V_{m} is the molar volume of water (a constant in the relevant range).

Using the fact that the solute concentration C is close to $(1 - X)/V_{\text{m}}$, the osmotic pressure simply reads:

$$\Pi = N_{\text{a}}k_{\text{B}}TC \tag{7}$$

where N_{a} is the Avogadro number.

This equation applies to dilute solutions and must be corrected for highly concentrated solutions by a virial-like expansion and also for complex solutions, those containing polymers for example. In the following, we will assume this equation to be valid in our range of parameters.

2.2. Osmotic Growth

In the following, the boundary which limits the cell or the physicochemical system is called interface for generality. It can be shown from entropy production that the flow J_{v} is proportional to the difference in water chemical potential between the two sides. Then, for very small values, the flow velocity J_{v} across the interface is proportional to $k_{\text{B}}TLnX$, the osmotic pressure Π. It is clear that when the pressure P_{l} increases, the flow velocity decreases linearly with the pressure. Finally, the relevant quantity is $\Pi - P$. The flow velocity through the interface is:

$$J_{\text{v}} = L_{\text{p}}\left[P_{\text{i}} - P_{\text{e}} - N_{\text{a}}k_{\text{B}}T\left(C_{\text{i}} - C_{\text{e}}\right)\right] \tag{8}$$

where the subscripts note the intracellular and extracellular media. J_{v} is positive when the flow goes out. The equation of water conservation at interface is simply:

$$\vec{U} - \vec{V} = -J_{\text{v}}\vec{n} \tag{9}$$

where $\vec{U}, \vec{V}$ and $\vec{n}$ are respectively the interface velocity, the hydrodynamic velocity and the normal unit vector pointing outwards. Let us emphasize that the interface velocity has two components, one tangential and the other normal to the membrane. Another equation on osmotic growth is the stress boundary condition at interface:

$$\left[\sigma_{\rm e}^{\rm hydro} - \sigma_{\rm i}^{\rm hydro}\right] \vec{n} = -(\delta F/\delta u)\vec{n} \tag{10}$$

$$F = \int \gamma \mathrm{d}S \tag{11}$$

where γ is the surface tension and δu a displacement of the interface. F is the surface energy of the membrane. The integral runs over the total membrane surface.

Two processes of growth can be considered. The first one is absorption of nutriments. Schematically, consider a close interface which bounds two aqueous solutions. By an active process, the cell acquires nutriments from the extracellular medium. By this means, the intracellular osmotic pressure increases. The resulting water flow penetrates the cell and expands it. From (8) and (9), it is easy to check that $U > 0$. The second one doesn't imply membrane flux. An intracellular increase of solute concentration by some chemical process generates an entering flux of water which generates turgor and expands the interface.

3. STABILITY DURING OSMOTIC GROWTH

3.1. Formulation of the Problem

In this paper, we focus on the first growth mechanism: absorption-induced turgor. The problem is to know whether the shape is globally modified or not. The method consists, first, in determining the stationary shape (without deformations) and second, in investigating the behavior of a shape fluctuation around the stationary shape [6]. If the fluctuation decreases to zero, the shape is stable.

Let us consider the simplest case, a flat water permeable interface which pumps actively solutes from one side (the extracellular medium) to the other (the intracellular medium) (Fig. 2). z is the normal coordinate. We assume, for simplicity, that the solute flux J at the interface is a constant and is equal to J_0. J_0 is negative. We note U the interface velocity. U must be positive.

We also need to write the dynamical evolution of such physical quantities as the solute concentration C, the pressure P and the flow velocity V. These equations are known and can be simplified by evaluating some adimensional quantities. The Reynolds number $\mathrm{Re} = UL\rho/\eta$ evaluates the ratio of the convective term over the viscous term. For $U = 0.1$ μm/s, a typical cellular size L 50 μm, ρ the mass water density and η the viscosity, $\mathrm{Re} \ll 1$. The non-stationary term can also be neglected. In these limits, the pressure P and the hydrodynamic flow satisfy on each side of the membrane:

$$\nabla.\vec{V} = 0 \tag{12}$$

$$\eta\Delta\vec{V} - \vec{\nabla}P = \vec{0} \tag{13}$$

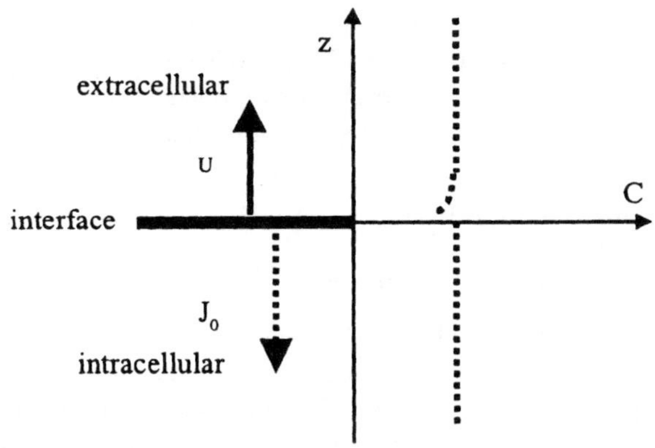

Fig. 2. — The absorption of solutes generates a concentration gradient at the interface.

The Peclet number $Pe = UL/D$ evaluates the ratio of solute transport by convection over the solute transport by diffusion. For a typical coefficient of diffusion D, $Pe \ll 1$. In this limit, the solute concentration C satisfies on each side of the membrane:

$$\partial C/\partial t = D\Delta C \tag{14}$$

As we will see in the following, we search solutions in a frame moving at velocity U with respect to the laboratory frame. Then, (14) becomes:

$$\Delta C + (U/D)(\partial C/\partial z) = 0 \tag{15}$$

The non-stationary term can also be neglected. To close the problem, we need some boundary conditions far from the interface:

$$C(z \to \pm\infty) = C_0 \tag{16}$$

$$P(z \to \pm\infty) = 0 \tag{17}$$

$$\vec{V}(z \to \pm\infty) = -\vec{U} \tag{18}$$

The conditions at the interface are provided by (9) and (10) with, in addition, the no-slip condition at the membrane.

3.2. The Planar Stationary Solution

It is straightforward to determine the physical quantities for the regime in which the planar interface grows at constant velocity U:

$$C_{\mathrm{i}}(z) = C_0 \tag{19}$$

$$C_{\mathrm{e}}(z) = C_0 + \frac{J_0}{L_{\mathrm{p}} N_{\mathrm{a}} k_{\mathrm{B}} T} \exp(-Uz/D) \tag{20}$$

$$U = -J_0/C_0 \tag{21}$$

As J_0 is negative, the interface velocity U is hopefully positive and doesn't depend on the hydraulic conductance. The flow velocity V is equal to $-U$ and the pressure is equal to zero. There is no tangential flow. We recall that the subscripts i and e are respectively for intracellular and extracellular media.

3.3. Linear Stability Analysis

3.3.1. Expressions of the Perturbations

In order to study the linear shape stability, we consider a small sinusoidal fluctuation of the shape interface $z = u(x,y,t) = u\exp(\omega t + jq_x x + jq_y y)$ where u is a constant ($j^2 = -1$). We determine the dispersion relation $\omega = \omega(q_x, q_y)$. If ω is positive, the fluctuation is amplified: the shape is unstable. If ω is negative, the shape returns to the flat state. Following the perturbation of the shape, fluctuations of the concentrations, pressures and hydrodynamic flows need to be determined. We only provide the results:

$$C_{\mathrm{i,e}} - C^{(0)}_{\mathrm{i,e}} = \delta C_{\mathrm{i,e}} = a_{\mathrm{i,e}} \exp(s_{\mathrm{i,e}} z) \exp(\omega t + jq_x x + jq_y y) \tag{22}$$

$$P_{\mathrm{i,e}} - P^{(0)}_{\mathrm{i,e}} = \delta P_{\mathrm{i,e}} = b_{\mathrm{i,e}} \exp(\pm qz) \exp(\omega t + jq_x x + jq_y y) \tag{23}$$

$$V_{z\mathrm{i,e}} - V^{(0)}_{z\mathrm{i,e}} = \delta V_{z\mathrm{i,e}} = \left[d_{\mathrm{i,e}}/b_{\mathrm{i,e}} + z/2\eta_{\mathrm{i,e}}\right] \delta P_{\mathrm{i,e}} \tag{24}$$

$$V_{x\mathrm{i,e}} - V^{(0)}_{x\mathrm{i,e}} = \delta V_{x\mathrm{i,e}} = (\pm j d_{\mathrm{i,e}}/b_{\mathrm{i,e}} + (1 \pm qz)/2\eta_{\mathrm{i,e}})\, \delta P_{\mathrm{i,e}} \tag{25}$$

where $q^2 = q_x^2 + q_y^2$, δC, δP and δV are the perturbations of the solute concentration, the pressure and the flow velocity. $a_{\mathrm{i,e}}$, $b_{\mathrm{i,e}}$ and $d_{\mathrm{i,e}}$ are some constants. We have used the boundary conditions at infinity. s_{i} and s_{e} are quantities coming from the resolution of (15):

$$s_{\mathrm{i}} = -U/2D_{\mathrm{i}} + \sqrt{q^2 + (U/2D_{\mathrm{i}})^2} \tag{26}$$

$$s_{\mathrm{e}} = -U/2D_{\mathrm{e}} + \sqrt{q^2 + (U/2D_{\mathrm{e}})^2} \tag{27}$$

3.3.2. Dispersion Relation

To determine the previous constants, one takes into account the boundary conditions at the interface. We obtain:

$$\omega = -(U^2/D_{\mathrm{e}} + \gamma q^2) - (\gamma q/(2\eta_{\mathrm{e}} + \eta_{\mathrm{i}}))(1+\alpha) \tag{28}$$

$$\alpha = L_{\mathrm{p}} N_{\mathrm{a}} k_{\mathrm{B}} T C_0 (D_{\mathrm{e}} s_{\mathrm{e}} - D_{\mathrm{i}} s_{\mathrm{i}})/(D_{\mathrm{i}} s_{\mathrm{i}} + U)(D_{\mathrm{e}} s_{\mathrm{e}} + U) \tag{29}$$

ω is always negative. This process of growth by nutrient absorption is stable.

4. CONCLUSION

Using a simplified model, we have shown that the mechanism of nutrient acquisition by the cell is stable for the shape. Considering a small perturbation of the shape, it is easy to see that the difference of concentration $\delta C_\mathrm{i} - \delta C_\mathrm{e}$ decreases at the tip of the perturbation (Fig. 2). This effect is due to the decrease of the stationary concentration C_e at the membrane. Now, let us consider the other process of growth. We assume for simplicity that the cell increases its intracellular solute concentration by a chemical process while the solute flux at the membrane is equal to zero. The interface velocity has the same sign but, now, it is the intracellular concentration which decreases at the interface. Then, at the tip of the shape fluctuation, $\delta C_\mathrm{i} - \delta C_\mathrm{e}$ increases and the perturbation is amplified. This mechanism is unstable.

In summary, once the mechanism of transport across the membrane are known, one can use the techniques of linear stability analysis to derive the behavior of the cell contour. Unstable modes are believed to generate, in the non-linear regimes, the complex features of cell shapes.

Acknowledgements

It is a pleasure to thank Vincent Fleury for stimulating discussions and Pierre Pelcé which introduced me to the problem of curved fronts.

REFERENCES

[1] W.K. Silk, *Ann. Rev. Plant Physiol.* **35** (1984) 479-518.
[2] L. Taiz, *Ann. Rev. Plant Physiol.* **35** (1984) 585-657.
[3] D. Cosgrove, *Ann. Rev. Plant Physiol.* **37** (1986) 377-405.
[4] J.A. Lockhart, *J. Theor. Biol.* **8** (1965) 264-275.
[5] N.P. Money, *Fungal Genet. Biol.* **21** (1997) 173-187.
[6] P. Pelcé, *Dynamics of curved fronts* (Academic Press, 1988).

CONTRIBUTED PAPER 3

On Transcellular Ionic Currents

M. Léonetti and E. Dubois-Violette

1. INTRODUCTION

Living organisms drive electric current through themselves. The current is a flow of ions in solutions which passes through the cell *via* specific proteins like pumps, symports, channels and others. There is evidence that such currents convey information in excitable cells like neurons [1], a fact which sounds quite likely. However, the role of stationary currents in the generation of spatial order or pattern formation in growing and developing organisms looks at first glance more questionable. These stationary ionic currents have been described in plant, animal and microbial systems, in individual cells of unicellular and multicellular organisms, and in both embryonic and mature cells [2–6]. They enter in one region, generate ionic gradients in the cytoplasm and then exit in another part of the membrane. The problem is to know whether these currents are just a passive consequence of an unrelated asymmetry of the cell or if they play by themselves a key role in the generation of those cellular asymmetries. In the context of morphogenesis of cells, one may expect that the self-organization of ionic currents may induce differential growth or deformation of membranes, and hence play a role in tip growth or branching of such cells, in the long run (non-linear regime).

The most convincing example of ionic currents is the well known phenomenon of band formation in Characean algae and notably, in *Chara corallina* a fresh-water green alga [7]. The internodal cell of Chara is cylindrical; it has a radius and a length respectively of the order of 0.5 mm and 6 cm. In 1828, Bischoff pointed out the existence of regularly spaced ring-shaped calcareous deposits along the cell. In 1908, Migura showed that these deposits depend on the extracellular calcium concentration, on the photosynthetic activity and, notably, on the light intensity. The symmetry-breaking is induced above a threshold of light intensity (Fig. 1). In the thirties, Arens established, with pH-indicating

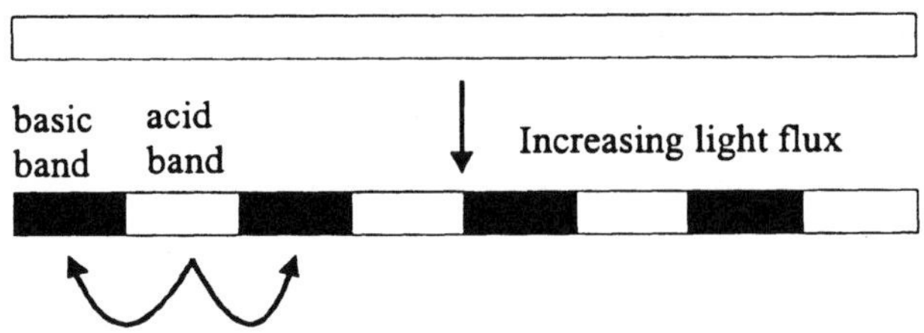

Fig. 1. — Symmetry-breaking in *Chara corallina*. Increasing light flux generates successive acid and basic extracellular pH bands along the cell. In an acid band (basic band), there is an efflux (influx) of protons pointed by the arrows.

dyes, that the deposits correlate with a local alkalination of the extracellular medium (even without calcium). The formation of bands has been intensively studied in the seventies and the eighties. Above some light intensity, basic and acid bands, regularly spaced, develop along the cell: Figure 1.

The wavelength is of the order of 2 cm. For a bath pH of 8.2, the extracellular basic and acid pH, close to the cell wall, are respectively of the order of 10.5 and 6.5 (probably less in the cell wall). In Figure 2, the variation of the pH along the cell has been measured with a pH electrode made at the laboratory (Fluka 95297 as ionophore [8]). This corresponds to a huge variation of the pH; four orders of magnitude on two centimeters. The excess (deficit) of protons in the acid (basic) band is produced by a transmembrane efflux (influx) of protons. Then, a ionic circulation and ionic currents are induced. The intensity and the characteristic time vary respectively from 8 to 80 μA cm^{-2} and from 5 to 60 minutes. However, the quantitative results must be taken with caution as the measurement of the extracellular current by a vibrating electrode is very perturbative.

2. THE THEORETICAL APPROACH

In the theoretical models, the symmetry-breaking which is characterized by the appearance of ionic currents, results fundamentally from a dynamical instability. The basis of these linear instabilities is a local positive feedback (destabilizing effect). For example, Toko *et al.* [12, 13] have assumed in their model of ionic currents in *Chara corallina*, that an intracellular increase of the proton concentration generates an influx of protons which, in turn, increases even more the proton intracellular concentration. However, two arguments show that the presence of a local positive feedback is not sufficient to generate ionic currents.

First, these models imply the diffusion of membrane proteins, ionic concentrations or membrane potential. Diffusion tends to homogenize these quantities inside the cells. This provides a general negative feedback (stabilizing effect). Then, the positive feedback must be more efficient than the negative feedback.

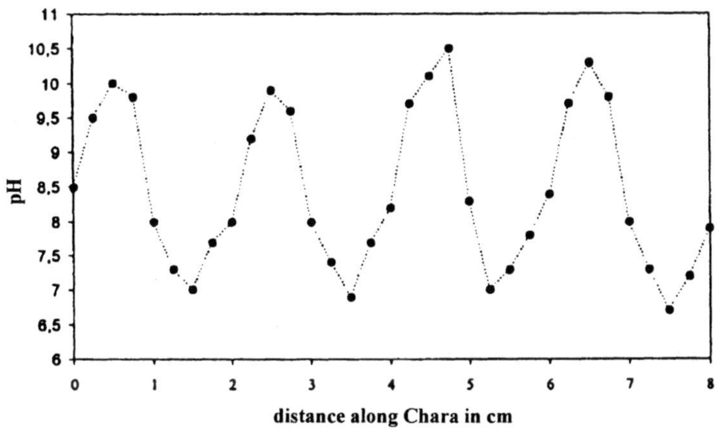

Fig. 2. — The pH is measured by a laboratory made pH electrode with a fine tip ($\sim$ 10 μm). The light flux is provided by a 150 W halogen. Bath: NaCl 1 mM, KCl 0.2 mM, $CaSO_4$ 0.2 mM and $NaHCO_3$ 1 mM. The bath pH is 8.2.

This gives a first criterion for the development of ionic currents. In the case of ionic currents, this criterion is actually on the cell size.

Second, it is possible to show that different modes can appear from dynamical instabilities. In order to perform a linear study of the stability, one introduces a spatial perturbation of the relevant physical quantities. For example, in the case of *Chara corallina*, Toko introduces a perturbation δC of the proton concentration of the form $\delta C = \mathbf{a}\cos(2\pi x/\lambda)\exp(\omega t)$ where $\mathbf{a}$ is a constant, x is the coordinate along the cell, ω is the growth rate of the perturbation and λ is the wavelength of the perturbation. The theoretical analysis of the stability derives the curve $\omega = \omega(1/\lambda)$. If ω is positive for a specific λ, this means that the perturbation characterized by this λ is unstable and its amplitude increases on a characteristic time scale $T = 1/\omega$: this simply means that ionic currents could appear on this time T. On the contrary, if ω is negative, this perturbation is stable and its amplitude decreases: ionic currents never appear by such a mechanism. In general, there exists, among all wavelengths, a band of wavelengths characterized by a positive ω. Let us emphasize the peculiar case of the homogeneous perturbation $\delta C = a\exp(\omega t)$ corresponding to an *infinite* wavelength λ. Infinite means in fact that λ becomes larger than the cellular size. In some mechanisms, this perturbation is the most unstable: it means that the maximum of ω is reached when $1/\lambda$ tends to zero. In this case, there are no ionic currents, but one finds an homogeneous absorption or desorption (uniform influx or outflux). Without other ingredients, the final state is characterized by a uniform concentration and/or membrane potential. We will see that this remark is crucial if we want to understand a few mechanisms which have been proposed in the literature.

In conclusion, what should be the predictions of the physical models of pattern formation of ionic currents? These models must provide a qualitative biophysical positive feedback, a quantitative criterion of instability, a range of unstable modes and a quantitative characteristic time of appearance of ionic currents. In the following, we propose three mechanisms in order to explain the appearance of ionic currents in cells where cellular growth is neglected (constant shape). See [9] for example. It is expected that, ultimately, these instabilities appearing on the non deformed shape will induce (in the relevant cases) undulations and deformations.

3. IONIC CURRENTS RESULTING FROM TURING INSTABILITY

In 1952, in a famous paper, Turing has provided a general mechanism of pattern formation in reaction-diffusion systems, taking the biological system as example [10]. The final structure is a stationary spatial modulation of concentrations. Almost forty years have passed before observing the first Turing pattern in a chemical system [11]. There are two main ingredients in this mechanism: an activator and an inhibitor, two diffusive species. For simplicity, we assume that the activator diffuses very slowly, conversely the inhibitor diffuses rapidly ($D_{\text{act}} < D_{\text{inh}}$ where D are the respective coefficients of diffusion). Activator means that a local small increase of the activator concentration generates an even larger value of that activator. This local activation is termed a positive feedback. However, this process is not sufficient. The linear study shows that the most unstable mode is the homogeneous perturbation: no spatial modulation of concentrations. The second ingredient is the inhibitor which is produced by the activator. Inhibitor means that a local increase of inhibitor concentration generates a local strong decrease of the activator concentration. If there is no diffusion, the system is stable because of this strong decrease which smooths out any perturbation. If diffusion is taken into account, the situation is completely modified: Figure 3. The inhibitor diffuses laterally very rapidly and two things happen. First, there is an inhibition of the activator, around a peak of activator concentration, and second, the activator dominates at the peak (see Fig. 3). This mechanism involves an interplay between local activation and lateral inhibition. It has been applied to the case of ionic currents in Chara and Fucus. Toko *et al.* have assumed that a local increase of intracellular proton concentration generates an intracellular production of protons [12,13]. More precisely, they assumed that the proton pumps are more active if the intracellular proton concentration decreases. To our knowledge, there are no data which support such a variation. However, we emphasize that this variation could be the result of several stages involving other molecules or organelles such as ATP and chloroplasts. The inhibitor is ADP, adenosinediphosphate which is assumed to decrease the activity of pumps. In Fucus, the local activation seems to have been measured. The influx of calcium is sevenfold greater at 1 mM than at 10 mM extracellular calcium concentration [14,21]. However, as in the case of Chara, the inhibitor is still elusive. Moreover, the experiment is done when currents are already developed.

The characteristic time is provided by the diffusion of ionic species:

$$T \sim \lambda^2/D^{\text{eff}} \tag{1}$$

where λ is about the characteristic wavelength of the currents. D^{eff} is the effective coefficient of activator diffusion close to the membrane. For a numerical evaluation, see the end of the part 5. The time of the third mechanism implies also the diffusion of molecules. For Fucus, T varies from a few seconds to a few minutes and for Chara, T varies from ten minutes to a few hours.

4. IONIC CURRENTS BY SELF-ORGANIZATION OF MEMBRANE PROTEINS

Let us consider the following naive, but instructive, example of a spherical cell with two types (# and §) of membrane proteins which transport transmembrane current. The concentrations and the currents are respectively $C_\#$, $C_\S$, $I_\#$ and $I_\S$. We also assume that the currents do not depend on ionic concentrations and membrane potential. We recall that the assumptions are just made in order to simplify the following discussion.

At equilibrium, the total transmembrane current I is equal to zero: $I = I_\# C_\# + I_\S C_\S = 0$. If a mechanism induces spontaneously a segregation of the two types of proteins, ionic currents develop: Figure 4.

Let us note that $I_\S$ and $I_\#$ are necessarily opposite. We choose $I_\# < 0$. In 1977, Jaffe *et al.* have shown that membrane proteins move under the action of an electric field [15,16]. For simplicity, we assume that only the proteins # move. Consider a region of membrane where proteins # are much more concentrated (as a consequence of a fluctuation or of a defect) and a protein # far from that region. Close to this zone, the current produced by the proteins # is not compensated by the static proteins § due to the excess of proteins #: Figure 5. Then, an extracellular E_e and an intracellular E_i electric field are generated which act on the charges of the protein # far from the aggregate and can move it. If the protein # is conveniently charged, this electric force tends to bring the previously far protein nearer to the concentrated part. Then, the concentration and the current generated this part increase. The feedback is positive. Let us emphasize that the total force F is $F = z_\text{e}E_\text{e} + z_\text{i}E_\text{i}$ which reduces to $(z_\text{i} - z_\text{e})E_\text{i}$ in the hypothetical case of a flat membrane. In the Figure 5, the feedback is positive if $z_\text{i} < z_\text{e}$. Obviously, if we had chosen $I_\# > 0$, the condition would be $z_\text{i} > z_\text{e}$. However, the electric force must be strong enough to overcome the Brownian motion which tends to homogenize the protein concentration. This fact provides the criteria which must be determined in any given geometry. For a spherical cell, the concentration is unstable and

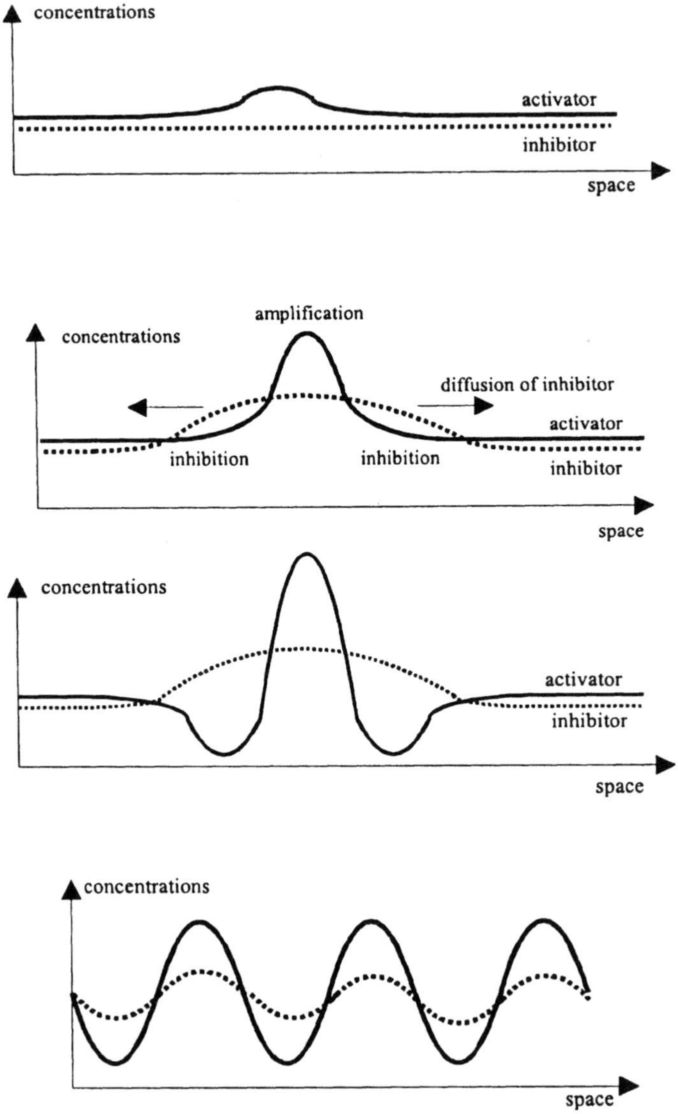

Fig. 3. — A local increase of activator concentration generates an activation of the activator production (local activation) and an increase of the inhibitor concentration. As the inhibitor diffuses laterally more rapidly than the activator, the inhibitor decreases the activator concentration around the local peak (lateral inhibition). At the peak, the inhibitor diffusion has sufficiently reduced its concentration that the activator dominates. The final pattern is a stationary spatial modulation of concentrations (similar instabilities are invoked for zebra stripes or leopard's spots).

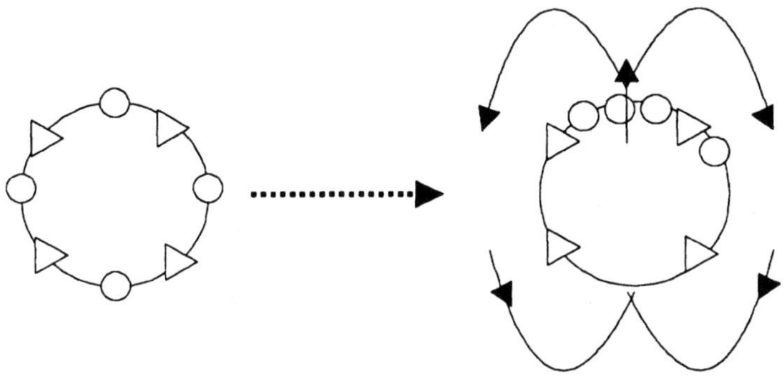

Fig. 4. — Initially, the concentrations of the two types of proteins are uniform. The current produced by one type is compensated by the other type: no ionic currents form. If one type of proteins aggregates spontaneously, there is a symmetry-breaking which generates ionic currents.

dipolar ionic currents appear if the size R of the cell satisfies [17–19]:

$$R > R_c = \left[(eI/\varepsilon k_B T)\left(2z_i/\chi_i^2 D_i^{\text{eff}} - z_e/\chi_e^2 D_e^{\text{eff}}\right)\right]^{-1} \tag{2}$$

where I is the current density produced by proteins #, z_i and z_e are respectively the intracellular and extracellular charges of proteins #, $\chi_{i,e}$ the inverse of the Debye length ($\chi_{i,e}^2 = \text{Fe} \sum z_j^2 C_{i,e}^j/\varepsilon k_B T$). $D_{i,e}^{\text{eff}}$ is the effective coefficient of diffusion close to the membrane. It means that it takes into account the chemical binding with other molecules (especially important for ions implied in internal chemical processes as calcium in Fucus, protons in Chara and methionine in Achlya). This effect is important (a few orders of magnitude); since it decreases the critical size, it increases the possibility for ionic currents to appear by such a mechanism.

The theoretical analysis of this mechanism predicts also the appearance of a quadrupolar mode if $R > 2R_c$. The quadrupolar mode could correspond, for example, to the formation of two rhizoids in Fucus, when deformation follows, after the instability has appeared on the non-deformed cell. In the case of Fucus, the model predicts that the quadrupolar mode will appear preferentially if the diffusion coefficient of the calcium is reduced after introduction in the bath and in the cell of some buffer of calcium. Unfortunately, as we will see, a similar effect occurs with the other mechanisms.

For a cylindrical cell such as Fucus, the wavelength of ionic currents along the cell is large compared to the cellular radius, the protein concentration is unstable if the length L of the cell satisfies [18–20]:

$$L > \lambda_c = \left[\left(e\left|I\right|/\varepsilon k_B T \chi_i^2\right)\left(z_i/R D_i^{\text{eff}}\right)\right]^{-1/2} \tag{3}$$

where R is the cellular radius. In this geometry, the internal electric field is

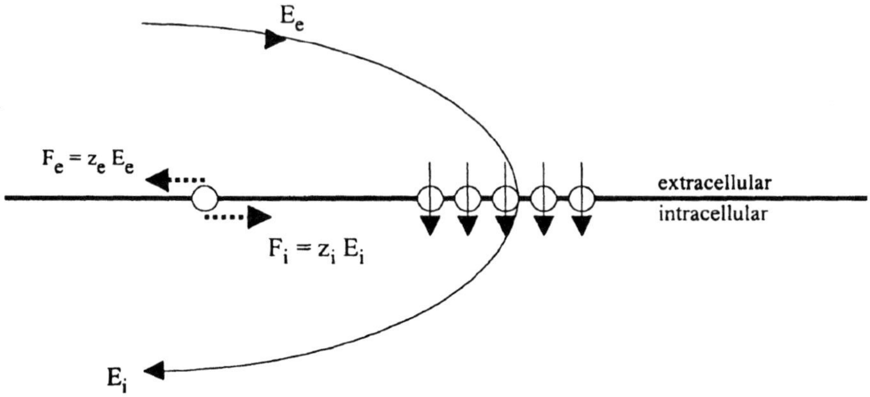

Fig. 5. — An aggregate of mobile membrane proteins produced by a local fluctuation generates an extracellular electric field E_e and an intracellular one E_i. The total force on proteins far from the aggregate is $F = F_e + F_i$, F. It is negative as we choose arbitrary $z_j < 0$. If the protein is conveniently charged, this electric force tends to bring the previously far protein nearer to the aggregate. The feedback is positive.

much larger than the extracellular electric field and it dominates the motion of mobile proteins.

The characteristic time T is provided by the limiting stage of the mechanism, the diffusion of the proteins along the membrane:

$$T \sim \lambda^2 / D_p \tag{4}$$

where λ is the radius for a spherical cell and λ_c for a cylindrical cell. D_p is the coefficient of diffusion of mobile proteins.

In the case of Fucus [21–24], the numerical value is easy to obtain. For $R = 25$ μm and $D_p = 10^{-13}$ m^2/s a typical value of protein diffusion coefficient, T varies approximatively from 30 minutes to two hours. In the case of *Chara corallina* [7], T is 10^6 s. This is larger than 10^3 s, the value actually observed in experiments. This evaluation shows that this mechanism cannot be applied to *Chara corallina* but, could still be correct for Fucus. Unfortunately, it is quite difficult to evaluate the criteria on sizes given by (1) and (2) because of lack of data and because of the difficulty to extract the values of the relevant quantities: type of ions, charge of proteins, value of the current I... We refer to the quoted articles for more information on these difficulties. But, we recall that a simple numerical evaluation of the characteristic time allows us to rule out this mechanism as the origin of ionic currents in *Chara corallina*.

5. IONIC CURRENTS BY NEGATIVE DIFFERENTIAL CONDUCTANCE

Consider now that the current I_j of the ion j only depends on the membrane potential V. This means that we assume the membrane protein concentration to be constant and the current I_j not to depend on concentrations, in a first approximation. We consider two ions 1 and 2, to make the discussion clear.

Let us describe first the local positive feedback. It is well known that a negative differential conductance may be the origin of instabilities in physical and biological systems: for example, the action potential in excitable cells. We recall that a membrane can be considered, with a good approximation, as a conductance in parallel with a capacitance. The origin of capacitance is the lipid bilayer which constitutes a good insulator while the proteins embedded in the bilayer transfer specifically ions from one side to the other side. The membrane potential V is the difference between the intracellular and the extracellular electric potentials. Schematically, a fluctuation δV generates an ionic current $\delta I_j = G_j \delta V$ where G_j is the differential conductance. If δV is positive and the total differential conductance $G_1 + G_2$ negative, the current δI is negative. This is what increases the intracellular charge $\delta\rho$ close to the membrane ($\delta\rho\chi^{-1}$ is the capacitance charge). By capacitive effect, the membrane potential increases: $\delta V \sim \delta\rho/\chi C_{\mathrm{m}}$. So, the feedback is positive. Then, if $G_1 + G_2$ is negative, the membrane potential is unstable. It is quite obvious that the uniform perturbation (δV uniform along the membrane) is the most unstable perturbation because there is no lateral leak of ions as in the Turing instability described in part 3. The dynamics is provided by the propagation of the membrane potential due to the limiting step which is the passage of ions through the membrane. It is important to note that there is no macroscopic motion of ions along the membrane (contrary to the case which we shall study below). The characteristic time is about $C_{\mathrm{m}}/(G_1 + G_2)$ where C_{m} is the capacitance of the membrane. This time is very small as compared to the characteristic time of establishment of the ionic currents. The final state is characterized by a uniform membrane potential. This mechanism is at the origin of action potential and fertilization potential. In this case, the transitory negative differential conductance is provided by the sodium and potassium voltage-gated channels. This instability is well-known. It has been described in great details using the cable model which is an analogy between a cell and an electric cable [1,25,26]. So, if $G_1 + G_2 < 0$, there are no ionic currents.

Now, we consider the case $G_1 + G_2 > 0$. The previous study of stability does not take into account the diffusion of ions or the macroscopic lateral motion of ions along the membrane. This point is specially important when the diffusion differs for the various ion species (note that it is not the case for sodium and potassium implied in the action potential). Indeed, in the bulk, where electroneutrality is more or less satisfied outside the Debye layer (charge of the capacitance), the electric current reads:

$$\vec{I} = -F(D_1 - D_2)z_1\vec{\nabla}c_1 + \gamma\vec{E} \tag{5}$$

where F is the Faraday constant and γ the conductivity. The second term of this expression is the Ohm relation and it corresponds to that taken into account in the cable model. The first term which emerges when $D_1 \neq D_2$ enhances the important contribution of relative ion diffusion.

Our goal in the following is to show that diffusion currents, induced by concentration gradients, may also strongly affect the electric stability of the membrane and finally, generate stationary ionic currents. We consider local perturbations of the concentrations δC_1 and δC_2 and of the membrane potential δV. We emphasize that these perturbations are not uniform (finite wavelength λ) to allow the lateral diffusion of ions along the membrane. δC_1 and δC_2 are proportional to each ionic current divided by the diffusion coefficient $z_1 F\delta C_1 \sim -\delta I_1/D_1$ and $z_2 F\delta C_2 \sim -\delta I_2/D_2$. Then, the intracellular density of charges $\delta\rho$ is $\delta\rho = F(z_1\delta C_1 + z_2\delta C_2) \sim -(G_1/D_1 + G_2/D_2)\delta V$. The fundamental fact is that the local charge $\delta\rho$ depends on the sign of $G_1/D_1 + G_2/D_2$ while the total current $\delta I = \delta I_1 + \delta I_2 = (G_1 + G_2)\delta V$ which passes through the membrane depends on $G_1 + G_2$. Now, and contrary to the previous case, $\delta\rho$ and δI can have independent signs. This was not possible in the case $G_1 + G_2 < 0$. If $G_1 + G_2 > 0$ and $G_1/D_1 + G_2/D_2 > 0$, the membrane potential and the concentrations are of course stable, ionic currents are absent. If $G_1 + G_2 > 0$ and $G_1/D_1 + G_2/D_2 < 0$, a positive perturbation of the membrane potential $\delta V(> 0)$ generates a positive charge perturbation $\delta\rho > 0$. By capacitive effect δV is modified so that $\delta V \sim \delta\rho/\chi C_\mathrm{m}$, this results in an increase of the initial perturbation of the membrane potential. The feedback is positive [26]. For a spherical geometry, the criterion is

$$R < R_\mathrm{c} = 2\varepsilon\chi^2 \left(G_1/D_1^\mathrm{eff} + G_2/D_2^\mathrm{eff}\right)^{-1} . \tag{6}$$

For a cylindrical geometry of radius R and length L,

$$L^2 < L_\mathrm{c}^2 = 2\pi^2\varepsilon\chi^2 R \left(G_1/D_1^\mathrm{eff} + G_2/D_2^\mathrm{eff}\right)^{-1} \tag{7}$$

where D_1^eff takes into account the chemical binding of ions to other molecules. This can have a very strong effect (few orders of magnitude) on the cell wall and the intracellular medium.

The dynamics follows the diffusion of ions. Then, the characteristic time is diffusive:

$$T \sim \lambda^2/D_1^\mathrm{eff} \tag{8}$$

where λ is R_c or L_c.

In Fucus, T is of the order of a few seconds, for $\lambda = 30$ μm and $D_1^\mathrm{eff} = 10^{-6}$ cm^2/s which is the value of the diffusion coefficient of calcium in water. However, in cells, just close to the membrane, this coefficient should be smaller and the characteristic time should be of the order of a few minutes. By this mechanism, the currents appear soon after the fertilization. One interesting point is the presence of calcium voltage-gated channels in the membrane of Fucus. These channels seem to be crucial for the appearance of ionic currents. In Chara, T is about thirty minutes for $\lambda = 1$ mm and $D = 10^{-5}$ cm^2/s a typical coefficient of diffusion in water in agreement with experiments. In this mechanism, there is a local activation and a lateral inhibition as in the Turing mechanism. However, there are also some differences which are not explained here, for example, in the dispersion relation.

6. CONCLUSION

We have recalled the different theoretical mechanisms proposed in literature to generate ionic currents. From these models two physical quantities emerge, a critical size and a characteristic time. While the time can be compared to experiments, the same is quite difficult for the critical size because of the lack of data (on protein charge, on chemical binding at membrane and on characteristics of the current produced by the relevant membrane proteins). Moreover, these studies are only linear and further studies would anyway be able to provide the variations of the pH along Chara close to the threshold only while all experiments are done far from it. One path of progress could be to make artificial cellular models based on vesicles, with embedded proteins. However, the main problems will still be to reach the critical size and to set the system out of equilibrium.

Acknowledgements

It is a pleasure to thank Vincent Fleury, Fabrice Homblé, Darryl Kropf and Pierre Pelcé for stimulating discussions. This work was supported by the CNRS program Physico-Chimie du Vivant.

REFERENCES

[1] B. Hille, *Ionic Channels of excitable membranes*, 2nd edition (Sunderland, 1992).
[2] R. Nuccitelli, *Modern Cell Biology* **2** (1983) 451-481.
[3] N.A.R. Gow, *Adv. Microb. Physiol.* **30** (1989) 89-123.
[4] F.M. Harold and J.H. Caldwell, *Tip Growth in Plant and Fungal Cells* (Academic Press, 1990) pp. 59-90.
[5] D. Kropf, M.D.A. Lupa, J.H. Caldwell and F.M. Harold, *Sci* **220** (1983) 1385-1387.
[6] L.F. Jaffe and R. Nuccitelli, *Ann. Rev. Biophys. Bioeng.* **6** (1977) 445-476.

[7] W.J. Lucas, *Oscillations and Morphogenesis*, edited by L. Rensing (Marcel Dekker NY, 1993) pp. 411-425.
[8] M. Cougnon, P. Bouyer, G. Planelles and F. Jaisser, *Proc. Natl. Acad. Sci.* **95** (1998) 6516-6520.
[9] L. Limozin, B. Denet and P. Pelcé, *Phys. Rev. Lett.* **78** (1997) 4881-4884.
[10] A.M. Turing, *Philos. Trans. R. Soc. Lond. B* **237** (1952) 37.
[11] V. Castets, E. Dulos, J. Boissonade and P. deKepper, *Phys. Rev. Lett.* **64** (1990) 2953-2956.
[12] K. Toko, H. Chosa and K. Yamafuji, *J. Theor. Biol.* **114** (1985) 125.
[13] K. Toko, M. Nosaka, T. Fujiyoshi, K. Yamafuji and K. Ogata, *Bull. Math. Biol.* **50** (1988) 225.
[14] P. Pelcé and M. Léonetti, *Europhys. Lett.* **30** (1995) 221-225.
[15] L.F. Jaffe, *Nat* **265** (1977) 600-602.
[16] M.-M. Poo and K.R. Robinson, *Nat* **265** (1977) 602-604.
[17] R. Larter and P. Ortoleva, *J. Theor. Biol.* **96** (1982) 175.
[18] M. Léonetti and E. Dubois-Violette, *Phys. Rev. E* **56** (1997) 4521-4525.
[19] M. Léonetti and E. Dubois-Violette, *Europhys. Lett.* **46** (1999) 107-113.
[20] P. Fromherz and W. Zimmermann, *Phys. Rev. E* **51** (1995) R1659.
[21] D. Kropf, *Microbiol. Rev.* **56** (1992) 316-339.
[22] D. Kropf, *Plant Cell* **9** (1997) 1011-1012.
[23] D. Kropf, in this book.
[24] D. Kropf, *Dev. Biol.* **165** (1994) 361-371.
[25] A.C. Scott, *Rev. Mod. Phys.* **47** (1975) 487.
[26] M. Léonetti and E. Dubois-Violette, *Phys. Rev. Lett.* **81** (1998) 1977-1980.

CHAPTER IV

Branched Patterns in Geology: Rivers and Other Systems

P. Meakin, T. Sun and T. Jøssang

1. INTRODUCTION

Branched patterns are common in geological systems. Perhaps the most ubiquitous example is dendritic patterns formed by the growth of mineral crystals under far-from-equilibrium conditions, and the most familiar example is branched drainage networks (and the associated drainage divide networks). The growth of minerals leads to a wide variety of branched (and unbranched) patterns, because of the wide range of mineral compositions, growth conditions and growth rates. Branched patterns can also be formed by a variety of other processes such as the dissolution or melting of porous materials by fluids that are injected into them, fracture processes and fluid-fluid displacement processes at high capillary numbers. Aggregates composed of small particles may also form extensively branched structures, and structures of this type are widespread in rivers, lakes, marine environments, and in planetary atmospheres [1]. Marine snow [2–4] formed from a variety of materials including living and dead organisms (such as bacteria and diatoms) fecal pellets, polysaccharides and inorganic particles is an important example. Other examples include micrometeorites [5] and aggregates formed in river estuaries from clay and other mineral particles, which are transported in great quantities by rivers. This important class of branched structures will not be discussed further in this review. A wide variety of less common, but interesting, examples such as fulgurites (branched patterns composed of fused sand formed as a result of lightning striking sandy soil [6]) have also been studied.

In some cases, branched patterns can be understood quite well in terms of simple computer models such as the diffusion-limited aggregation model [7, 8] or the Eden-tree model [8, 9]. However, in most cases, branched structures found in geological systems have complex and often poorly understood origins. Sometimes, the structure of a poorly understood branched pattern can be reproduced quite well by a simple model. This may provide valuable insight into the process(es) involved in the formation of the pattern. However, a superficial similarity between two patterns does not necessarily imply that there is any fundamental relationship between the processes that formed them. For the past two decades, considerable attention has been focused on simple models that exhibit complex structural and dynamic behavior. Very often, the similarities between the behavior exhibited by these models and the behavior of complex natural systems has been taken too seriously.

During the past two decades, it has become popular to attempt to describe complex patterns in terms of fractal geometry [10]. This has led to many valuable insights. However, many of the attempts that have been made to describe complex geological structures in terms of fractal geometry have led to results that are imprecise and/or misleading. It is widely recognized that many geological structures do not contain characteristic features that can be used to set a natural length scale. For this reason, an easily recognized object with a known size (such as a coin, lens cap, hammer or person) is often included in images of geological patterns. This motivates the idea that such structures may have a fractal geometry. However, in the analysis of geological structures, care must be taken to distinguish between systems that have structure over a wide range of length scales and systems that have the *same* structure over a wide range of length scales. Many attempts, based on many different approaches [11], have been made to characterize rivers using the concepts of fractal geometry. At the time of writing, there is no consensus concerning the value of fractal geometry for the quantitative characterization of branched rivers.

Rivers play an important role in all of our lives, and they have been studied since antiquity [12]. A very wide variety of patterns are formed by rivers [13]. Terms such as "meandering", "braided" (multichannel), "anastomotic" (a few interconnected channels), deltaic and branched are often used to describe and classify rivers. However, the appearance of a river may change rapidly (for example, a "braided" river can change into a single sinuous channel as the water level increases) and both transitional and mixed forms are common. In addition, the individual channels in a single branched river system may have a variety of forms.

One reason why rivers have become of interest to physicists as well as geologists is the large number of empirical scaling relations (simple power laws) that have been discovered. Some widely recognized examples include:

- Hack's law [10,14] – $\ell_{\rm m} \sim \mathcal{A}_{\rm b}^{\zeta}$, with $\zeta \approx 0.6$ – where $\ell_{\rm m}$ is the length of the "main channel" of a branched river and $\mathcal{A}_{\rm b}$ is the area of the river basin. This is consistent with the idea that the main channel is a self-similar fractal with a fractal dimensionality given by $D = 2\zeta \approx 1.2$. This interpretation of Hack's law is valid only if the overall shape of the basin is independent of its size. In general, both allometric and fractal effects may contribute to the effective value of the exponent ζ. However, direct analysis of river channels indicates that they are self-similar with fractal dimensionalities of about 2ζ [15].

- The average water velocity v, the river width w and the water depth d_w are related to the channel discharge Q (and to each other) by the power laws

$$w \sim Q^x, \quad d_w \sim Q^y \quad \text{and} \quad v \sim Q^z, \tag{1}$$

where "$\sim$" means "scales as". Since $Q \sim w \cdot d_w \cdot v$, $x + y + z = 1$. The exponents x, y and z have empirically determined values of $x \approx 1/2$, $y \approx 1/3$ and $z \approx 1/6$. The (small) positive value of z implies that water flows more rapidly in large lowland rivers than in mountain streams. This may appear to be contrary to common experience, but it is natural to subconsciously take into account the channel width as a length scale when the flow velocity is estimated. These relationships indicate that $v \sim d_w^{1/2}$. As early as the late 1800s, Kennedy [16] found a power law relationship between the mean velocity and depth, with an exponent of 0.64 instead of 0.5.

- The Lacey equations [17], which can be written in the form

$$p \simeq Q^{1/2} \tag{2}$$

and

$$s \sim \mathcal{F}^{5/3} Q^{1/6} \tag{3}$$

where p is the "wetted perimeter" of the channel cross section. In equation (3), s is the slope of the channel and $\mathcal{F}$ is the Froude number ($\mathcal{F} = v^2/gd_w$, where g is the acceleration in the Earth's gravitational field[1]).

- The area-slope-particle diameter relationship

$$s \sim (d/A)^{\varepsilon}, \tag{4}$$

where A is the area drained through a selected position along a channel, s is the average slope in the vicinity of the selected position and d is the median diameter of the particles on the channel bed, in the vicinity of the selected position. The exponent ε has a value of about 0.6 [14]. For many river systems, the relationship

$$s \sim (A)^{-\zeta} \tag{5}$$

or

$$s \sim (Q)^{-\zeta'} \tag{6}$$

has been proposed [18,19]. A quite wide range of values for the exponents ε, ζ and ζ' [20] have been reported.

[1] A different, but closely related, Froude number $\mathcal{F}_2$, defined as $\mathcal{F}_2 = v/\sqrt{(gd_w)}$, is more commonly used in the recent literature concerned with the hydraulic behavior of rivers.

- The discharge-drainage basin area relationship

$$Q \sim A^n. \tag{7}$$

  In computer models, it is often assumed that $n = 1$, but values in the range $0.7 < n < 0.9$ have been reported in the literature, and the value of n depends on the way in which Q is defined (mean annual discharge, mean annual flood discharge...).

Irrigation canals in the Indian subcontinent were studied extensively by British engineers during the late 19th century and the first half of the 20th century[2]. They found a large number of power law relationships called *regime equations* that characterize stable canals. These stable canals were said to be *in regime.* Canals that do not satisfy these power laws are not in regime, and they undergo undesirable processes such as silting bank erosion and migration. They evolve over periods of the order of a few decades until they are in regime. These regime equations were of considerable value in the design of canals [21].

River networks evolve in concert with the land surface that supports them. The structure and dynamics of river networks and topography cannot be fully understood in isolation. Some of the characteristic features of the Earth's surface can be reproduced by simple self-affine models. However, these surfaces contain far too many local minima that would fill with water to form ponds and lakes [22]. Under most conditions, small lakes and ponds have short lifetimes, and they are soon filled by sediment and dead vegetation. One of the motivations for developing erosion and erosion/deposition models is to generate surfaces that support realistic river networks and watersheds [22].

2. QUANTITATIVE ANALYSIS OF BRANCHED PATTERNS

A variety of approaches have been developed to characterize disorderly branched patterns in quantitative terms. Since statistical methods are almost invariably used, large data sets are required. Unfortunately, a detailed mapping of all of the streams in a river network is an arduous and expensive undertaking, even for quite small river basins. In addition, somewhat arbitrary criteria that may depend on the weather conditions prior to the study must be used to identify channels and distinguish between perennial and intermittent streams. In recent years, these difficulties have been alleviated by the use of digital elevation models obtained from aerial photography, satellite imaging or topographic maps. However, the channel network cannot be identified directly from digital elevation models. The river network obtained from digital elevation models depend on the resolution and quality of the digital elevation model, and the algorithm used to extract the network pattern [23, 24].

[2] Until the middle of the 20th century, the irrigation system in the Punjab was larger than in the United States.

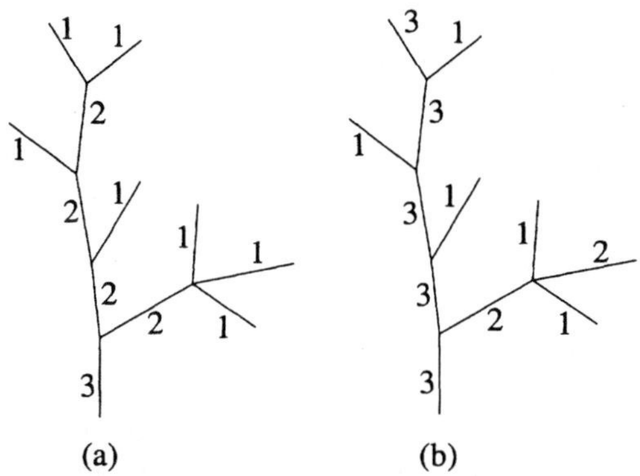

Fig. 1. — A small tree labeled according to the Strahler (part a) and the Horton (part b) schemes.

2.1. Stream order and Horton/Strahler ratios

Schemes that are now very widely used to characterize loopless networks or trees [25] were first developed for the study of branched rivers by Horton [26] and Strahler [27]. In topological terms, a tree is a distribution of nodes connected by line segments, with no loops. One of the nodes is selected to represent the "root". In the case of a branched river, it is natural to place the root at the outlet of the river. In the Strahler scheme, the tree is labeled starting with all of the terminal segments or links (segments that disconnect only one node from the root of the tree, if they are removed). These terminal segments are given the label "1". The rest of the tree is labeled by following the "downstream" paths from the terminal segments to the root. If only a single labeled segment contacts a node the "downstream", unlabeled segment is given the same label as the labeled, upstream segment[3]. If n labeled segments meet at a node, then the label given to the previously unlabeled, downstream segment is either $\sup\{l_i\}$, where $\{l_i\}$ is the set of "tributary" labels and $\sup\{l_i\}$ is the largest member of this set, or $\sup\{l_i\} + 1$, if more than 1 tributary has the same maximum value. A downstream segment can be labeled only after all of the upstream segments have been labeled. The link labels, obtained in this manner, are called the Strahler orders of the links or segments. A small tree labeled according to the Strahler scheme is shown in Figure 1a.

The earlier Horton scheme is more complicated. Whenever two or more tributary links, with the same maximum label $l_{\max}$ meet, then the down stream link is given the label $l_{\max} + 1$, just as in the Strahler scheme. However, once this has been done, all of the segments along one of the upstream paths, from the segment labeled $l_{\max} + 1$ up to and including one of the terminal segments,

[3] A node connected to only two segments is topologically equivalent to a single segment and it is usual to replace all such configurations by a single segment.

are relabeled $l_{max}+1$. This relabeling is propagated along an upstream path, always following the tributary link with the highest, previously assigned label. If two or more tributary links have the same highest order or label, the tie can be broken in several ways. In a real system, such as a river, the tributary with the longest length or the tributary draining the largest area could be chosen. One of the highest order tributaries could also be selected at random. After this relabeling has been completed, the labeling process proceeds in a downstream direction, by selecting nodes for which all of the tributaries have been labeled and labeling the downstream link or segment. Each time a node with two or more tributaries having the same highest label is identified, the upstream relabeling procedure, described above, is carried out. When the "root" of the tree is finally reached by all of the links contacting the root, the relabeling is carried out for a final time, if necessary. The labels on each of the links are then their Horton orders. When the labeling process is complete, a single path of links from the "root" to one of the terminal links will carry the same label, corresponding to the highest order. The order of an entire river or a tributary is equal to the highest order of any of its links. A contiguous set of links with the same order O is called a "stream of order O", and the links with the highest order form the "mainstream".

It is common to express the results of a Horton or Strahler analysis in terms of the ratios

$$r_{\mathrm{b}}(O)=\langle N_O\rangle/\langle N_{O-1}\rangle \tag{8}$$

and

$$r_{\ell}(O)=\langle \ell_O\rangle/\langle \ell_{O-1}\rangle, \tag{9}$$

where $\langle N_O\rangle$ is the number of streams of order O and ℓ_O is the mean length of streams of order O. A stream consists of all of those links of the same order along a path from the root to a terminal link, including the terminal link. The links in each stream are connected to form a topologically linear structure and the stream length is the length of this path[4]. For a tree with a simple internal scaling structure the ratios $r_{\mathrm{b}}(O)$ and $r_{\ell}(O)$ do not depend on the order O for essentially all orders O_i, so that

$$R_{\mathrm{b}}=r_{\mathrm{b}}(O_i)=r_{\mathrm{b}}(O_j) \tag{10}$$

and

$$R_{\ell}=r_{\ell}(O_i)=r_{\ell}(O_j), \tag{11}$$

where R_{b} is the branching ratio and R_{ℓ} is the length ratio. Many investigators have concluded that $r_{\mathrm{b}}(O)$ and $r_{\ell}(O)$ do not depend on the order O for natural rivers. A variety of other ratios, based on the dependence of quantities such as the basin areas and dimensionless groups formed from such quantities on

[4] In some cases, the stream is a fractal and the length measured along the stream must be distinguished from the end-to-end distance, measured in the embedding space.

the stream order [28] have also been studied. However, the maximum order of most streams that have been analyzed in these terms is small, so that finite size effects and statistical uncertainties are large. An additional problem arises because real rivers (in contrast to the branched patterns generated by most "river models") have inhomogeneous structures (the structure varies from region to region inside the basin supporting the entire network). The Horton and Strahler definitions of stream order have been compared, and some of the fundamental relationships between them described by Shreve [25]. Horton/Strahler ratios have been of little value for the purpose of distinguishing between rivers subjected to different geological controls or between rivers at different stages of development. With a few significant exceptions [29, 30], Horton/Strahler ratios have not been used in the quantitative analysis of branched structures generated by simple growth models such as the diffusion-limited aggregation model [7], discussed below.

Several other schemes have been devised for describing the topology of trees. For example, in his studies on the human lung, Weibel [31] simply started with the "trunk", which was given a label of "0" and increased the label by 1 following each branched node encountered along every outwards directed (upstream) path.

If the individual streams have a fractal dimensionality of 1, the fractal dimensionality of the tree is given by

$$D = \log R_{\mathrm{b}} / \log R_{\ell}, \tag{12}$$

where R_ℓ is measured using the distances between the stream ends, and if the individual streams have a fractal dimensionality of D_{s} the fractal dimensionality of the tree is given by

$$D = [\log R_{\mathrm{b}} / \log R_{\ell}] D_{\mathrm{s}}. \tag{13}$$

Equations (12) and (13) have been used to determine the fractal dimensionalities of branched rivers using values for R_{b} and R_ℓ obtained using both the Horton and Strahler labeling schemes, and a number of studies have led to the result that $R_{\mathrm{b}} \approx 4$ and $R_\ell \approx 2$ ($D = 2$).

Some investigators have claimed that branched rivers form self-similar fractal trees with fractal dimensionalities D smaller than 2 [32]. However, as Mandelbrot [10] has pointed out, if a tree consisting of channels of vanishing width is to drain a river basin, it must penetrate everywhere, and the tree must be plane filling ($D = 2$). It is a common experience that in most regions, a stream can always be reached by traveling through a short distance[5] of the order of 1 km. In some cases, the fractal dimension has been calculated from equation (12) instead of equation (13). If the correct equation is used [33], the fractal dimensionalities are often consistent with the idea that $D = 2$ on

[5] Karst regions, in which most of the water flows beneath the surface and desert regions are important exceptions. However, if the drainage channels formed by ephemeral streams are included, river networks are also space filling in desert regions.

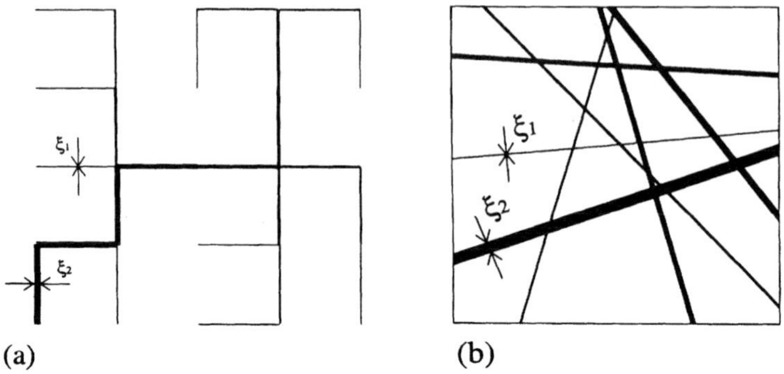

Fig. 2. — Examples of patterns containing linear elements with a distribution of widths. Part a) shows a "tree", that is similar to branched river networks, and part b) shows a network of intersecting lines that can be used as a crude model for fracture networks. The widths of the line segments w lie in the range $\xi_1 \leq w \leq \xi_2$ and the average distance between neighboring lines is ξ.

lengths scales greater than the crossover length ξ given by $\xi = 1/\rho_{\rm d}$, where $\rho_{\rm d}$ is the drainage density (total channel length/basin area) [26].

It has also been claimed that the boundaries of river basins are self-similar fractals. For example, Breyer and Snow [34] have measured fractal dimensionalities in the range $1.06 \leq D \leq 1.12$ for 12 drainage basins in the United States using the "walking yardstick", or "divider", method [10,35]. While the log–log plots are more or less linear over more than a decade of length scales for some river basin boundaries, the log–log plots appear to be continuously curved for others.

2.2. Crossovers

It is often instructive to study simple geometric models for complex natural structures. For example, Figure 2a can be considered to be a crude two-dimensional model for a branched river, and Figure 2b can be considered to be a simple two-dimensional model for a fracture network. If the widths of the lines in Figures 2a and 2b are ignored, it is apparent that the structure is linear ($D = 1$) on length scales much smaller than ξ and space-filling ($D = 2$) on length scales much larger than $\xi = \mathcal{A}/\mathcal{L}$, where $\mathcal{L}$ is the total length of the line segments in the total area $\mathcal{A}$. If these patterns are analyzed using a box counting method in which the pattern is covered with a square grid composed of grid elements with sides of length ℓ, and the number of grid elements $N(\ell)$ that are fully of partially occupied by the pattern are counted, then it will be found that $N(\ell) \sim \ell^{-1}$ if $\ell \ll \xi$ and $N(\ell) \sim \ell^{-2}$ if $\ell \gg \xi$. For a self-similar fractal, $N(\ell) \sim \ell^{-D}$, where D is the fractal dimensionality. In the crossover region, where $\ell \approx \xi$ a plot of $\log N(\ell)$ against $\log \ell$ will be curved, and in most cases, the *local* slope s will lie in the range $-2 < s < -1$. For the pattern shown

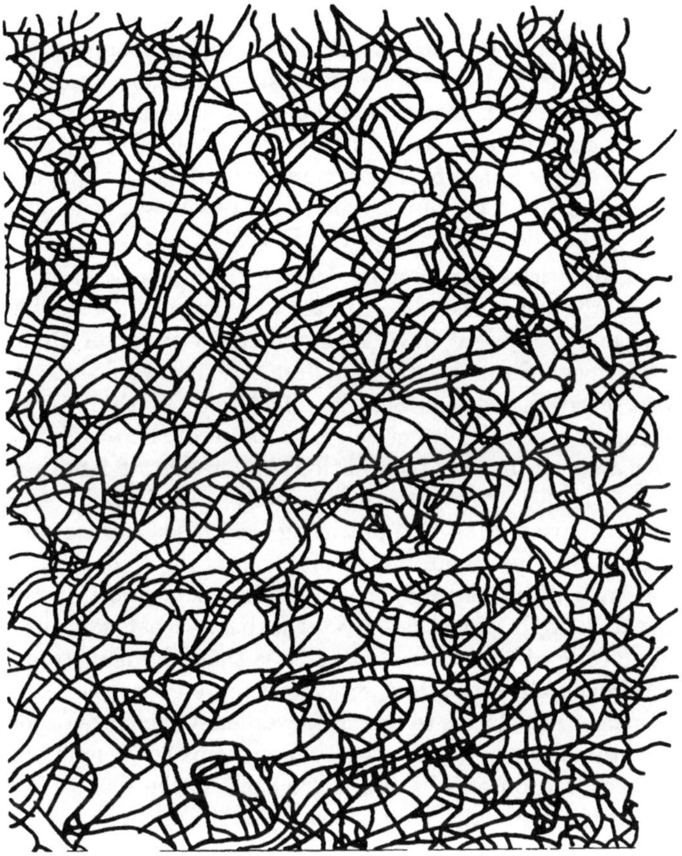

Fig. 3. — A digitized image of a 36 cm × 47 cm part of a highly tempered 86 cm × 100 cm glass sheet that was accidentally fractured. The original photograph was provided by P. D'Onfro and the digitized image was prepared by T. Walmann, using a high resolution CCD camera.

in Figure 2a, the crossover from the short length scale $D = 1$ regime to the long length scale $D = 2$ regime will take place over a quite short range of length scales, and the crossover from the $D = 1$ regime to the $D = 2$ regimes is often quite apparent from a plot of $\log N(\ell)$ against $\log \ell$. For the example shown in Figure 2b, the crossover extends over a wider range of length scales. Nevertheless, it should be possible to discern the linear regime, the space filling regime and the crossover, if the range of length scales examined ($\ell_{\rm min} \leq \ell \leq \ell_{\rm max}$) is centered on the crossover length ξ and extends over a few decades. In the fracture network pattern shown in Figure 3, the sizes of the unfractured regions range over more than a decade, and the crossover will extend over a somewhat larger range of length scales (at least two decades). In many natural systems, the pattern is not homogeneous. For example, in natural rivers, the drainage density may vary from region to region, depending on the climate and geological conditions. Under these circumstances, the crossover may extend over such a large range of length scales that the entire data set lies in or near the crossover regime, and the curvature in the plot of $\log N(\ell)$ against $\log \ell$ may not be obvious in the presence of fluctuations caused by the random nature of the pattern. This has led many investigators to report spurious values for the fractal dimension that lie in the range $1 < D < 2$.

Practical considerations conspire to deceive the unwary. For example, in the study of a large river like the Amazon or Congo, the smallest streams included in the analysis might have sizes similar to or larger than rivers like the Seine or Thames. Similarly the smallest streams would not be included in the analysis of rivers like the Seine or Thames. In the study of natural fracture patterns, it is usually practical to measure of the order of a thousand individual fractures, at most. In addition, the range of length scales used to represent and analyze the pattern is usually quite small. Consequently, almost the entire analysis is carried out using length scales that are within the crossover region. The investigator very often plots $\log N(\ell)$ against $\log \ell$ and either draws a straight line through the entire data set, ignoring the curvature, or calculates the straight line that gives the best "least squares" fit to the date. The magnitude of the slope of the line is then claimed to be the fractal dimensionality. Since all of the data sets cover more or less the same range of length scales, measured relative to the crossover length ξ, more or less the same slope may be obtained from a wide variety of systems covering a wide range of length scales measured on an absolute scale such as the meter. This leads to the seductive but incorrect idea that the slope of the straight line (the "fractal dimensionality") has a universal value. Fortunately, some investigators [36] have avoided this trap.

A simple model that can be used to illustrate the idea of a resolution dependent crossover [8] is based on the Eden growth model [37,38]. In the standard Eden model, unoccupied sites on the perimeter of a growing cluster are selected randomly and filled. In the Eden tree model, a bond is formed between the randomly selected unoccupied perimeter site that is filled at each stage in the growth process and one of its previously occupied nearest neighbor sites, which is selected at random[6]. In a completed cluster, each site at position $\mathbf{r}$ connects $n(\mathbf{r})$ other sites to the origin of the cluster *via* the tree of bonds ($n(\mathbf{r})$ sites would be disconnected from the main part of the cluster by removing the site at position $\mathbf{r}$). In Figure 4a, each of the sites in the cluster is represented by a circle with a radius proportional to $n(\mathbf{r})^{1/2}$. In terms of river networks, $n(\mathbf{r})$ can be thought of as the area of land surface that drains through the river channel

[6] In the original Eden model, an occupied perimeter site is selected at random and one of its randomly selected nearest neighbors is filled, at each stage in the simulation. In this model the bonds (between the occupied perimeter site and the newly occupied site) appear in a more natural manner.

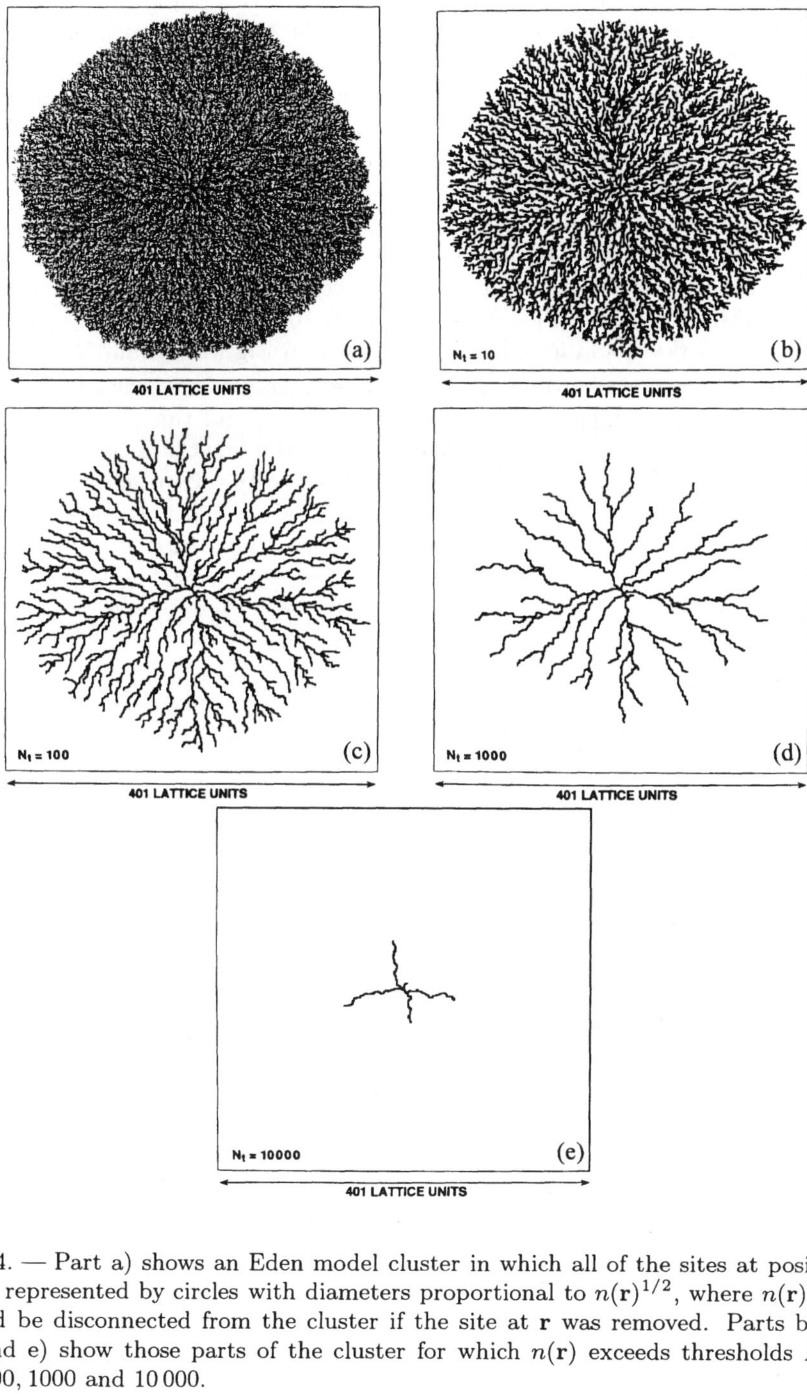

Fig. 4. — Part a) shows an Eden model cluster in which all of the sites at positions $\mathbf{r}$ are represented by circles with diameters proportional to $n(\mathbf{r})^{1/2}$, where $n(\mathbf{r})$ sites would be disconnected from the cluster if the site at $\mathbf{r}$ was removed. Parts b), c), d) and e) show those parts of the cluster for which $n(\mathbf{r})$ exceeds thresholds N_{t} of $10, 100, 1000$ and $10\,000$.

at position $\mathbf{r}$. In this case, the empirical data supports a power law relationship between the river width $w(\mathbf{r})$ and the flow $Q(\mathbf{r}) \sim n(\mathbf{r})$ ($w(\mathbf{r}) \sim Q(\mathbf{r})^x$ with $x \approx 1/2$ – Eq. (1)). Consequently, the "Eden tree" model, with branch widths at $\mathbf{r}$ proportional to $n(\mathbf{r})^{1/2}$, can be thought of as a primitive river model. Parts b), c), d), and e), in the same figure, show only those parts of the cluster in which $n(\mathbf{r})$ exceeds a threshold N_t. In parts b) and c), ($N_\mathrm{t} = 10$ and $N_\mathrm{t} = 100$) it is clear that those sites for which $n(\mathbf{r})$ exceeds the threshold form a network that is uniform on all but quite short length scales. The clusters shown in parts d) and e) ($N_\mathrm{t} = 1000$ and $N_\mathrm{t} = 10\,000$) look more "fractal". However, this is an illusion. If the growth process was continued to a much larger length scale and the sites with $n(\mathbf{r}) > 1000$ or $n(\mathbf{r}) > 10\,000$ were displayed, then the resulting networks would look like those shown in Figures 4b and 4c.

Another type of crossover is illustrated by meandering rivers such as the Mississippi. If the width of the channel is ignored, and the channel is represented by its center line. Then it is clear that the structure on scales much less then the crossover length, which corresponds to the size of a meander loop, is one-dimensional. Similarly, the structure on long length scales is also one-dimensional. If a box counting analysis is carried out, then it will be found that

$$N(\ell) = \begin{matrix} A_S \ell^{-1} & \text{if } \ell \ll \xi \\ A_L \ell^{-1} & \text{if } \ell \gg \xi, \end{matrix} \tag{14}$$

and that $A_S > A_L$. Similarly, if a walking yardstick analysis is carried out it will be found that the channel length L measured with yardstick lengths of λ is given by

$$L(\lambda) = \begin{matrix} L_S & \text{if } \ell \ll \xi \\ L_L & \text{if } \ell \gg \xi. \end{matrix} \tag{15}$$

where L_S/L_L is the sinuosity of the channel. Depending on the complexity of the meander pattern, it might be possible to represent the dependence of $N(\ell)$ on ℓ by a power law with an exponent other than -1 or represent the dependence of $L(\lambda)$ on λ by a power law over a small range of length scales [39]. However, this does not mean that the meandering channel has a self-similar fractal structure. If the meander loops have a complex structure, the crossover may extend over a relatively large range of length scales. This only increases the probability that the crossover will be missed, and that an invalid conclusion will be reached.

The analysis of natural trees and networks is often complicated by inhomogeneity. For example, the drainage density can vary from ≈ 1.0 km^{-1} in areas with high rainfall to ≈ 1000 km^{-1} in some badland areas [40]. Consequently, for very large river systems that drain regions with different climates and geologies, the crossover from $D = 1$ on short length scales to $D = 2$ on long length scales may extend over a length scale range of approximately 4 orders of magnitude (powers of 10). This implies that the data must cover more than

5 length scale decades to clearly see the nature of the crossover[7]. This can make it very difficult to distinguish between fractal patterns and inhomogeneous non-fractal patterns, particularly for cases in which the pattern is inhomogeneous on essentially all relevant scales. However, if ephemeral streams are included in dry regions, the range of length scales spanned by the crossover may be substantially reduced.

3. THE DIFFUSION-LIMITED AGGREGATION MODEL

In the original diffusion-limited aggregation (DLA) model of Witten and Sander [7], "particles", represented by lattice sites, are added, one at a time, to a growing cluster or aggregate of particles *via* random walk paths starting from outside of the region occupied by the cluster. In the most simple version of the model [7], a simulation is started by occupying a site in the center of a square or triangular lattice to represent the "seed", "growth site" or "nucleation site". A site far from the cluster is then selected and a random walk, consisting of steps between nearest neighbor sites, is started from the selected site. If the random walker moves too far from the growing cluster, it is terminated and a new random walk is started. If the random walker eventually reaches a site that is the nearest neighbor to a previously occupied site, the random walk is stopped and the unoccupied perimeter site (the last site occupied by the random walker) is filled to represent the growth process. The process of launching random walkers from outside of the region occupied by the growing cluster and terminating them when they wander too far from the cluster or "stick" to the growing cluster by reaching an unoccupied perimeter site is repeated many times to simulate the cluster growth process.

Because many random walks must be generated, this algorithm is very time consuming, and the largest clusters that can be grown, using reasonable amounts of computer time, contain only a few thousand lattice sites. The algorithm can be improved [41] by starting the random walk trajectories at a randomly selected point on a circle, called the launching circle, that just contains the cluster, and transferring the random walker to the nearest lattice site. This is justified, since a random walker started at a random position far outside of a circle will first cross the circle at random, with equal probability at all positions on the circle. For small clusters, it is satisfactory to terminate random walkers if they reach a "killing circle" centered on the original growth site (the cluster origin) and having a radius $2-3$ times the radius of the launching circle. This model is illustrated in Figure 5. Clusters containing a few tens of thousands of sites can be generated using this algorithm.

Diffusion-limited aggregation models can be made much more efficient by allowing the random walkers to take long steps when they are far from the cluster (including positions far from the cluster, but between the branches of

[7] In practice, the range of length scales covered by the crossover may extend to the size of the entire system.

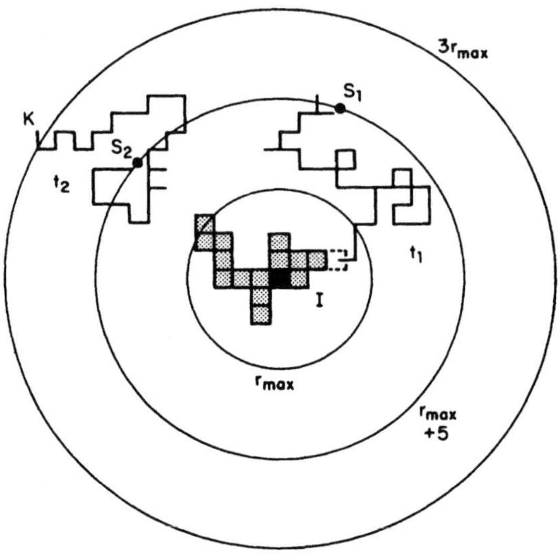

Fig. 5. — A simple, square lattice DLA model. The initial seed or growth site is shown in black. Sites that have already been added to the growing cluster are shaded. Two typical random walk trajectories, originating at random points on the launching circle with a radius of $r_{\max} + 5$ lattice units are shown, where $r_{\max}$ is the maximum cluster radius measured from the initial growth site. Trajectory t_1 starts from point S_1 and eventually reaches the unoccupied perimeter site, with dashed borders, which is filled to represent the growth process. Trajectory t_2, starting from point S_2, reaches the killing circle with a radius of 3 $r_{\max}$, before reaching the unoccupied perimeter, and it is terminated.

the cluster). This is justified, because if the randomly walking particle is at the center of a circular region with a radius of r (or at the center of a sphere of radius r, in the three-dimensional case) that contains no traps (adsorbing sites) then the entire path of the random walker, up to the point at which it emerges from the circle, can be replaced by a single step to a randomly selected position on the circle[8]. This, and other improvements on the basic DLA algorithm, has allowed very large clusters to be generated [8]. Figure 6 shows a 20 000 000 particle cluster generated using a two-dimensional off lattice model, and Figure 7 shows a 10 000 000 particle cluster generated using a three-dimensional off lattice model. In off lattice models, the particles are represented by disks or spheres, and the random walk is represented by a sequence of steps in randomly selected directions. The particles stick to the growing cluster in the position in which they first contact the cluster.

[8] This approach was first used by Ball and Brady [42] for square lattice DLA. In these simulations, they replaced the random walk from the center of a square to the perimeter of the square by a single step using the analytically calculated first exit probability distribution.

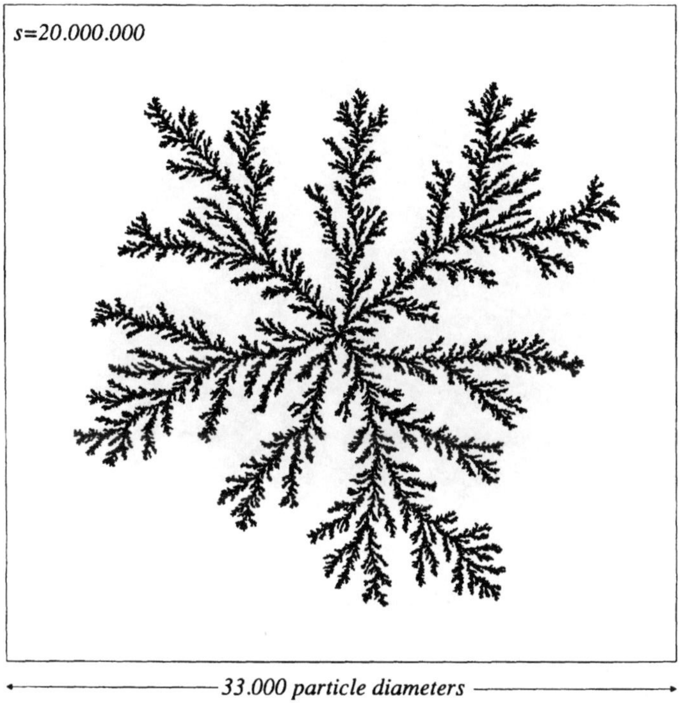

Fig. 6. — A cluster of 20 000 000 equal sized discs generated by a two-dimensional off lattice DLA model. This figure was provided by Thomas Rage.

In most studies of three-dimensional systems, laboratory samples or field specimens are examined *via* two-dimensional cuts, or cross-sections. As Figure 7b shows, a two-dimensional cut through a three-dimensional DLA cluster is a fractal dust that may be difficult to recognize. In contrast, two-dimensional DLA clusters have a striking, easily recognized appearance. For this reason, most studies have been concerned with systems that are essentially two-dimensional. Although, we live in a three-dimensional world, many systems are extremely anisotropic, and pattern formation processes are often controlled by essentially two-dimensional transport processes[9]. Nevertheless, three-dimensional DLA-like growth processes are probably very common, and they have been neglected by comparison with two-dimensional DLA.

The (DLA) model is one of the most striking examples of the generation of a complex disorderly pattern by a simple model or algorithm. It was developed at a time when interest in fractals was growing rapidly, and it soon became the

[9] The effects of anisotropy and inhomogeneity may also make it more difficult to recognize branched structures grown *via* DLA-like processes in three-dimensional systems.

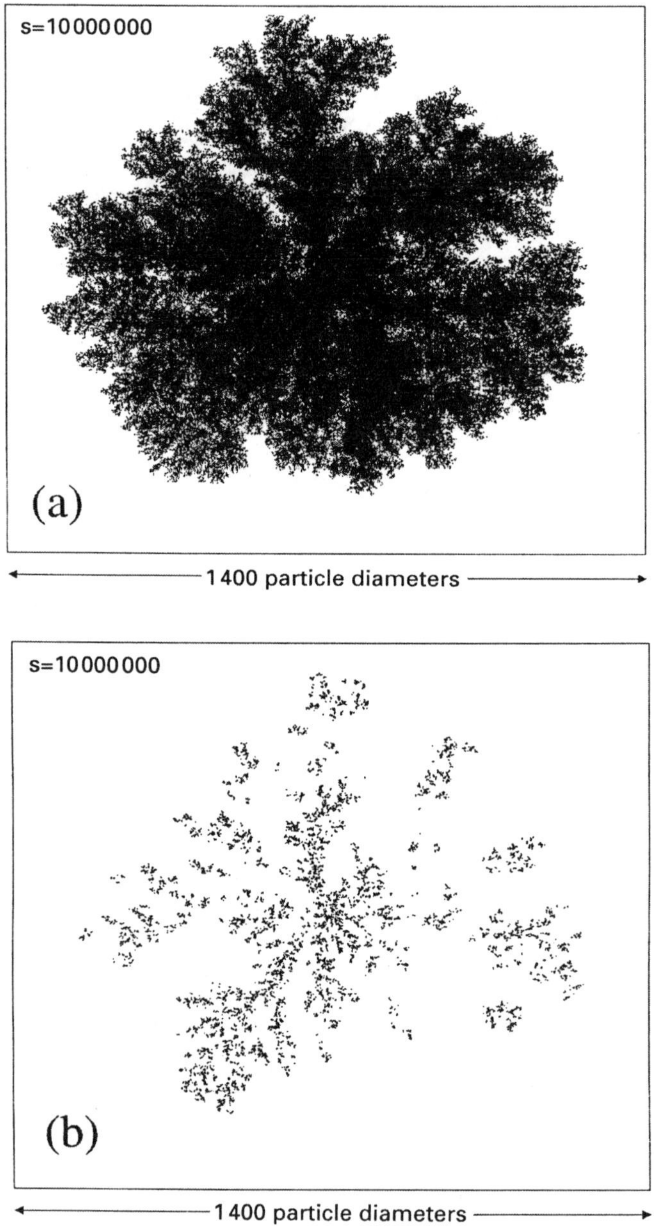

Fig. 7. — A cluster of 10 000 000 equal sized spheres generated by a three-dimensional off lattice DLA model. Part a) shows a projection of the cluster onto a plane, and part b) shows a cross-section through the origin. This figure was provided by Thomas Rage.

principal paradigm for pattern generation far from equilibrium. After almost 20 years, interest in this model is still at a high level. This interest is maintained by a wide range of applications. The DLA model also continues to present an important theoretical challenge that has not yet been fully met. Many aspects of the structure of DLA clusters can be described in terms of a statistically self-similar fractal scaling model. For example, the dependence of the radius of gyration of two-dimensional DLA clusters on the number of particles is a robust power law, corresponding to a fractal dimensionality of $D \approx 1.715$, over about 6 decades in the number of particles and about 4 decades in the radius of gyration. However, it is apparent that the structure of DLA may be more complex than that of a homogeneous, statistically self-similar fractal, such as a percolation cluster [8]. There is, at present, no systematic theoretical framework for the class of non-equilibrium, non-local growth problems that DLA represents.

Despite the lack of a fundamental understanding of DLA in more basic terms, the DLA model itself provides a basis for understanding a wide range of natural phenomena. The idea that understanding natural processes in terms of simple algorithms provides an alternative to understanding in terms of the analytical solution of equations is gaining acceptance. In some cases, the "quality" of understanding may be greater for the "algorithmic" approach, and simple algorithmic models frequently lead to a more intuitive understanding.

Very similar patterns can be generated using the "dielectric breakdown" model [43]. The dielectric breakdown model was motivated by the idea that in dielectric breakdown processes, dielectric material with a very low conductivity is converted into a material with a high conductivity (a plasma in the dielectric breakdown of a gas or carbonaceous material in the dielectric breakdown in an organic polymer, for example) and that the rate of dielectric breakdown depends on the electric potential gradient, in the dielectric material, at the interface with the conducting breakdown products. In this model, it is assumed that the rate of growth of the breakdown pattern is given by

$$V_{\mathbf{x}_s} = f(|\nabla_n \Phi_{\mathbf{x}_s}|), \tag{16}$$

where $V_{\mathbf{x}_s}$ is the growth velocity normal to the interface between the breakdown pattern and the surrounding material at position $\mathbf{x}_s$ on the interface and $\nabla_n \Phi_{\mathbf{x}_s}$ is the gradient of the electric potential Φ in the direction normal to the interface, measured in the dielectric material at $\mathbf{x}_s$, in a direction normal to the interface. Niemeyer *et al.* [43] simulated the growth of breakdown patterns using a model in which a discretized version of the Laplace equation, $\nabla^2\Phi = 0$, was solved numerically, at each stage in the simulation, with the boundary conditions $\Phi = \Phi_o$ at all occupied sites used to represent the breakdown region and $\Phi = 0$ on a distant surface enclosing the growing cluster of "breakdown" sites[10].

[10] Because of the linearity of the Laplace equation, Φ_o and Φ_∞ can be given any values, providing that they are not equal. In most simulations, Φ_∞ is set to zero, and Φ_o is given a convenient value such as unity.

On a square lattice, the discretized Laplace equation can be written as

$$\Phi(i,j) = 1/4[\Phi(i,j-1) + \Phi(i,j+1) + \Phi(i-1,j) + \Phi(i+1,j)], \qquad (17)$$

where $\Phi(i,j)$ is the "potential" at the lattice site with coordinates (i,j).

If the function $f(|\nabla_n \Phi_{\mathbf{x}_s}|)$ has the simple form $f(|\nabla_n \Phi_{\mathbf{x}_s}|) \propto |\nabla_n \Phi_{\mathbf{x}_s}|$ and appropriate boundary conditions are used at the surface of the growing cluster, then the dielectric breakdown becomes equivalent to the DLA model[11]. The close relationship between the DLA model and the dielectric breakdown model was anticipated by Witten and Sander [44], who pointed out an "electrostatic analogy" between the probability that a site at position $\mathbf{x}$ on the perimeter of a growing cluster will be the first perimeter site reached by a random walker and the growth probability or growth velocity at that site in a Laplacian growth process. Indeed, they commented on the similarity between DLA clusters and lightning. This can be seen by comparing equation (17) with the equation

$$P(i,j) = 1/4[P(i,j-1) + P(i,j+1) + P(i-1,j) + P(i+1,j)], \qquad (18)$$

for the probability that a random walker will be found in the site at position (i,j) in terms of the probabilities of finding it in the neighboring sites. This means that a scalar field Φ that obeys the Laplace equation can be simulated (statistically sampled) by a particle following a random walk with the same boundary conditions. The probability $P(i,j)$ in equation (18) is then proportional to $\Phi(i,j)$.

3.1. Examples of DLA-like patterns

One of the most familiar DLA-like patterns is the manganese/iron oxide "pseudo fossils" (MnO_2 dendrites) that are found at the joint surfaces in sedimentary rocks such as Solnhofen chalks[12]. However, the origin of these DLA-like patterns is uncertain, and it is not clear if the field that obeys the Laplace equation is a concentration field or a fluid pressure field [8]. However, a preponderance of the available information indicates that these patterns are formed by a viscous-fingering process driven by pressure gradients. A much better understood example is the formation of Lichtenberg figure electrical breakdown patterns in gases such as SF_6 [43]. In this case, fractal dimensionalities of $D \approx 1.7$ (in agreement with the fractal dimensionality of $D \approx 1.715$ obtained for off lattice two-dimensional DLA clusters) have been obtained by analyzing the branched patterns.

Lightning provides a familiar and important large scale example of a three-dimensional dielectric breakdown process. Apart from the branched discharge

[11] Except that the dielectric breakdown model correspond to a DLA model in which growth occurs in the last unoccupied site that the randomly walking particle takes, before it enters a previously occupied site and is terminated.

[12] The interesting history of work on this phenomenon is described in a recent book authored by V. Fleury [45].

patterns in the air, branched patterns can also be observed in the surfaces of golf courses that have been struck by lightning, and fulgurites also have a branched structure. In addition, quite intricate branched patterns have been observed in experimental electrical discharges in wet and dry sand [46]. In all of these examples, it is clear that the electric potential is the scalar field that obeys the Laplace equation.

The experiments of Daccord and Lenormand [47, 48] on the dissolution of plaster of Paris ($CaSO_4 \cdot 0.5H_2O$) provide some of the most spectacular examples of DLA-like patterns. In these experiments, water was slowly injected into water saturated plaster of Paris, which is a porous material with a small solubility in water. Despite the relatively small solubility of $CaSO_4$, water in contact with plaster of Paris soon becomes saturated with $CaSO_4$ and will dissolve no more. Consequently, the dissolution of the plaster of Paris takes place near to the interface between the water filled cavity formed by dissolution of the plaster of Paris and the surrounding water saturated plaster of Paris. The rate of dissolution (the velocity of the interface) at position $\mathbf{x}$ on the interface is proportional to the flux density of water through the interface at $\mathbf{x}$, if the injection rate is neither too fast nor too slow[13]. The flow of water through the interface is controlled by the pressure field inside the porous plaster of Paris, since there is essentially no pressure drop inside the cavity in the porous medium, etched by the water.

Slow flow inside a porous medium can be described by Darcy's equation,

$$\mathbf{V}(\mathbf{x}) = -m\nabla\Phi(\mathbf{x}), \tag{19}$$

where $m = k/\mu$ is the mobility (μ is the fluid viscosity and k is the permeability of the porous medium), $\mathbf{V}(\mathbf{x})$ is the fluid velocity at $\mathbf{x}$ and $\Phi(\mathbf{x})$ is the hydraulic potential ($\Phi(\mathbf{x}) = P(\mathbf{x}) + \rho g h(\mathbf{x})$, where P is the fluid pressure, ρ is the fluid density, g is the acceleration due to gravity and $h(\mathbf{x})$ is the distance in the direction of the gravitational field at position $\mathbf{x}$). For incompressible fluids, $\nabla \mathbf{V} = 0$, so that Darcy's law implies that $\nabla^2\Phi = 0$, and the hydraulic potential field driving the fluid through the porous medium obeys the Laplace equation. Consequently, the experiment can be described in terms of the random propagation of an interface controlled by a field that obeys the Laplace equation (the pressure field inside the porous plaster of Paris) and the dissolution process can be represented quite well by the DLA model.

Experiments were carried out with several different boundary conditions including injection into a "radial" Hele-Shaw cell consisting of two parallel circular glass sheets that confined a thin layer of plaster of Paris. In these experiments, the water entered through a small hole in the center of one of the glass sheets. The glass sheets had diameters of 250 mm and were separated by

[13] If the flow rate is too fast, then the water does not become saturated with $CaSO_4$ before it penetrates too far into the porous plaster of Paris. If the flow rate is too slow, the rate of dissolution is controlled by the diffusion of $CaSO_4$ in the etched cavity instead of the flow of water through the water/plaster of Paris interface.

Fig. 8. — A Wood's metal casting of a cavity etched into a 1 mm thick layer of plaster of Paris, in a radial Hele-Shaw cell. This figure was provided by G. Daccord and is reproduced with the permission of Dowell-Schlumberger.

a gap of 1 mm (the porous layer thickness). The patterns produced in these experiments closely resembled those generated in corresponding DLA model simulations. Figure 8 shows the dissolution pattern obtained by injecting water at a rate of 48 cm^3/hour into plaster of Paris with a porosity of 60% and a permeability of 60 μm^2, where 1 μm^2 = 1.013 Darcy. After the water had been injected, the plaster of Paris was dried and Wood's metal was injected into the cavity etched into the plaster and the remaining plaster of Paris was then dissolved[14]. Figure 8 shows one of these Wood's metal castings. At lower injection rates, the branches were broader and less ramified.

Daccord and Lenormand [47,48] also carried out experiments in which water was injected into cylindrical blocks of plaster of Paris, *via* a hole drilled along the axis of each cylinder and experiments in which water was injected into pipes filled with plaster of Paris (through one of the exposed ends). Woods metal castings of the cavities formed in some of these experiments are shown in Figure 9. Similar patterns can be generated by three-dimensional DLA model simulations [49] with the appropriate boundary conditions. The experiments in which the water was injected through a hole drilled into the plaster of Paris can be represented by a DLA model in which particles are deposited onto a line

[14] Because of the small pore size, the molten Wood's metal does not penetrate into the dried plaster of Paris at the relatively small pressures used in these experiments.

Fig. 9. — Wood's metal castings of cavities etched into plaster of Paris. This image shows results from an experiment in which water was injected through a 2 mm diameter hole drilled along the center of a plaster of Paris cylinder, and the image on the next page shows the cavities formed when water was injected into a pipe filled with plaster of Paris. This figure was provided by G. Daccord and is reproduced with the permission of Dowell-Schlumberger.

Fig. 9. — Continued.

(or onto previously deposited particles) *via* random walk paths that start far from the line or growing deposit. In the experiments, the characteristic radius

Fig. 10. — Branched patterns formed by the invasion of relatively warm water through holes in "black ice" into a layer of overlying slush. After this event, the slush froze to form "white ice" and the melted regions formed black ice dendrites within the white ice layer. This figure was provided by C.A. Knight.

R of the pattern, measured *via* the pressure drop at a constant injection rate was found to grow as $R \sim t^{\gamma}$, with $\gamma = 0.65 \pm 0.07$. This value for the exponent γ is in good agreement with the value of $\gamma \approx 0.66$ expected for a structure with the fractal dimensionality of three-dimensional DLA ($D = 2.49 \pm 0.01$ [49]) and with the exponent describing the dependence of the radius of gyration on the number of particles deposited, measured directly from the simulations of deposition onto a line in three-dimensional space [49].

The work of Daccord and Lenormand was partially motivated by applications in enhanced oil recovery (a processes called "wormholing"). Similar patterns should be formed when fluids are forced through other soluble porous media, at appropriate flow rates, and they may be common in nature.

A phenomenon closely related to the dissolution of porous materials has been described by C.A. Knight [50]. This is the formation of branched patterns (Fig. 10) in "white ice" (opaque ice) when snow is deposited on top of a layer of clear "black ice" (transparent ice) on a lake. This phenomenon is quite common. I have observed these patterns many times on ponds in Southeast Pennsylvania. The overall size of the patterns ranges from $\approx$20 cm to >10 m, and these branched patterns often emanate from the edge of the pond. If sufficient snow is deposited on the ice, water may flood the surface of the ice, forming a layer of slush that later freezes. The pattern consists of branches of clear ice within the layer of white ice (frozen slush). The pattern originates at a hole in the underlying layer of black ice. Relatively warm water flows through this hole into the layer of slush on the ice. The water melts some of the slush, but is soon cooled to the freezing point, because the water temperature is near to the freezing point and the latent heat of melting is large for water. The cooled water then penetrates through the slush layer. This causes no further melting, but the flow of water through the slush, which can be described by Darcy's equation, controls the flow of warm water to the water/slush interface. Consequently, all the ingredients of DLA (a random growth process controlled by a scalar field that obeys the Laplace equation) are present. Other investigators [51] have suggested that these patterns are related to cracking of the ice. This may take place under some circumstances. However, the generation of similar patterns in laboratory experiments and experiments in which water slightly above the freezing point was poured onto lakes with snow covered ice [52] lends strong support to the "flow controlled melting" scenario.

The patterns formed by some "river" systems, particularly drainage channels in dry climates, are quite similar to two-dimensional DLA clusters. In some cases, there is reason to believe that this similarity in form may be more than superficial. The importance of mechanisms such as "piping" and "sapping" in the initiation and extension of streams has been recognized for some time [53–55]. In piping and sapping, accelerated erosion takes place when ground water flow emerges to generate overland flow or flow in an underground channel, which usually collapses relatively quickly to form an open channel. The ground water flow can be described in terms of the Darcy equation. This process is similar to the dissolution and melting processes discussed above, except that the direction of the water flow is reversed. If the rate of erosion at the water saturated soil/water interface is proportional to the outflow flux density, then a DLA model would be a good starting point for simulating these phenomena. In practice, the dependence of the rate of erosion on the flux density is likely to be complex and non-linear, but it may be approximated by a power law, suggesting that a dielectric breakdown model in which the function $f(|\nabla_n \Phi_{\mathbf{x}_s}|)$ in equation (16) has the form $f(|\nabla_n \Phi_{\mathbf{x}_s}|) \propto |(\nabla_n \Phi_{\mathbf{x}_s})|^{\eta}$ might be used.

Cerasi *et al.* [56] have carried out experiments in which water/glycerin was withdrawn from the center of a radial (circular) Hele-Shaw cell filled with Fontainebleau sand that was saturated with the water/glycerin mixture. The gap between the plates of the Hele-Shaw cell was about 300 μm, and the diameters of the sand grains were in the 100 μm to 300 μm range. Experiments were carried out using a range of pumping rates. At low flow rates, the fluid flowed through the cell with little or no erosion. At a threshold flow rate $Q_c^{(1)}$, localized erosion began, and an unstable erosion front propagated through the unconsolidated porous medium. At a second threshold flow rate $Q_c^{(2)}$, fluidization took place with motion of the entire porous medium. Branched erosion channels, like those shown in Figure 11a, with fractal dimensionalities of $D \simeq 1.70$

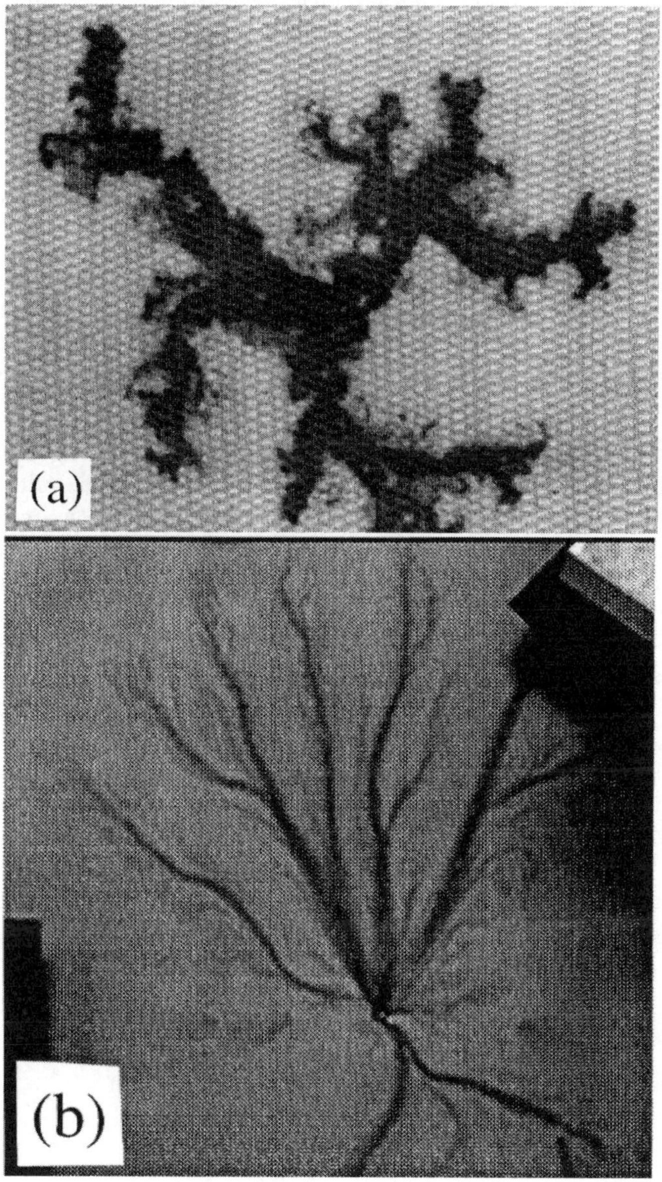

Fig. 11. — Branched patterns formed by erosion of unconsolidated materials in a Hele-Shaw cell. Part a) shows erosion channels obtained using Fontainebleau sand and part b) shows a pattern obtained in a similar experiment using glass spheres with diameters in the 5 μm to 60 μm range. These figures were provided by P. Mills and P. Cerasi.

(consistent with the fractal dimensionality of two-dimensional off lattice DLA) were formed under a quite wide range of conditions, with flow rates in the $Q_c^{(1)} < Q < Q_c^{(2)}$ range. Under some conditions, less well defined channels with lower fractal dimensionalities were obtained. Similar experiments were carried out using glass beads in place of the sand (P. Cerasi, private communication). Branched patterns like that shown in Figure 11b were formed at intermediate flow rates. Although the effective fractal dimensionality is also about 1.7, the branched morphology seems to be quite different from that obtained with sand.

Another important example of DLA-like patterns is the growth of mineral dendrites. In this case, it is the concentration field that obeys the Laplace equation (in the large diffusion length limit). For example, Fowler *et al.* [57] have used a simple modification of the DLA model to simulate the growth of olivine dendrites formed at the margins of archean lave flows.

Masek and Turcotte [58] have developed a diffusion-limited aggregation model for the evolution of river networks. In this model, particles are added randomly at any position in the system (including positions that are between the branches of the growing patterns). The deposited particles then undergo random walks until they reach a site that is a nearest neighbor to an already occupied site (either a "seed" site or a site that is part of the growing pattern). An additional modification of the standard square lattice DLA model is that the random walk is terminated, but growth does not take place, if the random walk enters an unoccupied perimeter site with more than one occupied nearest neighbor. A fractal dimensionality of $D = 1.87$ (determined using Eq. (12)) was reported by Masek and Turcotte [58]. However, this type of DLA model generates a pattern that has a fractal dimensionality of $D \approx 1.7$ (the fractal dimensionality of two-dimensional DLA) on short length scales, and a fractal dimensionality of $D = 2$ on long length scales. The purported fractal dimensionality of $D \approx 1.87$ is yet another example of an *effective* fractal dimensionality, with no fundamental significance, generated by a crossover.

4. BRANCHED RIVERS

Branched rivers are one of the most striking features associated with erosion and sediment transport due to overland flow. However, it is important to recognize that branched rivers are not purely erosional features (though they may be dominated by erosion processes, under some conditions) and that branched rivers evolve in concert with the supporting topography and the underlying soil and porous rocks. Many attempts to model the structure and dynamics of rivers focus attention on only a few physical processes, and simple initial conditions and boundary conditions are used. However, real rivers are influenced by a large number of coupled physical, chemical and biological processes, and simple well defined initial conditions are rare[15]. In general, fluvial systems evolve continuously over very long periods of time. They may be considered to have no well defined beginning or end.

Under some circumstances, the evolution of branched rivers is dominated by overland flow, erosion and sediment transport. If it is assumed that once erosion has occurred, the eroded material is immediately removed from the system, then attention can be focused on the coupling between erosion and overland flow. It is clear that uniform flow over a smooth erodible surface is an inherently unstable process. An infinitesimal fluctuation in the smooth surface will begin to focus the flow onto preferred pathways, and these preferred pathways will become even more preferred due to enhanced erosion along them.

In a "mature" branched river system, the landscape can be divided into hillslopes and channels. In general, the sediment transport and erosion processes are qualitatively and/or quantitatively different in the channels and hillslopes. However, the division into channels and hillslopes is ambiguous, and the transition between hillslopes and channels is poorly understood.

On the hillslopes, the overland flow is relatively uniform (it increases with increasing distance from the ridge crests) and sediment transport is dominated by *soil creep* processes. Soil creep is a result of internal reorganization of the soil structure due to processes such as freeze-thaw cycling, wetting and drying, the growth and the decay of plant roots, and the effects of burrowing animals. Processes that take place at or near the surface of the soil, such as rain splash, the combined effects of erosion, transport and deposition by overland flow, hoof impact by mammals and the uprooting of trees also contribute to the evolution of the topography. Vegetation plays an important role in sediment transport on hillslopes, and the vegetation is influenced by factors such as fires, climate change, disease and grazing. All of these processes result in a down-slope transport of sediment, driven by the effects of gravity.

In a variety of models, it is assumed that the effects of hillslope processes on the topography can be represented by the equation[16]

$$\partial h(\mathbf{x},t)/\partial t = \mathcal{D}\nabla^2 h(\mathbf{x},t), \tag{20}$$

where $\mathcal{D}$ is the transport coefficient that relates the sediment flux density to the

[15] Reasonably simple well-defined initial conditions may occur under artificial conditions such as waste disposal sites that have been covered by a layer of soil with a flat surface [59] or after a large scale catastrophic event such as a lava flow has occurred.

[16] This equation is based on conservation of the sedimentary material, which requires that $\partial h(\mathbf{x},t)/\partial t = -\nabla \mathbf{J}_s(\mathbf{x})$, where $\mathbf{J}_s(\mathbf{x})$ is the sediment flux density, and the idea that $\mathbf{J}_s(\mathbf{x}) = -\mathcal{D}\nabla h(\mathbf{x},t)$. The idea that the sediment is transported in the steepest down slope direction with a flux density that is proportional to the slope is reasonable for processes such as soil creep and rain splash. However, this relationship must be, at best, a reasonable approximation, and there is little or no field data to support it [40].

the surface inclination. In this equation, $h(\mathbf{x}, t)$ is the height of the land surface at position $\mathbf{x}$ in the horizontal plane. This diffusion-like process makes the topography smoother and fills in incipient channels. Channels are formed when this healing process can no longer keep up with erosion. The processes described by equation (20) transports sediment from the hillslopes to the channels, in which the fine-grained components of the sediment are quickly removed by the flowing water. In mature river systems, this loss of soil to the channels is compensated for by the generation of new soil from the underlying rock, and the total amount of soil is more-or-less conserved, provided that conditions such as the climate and vegetation do not change.

If the rate of channel growth depends only on the slope $\nabla_n h(\mathbf{x}_\mathrm{s}, t)$ at the channel boundary, where $\mathbf{x}_\mathrm{s}$ is a position at the channel boundary and $\nabla_n = \mathbf{n} \cdot \nabla$ is the gradient along the vector $\mathbf{n}$, perpendicular to the channel boundary, then the use of a DLA-like model is motivated. However, this approach is also based on the implicit assumption that sediment transported to the channel network is removed rapidly from the system and that the slope along the channel is much smaller than slope on the adjacent land surface. In such a model, uplift can be represented by a uniform "rain" of random walkers, and erosion without uplift can be represented by an initial "sea" of random walkers, which is depleted as the simulation proceeds. Both models would lead to a pattern that is fractal on short length scales, $\ell \ll \xi$ ($D \approx 1.715$ for an off-lattice model), and uniform ($D = 2$) on long length scales, $\ell \gg \xi$. The crossover length ξ would depend on the initial height of the land surface with respect to the base level or the amount of uplift.

On short length scales, sediment transport is dominated by the diffusion-like hillslope processes, and on long length scales sediment transport is dominated by advective channel processes. This is similar to the behavior found in many systems in which diffusive and advective processes act independently. In this systems, diffusive transport dominates on scales shorter than the Peclet length (the length at which the Peclet number is unity) and advective processes dominate on longer length scales. However, the analogy is weak, because the diffusive and advective processes are not independent, and the transport processes in the channels are much different from the hillslope transport processes. Fluvial channels differ substantially from hillslopes. For example, the particles in the channels are, on average, much coarser than the soil particles, and the vegetation in the channels is either much different from the hillslope vegetation or is essentially absent. Equation (20) is consistent with the convex form of hillslopes in mature river basins, which contrasts with the concave form of the height profiles measured along river channels [14][17].

[17] These features were described and explained in terms of soil creep (including freeze–thaw, heat-cool and wet–dry cycling) on hillslopes and the dependence of sediment transport on channel slope and channel flow, before the end of the 19th century [60–62].

In most natural drainage channels, particularly large channels, the channel width w is much larger than the water depth d_w (despite the illustrations used in many text books). Consequently, most of the sediment transport takes place on the bed of the channel, away from the banks. However, bank erosion is an important process in the evolution of rivers. In many computer modeling studies, the sediment transport in the channels Q_s is represented by the simple form [20, 40].

$$Q_s = C_b Q^m s^n, \tag{21}$$

where s is the bed slope. Assuming that $w \gg d_w$, and that $w \simeq Q^{1/2}$ this equation can also be written in the form

$$J_s = C_j J^{2m-1} s^n, \tag{22}$$

Where J is the flow density (flow rate per unit channel width) and J_s is the sediment flow density in the channel. Similar equations, such as

$$E = C_b^* Q^{m^*} s^{n^*}, \tag{23}$$

have been used for the erosion rate E in erosion models. The exponents and sediment transport (or erosion) coefficients in these equations depend on the sediment transport (or erosion) processes, the composition of the channel bed and other factors. At best, these equations provide only an approximate description of the dominant process[18].

Since a wide variety of different models lead to realistic river network patterns, it is not possible to evaluate the validity of the sediment transport equations used in a particular model by comparing model channel networks with real systems. In fact, the main characteristics appear to be determined by fundamental constraints such as connectivity, randomness, space filling and conservation of sediment and/or water.

The development of river networks can also be influenced by sudden events such as earthquakes, landslides, volcanic lahars and jökulhlaups (glacial outburst floods[19]). In some areas, such as the scablands in the Northwestern United States, the landscape is still dominated by a single, albeit relatively recent, event (in this case the release of a large amount of water due to the collapse of an ice dam near the end of the last ice age[20]). In the past, there was an

[18] In some models, it is assumed that the erosion rate is zero if $\tau \le \tau_c$, where τ is the shear stress exerted by the flow of the water on the channel bed and τ_c is the critical shear stress, below which erosion is negligible. In these models, the erosion rate is assumed to have the form $f(\tau - \tau_c)$ if $\tau > \tau_c$, where $f(x)$ is a simple function of x, such as a power law. The shear stress is given by $\tau \approx gds$, where d is the water depth, g is the acceleration in the Earth's gravitational field and s is the bed slope.

[19] In many regions at high latitudes, the topography is dominated by the effects of active glaciers or glaciers that were formed during the recent ice ages. In addition, sediment generated by glaciers may play an important role in some fluvial systems.

[20] There may have been a series of such events. It has been estimated that during the largest event, water was released from Lake Missoula at an average rate of about 10^7 m^{3}s^{-1} for a period of several days.

intellectual battle between *gradualists* and *catastrophists.* The catastrophists believed that the Earth's topography was dominated by large scale events. In the most extreme form of catastrophism, it was claimed that the topography in a particular region is dominated by the largest event in the past and that this topography will remain more or less unchanged until an ever larger event occurs. The battle was eventually won by the gradualists. However, the emphasis on gradual processes has led to the neglect of events such as landslides and earthquakes in almost all river models[21].

4.1. Optimization models

One approach to the modeling of river networks is based on the idea that they are "optimal collecting networks" [65]. This approach can be used to model the structure of river networks but not their dynamics[22]. It is clear that such models can be applied only under very restrictive conditions, under which the structure of the river network has reached an essentially stationary form. This could occur under a period of more-or-less constant uplift, when erosion and uplift come into balance.

Optimization concepts have been used in an attempt to understand rivers for more than a century. For example, Gilbert [61, 62] used the idea that... *a stream automatically adjusts slope to volume in such a way as to equalize its work of transportation in different parts.* In an influential publication [66], with the title "The concept of entropy in landscape evolution" Leopold and Langbein proposed that the "principle of least work" can be applied to river systems and that *the distribution of energy in a river system tends towards the most probable state.* They also stated that... *the most probable distribution of energy exists when the rate of gain of entropy in each interval of length along the river is equal.* They argued that (if there is no constraint on the channel length) the elevation with respect to sea level (or base level) plays the same role as temperature in classical thermodynamics, and that the most probable state is the one in which the energy expenditure per unit mass of fluid is proportional to the height above base level. However, these ideas are based on quite unconvincing analogies between poorly understood far from equilibrium fluvial systems and equilibrium systems.

[21] Leheny and Nagel [63,64] have included "avalanching" in a simple erosion model by removing material to reduce the slope if the slope ∇h exceeds a critical value M. If the surface height difference between nearest neighbors exceeds M, then the height of the higher site is reduced by $\nabla h/4$, and this process is repeated until all local slopes are less than or equal to M. This results in well defined hillslopes with slopes of $3M/4$. The avalanche size distribution was not determined. Some of the avalanches might be quite large. However, no single avalanche has a large effect on the overall topography and there are no catastrophic events in this model.

[22] However, the evolution of branched rivers in response to slowly changing conditions could be modeled.

The ideas proposed by Leopold and Langbein were influenced by the minimum entropy production concept promoted by Prigogine [67] to explain the behavior of non-equilibrium systems. This approach to the development of a better understanding of non-equilibrium systems was popular during the 1960s and 1970s. However, more recent work has shown that this and related optimization concepts have no general validity for non-equilibrium systems, despite their intuitive appeal and many ostensible successes.

If small changes in the average water velocity along the channel are ignored, then the rate of energy dissipation (the power) in a channel segment of length ℓ is given by

$$\dot{E} = \ell s Q, \tag{24}$$

and if $s \sim Q^{\beta}$ then

$$\dot{E} \sim \ell Q^{1+\beta} \sim Q^{\alpha}. \tag{25}$$

This idea has found a large number of applications. For example, in 1971, Howard used the minimum power dissipation hypothesis to calculate the angles between the channels at a junction [19]. Almost two decades later, Howard [68] extended this approach to a minimum energy dissipation model for an entire system of river networks draining a large region. Simulations were carried out using several values for the exponent β in the range $-1.0 \leq \beta \leq -0.15$ ($0 \leq \alpha \leq 0.85$).

Rodriguez-Iturbe *et al.* [69–71] have extensively studied a river network model based on the idea that the channel network is both locally and globally optimized. This model is based on the following "principles":

- The energy expenditure (energy dissipation rate) in any link in the network is minimal;
- The energy expenditure per unit area of channel is equal everywhere in the network;
- The energy expenditure in the entire network is minimal.

The first two principles imply that the energy dissipation rate in the ith link of length ℓ_i is given by

$$\dot{E}_i = \ell_i Q_i^{0.5}, \tag{26}$$

and it is assumed that $Q_i \propto A_i$, where A_i is the area drained through the ith link. This model is a simplified version of that developed by Howard [68], and it has been much more extensively investigated.

In a typical simulation, carried out on a square lattice, the initial channel network consists of a single branched network composed of links that connect the centers of nearest neighbor sites. The initial network connects the centers of all the lattice sites to one of the corners of the lattice, where the entire discharge leaves the system. Starting with this configuration, the total energy dissipation rate $\dot{E}_{\Omega} = \sum_i Q_i^{0.5}$ ($\ell_i = 1$) is then minimized, by changing the structure of the network (the network configuration Ω) while the connectivity between all of the lattice sites and the corner through which the channel network discharges is maintained[23]. A configuration that is *representative* of the global minimum[24] (or degenerate global minima) can be found efficiently by simulated annealing. After the initial network has been set up, a simple simulated annealing algorithm consists of the following steps:

1. A link is randomly selected and its direction is randomly changed;
2. The new configuration is tested for connectivity. If the new configuration does not connect all of the lattice sites to the exit site, the original configuration is restored and a new stage in the simulation begins with step 1;
3. The change in energy dissipation rate $\delta\dot{E} = \dot{E}_n - \dot{E}_o$ is calculated, where $\dot{E}_n$ is the energy dissipation rate in the new configuration and $\dot{E}_o$ is the energy dissipation rate in the original configuration;
4. If $\delta\dot{E} \leq 0$, the new configuration, which becomes the original configuration in the next stage of the simulation, is accepted and a new stage in the simulation begins with step 1. If $\delta\dot{E} > 0$, a random number x uniformly distributed over the range $0 \leq x \leq 1$ is generated. If $x < e^{-\delta\dot{E}/T}$, the new configuration, which becomes the original configuration in the next stage in the simulation, is accepted and step 1 is repeated.
   If $x \geq e^{-\delta\dot{E}/T}$, the system is restored to the original configuration and a new stage in the simulation begins with step 1.

This procedure is repeated many times as T is slowly lowered from a large value to a value that is low enough to ensure that only transitions between states with the same values of $\dot{E}$ can occur. In many physical processes that are modeled using simulated annealing, T represents the temperature and $\dot{E}$ is replaced by E, the energy of the system in units in which k_B, the Boltzmann constant, is unity. Under these circumstances, the gradual lowering of T can be thought of in terms of a physical annealing process. In the modeling of river networks, no such interpretation is possible, and the gradual lowering of T is merely a means of generating a configuration with a low energy dissipation rate, which is representative of the global minimum. This algorithm generates branched networks that resemble natural fluvial systems with branching ratios of $R_b \approx 4$ and length ratios of $R_\ell \approx 2$. The algorithm generates a branched channel network that connects all of the lattice sites. Such a structure is called a

[23] In principal, all configurations that drain the entire area through the corner site could be generated, and the unique configuration with the lowest total energy dissipation rate or the set of configurations with the lowest total energy dissipation rate could be selected. However, the number of configurations grows faster than any power of the number of links in the network, and this procedure is completely impractical for all but the very smallest systems.

[24] It appear that many configurations are statistically indistinguishable from the global minimum, and that the statistical properties of the configuration found by a minimization algorithm are insensitive to both the algorithm details and the starting configuration.

spanning tree, and randomly generated spanning trees have been used to model branched rivers [72]. The simulated annealing Monte Carlo algorithm can also be run at a constant non-zero temperature (or the temperature can be lowered to a non-zero value). This generates "hot" optimal channel networks [73]. In the high temperature limit, this algorithm will generate a random spanning tree if the system size is finite.

At the end of a minimum energy dissipation simulation, the energy in each of the links is known. This allows the topography of the drainage basin(s) to be constructed [74]. If the energy dissipation in the ith link is given by $\dot{E}_i = Q^\alpha$ then

$$\dot{E}_i = s_i \ell_i Q_i = Q_i^\alpha, \tag{27}$$

and the height difference across the ith link is given by

$$\delta h_i = s_i \ell_i = \dot{E}_i / Q_i = Q_i^{\alpha - 1}. \tag{28}$$

Consequently, the height can be calculated anywhere in the channel network by simply summing the height across all of the links along the path leading to the outlet (the base level or sea level). This leads to a rough surface without any discontinuities across the ridge crests separating adjacent basins. For small values of the exponent α ($\alpha \le 0.25$), the topography is extremely rough. While for large values of α ($\alpha \ge 0.75$), the topography consists of hillslopes with a more or less constant inclination. Values of α in the range $0.25 \le \alpha \le 0.75$ give the most realistic looking landscapes[25].

If rivers are allowed to develop in an extended area, and there is no restriction on where they can exit from this area, then the area can be divided into a large number of basins, each drained by its own river. For a variety of simple models [75], the distribution of basin areas has the form

$$N_A \sim A^{-\tau}, \tag{29}$$

where $N_A \delta A$ is the number of basins with areas in the range $A - \delta A/2$ to $A + \delta A/2$, in the limit $\delta A \to 0$. If the basin area distribution is indeed a power law, the exponent τ is determined by the requirement that the individual river basins must fill the entire area. For a system of size L, the total area $\mathcal{A}$ is proportional to L^2, and the length of the largest basins is proportional to L. In most simple models, the basin width w_b and basin length ℓ_b are related by a simple power law, so that the basin area is proportional to ℓ_b^γ and the area of the largest basins is proportional to L^γ. The space filling requirement can be represented by the equation

$$\mathcal{A} = N_\circ \int_0^\infty A \cdot A^{-\tau} f(A/L^\gamma) \mathrm{d}A \propto L^2, \tag{30}$$

[25] Simulations have been carried out for $0 < \alpha < 1$. If $\alpha < 1$, the river will consist of a single unbranched sinuous channel that connects all of the sites. If $\alpha > 1$, the river will consist of long straight channels with few branches.

where the function $f(x)$ has the form $f(x) = \text{const.}$ for $x \ll 1$, and for $x \gg 1$, $f(x)$ decreases with increasing x faster than any power of x. In this equation, $N_\circ$ is proportional to the length of the perimeter of the total area, and equation (30) can be written in the form

$$\mathcal{A} = L \int_0^\infty A \cdot A^{-\tau} f(A/L^\gamma) \mathrm{d}A \propto L^2. \tag{31}$$

Making the substitution $x = A/L^\gamma$, equation (31) becomes

$$\mathcal{A} = L^{1+\gamma-\tau\gamma+\gamma} \int_0^\infty x^{1-\tau} f(x) \mathrm{d}x \propto L^2. \tag{32}$$

Since the integral has a finite and non-zero value, the exponents τ and γ are related by

$$\tau = 2 - 1/\gamma. \tag{33}$$

If the basin lengths are proportional to their widths, $\gamma = 2$ and $\tau = 3/2$. On the other hand, if the basin lengths l_b and widths w_b are given by $l_\mathrm{b} \sim A^{\nu_\parallel}$ and $w_\mathrm{b} \sim A^{\nu_\perp}$

$$\tau = 2 - \nu_\parallel. \tag{34}$$

It appear that $\nu_\parallel \approx \nu_\perp \approx 1/2$ for real river basins [76]. If the boundary of the entire system is a self-similar fractal, with a fractal dimensionality of D, then the number of rivers is proportional to L^D, and equations (31) and (32) must be replaced by

$$\mathcal{A} = L^D \int_0^\infty A \cdot A^{-\tau} f(A/L^\gamma) \mathrm{d}A \propto L^2. \tag{35}$$

Making the substitution $x = A/L^\gamma$, equation (35) becomes

$$\mathcal{A} = L^{D+\gamma-\tau\gamma+\gamma} \int_0^\infty x^{1-\tau} f(x) \mathrm{d}x \propto L^2. \tag{36}$$

so that

$$\tau = 2 - (2 - D)/\gamma \tag{37}$$

and

$$\tau = 1 + D/2 \tag{38}$$

for $\gamma = 2$. Computer simulations using a minimum energy dissipation model with fractal drainage area boundaries lead to a power law distribution of basin areas in good agreement with equation (38) [77]. Figure 12 shows some results from a minimum energy dissipation rate model for an "island" with a fractal shore-line.

An energy dissipation rate function of the form

$$\dot{E} \propto Q^\alpha \tag{39}$$

with $0 < \alpha < 1$ corresponds to a concave height profile along the channel. Energy dissipation rate optimization models based on these energy dissipation

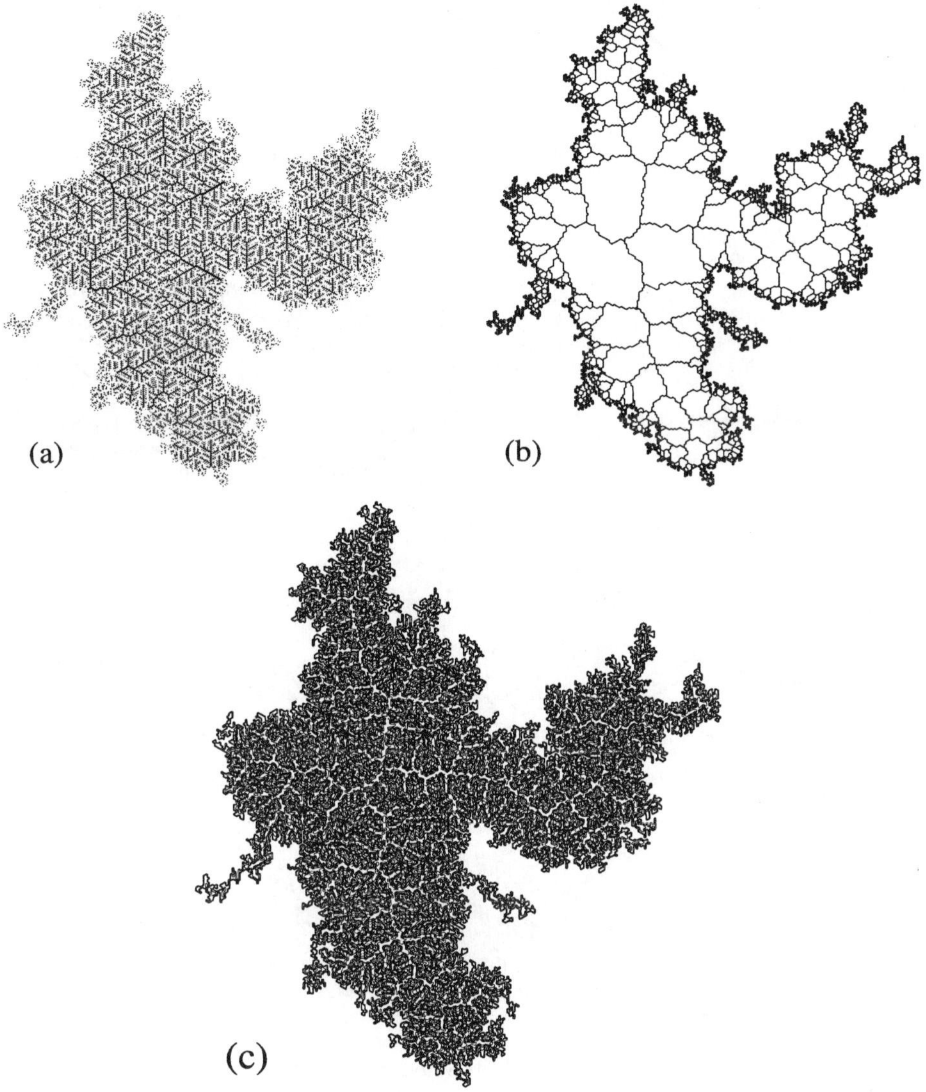

Fig. 12. — Some results from a minimum energy dissipation rate model simulation for an "island" with a fractal shore-line (the external perimeter of a percolation cluster, with a fractal dimensionality of $D = 4/3$). Part a) shows the drainage channels, part b) shows the basin boundaries and part c) shows the contours of the island topography. The distribution of the basin areas is in good agreement with the idea that the distribution is a power law with an exponent given by equation (38).

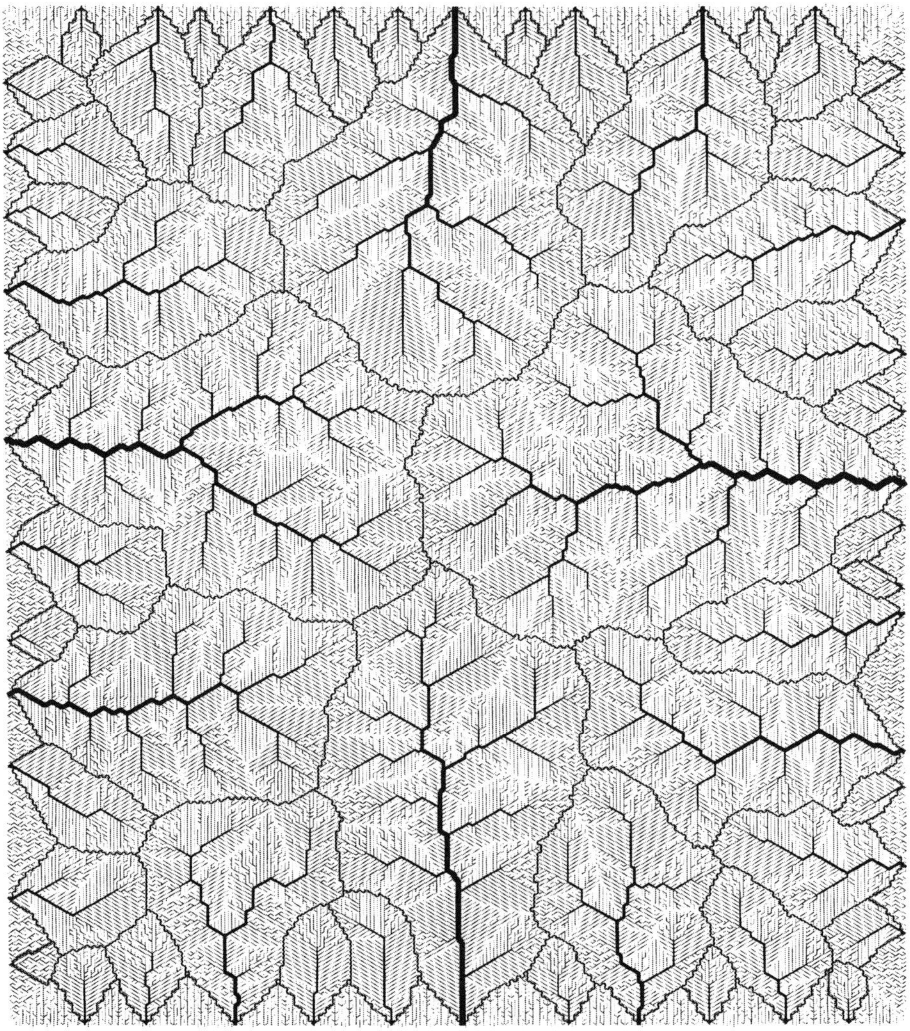

Fig. 13. — A drainage channel pattern obtained from a minimum energy dissipation model that includes distinct hillslope and channel regions. The simulation was carried out using a triangular lattice, and the energy dissipation rate exponents were $\gamma = 1/2$ and $\omega = 1$. The channels are shown in blue, the drainage directions on the hillslopes are indicated in green, and the basin boundaries are shown in red.

rate functions are representative of the channel network, but not the hillslopes. These models can be extended to include both channels and hillslopes by using a more general dissipation function of the form [78]

$$\dot{E} = \begin{array}{ll} \beta Q^{\alpha} & \text{if } Q \geq Q_{\mathrm{c}} \\ s_{\mathrm{h}} Q^{\omega} & \text{if } Q < Q_{\mathrm{c}}, \end{array} \tag{40}$$

where s_{h} can be interpreted to be the inclination in the hillslope regions and Q_{c} is the discharge at which channel initiation occurs[26]. The critical discharge Q_{c} can be used as a model parameter, or it can be calculated from a channel initiation criterion such as $Q^m s_{\mathrm{h}} < a_{\mathrm{c}}$ for hillslope regions, so that $Q_{\mathrm{c}} = (a_{\mathrm{c}}/s_{\mathrm{h}})^{1/m}$. In practice, the time required to fill channels that do not satisfy the channel initiation criterion (due to changes in the climate or topography) may be long compared with the time required to form new channels or extend preexisting channels when the channel initiation criterion is satisfied. Under these conditions, a stationary state may not be reached [76], and optimization models may not be appropriate. Figure 40 shows the channel network obtained from a simulation carried out with $\gamma = 1/2$ and $\omega = 1$. A value of unity for the exponent implies that the slope of the hillslope links is independent of the discharge. This is consistent with the topographies reconstructed from minimum energy dissipation rate channel network models with $0 \leq \gamma \leq 1$. For large values of γ in this range, "V"-shaped valleys are formed. A value for ω larger than unity would correspond to a convex hillslope topography. Similar approaches, based on channel initiation functions, can be used to include both hillslope and channel regions in erosional models and sediment transport models [79].

Sinclair and Ball [80] have studied a pure erosion model that can be described by the equation

$$\partial h/\partial t = Q^m s^n. \tag{41}$$

Starting with a horizontal or inclined surface with random initial height fluctuations and a uniform precipitation, the drainage network evolved towards an invariant form, while the surface heights continued to decrease. Assuming that the height field $h(\mathbf{x}, t)$ can be separated into time dependent and position dependent components

$$h(\mathbf{x}, t) = g(t)\eta(\mathbf{x}). \tag{42}$$

They showed that the function $\eta(\mathbf{x})$ corresponds to minimization of

$$\sum_i Q_i^{(n-m)/n} \text{ or } \sum_i Q_i^{1-m/n} \tag{43}$$

at a particular time t in the stationary asymptotic regime. A visual comparison between river networks generated using the minimum energy dissipation model and the erosion support the idea that the networks are statistically equivalent if the energy dissipation rate exponent α is related to the erosion exponents m and n by $\alpha = 1 - m/n$.

[26] For $Q < Q_{\mathrm{c}}$, there is no channel, and Q is the flow through the region represented by a lattice cell. Channel initiation occurs when the flow density J exceeds $J_{\mathrm{c}} = Q_{\mathrm{c}}/a$, where a is the length of a side of the area represented by a lattice cell.

5. DISCUSSION

Drainage channels form branched patterns that are both beautiful and familiar. These structures appear to have simple scaling properties that can be more or less reproduced by a variety of simple models. However, branched rivers are complex inhomogeneous structures that are subject to a variety of controls and perturbations. The fact that so many simple and unrealistic models generate patterns that are similar to branched rivers is at least in part a consequence of fundamental constraints such as connectivity, continuity of the supporting topology, and conservation of water and sediment. However, natural rivers form a very wide variety of branched patterns that can be easily distinguished by eye. Unfortunately, most simple statistical measures are insensitive to the differences between these patterns. This problem is exacerbated by the difficulties that are encountered when attempts are made to characterize branched rivers in quantitative terms, as well as the poor statistics and small range of length scales associated with most field studies.

Consequently, the "good agreement" that is reported when statistical measures such as the Horton/Strahler ratios and fractal dimensionalities obtained from simple computer models and real rivers are compared have little or no fundamental significance. These models do not contribute much to the development of a better understanding of natural rivers on a detailed level. However, they do demonstrate that many of the most fundamental properties of branched rivers are not sensitive to the details associated with erosion, sediment transport and other processes.

To develop a better understanding of branched rivers it will be necessary to study large river networks in more detail, develop new ways of characterizing branched patterns in quantitative terms and develop more realistic computer models. In the past, the focus on simple models (often models that are too simple) has been justified in part by the limited computer resources that are available. As faster computers become available, it will be possible to develop more detailed models that will still enable large networks (networks covering a large range of length scales) to be generated.

Most computer modeling has been concerned with branched rivers that are dominated by erosion processes. However branched channel are also prominent on fans and fan deltas, which are formed primarily by depositional processes. Branched channels are also formed in tidal regions, where there is little net deposition or erosion.

REFERENCES

[1] P. Meakin, *Rev. Geophys.* **29** (1991) 317-354.

[2] S.W. Fowler and G.A. Knauer, *Progr. Oceanogr.* **16** (1986) 147-194.

[3] A.L. Alldredge and M.W. Silver, *Progr. Oceanogr.* **20** (1988) 41-82.

[4] A.L. Alldredge and C. Gotschalk, *Limnol. Oceanogr.* **33** (1988) 339-351.

[5] D.E. Brownlee, Interplanetary dust: Possible implications for comets and presolar interstellar grains, *Protostars and planets*, edited by T. Gehrels (University of Arizona Press, Tuscon Arizona, 1978) pp. 134-150.
[6] W.R. Newcott, *Nation. Geogr.* **184** (1993) 80-103.
[7] T.A. Witten and L.M. Sander, *Phys. Rev. Lett.* **47** (1981) 1400-1403.
[8] P. Meakin, *Fractals, scaling and growth far from equilibrium* (Cambridge University Press, Cambridge, U.K., 1998).
[9] P. Meakin, *Phys. Scr.* **46** (1992) 69-74.
[10] B.B. Mandelbrot, *The fractal geometry of nature* (W.H. Freeman and Company, New York, 1983).
[11] T. Beer and M. Borgas, *Water Resour. Res.* **29** (1993) 1475-1487.
[12] Tao Sun, *Computer simulations of meandering rivers, braided rivers and fluvial fan-deltas*, Dr. Philos. Thesis (University of Oslo, 1998).
[13] S.A. Schumm, *Ann. Rev. Earth Planet. Sci.* **13** (1985) 5-25.
[14] J.T. Hack, *U.S. Geolog. Survey Prof. Paper* **294b** (1957) 55-97.
[15] A. Robert and A.G. Roy, *Water Resour. Res.* **26** (1990) 839-842.
[16] R.G. Kennedy, *Proc. Inst. Civil Eng.* **119** (1895).
[17] G. Lacey, *Proc. Inst. Civil Eng.* **229** (1929-1930) 259-384.
[18] L.B. Leopold, *Am. J. Sci.* **251** (1953) 606-624.
[19] A.D. Howard, *Water Resour. Res.* **7** (1971) 863-873.
[20] D.G. Tarboton, R. Bras and I. Rodriguez-Iturbe, *Geomorphology* (1992) 59-76.
[21] T. Blench, *Regime behavior of canals and rivers* (Butterworth Scientific Publications, London, 1957).
[22] H. Takayasu and H. Inaoka, *Phys. Rev. Lett.* **68** (1996) 966-969.
[23] L.E. Band, *Water Resour. Res.* **1** (1986) 15-24.
[24] L.W. Martz and J. Barbrecht, *Comp. Geosci.* **18** (1992) 747-761.
[25] R.L. Shreve, *J. Geology* **74** (1966) 17-37.
[26] R.E. Horton, *Bull. Geol. Soc. Am.* **56** (1945) 275-370.
[27] A.N. Strahler, *Bull. Geol. Soc. Am.* **63** (1952) 1117-1142.
[28] J.S. Smart, *Water Resour. Res.* **8** (1972) 1487-1496.
[29] J. Vannimenus and X.G. Viennot, *J. Stat. Phys.* **54** (1989) 1529.
[30] E.L. Hinrichsen, *Structure and geometry of two dimensional random systems*, Ph.D. Thesis (University of Oslo, 1989).
[31] E.R. Weibel, *Morphometry of the human lung* (Springer-Verlag, Berlin, 1963).
[32] P. La Barbera and R. Rosso, *Water Resour. Res.* **25** (1989) 735-741.
[33] D.G. Tabboton, R.L. Bras and I. Rodriguez-Iturbe, *Water Resour. Res.* **26** (1990) 2243-2244.
[34] S.P. Breyer and R.S. Snow, *Geomorphol.* **5** (1992) 143-157.
[35] L.F. Richardson, The problem of contiguity: An appendix to statistics of deadly quarrels, edited by A. Rapoport, L. von Bertalanfly and R.L. Meier, *General systems: yearbook of the Society for General Systems* **6** (The Society for General Systems, Ann Arbor, Michigan, 1961) pp. 139-187.

[36] A.A. Beauvais and D.R. Montgomery, *Geology* **25** (1997) 1063-1066.

[37] M. Eden, A probabilistic model for morphogenesis, edited by H.P. Yockey, R.L. Platzman and H. Quastler, *Symposium on information theory in biology* (Pergamon, New York, 1958. Gatlinburg, Tennessee, October 29–31, 1956) pp. 359-370.

[38] M. Eden., A two-dimensional growth process, edited by J. Neyman, *4th Berkeley symposium on mathematics, statistics and probability* (The University of California Press, Berkeley, 1961) pp. 223-239. Volume IV, Biology and the problems of health.

[39] R.S. Snow, *Pure Appl. Geophys.* **131** (1989) 99-109.

[40] M.J. Kirby, A two-dimensional simulation model for slope and stream evolution, edited by A.D. Abrahams, *Hillslope processes* (Boston: Allen & Unwin, 1985) pp. 203-222. The 16th Bingham symposium on geomorphology.

[41] P. Meakin, *Phys. Rev. A* **27** (1983) 1495-1507.

[42] R.C. Ball and R.M. Brady, *J. Phys. A* **18** (1985) L809-L813.

[43] L. Niemeyer, L. Pietronero and H.J. Wiesmann, *Phys. Rev. Lett.* **47** (1984) 1033-1036.

[44] T.A. Witten and L.M. Sander, *Phys. Rev. B* **27** (1983) 5686-5697.

[45] V. Fleury, *Arbres de Pierre* (Flammarion, Paris, 1998).

[46] V.M. Cabrera, *J. Electrostat.* **30** (1993) 47-56.

[47] G. Daccord, *Phys. Rev. Lett.* **58** (1987) 479-482.

[48] G. Daccord and R. Lenormand, *Nat* **325** (1987) 41-43.

[49] P. Meakin, *Phys. Rev. A* **27** (1983) 2616-2623.

[50] C.A. Knight, Slush on lakes, *Structure and dynamics of partially solidified systems*, edited by D.E. Loper (Martinus Nijhoff, Dordrecht, 1987) pp. 455-465. NATO ASI Series E125.

[51] A.J. Gow and J.W. Govoni, Ice growth on Post Pond, 1973-1982, Technical report, United States Army Cold Regions Research and Engineering Laboratory (Hanover, New Hampshire, 1983). CREEL Report 83-4, 25pp.

[52] A. Tokairin, *Lake Ice* (Kodansha, Tokyo, 1977).

[53] C.G. Higgins, Piping and sapping: Development of landforms by groundwater outflow, edited by R.C. La Fleur, *Groundwater as a geomorphic agent* (Allen and Unwin, Boston, 1984) pp. 18-58.

[54] T. Dunne, *Progr. Phys. Geol.* **4** (1980) 211-239.

[55] J.A.A. Jones, *Brit. Geomorpholog. Res. Group, Res. Monogr.* **3** (1980) 602-610.

[56] P. Cerasi, P. Mills and S. Fautrat, *Europhys. Lett.* **29** (1995) 215-220.

[57] A.D. Fowler, H.E. Stanley and G. Daccord, *Nat* **341** (1989) 134-138.

[58] J.G. Masek and D.L. Turcotte, *Earth Plan. Sci. Lett.* **119** (1993) 379-386.

[59] S.A. Schumm, *Bull. Geol. Soc. Am.* **67** (1956) 597-646.

[60] W.M. Davis, *Sci* **20** (1892).

[61] G.K. Gilbert, *Geology of the Henry Mountains* (U.S. Geographic and Geological Survey, Wahington D.C., 1877).

[62] G.K. Gilbert, *J. Geology* **17** (1909) 344-350.
[63] R.L. Leheny and S.R. Nagel, *Phys. Rev. Lett.* **71** (1993) 1470-1473.
[64] R.L. Leheny, *Phys. Rev. E* **52** (1995) 5610-5620.
[65] J. Wejchert, *Europhys. Lett.* **9** (1989) 503-508.
[66] L.B. Leopold and W.B. Langbein, *United States Geolog. Survey Prof. Paper* **500-A** (1962) 1-20.
[67] I. Prigogine, *Introduction to thermodynamics of irreversible processes*, edited by C.C. Thomas (Springfield, Illinois, 1955).
[68] A.D. Howard, *Water Resour. Res.* **26** (1990) 2107-2117.
[69] I. Rodriguez-Iturbe, A. Rinaldo, R. Rigon, R.L. Bras, E. Ijjasz-Vasquez and A. Marini, *Geophys. Res. Lett.* **19** (1992) 889-892.
[70] I. Rodriguez-Iturbe, A. Rinaldo, R. Rigon, R.L. Bras, E. Ijjasz-Vasquez, A. Marini and E. Ijjasz-Vasquez, *Water Resour. Res.* **28** (1992) 1095-1103.
[71] A. Rinaldo, I. Rodriguez-Iturbe, R. Rigon, E. Ijjasz-Vasquez and R.L. Bras, *Phys. Rev. Lett.* **70** (1993) 822-825.
[72] S.S. Manna and B. Subramanian, *Phys. Rev. Lett.* **76** (1996) 3460-3463.
[73] A. Rinaldo, A. Maritan, F. Colaioro, A. Flammini, R. Rigon, I. Rodriguez-Iturbe, and J. Banavar, *Water Resour. Res.* **76** (1992) 3364-3367.
[74] T. Sun, P. Meakin and T. Jøssang, *Water Resour. Res.* **30** (1994) 2599-2610.
[75] P. Meakin, J. Feder and T. Jøssang, *Physica A* **176** (1991) 400-429.
[76] D.R. Montgomery and W. Dietrich, *Sci* **255** (1991) 826-830.
[77] T. Sun, P. Meakin and T. Jøssang, *Phys. Rev. E* **51** (1995) 5353-5359.
[78] T. Sun, P. Meakin and T. Jøssang, *A minimum energy dissipation model for drainage basins that explicitly differentiates between channel networks and hillslopes.*
[79] G. Willgoose, R. Bras and I. Rodriguez-Iturbe, *Water Resour. Res.* **27** (1991) 1671-1684.
[80] K. Sinclair and R. Ball, *Phys. Rev. Lett.* **76** (1996) 3360-3363.

CHAPTER V

Neuronal Arborization

J.-P. Ternaux

1. INTRODUCTION

Neural and glial cells are the fundamental cellular components of brain tissue. When compared to cells which form the majority of the body tissues, neural cells appear as characterized by their shapes, which exhibit a large variety of branched geometries (Fig. 1). Such a branched architecture is observed in the whole nervous tissue, but shape specificities can be noticed from one brain region to another. Nevertheless, among this diversity, similar geometries are found in the same region of different brains of the same species or in the same region of various brains selected, as example, in the vertebrate phylogenetic scale. These observations suggest that neural branched geometry is certainly related, in part, to the expression of genetic factors which are preserved during phylogenesis.

In adult brain, neural cell morphology is characterized by spherical or stellar cell bodies, with diameters ranging from 5 to 50 μm, giving rise to long cytoplasmic branched processes (from several μm up to several mm): the axon and the dendritic tree. The dendritic arborization, which represents about 90% of the membrane surface of the neuron, receives afferent connections from other neural cells (essentially axon terminals) at the level of specialized microscopic structures named synapses. Synapses are distributed all along the dendritic branches in varying densities. At presynaptic level, the electrical activity of the afferent neuron (the action potential), which is conveyed from the cell body to the axon terminal, induces the release of specific molecules which can act at the postsynaptic membrane, *i.e.* on proximal or distal dendrites. Neurotransmitters, which are released at the synapse, bind to specific receptors, at postsynaptic membrane, inducing, by various mechanisms, changes in the ionic membrane permeability (opening and closing of ion channels). They may consequently

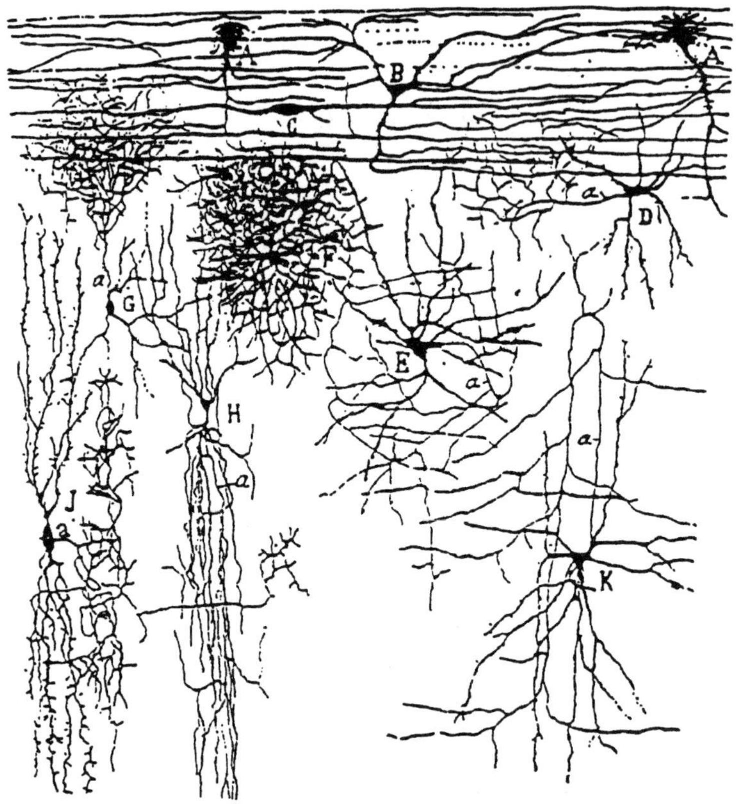

Fig. 1. — Various types of neuronal shapes in the motor cortex of human aged of one month. Staining of neuronal cells is obtained using the Golgi's method. A, B, C: horizontal cells of the plexiform layer. D, E, F: neurons with short axons. G: neuron with ascendant axon towards the superficial layer. H, I: cells with a double dendritic arbor. K; big neurons with short axons from the third layer. After Ramón y Cajal [1].

generate potentials in the dendritic branches. Classically, such synaptic currents are conveyed passively to the cell body. When the threshold of depolarization is reached at somatic level, an action potential is generated at the initial segment and propagated in the axon, towards a target cell, *via* a new synapse. These mechanisms (although schematically described) are involved in the processing of neural information.

According to these data, which were accumulated essentially during the second half of the last century, it can be assumed that the processing and transmission of neural information in the neuron are linked, on one hand, to the efficiency of the synapses which represent the key points of the neural networks and, on the other hand, to the morphological properties of the dendritic arborization. As the propagation of electrical events in the dendrites is essentially passive, location of the synapses on dendritic branches, diameter and length of the dendrites, number and asymmetry of the bifurcations are relevant parameters which must be considered in order to understand how neural information is processed at the cellular level.

The relationship between structure and function of the neural (branched) structure can be studied experimentally by means of the various methods which were developed in the field of neurobiology. Questions asked about the involvement of the neuronal architecture in the processing of neural information can also be answered using biophysical approaches, combining modelisation and simulation. In the present chapter, we will successively examine the biological processes which are involved in the acquisition of neural cell branched morphology during brain development and the involvement of dendritic geometry arborization in the neural information processing.

1.1. Historical Orientation

The first description of neuronal cells as branched structures and independent units in cerebral tissue was achieved at the end of the nineteenth century. At that date, it was not possible to determine, with the available methods, whether the branches emerging from the nerve cells formed a continuous net or whether they retained some independence. The idea that the nervous system was made of anatomically distinct units began with improved microscopy, resulting largely from the achromatic objective developed in 1826 and from silver nitrate staining which was put forward by Camillo Golgi [2]. Surprisingly, the procedure used by Golgi did not stain all nerve cells but occasional cells that stain black against a yellow background. Under these conditions, cell body, axon and dendrites were observed, but Golgi, in his initial report, assigned a nutritive role to the dendrites and firmly believed that the axons formed a dense fused reticulum. That affirmation was broken down in 1886 when Golgi concluded: "... transmission of a stimulus without direct continuity is possible". This idea of contiguity, named "cell theory" against continuity promoted in the previous "anastomosis theory" or "reticular theory" was reinforced by the work of Santiago Ramón y Cajal (1852-1934) who showed, using a slightly modified Golgi staining method, that nerve fibers are very intricate in the brain tissue but that no evidence could be found for either axons or dendrites undergoing anastomosis and forming nets. Following many observations in fixed brain tissue from various regions and from various vertebrate species, Cajal argued that nerve cells, *i.e.* neurons, were independent elements:

"Des individualités en nombre immense, les neurones, complètement indépendants, simplement au contact les uns avec les autres, constituent, nous

l'avons démontré, le système nerveux. Mais ces neurones, à cause de la diversité de leurs positions et de leurs rapports, sont traversés par des courants de nature et de qualité différentes; leur variété de forme influe aussi sur l'intensité, la direction et le mode de distribution de l'onde nerveuse. Et cependant, malgré cette diversité, les neurones possèdent, à coup sûr, une physiologie qui leur est commune, et cela, en raison de leur composition anatomique semblable et de leurs rapports analogues..." [3].

The viability of the neuron doctrine was also defended by Waldeyer (1836-1921) who claimed that the neuron is the anatomical and physiological unit of the nervous system. In this context, Cajal suggested that the diversity of shapes observed in dendritic arborizations of neurons, from different regions of the brain, was certainly in close relationship with the intensity and direction of the currents which are conveyed in the branches. Consequently, as early as the start of the twentieth century, both the idea of polarity in nerve cell and the relationship between geometry and information processing in neuronal branches, were suggested. Cajal also showed how axons grow from the growth cone found at the tip of neuritic branches of embryonic neurons, thus establishing more firmly the veracity of the neuron theory. But at that date, a fundamental question had no answer: if neurons do not fuse to one another, how do they communicate? Do they just touch each other or is it something else? The solution to this problem came largely from the work of Sir Charles Scott Sherrington (1857-1952) who recognized the gap between neurons and muscular cells effectors and between one neuron and another. In 1897, Sherrington called this "junction": synapse [4]. Although Sherrington developed the concept of synapse, he was not the first to propose chemical transmission as an explanation of neuron communication. Such a chemical transmission of the neuron messages was previously suggested by Dubois-Raymond in 1870 [5]. Nevertheless, the demonstration of electrically evoked release of substances from nerve endings, the neurotransmitters, was achieved during the mid twentieth century [6]. Later on, synapses and contiguity were clearly identified and found support in electron microscopy pictures of brain tissue. Electron microscopy observations of synapses made with brain tissue fixed in osmium tetroxide, embedded in methacrylate and sectioned with glass knives, showed the presynaptic and postsynaptic membranes separated by a synaptic cleft about 20 nm wide [7,8]. Synaptic vesicles which characterized the presynaptic endings [8,9] were immediately recognized as possible storage sites for chemical neurotransmitters [10]. The crucial physiological evidence proving the existence of chemical synaptic transmission was obtained by means of intracellular microelectrode recordings, using glass pipettes at neuromuscular junctions [11,12], in the spinal cord of vertebrates [13] and in the abdominal ganglion of aplysis [14]. Intracellular microelectrode recordings showed that the excitatory postsynaptic potential is thousands of times greater than can be accounted for by electrotonic transmission from neuron to neuron, thus making it certain that the observed amplification is mediated by chemical synaptic transmission.

2. BIOLOGICAL EVENTS INVOLVED IN THE GROWTH OF NEURONAL ARBORIZATIONS

According to anatomical data, the adult brain tissue is composed of branched neuronal cells, placed in contiguity, which form neuronal networks. Neurons are surrounded by glial cells which assume respiratory, nutritive and catabolic cellular functions for the neurons, buffering of the extracellular fluid as well as providing mechanical support. The adult neuron, in terms of its cellular architecture which can be also named "macrostructure", results from growth mechanisms which take place during embryonic and post-natal periods. At the molecular scale, the macrostructure is dependent on the expression and distribution of various molecules which constitute the "microstructure". During growth, maturation and differentiation of nerve cells, both the macrostructure and microstructure are closely dependent on genetic and epigenetic factors.

2.1. Neuronal Morphogenesis

In all living species, there are specific genes which are responsible for the existence of organized cellular interactions and for the maturation of cellular and organ shapes. The discovery of these genes results essentially from experimental work done in a small fly: the drosophila. For example, with the same chemical components, three types of cells such as fibroblasts, myoblasts and epithelial cells can be combined to build a limb or an antenna. Even if the cells are identical, the organ formed is different. The design of these organs is under the control of genetic elements: the homeotic genes which contribute to the organogenesis. The shape of a cell and the shape of an organ are questions which can be either separated or either related: relations and interactions between the shape of the cellular components and the shape of an organ which is built with these components, clearly occur during development.

The shape of a cell results from interactions between its intrinsic properties and the structure of its external environment. The three-dimensional shape of a cell and the involvement of the intracellular components which are implied in its expression, is far from being decrypted. To day, in a first step, membrane molecules which could be involved in the dynamic properties of living cells have been classified. The cytoskeleton, as we will see later, is certainly one of the main support of cell geometry and the main actor involved in morphogenic processes.

The shape of cells in living tissue is not only a morphological concept, but is also related to functional and physiological rules. Molecules, enzymes and water molecules are not randomly distributed among the cytoskeleton matrix but are specifically addressed in the cell, giving morphological asymmetries which are able to determine morphological properties and specific functional mechanisms.

The neuron, as other cell, exhibits these properties too, but with much more accuracy. The neuron is a polarized cell able to communicate with its neighboring elements *via* specialized structures: the synapses. When looking at brain tissue, neurons seem to be morphologically similar to each other with cell bodies and long branched cytoplasmic processes. Nevertheless, when observed in detail, it is clear that neuronal shapes vary from one region to another, showing, for example, that the size of the cell bodies, the length and the diameter of processes and the number of bifurcations are variable and can be associated in a large scale of combinations. Nevertheless, the observations of a neuronal type of the same region in different brains of the same species or in species distributed all along the phylogenetic scale, clearly demonstrate that neurons located in a specific area always possess the same morphological features.

According to this assessment, neuronal morphogenesis is not only due to the expression of a genetic pre-programme, encoded in the neuroblasts, its precursor cells, but results, in part, from the influences of its external field. The epigenetic factors are structurant for neuronal shape.

Epigenetic factors which are involved in the control of neuronal shape during development can be classified in three categories: mechanical factors, electrical factors and biochemical factors. Each of them were demonstrated to act on neuronal cytoskeleton structure and consequently on the shape of the branched nerve cell, during ontogenesis.

2.2. Cytoskeleton and Neuronal Development

The network of proteins, microfilaments and microtubules in any cell and esp. in the neuron is termed the cytoskeleton. The structure of the cytoskeleton includes actin filaments [15], intermediate filaments, mainly neurofilaments [16] and microtubules [17] distributed in longitudinal arrays and bundles in the cytoplasmic branched processes of the nerve cell. Polymerized cytoskeletal proteins are in equilibrium with cytoplasmic pools of unpolymerized subunits [18]. The cytoskeletal proteins are linked together by a large variety of associated proteins [19] which dynamically interact with the cytoskeleton, inducing changes in cell shape, but also in cell movements, exocytosis and endocytosis of materials such as synaptic vesicles, movements of organelles and intracellular anterograde or retrograde transports of various molecules.

The cytoplasmic membrane with its bilayer of lipids and its constitutive proteins and the cytoskeletal proteins, which are all synthesized at cell body level and transported in both axon and dendrites, are the two main components involved on one hand in the shape maintenance and on the other hand in morphological plasticity. Growth of neuritic processes from neuroblastic precursor cells as well as changes in the geometry of adult neurons are closely related to the modifications occurring at cytoskeletal level. Such modifications are, in part, dependent on the intrinsic mechanical properties of the microtubules and filaments, bringing limits to the adaptability of the cytoskeleton, and consequently to the possibilities of geometrical changes, particularly in adult brains.

a. Microtubules

Microtubules are very abundant in both neurons and glial cells. They have two main functions. They provide mechanical stability to the cell and the transport of various proteins and organelles along them. Microtubules are found in cilia and mitotic spindles and originate from centrosomes [20]. Microtubules are built up of tubulin dimers and form a polymer of indefinite length, with an external diameter of 24 μm, a wall 5 nm thick and a hollow core 14 nm in diameter. Cytoplasmic microtubules, typically have 13 protofilaments coiled helically. The tubulin dimer is a 100 KDa complex of two nearly identical proteins called α and β tubulin. In a majority of cell types, the dimers are oriented in the same direction in the microtubule, giving it an intrinsic structural polarity, with plus and minus ends where polymerization and depolymerization of tubulin occur, respectively [21]. In the neuron, bundles of microtubules are organized differently in the axon and dendrites [22]. Axonal microtubules are parallel with their plus ends oriented towards terminals whereas they exhibit antiparallel organization in the dendrites. Microtubules are in condition of dynamic instability, undergoing phases of elongation and rapid shortening, particularly during growth of neuritic processes. They are stabilized by an energy-dependent mechanism involving hydrolysis of GTP and by interactions with Microtubules Associated Proteins (MAPs). Equilibrium between the tubulin cytoplasmic free pool and microtubules is affected by many factors, especially the intracellular calcium concentration in the presence of calmodulin. At neurite tip and growth cones, microtubules are oriented in the same way, with the plus end pointing in the direction of growth. Distribution of microtubules in axon and dendrites is one of the most important morphological support of the neuronal polarity (see reviews [23, 24]).

b. Microtubules Associated Proteins (MAPs)

A significant difference between cells that have long processes, such as neurons and astrocytes, and those that do not have them is that the former have MAPs whereas the latter lack MAPs. MAPs are polypeptides that are associated with microtubules in neurons and glial cells where they form molecular crossed bridges, playing an important role in the assembly and stabilization of microtubules. Long MAPs [25] and smaller Tau proteins [26] which stabilize microtubules in the cytoplasmic processes are essential during development and for the acquisition of the adult neuronal architecture.

Antibodies raised against the different forms of MAPs have shown that they are intracellularly compartimentalized and developmentally regulated. MAPS also bind to neurofilaments and actin filaments. Binding to actin is regulated by phosphorylation of MAPs which, in the same time, reduces assembly of microtubules. MAP 1B is abundant in the newborn rat brain and then decrease in function of age whereas MAP 1A increase during post natal period. MAP 1B is abundant in the growing axons and MAP 1A is characterized in growing and mature dendrites. They are also both expressed in glial cells. MAP 2 appears to be a component of cross-bridge interactions between microtubules and neurofilaments. A high molecular weight form (280 KiloDalton (KDa)) expressed in developing dendrite and a low molecular weight form (70 KDa) expressed exclusively in adult neurons, are encoded by the same gene. Recently a MAP 2D form [27] was characterized in the dendrites of developing and adult rat hippocampal neurones. MAP 3 (180 KDa) as well as MAP 4 (210 KDa) occurs in astrocytes but not in neuron. MAP 5 was exclusively found in neurons in cell bodies, axons and dendrites and was ten times more abundant in the newborn than in the adult.

Tau is a protein which is found only in neurons and is mainly localized in the axon, in association with the microtubules. Tau promotes the polymerization of tubulin. Tau proteins exist in four forms which result from post translational modifications. During development of the axon an exchange between the juvenile Tau protein and the adult form is observed.

c. Microfilaments

Actin microfilaments, 6 to 8 nm in diameter, are ubiquitous cytoskeletal components of neuroepithelial cells, glial cells and neurons. They function with other cytoskeletal elements to determine the shape and the cell movements. During development, actin microfilaments are particularly involved in the growth cone motion an orientation, as well as in the initiation of branching processes [28].

Globular actin (Actin G) subunits form a cytoplasmic pool which is in dynamic equilibrium with actin polymers, which form double helical filaments. These filaments are organized into bundles and networks by means of cross-linking proteins, the most important of which is filamin. The resulting protein complex has the viscoelastic properties of a gel. Such properties are illustrated by the ability of microfilaments to recoil rapidly in response to a large transiently applied force and to deform gradually in response to a small but steadily applied force. These phenomenon can be observed at neuritic tips and growth cones during growth and branching.

d. Intermediate Filaments

Intermediate filaments have a diameter of 8 to 10 nm and are composed of protein subunits (40 to 200 KDa). They are encoded by a large multigene family with specific patterns of expression in different tissues: neurofilaments (neurons), glial filaments (astrocytes), vimentin (cells of epidermal origin, immature astrocytes), desmin (muscle) and keratin (epithelial cells). With exception of keratins, all other types of intermediate filaments have substantial similarities in their amino-acids sequences.

Neurofilaments are intermediate filaments which are only found in the neurons. They are distributed in cell bodies, axons and dendrites in longitudinal arrays of single neurofilaments or of neurofilaments bundles linked by cross-bridges. Proteins which form the neurofilaments can first be detected in post-mitotic neurons and are absent in dividing neuroepithelial cells, in which vimentin is the main intermediate filament protein [29].

2.3. Transport of Materials in the Neuron

During growth, synaptogenesis and functioning of the adult neuron, the neuronal cell must produce very large quantities of proteins. According to the architecture of the neuron, substances which are synthesized in the cell bodies must be transported to their specific cytoplasmic and membrane sites which can be far from the producing machinery. Growth of the axons and dendrites, as well as branching are dependent on both the synthesis and transport of various molecules originating from the cell bodies. According to the volume of the axon and dendritic tree (which is thousands of times greater than the cell body), a permanent synthesis of materials must be maintained at the cell body level. These dependences of the axon and of the dendrites on the cell body starts with the initial outgrowth of neurites from the neuroblasts and continues throughout life.

Even at the fastest rate of anterograde axonal transport (50 to 400 mm/day) materials produced in the cell body may need several weeks to reach the presynaptic terminals. In the same manner, materials which are transported retrogradely may journey for days or weeks from peripheral nerve terminals to the cell body. These delays set severe limits on the participation of the cell body in the regulation of growth cone activities during elongation and branching processing in axons and dendrites which possess specific morphological postsynaptic structures such as spines.

In the dendrites, transport of materials also occurs in both directions, but the distances are generally shorter, and so are the delays. Local synthesis of various substances was demonstrated in dendritic spines where the synthesizing machinery is close to the synaptic cleft, supporting rapid responses to synaptic activation.

During ontogenetic and post-natal maturation and differentiation of the nerve cell, transport of materials in the elongating processes is a way to provide new materials to the membrane and to renew various components whose lifespan is short. Such a material is also necessary to the formation, maintenance and functioning of synapses. Finally, transport is also involved in the conveying of trophic molecules from the extracellular environment at the tips of the axons and from the axons postsynaptic targets.

Axonal transport was shown by various methods including light microscopic cinematography, enhanced video differential interference microscopy, accumulation of materials following ligatures, transport of various markers such as viruses, horseradish peroxidase, tetanus toxin, Nerve Growth Factor (NGF)..., radiolabeling with isotopic precursors such as amino acids. Using these last technique, two main rates of axonal transport were defined: slow and fast

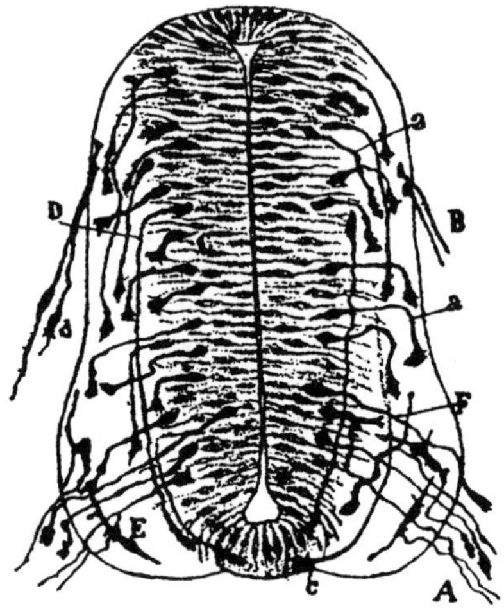

Fig. 2. — Growth cone in the spinal cord of a three day old chick embryo. A: ventral spinal nerve fibers; a: axon; B: dorsal root fibers; C: Growth cone on a very short axon; c: growth cone on a commissural fiber; d: dorsal root ganglion cell; E: ventral horn motor nerve cell; F: fiber of the lateral motor nerve cell. From Ramón y Cajal [30].

axonal transports. Slow axonal transport includes all the proteins of the cytoskeleton and associated proteins of the axoplasmic matrix. Plasma membrane proteins such as sodium and potassium ATPases, acetylcholinesterase, components of the mitochondries, membrane organelles as well as synaptic vesicles are transported with the fast axonal flux. Fast transport which is characterized by rates in between 50 to 400 mm/day occurs in both directions in the axon. It conveys materials from cell body to axon terminal and, retrogradely, materials entering axon terminals by endocytosis. Slow axonal transport is exclusively anterograde and exhibit two main peaks of velocities which are termed a and b components, with velocities ranging from 0.2 to 1 mm/day and 1 to 2 mm/day, respectively. Quantitatively, the flux of materials transported in the slow components is about five times higher than that transported in the fast components (see review [31]).

2.4. The Growth Cone

During development, elongation of neurites occurs exclusively at their growth cones which are the actively growing tips of all branches of axons and dendrites (Figs. 2 and 3). Growth of plasma membrane is achieved by insertion of membrane components at the growth cone. Growth of the axonal and dendritic cytoskeleton also occurs by assembly in the growth cone. These components, as seen previously, are delivered from the cell body to the growth cones by

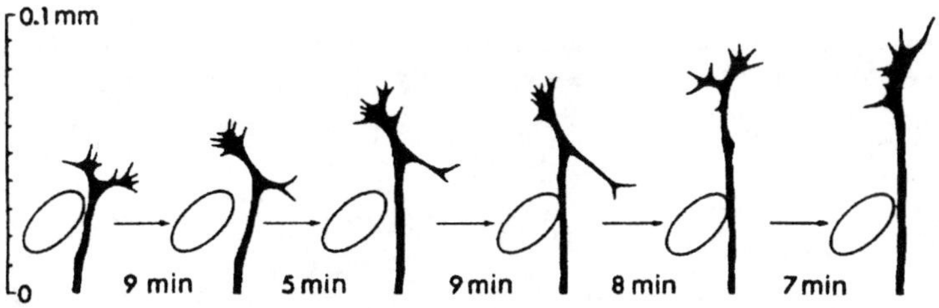

Fig. 3. — Successive views of a neuronal growth cone of a frog ectodermal sensitive neuron. The oblate corpuscule outside of the growth cone is fixed. The average rate of elongation of the neurite is about 1 μm per minute. The total length of the nerve fiber is 800 μm. After Harrison [34].

axonal transport. Nevertheless, synthesis of various materials also takes place at growth cone level. In fact, delivery of cytoskeleton and associated proteins by slow axonal transport sets a limit to the rate of axonal elongation and regeneration (Fig. 4). Growth cone movements, produced by actin filaments, determine the direction of growth of the axons and probably also of dendrites. The axonal growth cone functions are ultimately under control of the cell body which is the source of all the components of the growth cone except those obtained by local synthesis and endocytosis from the environment (Fig. 5).

Therefore, the activity of the growth cones is regulated by local conditions. As we will see below, the direction of elongation of the growth cone can be influenced by mechanical guidance by the substratum, extracellular matrix molecules, diffusible extracellular growth factors and by electrical fields. During development, growth cones are structurally and functionally different from the rest of the neuron. Most of the interactions of the neuron with its environment take place in this region. When the axon and the dendrites elongate, the cell body and the proximal segments of the axon and dendritic tree are attached and stationary whereas only the growth cone moves. The external morphologies of growth cones vary from a relatively simple pointed paintbrush ending with a single thin finger-like extension called fillopodium, to an elaborate fan-like expansion bearing a profusion of filopodia and broad and flat processes named lamellipodia. Ultrastructural organization of the growth cone, and particularly of its cytoskeleton, is correlated with its functional activities. The latter are involved in the growth and elongation of the neurites, motility and active exploration of the environment, uptake of materials for local synthesis, secretion of enzymes, neurotransmitters and trophic substances (Fig. 6). Such mechanisms require energy which is provided by mitochondries, located in the central part of the growth cone. The morphology of the growth cone can be rapidly modified under the influence of both intrinsic and extrinsic factors. The type of neuron, the stage of development as well as the contact with various molecules present on the substratum and in the environment will promote or

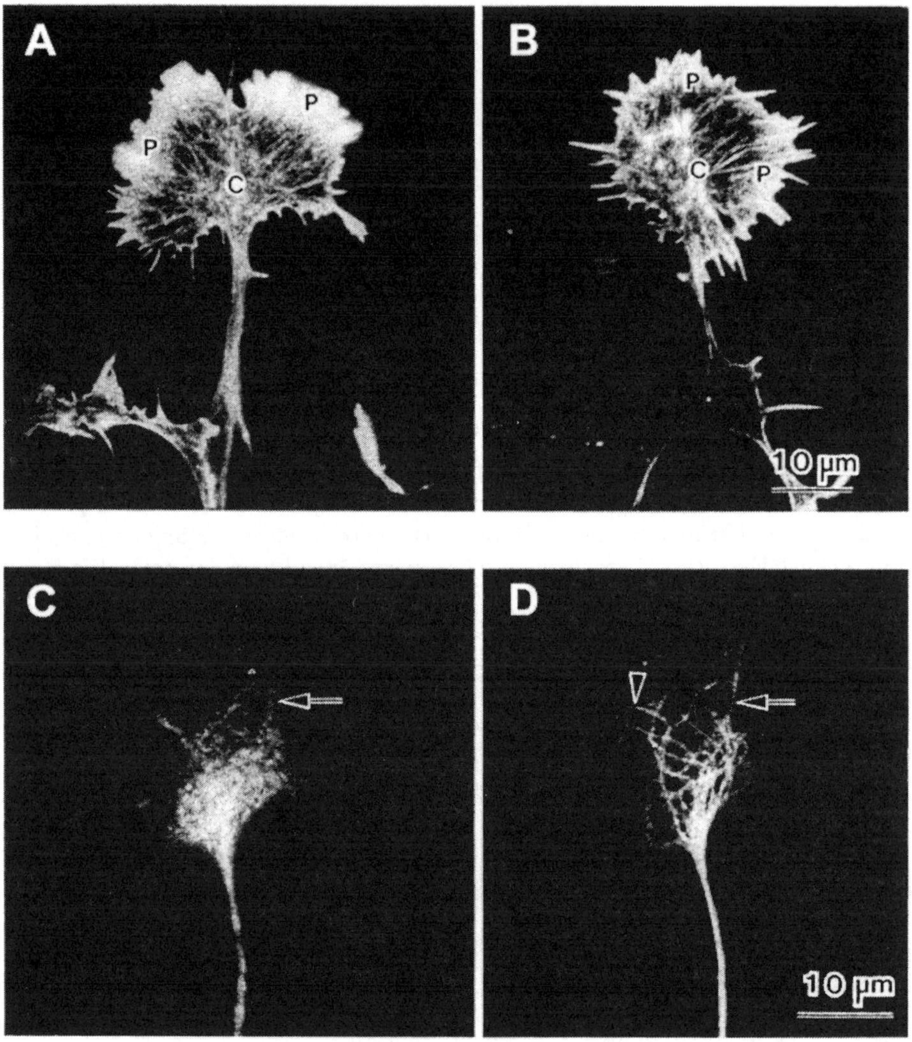

Fig. 4. — Distribution of growth cone actin filaments and microtubules. A and B) fluorescent antibody raised against actin; the relatively thin peripheral domain (P) exhibits high concentration of actin compared to the central domain (C). Lower part, A) fluorescent image of a growth cone stained by a carbocyanin DIO C6 and B with an antibody raised against tubulin. The carbocyanin stains the microtubules and sometimes their distal segments (arrow). Fluorescence corresponding to tubulin is only detected in the central domain. From Bridgman and Dailey [35] and Dailey and Bridgman [36].

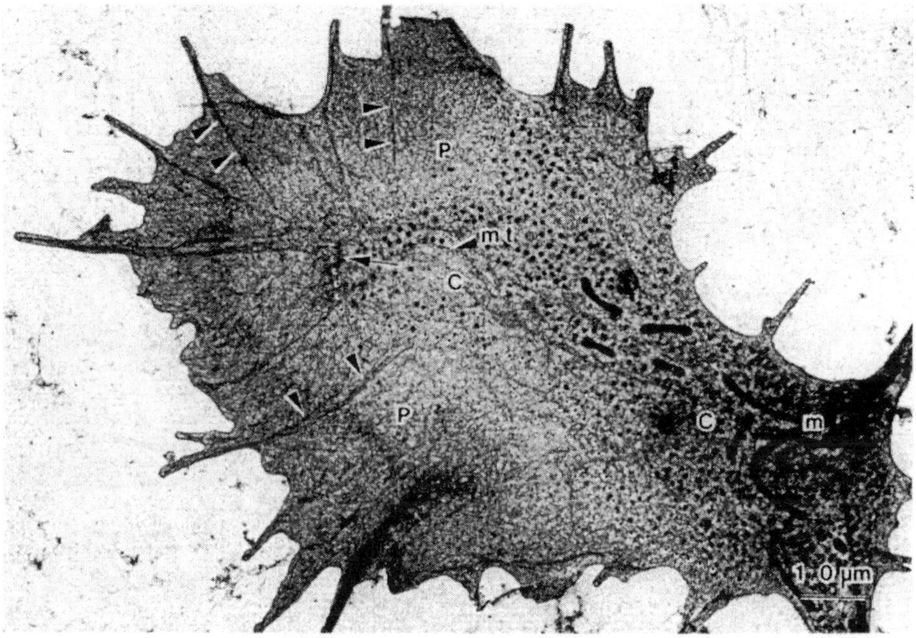

Fig. 5. — Electron microscopic image of a growth cone. C: central domain; mt: microtubules; m: mitochondries. The central domain also contains numerous dense core vesicles. P: peripheral domain with bundles of microfilaments directed in the filopodia (arrowheads). From Bridgman and Dailey (1989) [35].

reduce the neuritic elongation. Trophic factors such as Nerve Growth Factor (NGF), neurotrophins such as Brain Derived Neurotrophic Factor (BDNF), Neurotrophin 3 et 4/5 (NT 3, NT4/5), Ciliary Neurotrophic Factor (CNTF), Fibroblastic Growth Factor (FGF), Insulin Growth Factor (IGF I et IGF II), Glial Derived Neurotrophic Factor (GDND)..., but also molecules of the extracellular matrix such as laminin, fibronectin, or adhesive molecule such as Neural Cell Adhesive Molecule (NCAM) were shown to influence the neuritic elongation (Fig. 7).

Whatever the substance involved, growth cone morphology and motility changes are related with changes in the extracellular concentration of calcium ions. These variations induce intracellular modifications which regulate the microtubule assembly and thus influence neuritic elongation. The first mechanism involved is *via* calcium dependent kinases that can phosphorylate cytoskeletal proteins. The second one is through a direct effect of calcium on tubulin polymerization. The third is *via* the calcium binding protein calmodulin which interacts with Tau and MAP2 proteins which stabilize microtubules in the axons by forming cross-bridges between them (Fig. 8). In addition, rapid changes

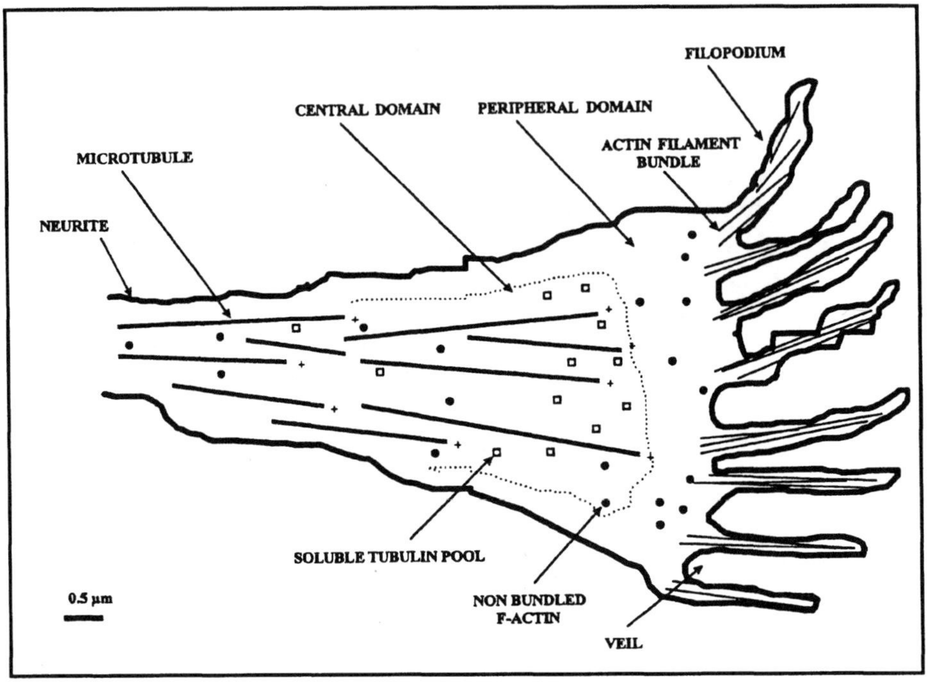

Fig. 6. — Schematic drawing of a growth cone showing the cytoskeleton organization. The squares indicate the free pool of soluble tubulin, and black circles the free soluble F-actin. Doted line indicates the limit of the central domain. Microtubules are oriented towards the peripheral domain with their plus ends in the direction of the filopodia.

in the morphology and motility of the growth cones can also result from changes in the actin filaments polymerization which is influenced by the intracellular calcium concentration. The motility of the growth cone, during the growth process, requires also to make and break adhesive contacts which the components of the extracellular matrix. When neurons migrate during ontogenesis to reach their position in the adult brain, the growth cones secrete proteases and plasminogen activators which result in activation of proteolytic enzymes which are able to break down the components of the extracellular matrix. In addition the growth of neuritic branches is also modified by the secretion of protease inhibitors from both glial cells and fibroblasts.

In vitro cultures of embryonic neurons of various types and various species of vertebrates, have shown that growth cone motility, shape and direction of the neurites can be regulated by various factors such as neurotrophic and/or adhesive molecules, neurotransmitters... In culture, axons of sympathetic neurons or of dorsal root ganglia respond to the addition of NGF in the culture medium

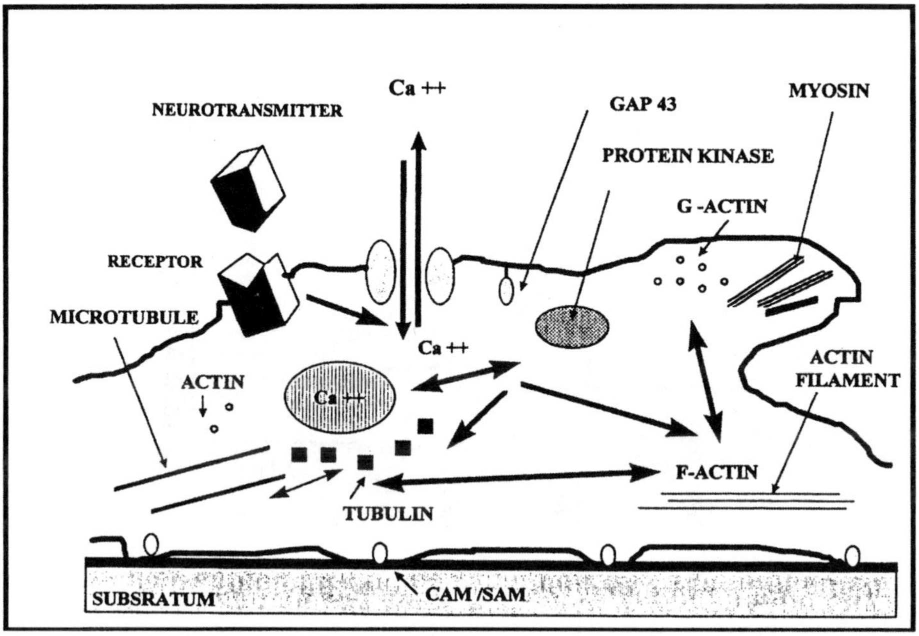

Fig. 7. — Schematic diagram of the sensory-effector machinery of growth cone navigation under the influence of a neurotransmitter. In order to simplify the drawing only the calcium second messenger system is shown. Arrows indicate the regulatory relationship between calcium and the different elements of the cytoskeleton. Modifications of intracellular calcium induce and regulate the assembly and disassembly of both the actin and tubulin. The movements of calcium between the cytoplasmic compartment are reversible and regulated. Calcium is also suspected to be involved in the binding of adhesion receptors to Cell Adhesion Molecules (CAMs) and substratum adhesion molecules (SAMs).

or to NGF bound to the substratum. Neuritic growth can also be promoted by laminin, a glycoprotein of the extracellular matrix. This molecule is acting on all types of peripheral and central nervous system neurons present in embryonic vertebrate brain. *In vitro*, neurons respond to laminin only when the molecule is bound to the substratum [32]. Under culture conditions, laminin has a powerful effect on neuritic growth but does not promote the survival of neurons [33]. Fibronectin, a glycoprotein located in the basement membrane is transiently expressed during development and it affects axon growth of various types of neurons. The adhesive interactions between growth cone and the extracellular matrix can be modulated during several days all along the development. Changing the distribution and number of integrin receptors and cell adhesion molecules such as NCAM can alter the strength of adhesion and consequently the rate of growth of neuritic processes. Postranscriptional

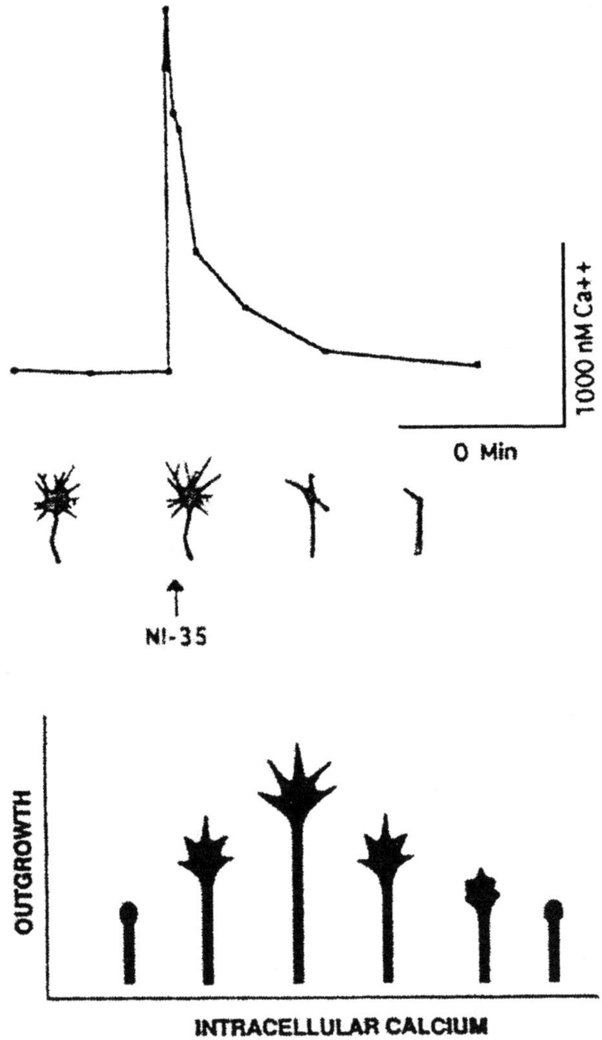

Fig. 8. — Diagrammatic representation of the effect of NI-35, a selective compound which induces calcium rise and subsequent collapse of the growth cone of dorsal root ganglion neurons. Application of a specific antibody able to block the NI-35, prevents both the calcium rise and all the effects observed on growth cone morphology and motility. After Bandtlow *et al.* [37]. Lower part: schematic diagram of the calcium hypothesis of growth cone regulation. Growth cones and their filopodia are displayed in graphic form in order to indicate the degree of outgrowth. Neuron will initiate neuritic branches with a particular intracellular concentration of calcium which induces the highest rate of growth. The growth is completely arrested with either low or high calcium concentrations.

modifications, especially phosphorylation, can provide another means of regulating integrins and of affecting neuritic growth process. (For a review concerning the growth cone, see [38]).

a. *Axonal Outgrowth*

As previously described, the nerve cell is highly polarized. It generally exhibits a long axonal process and a branched dendritic tree. Such a structure is supported by a specific organization of the cytoskeleton with parallel microtubules orienting their tubulin polymerizing ends towards the targets in axon, and with an antiparallel distribution in the dendrites. The single major feature that distinguishes neuron differentiation and growth from that of other cell types is outgrowth of the axon from the cell body of the embryonic neurons (neuroblasts) in a specific direction, along a specific pathway to form synaptic connections with specific targets (muscles, glands, other neurons). Functioning of the nervous system is absolutely dependent on the outgrowth of the axon making connections with the correct postsynaptic targets. The invariance of the initial growth trajectory of the axon is an essential feature of the axonal development during ontogenesis. The final pattern of neural connections, which is present in adult networks, is the result of different processes which take place in a coordinated pre-program during both pre- and post-natal development. The different steps of this program (which are dependent on gene expression and also on epigenetic control), can be summarized as follows: – outgrowth of the axons and selection of the pathways to their appropriate destinations, – dendritic outgrowth and formation of a specific dendritic tree morphology, – selection of specific targets by axon terminals, – elimination of incorrect and redundant synapses, – functional refinement of the final pattern of synaptic connections (Fig. 9).

During *in vivo* physiological development of the neuron, initiation of axonal and dendritic growth are influenced by non uniform extrinsic conditions. Mechanical constraints, electrical fields as well as a lot of various soluble and bound molecules are involved in the control of neuritic growth. This latter category of growth controlling factors includes mainly adhesive and extracellular matrix molecules, growth factors produced by neurons, glial cells and target tissues. All these factors should be taken into account in order to understand how neuritic processes are growing. The growing axon must be considered in its epigenetic environment, navigating through various parts of the space where some molecules are attached, others are diffusible, transiently present in homogeneous concentrations or distributed in gradients. Some of them can be permissive for growth, others can inhibit neuritic elongation and branching. Distribution and concentration of these molecules can be crucial in the initiation of an axon or a dendrite. Such a phenomenon can be observed *in vitro* with adhesive molecules: weak adhesion of the neurons on the substratum favors axonal outgrowth whereas strong adhesion favors dendritic outgrowth [32].

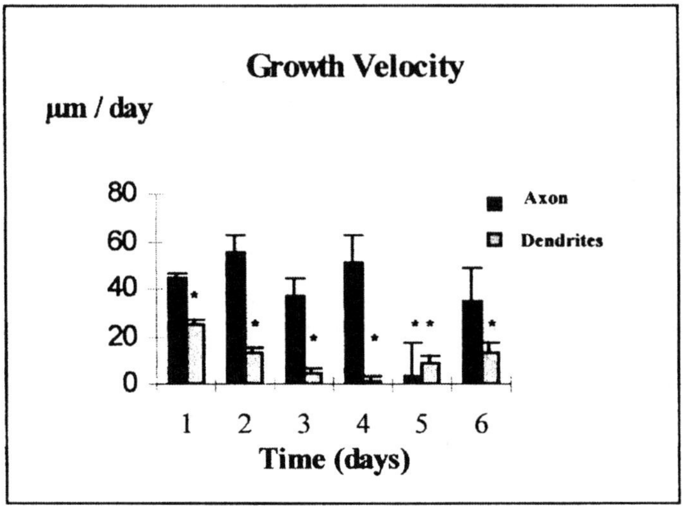

Fig. 9. — Growth velocity of axon and dendrites of embryonic rat spinal purified motoneurones (sampled at day embryonic 15) and maintained in culture during 6 days. Axon growth velocity is particularly high during the 4 first days compared to dendrites. Results are expressed as the mean ± SEM in μm per day. Each value is the mean of 120 determinations from different cultures. * Significantly different from the value measured at day 1 ($P < 0.001$). After Bataillé *et al.* [48].

b. Role of Electrical Currents in Neuritic Growing

The influence of electrical currents in the growth of neuritic processes was only demonstrated under *in vitro* conditions. Consequently, there is no good evidence that extracellular electrical fields can guide the growth and orientation of neurites. Nevertheless, the fact that the magnitude of experimentally applied current in neuronal cultures is in the range of currents measured *in vivo*, make it plausible that electric fields have a role in normal development of neuronal processes. Such an hypothesis is also supported by the fact that during neuronal development ionic channels, such as sodium and potassium ones, which are responsible for action potential onset, are progressively expressed at specific sites of the membrane, particularly at the initial segment, in parallel with an increase in electrical activity [39].

Assessing the effect of electrical fields on growth has been made possible by the use of linearly vibrating electrodes and circulatory vibrating microelectrodes. Under these conditions, it was shown that neurites extending out of explanted pieces of chick dorsal root ganglia grow faster in a weak electrical field towards the cathode. Changes in the growth of neuritic branches from embryonic dissociated neurons (from xenopus) in the direction of the cathode were also observed.

Using focal electric field, positive currents were shown to slow the rate of growth cone extension, whereas negative currents increase the growth of neurites. In these experiments, the magnitude of the currents required to produce an effect on neuritic growth was in the order of a few pico-amperes per square micrometer which is in the range of naturally occurring currents in *in vivo* physiological conditions. Currents measured from growth cones are in the range of 10 to 100 nA/cm^2 and appear to flow into the filopodial tip and out at its base [40]. Ion substitution experiments show that this current is carried by Ca^{2+} channels. This calcium current is also related to release of neurotransmitters from the growth cone [41]. Consequently, electrical currents can alter nerve growth through calcium influx which can change polymerization of microfilaments, act directly on tubulin or act *via* calmodulin or MAPs and Tau proteins which stabilize microtubules by linking them together by crossed-bridges.

c. *Chemotropism*

The response of the growth cone to chemotactants was first proposed as an hypothesis by Ramón y Cajal (1898) who provided a lot of circumstantial evidences but no direct proof of neurotropism hypothesis.

Under the rubric chemotropism we put together many phenomena that seem to show that nerve fibers grow up a concentration gradient towards the source of some diffusible substances, exhibiting positive chemotaxis. Chemotaxis involves action at a distance and should therefore be distinguished from chemoaffinity, which also occurs during growth of neuronal networks. Chemoaffinity requires contact between cells. Demonstration of stimulation of growth cones by diffusible molecules requires that the molecule forms a gradient extending to the outgrowth axons with a peak at the site of release from the targets, that the growth cone possesses receptors to these molecules and that they stimulate axonal elongation and change in direction of growth. Such conditions are found in the experimental work of Zheng *et al.* [42] who described a turning effect of nerve growth cones from axonal branches of xenopus motoneurons, induced by acetylcholine. This substance is involved, in the adult, in the transmission of the neural impulse at neuromuscular junction. The turning effect is mediated through nicotinic autoreceptors and needs a local increase of intracellular calcium concentrations, located in the regions of the growth cone oriented towards the highest concentration of acetylcholine. During growth of motoneurons such a phenomenon can be involved in the initiation of neurites from the embryonic motoneuron [44] and in the axonal targeting towards muscles. Both secretions of acetylcholine from embryonic motoneurons themselves and from developing myoblasts [45] are involved in these mechanisms. Other neurotransmitters, such as serotonin and glutamate, and other trophic factors can be responsible for chemotropism.

d. Mechanical Constraints and Growth of Neurites

During brain development, neurons which are generated from stem cells located in the embryonic neural tube, stop dividing rather early and then migrate to their final position before elongation of their neurites and differentiation. Growth of the axons and of the dendritic branches are not synchronous in the various parts of the future adult brain. That is to say, some neuronal pathways formed by bundles of axons emerging from several soma (brain nucleus) are still present when others are just starting to grow. Under this situation, it is clear that some elongating axons will meet, on their way to their targets, bundles of established axons, over which they have to climb, in order to reach their specific target. Apart from the fact that this dynamic process needs energy, and a large synthesis of various molecules, both at membrane and cytoskeleton systems, it is limited by visco-elastic properties of the various molecules which compose the membrane and, essentially, the cytoskeleton. Such mechanical constraints, occurring *in vivo* during development, can be mimicked *in vitro*, when culturing cells (fibroblasts, PC12 or neurons) on various substrates which present physical obstacles such as steps of various controlled hights or slip plans of increasing slopes. Under these experimental conditions crossing over the obstacles can only be performed when steps do not exceed a few μm high and with slopes below a few degrees [46, 47]. These results show that the molecules which are involved in the constitution of the cytoskeleton, and particularly the microtubules which are stabilized by MAPs (and consequently present the highest rigidity), can be submitted to limited mechanical constraints during building of brain tissue.

2.5. Formation of Dendritic Trees

a. Dendritic Development

In a majority of neurons, outgrowth of dendrites generally occurs after the growth of the axon. In most cases it is fairly certain that the axon has formed connections with its target before the differentiation of dendrites starts, suggesting that an axonal retrograde message to the cell body is necessary to start dendritic elongation and branching.

After the neurons have migrated to their final position in the brain, there is often a long delay before full differentiation of the dendrites occurs. Outgrowth of dendrites after growth of the axon is also observed *in vitro*, in primary cultures of neurons [48]. The young neuron is generally characterized by short and thick dendrite processes. Following axonal differentiation and contact with the target, the neuron will develop a complex system of branches which resembles the branching of multiaxiate plants. It is not certain whether branching occurs only at the tip of the existing neurites or whether branches also emerge preterminally. Branching results in a great extension in the surface area of the dendrites which form more than 90% of the post-synaptic membrane surface of the neuron.

When neurons are observed on histological sections impregnated with the Golgi method, the size, the shape and branching pattern of the dendritic arborization are characteristic of each type of nerve cell. There are differences in proportion and structure of different types of neurons and particularly in the pattern of dendritic branching on which a system of classification can be based.

b. Dendritic Outgrowth and Branching

As soon as the start of the twentieth century, Santiago Ramón y Cajal observed that the dendrites of spinal motor neurons and cerebellar Purkinje cells undergo regressive changes, analogous to pruning of a tree. He has distinguished an initial phase of outgrowth of an excessive number of dendritic branches, followed by a phase of regulation and resorption of the redundant dendrites. He proposed that a selection among the dendritic outgrowths takes place during development, presumably based on functional processes. Such an idea is expressed by Ramón y Cajal, as followed:

"what mysterious forces precede the appearance of the processes, promote their growth and branching... and finally establish those protoplasmic kisses, the intracellular articulations, which seems to constitute the final ecstasy of an epic love story?..."

Dendrites differentiate in conjunction with the specific axonal terminals with which they will formed synaptic connections. This suggests that the afferent axonal terminals might stimulate the elongation of dendritic branches. There is a large body of evidence showing that the removal of afferents or the blockade of afferent excitatory synaptic activities result in functional and structural changes in the dendrites of the postsynaptic neurons. The number of afferents is an important factor which determines the size and complexity of the dendritic arborizations. In some areas of the developing brain, dendrites may compete for the available afferents and that appears to be one of the ways in which the sizes of the dendritic trees are adjusted to fit into in the available space.

Dendritic growth cones were identified in Golgi preparations and also using electron microscopy. The initial contacts between dendritic and axonal growth cones result in the formation of immature synapses. Development of dendritic branches occurs at the same time as the development of axo-dendritic and dendro-dendritic synapses. As suggested by Barron [49], the development of the dendrites of some neurons is delayed until the axons have reached their targets and a retrograde signal, transmitted from the axon terminal to the soma, is required to initiate the dendritic growth.

The correct functioning of the dendrites depends on their coming into proper relationship with the axons which are destined to form effective working synapses on them. Dendritic spines are the main postsynaptic sites on most neurons. These specialized dendritic structures receive inhibitory as well as excitatory inputs which are often spatially segregated on different regions of the dendrites. For this reason, dendritic spines are important for integration of synaptic inputs. Nevertheless, cellular and molecular mechanisms involved in the development of the spines on dendrites remain unclear. How the spines develop in the correct position, in the correct numbers, how they are maintained and modified, are important questions to answer. What is the involvement of the afferents in the development and maintenance of the spines? To what degree the spine itself is responsible of its own functioning? It is not clear why synaptic junctions form early during development of dendritic spines in some types of neurons while in others the spines form first and synaptic junctions form later. Development of dendritic spines usually occurs after the formation of axo-dendritic synapses, but the spines can develop and persist in the absence of synapses on them. There is evidence that spines can develop in the absence of presynaptic terminals after experimental removal of the afferences, in some mutations, as well as during ontogenesis. There is also increasing evidence showing that changes in the size and shape of dendritic spines and their post-synaptic specializations play an important functional role during development and in the physiological maturation of the brain.

2.6. Mechanisms of Neurite Initiation and Branching. Biophysical Considerations

As previously described, growth of axonal and dendritic processes from the neuroblast is under the dependence of the expression of genetic as well as epigenetic factors. A lot of epigenetic factors including bound and soluble molecules are involved in morphogenesis. Development of the adult shape of the neurons implies different growth steps which occur during ontogenesis and post-natal period. During this period several successive phenomena might be observed: initiation of primary neurites from the soma, growth of the axon, elongation, branching and differentiation of the dendritic tree, combined with synaptogenesis. Such a period is characterized by the onset of various intracellular mechanisms, dependent on the local environment (epigenetic factors) which lead to the expression of specific morphological features in close relationship with functional neuronal properties. Presence of trophic or repulsive soluble molecules, specific adhesive molecules, extracellular matrix components, and continuous as well discontinuous gradients around the embryonic neurons, initiate growth, changes in intracellular calcium concentrations, membrane instabilities, modifications of cytoskeleton elements which are responsible for shape changes, elongation and branching of the primary neurites. Most of the changes occurring during brain development take place in the growth cones. Growth of neurons can be studied by using various experimental methods both *in vivo* and *in vitro*. In addition to the neurophysiological approach, such mechanisms can also be clarified by using modelisation and simulation. In this context, the biophysical approach of neuronal growth mechanisms suggests that generic rules concerning the growth of non-living or other living branched structures are also applicable to neuronal morphogenesis.

a. *Gradients of Soluble Molecules. Effect of Acetylcholine on Motoneuron Growth*

Cells, like neurons, which present tip growth are submitted to diffusion phenomena around their processes [50]. These mechanisms can be quantitatively studied. Two types of extracellular gradients can be distinguished: – external gradients settled down around natural sources, such as target cells, or artificially established, for example around a pipette containing a chemotropic molecule and releasing it in the extracellular medium; – natural gradients resulting from the absorption or the secretion of solutes by the cell itself which, by exchanges with the medium, give it anisotropic properties. During growth, the nerve cell is particularly sensitive to a lot of extracellular signals which determine *in vivo* the elongation of neurites, their directions and the establishment of synaptic contacts with target neurons or tissues. These signals can be experimentally reproduced or simulated under *in vitro* conditions, in embryonic neuronal primary cultures where extracellular conditions can be relatively well controlled. Chemotropism experiences show that processes, starting from embryonic motoneurons of the xenopus, are oriented by a chemo-attractive gradient of acetylcholine formed around the tip of a micropipette introduced in the culture medium, in the vicinity of the growing neurons [42]. Some substances are also known for their autocrin actions. That is to say, they are secreted by the cells and then act on them from the extracellular environment. Such influences where described during neuronal growth with Brain Derived Neurotrophic Factor (BDNF, [51]), Adenosine Tri Phosphate (ATP, [52]) and acetylcholine (Ach, [53]). Such autocrin modulations of neuronal growth can be observed if the secreted substances which form the extracellular gradient possess morphogenic properties, as it is the case for the previously cited molecules.

Control by acetylcholine (ACh) of neuritic growth from embryonic motoneurons, is certainly one of the mechanisms which was recently studied in details. The ACh, in the adult nervous system is a neurotransmitter involved at the neuromuscular junctions as well as at the central synapses. During ontogenesis, ACh is known to act as a chemotropic factor and a modulator of growth [54]. Embryonic rat spinal motoneurons, purified and maintained in culture, secrete ACh from their cell bodies at times where neurites just initiate from the somatic membrane [53]. Secretion of ACh was also observed from the growth cones of growing xenopus motoneurons in culture [55]. Secreted ACh in the extracellular space is hydrolyzed by acetylcholinesterase (AChE), its specific catabolic enzyme, which is present at membrane level and also released [56]. Consequently, the extracellular ACh concentration around the growing motoneuron is given by the net flux of secreted ACh minus the amount of hydrolyzed molecule [48]. Quantities of secreted ACh, as well as activity of AChE can be experimentally measured by using chemiluminescent assays [57,58]. Modelisation of these phenomena predicts that, at short times following plating of dissociated embryonic motoneurons, cell bodies are surrounded by a gradient of ACh which influences

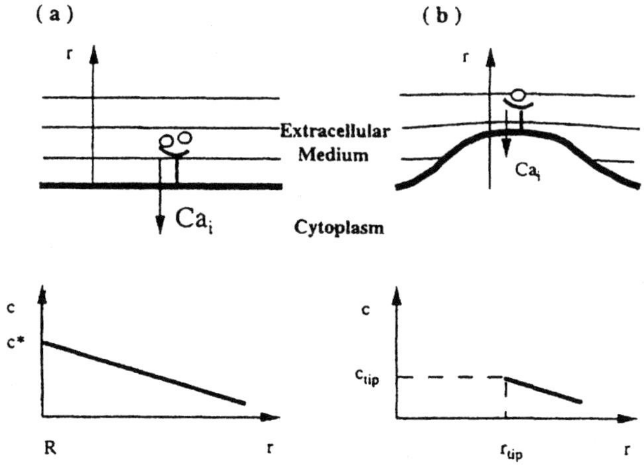

Fig. 10. — Schematic illustration of neurone tip growth mechanism due to extracellular ACh gradient near the membrane. This model stems from experimental results obtained at short times following plating of purified rat spinal embryonic motoneurones on glass coverslips and maintained in a defined medium without trophic factors. a) planar and b) deformed neuronal membrane. The ACh iso-concentration curves are indicated by solid lines lying ahead of the membrane. Since ACh concentration near the deformed membrane Ctip is less than C*, Ca^{2+} influx is reduced and tip growth rate is enhanced, leading to an amplification of the deformation. After Limozin *et al.* [53].

the iniation of primary neurites [53]. When neurites are already developed, the ACh gradient is restricted to their tips, at the growth cone level (Fig. 10).

The motoneuron responds to this extracellular field of ACh by means of specific receptors located on the membrane. Both nicotinic and muscarinic receptors which are early expressed during motoneuron development, seem to be involved in the morphogenetic response to ACh, but nicotinic one's are certainly more efficient in the regulation of neuritic elongation and turning effect [42,59], (Fig. 11). As previously shown at the level of the adult neuromuscular junction, nicotinic receptors are desensitized following a long application of ACh. In mice neuromuscular junction, one half of the nicotinic receptors are desensitized in presence of 0.3 μM ACh and only 10% of the receptors can be activated with 1 μM ACh [60]. The secretion of ACh by the developing motoneuron can affect its interaction with the extracellular environment and this behavior is particularly important when the axonal growth cone is reaching its muscular target, just before synaptogenesis [55]. At this stage, the growth cone will be submitted to the influence of a secreted gradient of ACh produced by the muscular fibers [45].

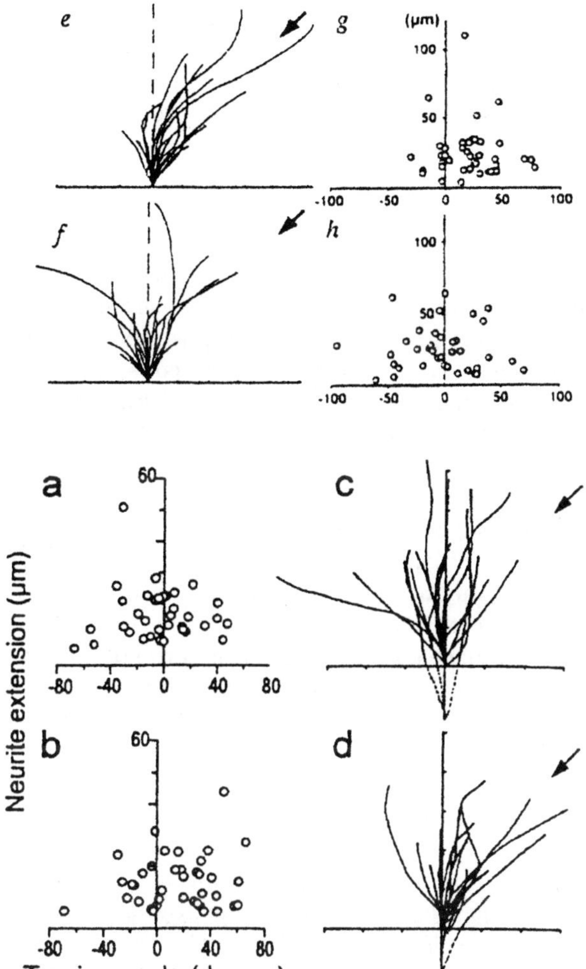

Fig. 11. — Turning of nerve growth cone induced by neurotransmitters. Top, turning response of Xenopus spinal neurons in presence of ACh gradient. The gradient is created by a pulsatile ejection of picolitres of 100 nM ACh solution through a micropipette. e and f) composite drawings of the path of neurite extension during a 2 hours period for a population of neurones plated in absence e) or presence f) of d-tubocurarin (0.1 mM). In g and h), each point represents the result of the turning angle and the net neuritic extension in the absence g) or presence h) of D-tubocurarin. (abscissa in degrees and ordinate in μm). Lower part, results obtained with 0.5 mM glutamate in the culture medium. a and c) in the presence of glutamate, b and d) in the absence of glutamate. Significant turning effects were observed with glutamate concentrations ranging from 0.05 and 0.5 mM. After Zheng *et al.* [42, 43].

Binding of ACh to the nicotinic receptor which is a typical pentameric channel protein induces an inward cationic flux through the channel and an increase of calcium cytoplasmic concentrations. Numerous data indicate that this increase in calcium concentrations, mediated by ACh, can be associated with the changes occurring during motoneuron growth. ACh inhibits neurite elongation [48, 54, 61] and this effect is probably due to the breakdown of the actin network which is involved in the motility of the growth cone. ACh is also an attractive molecule which promotes an increase of filopodial protrusions from the growth cone membrane [42]. Implication of cytoplasmic calcium concentrations in the effect of ACh during development of the motoneuron was recently demonstrated in the turning behavior of neurite submitted to a gradient of ACh. Zheng [62] has recently shown, in xenopus motoneuron, that the local photoactivation of caged calcium in the growth cone by a laser beam is able to produce a restricted increase of cytoplasmic calcium (1 to 2 μm^3) which is in close relationship with turning of the neurite in the direction of the highest calcium amount.

With high concentration of extracellular AChE, the extracellular content of ACh is highly reduced, suggesting that AChE is able to attract neurites towards the regions of lowest ACh concentrations [63]. The natural ACh concentration difference between the tip and the base is hardly affected, but the growth cone is more sensitive to the gradient of ACh. The variation of ACh content at the growth cone can influence its own behavior through an autocrin effect generating an increase of the intracellular calcium concentration. The screening length, which represents the distance between motoneurons maintained in culture, plays some role in the involvement of ACh during growth. When this distance is reduced and reaches a length smaller than the length of the neurite, this process is only submitted to its own field of ACh and the autocrin effect is great.

Involvement of ACh and nicotinic receptors in the morphological properties of growing motoneurons was recently confirmed by using antisense strategy in culture. Blockade of the synthesis of $\alpha 7$ subunit of the nicotinic receptor, using antisense oligonucleotides, leads to a reduction of axon length and spreading index of embryonic rat spinal motoneurons [64]. Such an inhibitory effect is certainly in relation with changes in the architecture of the cytoskeleton and particularly of actin filaments which seem to be involved in the clustering of $\alpha 7$ subunits [65].

b. Growth Cone, Elongation and Branching

As shown previously, mature neurons are characterized by dendrites and axons with complex and widely varying shapes. These structures often exist as branched trees, ranging from simple, as in ganglionic neurons, to more elaborate, as in pyramidal cells, and even very complicated shapes, such as Purkinje cells. Many studies have been concerned with the description of dendritic and axonal morphologies, starting from the famous work of Santiago Ramón y Cajal [3] to quantitative metric [66–68] and topological descriptions [69–71]. The nerve growth cone and the cytoskeleton play a key role in the formation of dendritic and axonal branching patterns. The growth cone is a mobile structure at the tips of the outgrowing axons and dendrites and is thought to act as the "eyes" and "ears" of a neurite [72, 73]. Growth cones directly contact the immediate environment of the neuron and this structure is assumed to sense other cells, such as glial and neuronal cells, and gradients of soluble and membrane bound chemical compounds. The growth cones also adhere to the surrounding substrate, establishing close contacts between the neurites and their environment. They are involved in guidance, elongation and branching of the neurite. The plastic properties of the cytoskeleton give it a crucial role in the maintenance of the shape [74]. Such a role is demonstrated by the fact that disrupting the network of actin by chemical compounds such as cytocalasin, results in a much slower elongation of the neurites and a lower probability of branching [75]. Disrupting the microtubules will result in retraction of the neurites [76] and blockade of tubulin polymerization stops elongation [77].

As described by Letourneau *et al.* [73], the growth cone is the structure where much of the growth activity occurs. The growth cone possesses two microscopically distinguishable domains. First, the central domain or C-domain directly connected to the neuritic shaft where microtubules end. This domain is rich in cell organelles. Second, the peripheral domain or P-domain which is flat, where organelles are absent and microtubules occasionally protude [78]. The P-domain can be divided into two parts: the filopodia and (proximal to the filopodia), a basal part containing lamellae and veils, with a high actin content, showing pseudopodal movements [79] (Fig. 5). Filopodia with their aligned actin bundles [80, 81] exhibit a high motility. They can extend and retract within a few minutes [82]. They adhere to the environment by forming bonds with Cell Adhesion Molecules (CAMs) such as NCAM, N-Cadherin, L1 and the integrin receptor family [83]. As shown by several authors (see review of Van Veen and Van Pelt [84]) the filopodia are involved in the guidance of neuritic processes using a "stepping mode". These observations which were obtained under various experimental conditions suggest that the filopodia are probing the environment and transduce a signal which will be received by the basal part of the P-domain. The translation from filopodia signals to growth cone displacement could result from interferences with the formation of lamellae and lamellipodia which are determined by the state of actin cortex inside the growth cone [79, 85–87]. A strong signal from the filopodia is able to affect the actin cortex and to trigger actin disassembly, leading to protrusion of lamellipodia and motion of the growth cone in the direction of the protusions [79, 86, 87]. Inhibitory signals would have an opposite effect inducing retraction of the growth cone [88–90].

Filopodia were also suspected to play an important role in the neuritic branching. The first attempt to assign a role to filopodia was made by Vaughan *et al.* [91] who suggested that branching would occur when two filopodia simultaneously contact synapses. This hypothesis was then generalized by

Berry *et al.* [92] who stated that the simultaneous formation of strong adhesive bounds with the environment of two filopodia will lead to branching. Both the angular distribution and the adhesive properties of the filopodia are involved in branching [93, 94]. Observations of spontaneous branching growth cones of cultured neurons show that branching probability is increased when filopodia are directed to the sides of the growth cone [95]. Such a phenomenon is also dependent on the interplay between filopodia and the bundles of microtubules present in the C-domain of the growth cone. Branching would occur when the filopodia orient microtubules into distinct clusters and this is favored when filopodia are oriented in distinct directed clusters located on the sides of the growth cone [72, 96].

According to these experimental data, Van Veen and Van Pelt [97, 98] have modeled the growth cone as a structure composed solely of filopodia which are present in a limited area near the basis of the growth cone. Both the number and the directions of the filopodia were assumed to follow a stochastic distribution. These distributions were generated in a model by scattering discrete adhesion sites on the substrate on which the modeled growth cone was positioned (Fig. 12). Filopodial vectors were projected towards all adhesion sites within the detection area of the growth cone. Filopodia vectors represent the filopodia and the detection area corresponds to the region around the growth cone which can be reached by the filopodia. Adhesion sites are used to generate stochastic distributions for the number and direction of the filopodia. The direction of growth shift can be defined by the sum vector of the filopodial vectors. In this model, the direction of each filopodial vector accounts for the direction, while the weight factor which cannot be measured experimentally either in terms of force of intracellular signals, accounts for the strength of the filopodial action on the growth cone. This model predicts that the growth cone will turn left or right if there are more filopodia pointing to the left or the right of the growth cone. This prediction is supported by experimental evidence suggesting that the growth cone will branch when the filopodia are directed mainly towards the sides of the growth cone [93, 95].

Once neurites are formed from the growth cone, the process must be maintained further. The main cellular components responsible for this maintenance are the cytoskeleton proteins. As shown by the pharmacological disruption of microtubules (which generally results in a retraction of the neurites [76]) or by blockade of tubulin polymerization (which stops outgrowth [77]), microtubules form a rigid continuous core all along the neurites and can be considered as one of the main factor involved in neuronal shape maturation and differentiation. Microtubules are chemically polarized with positive and negative ends where soluble tubulin can respectively be polymerized or depolymerized [99–101]. In addition, parallel organization of the microtubules, with their plus ends directed towards the growth cone, can be observed in the axons whereas antiparallel distribution occurs in dendrites [24]. The rate of dynamic assembly of the microtubules was shown to be linear, over a time scale of hours and depending on the concentration of soluble tubulin and a constant disassembly rate [99, 100].

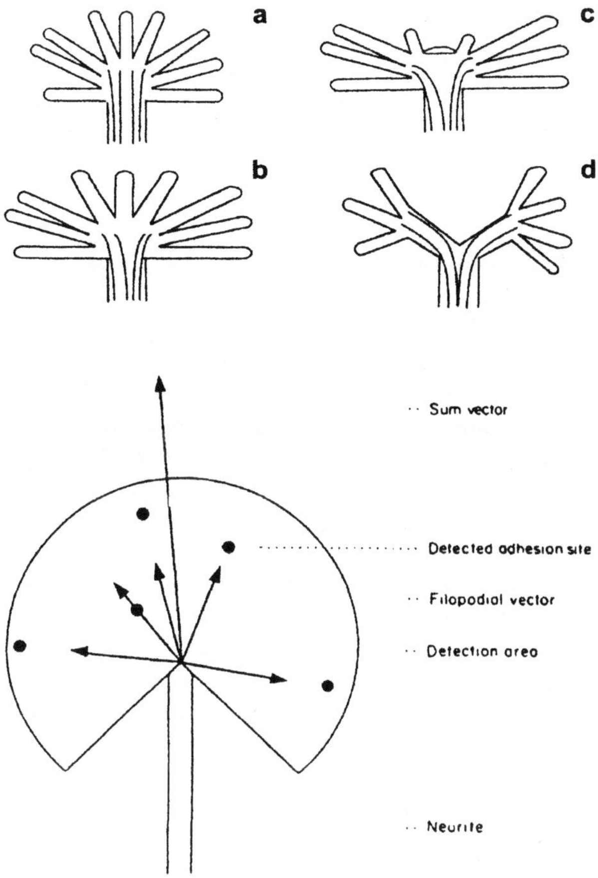

Fig. 12. — a, b, c, d) interplay between the filopodia and the microtubules in the growth cone leading to branching of the neurite. After Letouneau *et al.* [96]. Model of growth cone movement proposed by Van Veen and Van Pelt [84]. In the model, filopodia are represented by filopodial vectors and the area near the growth cone is described as a detection area which is submitted to various extracellular factors. The filopodial vectors are directed towards discrete adhesion sites which are randomly distributed on the surface of the central domain. The mean direction of the filopodia is given by their sum vector.

Such a linear dynamic suggests that a steady elongation of neuritic processes occurs. But, on a time scale of few seconds, microtubules can elongate or shrink, following a dynamic instability which can be described by the frequencies of the transitions between elongation and shrinking states [102–104]. Length of the microtubule, length of the tubulin dimer, concentration of soluble tubulin, assembly as well as disassembly constants interplay to determine if the average elongation rate will be positive, negative or equal to zero. Both the proximal and distal regions of the neurite are involved in polymerization of the tubulin [74]. Nevertheless, inhibition of tubulin polymerization at the proximal side of a neurite does not affect elongation rate whereas it stops immediately elongation when polymerization is blocked at distal parts [105].

Tubulin is produced in the soma and the half life of tubulin mRNA is one to two hours only [106]. This observation suggests that tubulin must be transported to the neurite tips in order to be available for assembly. Two kinds of mechanisms were suggested: tubulin is passively transported along the neurite in the form of tubulin dimers [107] or a large pool of soluble tubulin is present inside the neurite and can diffuse [108]. According to Van Veen and Van Pelt [84] a constant transport rate of tubulin over long distances can be obtained when the gradient of soluble tubulin reaches a constant slope, with the cell body forming a tubulin source and the neurite tips being the main polymerization sites. Under this configuration, mathematical models taking into account the tubulin synthesis, the diffuse transport and the polymerization can be used to determine the elongation rate of the microtubules [84]. These models predict that a constant elongation rate of the neurite is associated with a constant tubulin production at the somatic level.

In the models which predict that the growth cones are the main polymerization sites of tubulin, tips of neurites which are formed at branching points will compete for soluble tubulin which is present at a constant concentration at the branching points. Competition between the different tips will induce tips with low concentration of tubulin. This establishes a steep tubulin gradient where tubulin diffusion constants are of minor importance with respect to elongation rate [84]. Competition for soluble tubulin can also result in delayed outgrowth or retraction of tips which present low polymerization or high depolymerization rates. As shown by Faivre *et al.* [109] the stability of microtubules increases as a function of maturation. In mature Purkinje cells, which exhibit a high branched dendritic tree, a sudden cooling (which normally induces a disassembly of the microtubules), does not affect their shapes. Instability of microtubules in the early development can be related to the maturation of MAPs which play an important role in the distribution and stability of the microtubules [101, 110].

Genesis of neuronal shape is dependent on branching probability, direction of elongation, and rate of outgrowth. These parameters are dependent on various extra and intracellular factors which affect the distribution of cytoskeleton proteins which are themselves modulated, during *in vivo* development, by the properties of surrounding cells: *i.e.* other neurons and glial cells. As shown previously, elongation is in close relationship with changes occurring in the polymerization of microtubules. Microscopic observations of growing neurons in culture, show that neurites have elastic properties which are limited by the degree of elasticity of both microtubules and actin filaments. The elastic force of a neurite follows Hook's law, relating force to stretch [111, 112]. During elongation the growth cone exerts a permanent force on its neurite and the neurite is always in a stretched state. This stretching effect on the neuritic elongation rate might partially explain towed growth, as it was shown by Bray [113] and Zheng *et al.* [114]. Such a stretching of the neurite can be translated and it induces a release of compression of the microtubules, giving rise to a larger polymerization rate [115].

The interaction between tension generated by the growth cone and the polymerization of tubulin in the central domain combines the push or pull hypothesis for neuritic elongation which were respectively postulated by Bray [116] and Golberg and Burmeister [117]. They suggested that growth cones pull their neurite forward or that the growth cone is pushed forward by material which is transported within the central domain. In fact, both the tension generating pull of the growth cone and the input of polymerized tubulin into the growth cone are involved in elongation of neuritic branches.

As it was previously demonstrated, branching of neurites is correlated with the angular spread of filopodia, but this parameter does not give any information about the probability of branching. Using topological approach, where neuronal trees are considered as a collection of segments connected by branch points, outgrowth models can be developed [69, 70, 118–120]. Using two parameters: the branch probability and an exponential function defining the dependence of the branch probability on the centrifugal order of a segment (number of branch points between the considered segment and the proximal basis of the tree), Van Pelt *et al.* [119] concluded that the branching probability decreases with increasing centrifugal order, towards the top of the tree of cortical pyramidal as well as cerebellar Purkinje cells, whereas multipolar non pyramidal neurons exhibit a constant branching probability all along the tree. Using a stochastic model which takes into account the probability of branching and the lengths of the segments, Van Veen and Van Pelt [98] predict that terminal segments are expected to be longer than intermediate ones and that both types decrease in length with centrifugal order. Such a prediction agrees with measurements of neocortical rat pyramidal cells [121]. In this model, the branching probability was shown to be correlated to the number of filopodia, with very small probabilities of branching with low and high filopodial number and larger probability at intermediate numbers.

Outgrowth of neurite is activity-dependent. Electrical stimulation or application of neurotransmitters are able to stop or reverse outgrowth [122–126]. Effects of neurotransmitters are linked to changes in internal calcium concentration [126] and comparable with the effects of evoked action potentials [127–129]. Increase in intracellular calcium concentration is followed by a cascade of events which affect neuritic elongation and branching. Elevated calcium concentrations interfere with the formation of filopodia and veil protusions [126, 130] through actin polymerization [79] and microtubule disassembly [78, 131, 132]. Neurite outgrowth occurs at 35 nM cytosolic free calcium. During maturation and differentiation of neuronal networks, functional contacts are established and calcium concentrations become elevated. This induces changes in neuritic

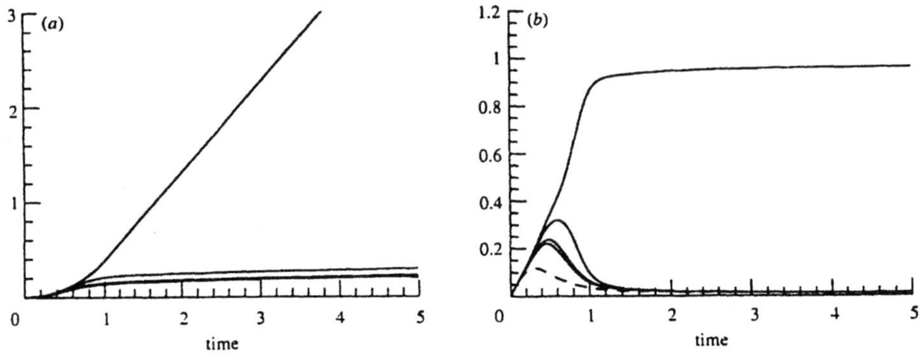

Fig. 13. — A model of axon formation and origin of neuronal polarization. After [103]. Typical simulation of the growth of a neurone bearing four neurites. Length of the neurite a) and concentration of morphogen at their tips and in the soma (dashed line, b) are plotted against time. The heavy lowest line in a) shows the growths of the two smallest neurites which are superimposed. All values are dimensionless. One neurite, which is the axon, grows more rapidly than the others.

morphologies. Such mechanisms are also observed in the adult brain where neurons extend and retract their dendritic branches [133, 134].

The concept that shape of neuronal cells is controlled by the local submembrane concentration of calcium ion is consistent with the mechanisms involved in growth of dendritic trees. Hentschel *et al.* [135], described a class of dynamic instabilities which are involved in the generation and branching of dendrites. Such instabilities are based on diffusion-controled growth through a biologically active membrane (Fig. 13). The interaction between diffusion and active transport of a morphogen whose concentration at the tip of a neurite regulates the growth rate, can lead to an instability forming a single quickly growing axon and inhibiting the growth of dendrites. Taking into account the consequences, in the case where calcium ion is a morphogen, for membrane growth, Hentschel *et al.* [135] demonstrated that the patterns of dendritic and axonal differentiation obtained by simulation were comparable to those observed on real neurons. Diffusion-regulated control by calcium (morphogen) of dendritic growth is linked to changes in the intra-cytoplasmic concentration of calcium which result from influx and active extrusion across the cell membrane. As shown by Hentschel and Fine [136] with simulations, nonlinearities in voltage-dependent ionic permeabilities enhance unstable growth, so that dendritic branching results from self-sustaining internal gradients of morphogen. Simulations predict that complex patterns of branching growth can be modified by changes in membrane conductance, galvanotropism and chemotropism. During development most of the neurons become polarized as soon as one emerging neurite, generally the longest, becomes the axon and the others the dendrites. The physical mechanism responsible for such length-related differentiation is not yet established. Recently, Samuel *et al.* [137] have built a model of neuronal polarization based on the existence of a chemical determinant whose concentration at the tip of a neurite is able to influence the growth rate. Equations taking into account the formation, the transport and the consumption of the chemical and the resulting neurite growth undergo a winner-take-all instability which leads to the rapid growth of one neurite which is the axon and the reduction of elongation rate of the others: the dendrites.

3. MORPHOLOGICAL PROPERTIES OF NEURONAL CELLS AND NEURONAL INFORMATION PROCESSING

The knowledge of operating modes in the dendritic trees of neuronal cells are mainly related to mathematical and physical models which mimic the behavior of neuronal cells, considered as the basic computational element of a neuronal network or the network itself. Since the first model of formal neuron proposed by McCulloch and Pitts [138] the field of computational neurosciences has exploded during the last decade [139–142]. Quantitative models were designed to simulate the traffic of electrical currents in dendritic branches of various neurones with different shapes [143–152]. These models try to define the type of dendritic computation, to determine the role of nonlinear processors and to evaluate the nature of interactions between various types of voltage-dependent channels. The presence of nonlinear conductances at the level of dendritic spines can lead to complex boolean logic [153, 154]. The presence of nonlinear interactions between different dendritic process should considerably influence the computational power of a neuron [155].

One may approach this field from modeling and simulation of information processing by neurons, but a close relationship to the biological reality is necessary in order to make progress [156]. Consequently, introduction in the models of parameters measured on living neurons is necessary to go further. Electrophysiological properties of the neurons as well as quantitative morphological descriptions, topological distribution of synapses are relevant parameters to be used for simulation. Predictions of the models are useful if they can be experimentally tested, giving rise to new hypotheses.

Dendritic branched trees can be analyzed at two levels of complexity. The "macrostructure" can be defined by the three-dimensional geometry of dendritic trees. This description encapsulates the different branch lengths, their local variations in diameter and the topology of their branching points. The "macrostructure" itself is composed of a "microstructure" which is defined by the spatial distribution of various populations of ionic channels, of molecular pumps, and receptors located at the membrane. These two components are involved in the dynamics of ionic currents generated at the synapses.

Using a simple model Bras *et al.* [157] have shown that the electrotonic structure of the dendritic tree is directly linked to its stochastic geometry (Fig. 14), when dendrites are considered as electrically passive. Under these

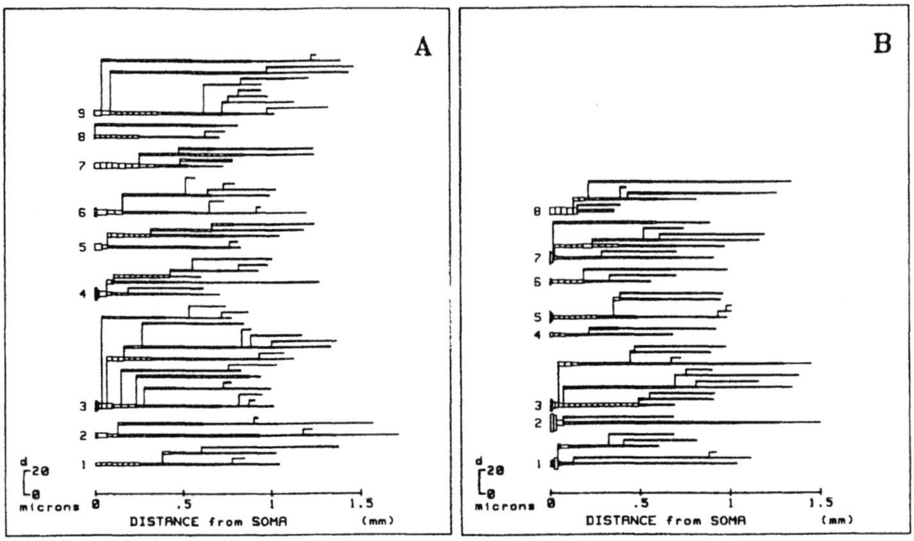

Fig. 14. — Topographical representation of the dendritic arborization of an abducens A) and a laryngeal B) motoneuron. Motoneurons where identified by electrophysiology (orthodromic stimulation of the nerve) and electrophysiological properties were determined by intracellular microelectrode recordings. Following electrophysiology, the soma was ionophoretically injected with horseradish peroxidase in order to stain the axon and the dendrites. Following fixation of the brain and serial sections, the stained neuron was reconstructed in three dimensions using a computer-aided microscope to obtain a high resolution. Each dendrite is coded by a number. Dendrites are represented by contiguous horizontal segments. The length of the segment represents a dendritic length of 30 μm. The width of the segment is proportional to the dendritic diameter given by the scale (d) at the lower corner of the diagrams. The distance from the soma is displayed on the abscissa. The branches are connected to the branch point by a thin vertical line and separated from each other by a constant distance. After Bras *et al.* [163].

conditions, using the geometrical description, at high spatial resolution, of the dendritic branches, the somatofugal voltage attenuation along the branches can be described with high precision. This somatofugal attenuation of the voltage is comparable to the somatopete electrical charges transfer [158]. The electrotonic architecture of reconstructed identified neurons shows that dendritic trees contain families of dendritic branches with statistically comparable somatofuge voltage and mean voltage gradients (Fig. 15). These branches families are distributed in the three dimensional space of the neuron and are composed of different dendritic branches belonging to different dendrites [159]. Such an approach allows a comparison between different dendritic arborizations of different regions of the brain of the same species or between the neurons

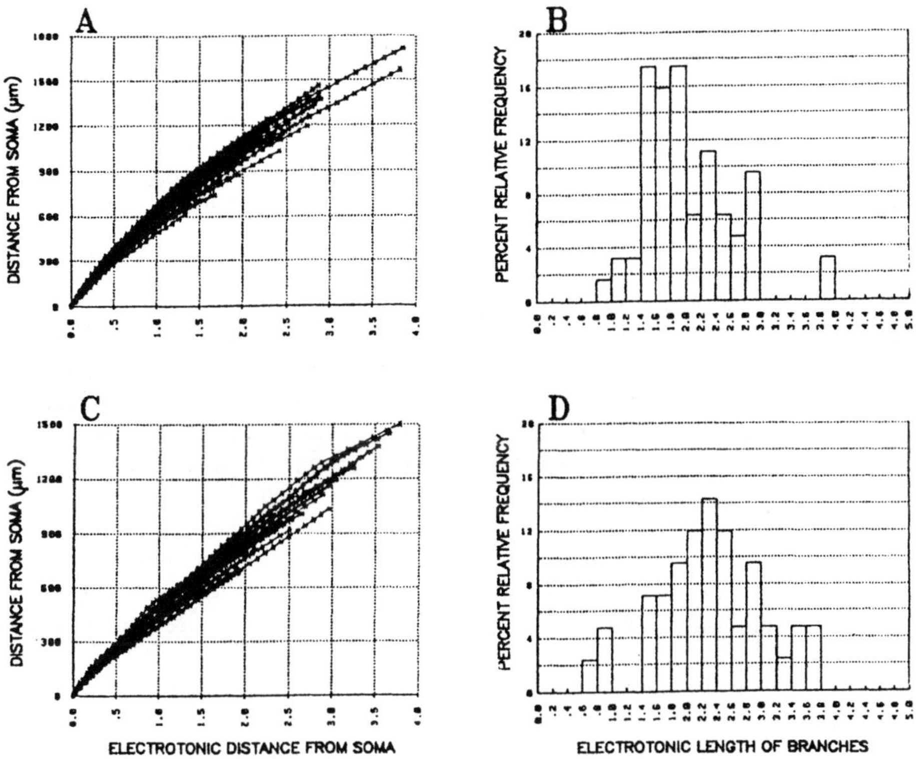

Fig. 15. — Electrotonic distances calculated for every dendrite plotted as a function of the geometrical distance from the soma. Neurons in this analysis correspond to those described in Figure 14. A) abducens motoneuron, C) laryngeal motoneuron. Data points represent successive sites every 30 μm along the dendritic paths. Histogram, on the left, showing the distribution of the total electrotonic length of all the dendritic branches. B) abducens motoneuron, D) laryngeal motoneuron. After Bras *et al.* [163].

of the same brain region in different species, located at various steps on the phylogenetic scale. In addition, this approach can be useful in order to predict the electrical behavior of various different arborizations and to demonstrate that the electrical complexity is different from the topological complexity [160].

When dendrites are considered in their active configuration, models can be helpful to determine the influence of the geometry on synaptic transfers. In a simple model, the effect of synapses homogeneously distributed on the surface of a dendrite, was considered. The conductances are activated by the voltage or by a ligand and the potentials are maintained constant and uniform. When the local potentials are different in various sites and when the channel densities are heterogeneous, axial currents are initiated and their direction is determined by the voltage gradient. With an homogeneous distribution of the excitatory dendritic conductances, the voltage profiles, along the different somatopete paths, are significantly modified when branches are asymmetric. The model also predicts that, according to the geometry of the dendritic branch, the somatofuge invasion of the dendrite will modulate the state of ionic channels which depends on the neuronal activity, thus influencing the efficiency of the synapses located on the dendritic branch [157]. This auto-regulatory phenomenon is dependent on the immediate electrical story of the neuron and could be involved in postsynaptic plasticity [159].

Concerning the microstructure, patch-clamp recordings of neuronal cells have shown that the ionic channels are stochastic elements which can be opened or closed at random. Theoretical studies have shown that an electrical capacitance coupling of the membrane (which implies large membrane domains) allows stochastic population of ionic channels to produce various behaviors such as: resting potential, action potentials, spontaneous discharges... [161]. The combined approaches of molecular properties of ionic channels (microstructure) and description of geometrical properties of the neuron (macrostructure) is certainly a key to understanding the relationship between structure and function of the neuron. Such a study can be experimentally approached by using electrical membrane fields imaging, which uses sensitive fluorescent potential dyes in cultured neurons [162]. Under these conditions, local changes of the fluorescence can be detected, at various sites of the neuronal membrane, during the action potential. These pictures were interpreted as resulting from the variations of the local electrical field. These are produced by charges transfer through ionic channel proteins. The large variation of topographical distribution of the changes, from one excitation to another, is not in contradiction with the fact that the macroscopic events (*i.e.* the action potentials) remain similar. During excitation channels open, but their stochastic behavior could explain the changes in the distribution of the depolarized membrane sites. Such an hypothesis is currently being tested using modelisation and simulation.

REFERENCES

[1] Ramón y Cajal (1909-1911), *Histologie du système nerveux de l'homme et des vertébrés*, 2 Vols., L. Azoulay (Trans) reprinted by Instituto Ramón y Cajal del C.S.I.C., Madrid (1952-1955).

[2] C. Golgi, *Gazz. Med. Ital. Lombardo.* **6** (1873) 41-56.

[3] S. Ramón y Cajal, *Histologie du système nerveux de l'homme et des vertébrés* (CSIC., Madrid, 1972) p. 106.

[4] C.S. Sherrington, *A text book in physiology* (7th edn. Macmillan, London, 1897).

[5] E. Du Bois-Raymond, *D. Allgem. Muskel und Nervenphysik* **2** (1877) 700.

[6] J.C. Eccles, *The physiology of the synapse* (Springer Verlag, Berlin, 1964).

[7] J.D. Roberston, *Proc. Soc. Exptl. Biol. Med.* **82** (1953) 219-223.
[8] G.E. Palade and S.L. Palay, *Anat. Record* **118** (1954) 335-336.
[9] E. De Robertis and H.S. Bennett, *J. Biophys. Biochem. Cytol.* **1** (1955) 47-58.
[10] J. Del Castillo and B. Katz, *J. Physiol. London* **124** (1954) 574-585.
[11] P. Fatt and B. Katz, *J. Physiol. London* **117** (1951) 109-129.
[12] P. Fatt, *Physiol. Rev.* **34** (1954) 674-710.
[13] L.G. Brock, J.S. Coombs and J.C. Eccles, *J. Physiol. London* **117** (1952) 431-460.
[14] L. Tauc, *J. Physiol. Path. Gen.* **47** (1955) 769-792.
[15] T.D. Pollard and J.A. Cooper, *An. Rev. Biochem.* **55** (1986) 987-1035.
[16] P.M. Steinert and D.R. Roop, *Ann. Rev. Biochem.* **57** (1988) 593-625.
[17] A.P. Dustin, *Microtubules* (2nd. edn. Springer Verlag, New-York, 1984).
[18] S. Okabe and N. Hirokawa, *Nat* **343** (1990) 479-482.
[19] A. Matus, *Ann. Rev. Neurosci.* **11** (1988) 29-44.
[20] A.S. Kashina, G.C. Rogers and J.M. Scholey, *Biochim. Biophys. Acta* **1357** (1997) 257-271.
[21] P.W. Baas, *J. Chem. Neuroanat.* **14** (1998) 175-180.
[22] L. Ferhat, C. Cook, M. Chauvière, M. Harper, M. Kress, G.E. Lyons and P.W. Baas, *J. Neurosci.* **18** (1988) 7822-7835.
[23] R.A. Nixon, *Bioassays* **20** (1998) 798-807.
[24] P.W. Baas, *Neuron* **22** (1999) 23-31.
[25] D.B. Murphy and G.G. Borisy, *PNAS* **72** (1975) 2696-2700.
[26] D.W. Cleveland, S.Y. Hwo and M.W. Krirschner, *J. Mol. Biol.* **25** (1977) 207-225.
[27] L. Fehrat, A. Represa, W. Fehrat, Y. Ben-Ari and M. Khrestchatisky, *Eur. J. Neurosci.* **10** (1998) 161-171.
[28] F. Bradke and G.G. Dotti, *Neuron* **19** (1977) 1175-1186.
[29] E. Fuchs and I. Hanukoglu, *Cell* **34** (1983) 332-334.
[30] Ramón y Cajal, *Anat. Anz.* **5** (1890) 85-95, 111-119, 609-613, 631-639.
[31] H. Thoenen and Y.A. Barde, *Physiol. Rev.* **60** (1980) 1284-1335.
[32] B. Chamak and A. Prochiantz, *C.R. Acad. Sci.* **308** (1989) 353-358.
[33] M. Stahlhut, V. Berezin, E. Bock and J.P. Ternaux, *J. Neurosc. Res.* **48** (1997) 112-121.
[34] R.G. Harrison, *J. Exp. Zool.* **9** (1910) 787-846.
[35] P.C. Bridgman and M.E. Dailey, *J. Cell Biol.* **108** (1989) 95-109.
[36] M.E. Dailey and P.C. Bridgman, *J. Cell Biol.* **107** (1989) 39a.
[37] C. Bandtlow, T. Zachelder and M. Schwab, *Neuron* **8** (1990) 831-842.
[38] J.P. Letourneau, S.B. Kater and E.R. Macagno, *The nerve growth cone* (Raven Press, New-York, 1991).
[39] N. Alessandri-Haber, C. Paillart, C. Arsac, M. Gola, F. Couraud and M. Crest, *J. Physiol. London* **518** (1999) 203-214.
[40] J.A. Freeman, P.B. Manis, G.J. Snipes, B.N. Mayes, P.C. Samson, J.P. Wikswo Jr. and D.B. Freeman, *J. Neurosci. Res.* **13** (1985) 257-283.
[41] H.S. Young and N.M. Poo, *Nat* **305** (1983) 634-637.

[42] Q.J. Zheng, M. Felder, A.J. Connor and M. Poo, *Nat* **368** (1994) 140-144.
[43] Q.J. Zheng, J.J. Wan and M. Poo, *J. Neurosci.* **16** (1996) 1140-1149.
[44] M. Barthélémy-Requin, P. Portalier, M.C. Chamoin and J.P. Ternaux, *Neurochem. Res.* **25** (2000) 377-384.
[45] M. Hamann, M.C. Chamoin, P. Portalier, L. Bernheim, A. Baroffio, H. Widmer, C.R. Bader and J.P. Ternaux, *J. Physiol. London* **489** (1995) 791-803.
[46] P. Clark, P. Connolly, A.S.G. Curtis, J.A.T. Dow and C.D.W. Wilkinson, *Development* **108** (1990) 635-644.
[47] A.S.G. Curtis and P. Clark, *Crit. Rev. Biocompatibility* **5** (1990) 343-362.
[48] S. Bataillé, P. Portalier, P. Coulon and J.P. Ternaux, *Eur. J. Neurosci.* **10** (1998) 560-572.
[49] D.H. Barron, *J. Comp. Neurol.* **78** (1943) 1-26.
[50] I.B. Heath, *Can. J. Bot.* **73** (1995) S131-S139.
[51] A. Acheson, W. Vogl, W.B. Huttner and H. Thoenen, *Eur. Mol. Biol. Org.* **5** (1986) 2799-2803.
[52] Y. Wang, R. Roman, S.D Lidofsky and J.G. Fitz, *PNAS* **93** (1996) 12020-12025.
[53] L. Limozin, P. Pelcé, L. Savtchenko, M.C. Chamoin and J.P. Ternaux, *Neuroreport* **8** (1997) 2957-2960.
[54] S.A. Lipton and S.B. Kater, *Trends Neurosci.* **12** (1989) 265-270.
[55] S.H. Young and N.M. Poo, *Nat* **305** (1983) 634-637.
[56] S.A Greenfield, U.P. Hansen, W.S. Long, C.L. Slayman and J. Warncke, *J. Memb. Biol.* **39** (1989) 333-367.
[57] M. Israel and B. Lesbats, *J. Neurochem.* **37** (1981) 1475-1483.
[58] J.P. Ternaux and M.C. Chamoin, *Chemilumin.* **9** (1994) 65-72.
[59] L.W. Role and D.K. Berg, *Neuron* **16** (1996) 1077-1085.
[60] J. Bufler, C. Franke, H. Parnas and Dudel, *Eur. J. Neurosci.* **8** (1996) 677-687.
[61] M.P. Mattson and S.B. Kater, *Neurosci.* **7** (1987) 4034-4043.
[62] J.Q. Zheng, *Nat* **403** (2000) 89-93.
[63] P.G. Layer, *Bioassays* **12** (1990) 415-420.
[64] C. Catone, P. Portalier and J.P. Ternaux, *Annual congress of the Society for Neurosciences* (New Orleans, 2000).
[65] R.D. Shoop, N. Yamada and D.K. Berg, *J. Neurosci.* **20** (2000) 4021-4035.
[66] H.B.M. Uylings, Ph.D. Thesis, Uva, Amsterdam, 1977.
[67] H.B.M. Uylings, J. Van Pelt, R.W.H. Verwer and P. McConnell, *Statistical analysis of neuronal populations. Computer techniques in Neuroanatomy*, edited by Capowski (Plenum, New-York, 1989).
[68] R.E. Burke, W.B. Marks and B. Ulfhake, *J. Neurosci.* **12** (1992) 2403-2416.
[69] W. Kliemann, *Bull. Math. Biol.* **49** (1987) 135-152.
[70] K. Horsfield, M.J. Woldenberg and C.L. Bowes, *Bull. Math. Biol.* **49** (1987) 413-429.

[71] J. Van Pelt, H.B.M. Uylings, R.W.H. Verwer, R.J. Pentney and M.J. Woldenberg, *Bull. Math. Biol.* **54** (1992) 759-784.
[72] P.C. Letourneau, T.A. Shattuck and A.H. Ressler, *Cell Motil. Cytoskel.* **8** (1987) 193-209.
[73] S.B. Kater, R.W. Davenport and P.B. Guthrie, *Prog. Brain Res.* **102** (1994) 49-60.
[74] M.M. Black, *Prog. Brain Res.* **102** (1994) 61-77.
[75] L. Marsh and P.C. Letourneau, *J. Cell Biol.* **99** (1984) 2041-2047.
[76] K.M. Yamada, B.S. Spooner and N.K. Wessels, *PNAS* **66** (1970) 1206-1212.
[77] J.R. Bamburg, D. Bray and K. Chapman, *Nat* **321** (1986) 788-790.
[78] P.R. Gordon-Weeks, *Bioassays* **13** (1991) 235-239.
[79] M.A. Lewis and J.D. Murray, *J. Math. Biol.* **31** (1992) 25-71.
[80] P.C. Letourneau and A.H. Ressler, *J. Cell. Biol.* **98** (1984) 1355-1362.
[81] T. Mitchinson and M. Kirschner, *Neuron* **1** (1988) 761-772.
[82] D. Bray and K. Chapman, *J. Neurosci.* **12** (1985) 3204-3213.
[83] J. Dodd and T.M. Jessell, *Sci* **242** (1988) 692-699.
[84] M.P. Van Veen and J. Van Pelt, *Prog. Brain Res.* **102** (1994) 95-108.
[85] G.F. Oster, Cell motility and tissue morphogenesis, in *Cell shape determinants, regulation and regulatory control*, edited by W. Stein and F. Bronner, (Academic Press, New-York, 1988).
[86] W. Alt, in *Biomechanisms of active movement and deformation of cells* (NATO ASI Series, 42, Springer Verlag, Heidelberg, 1990).
[87] T.P. Stossel, *Sci* **260** (1993) 1086-1094.
[88] J.P. Kapfhammer and J.A. Raper, *J. Neurosci.* **7** (1988) 201-212.
[89] M.E. Schwab, *Trends Neurosci.* **13** (1990) 452-456.
[90] C.E. Bantlow, M.F. Schmidt, T.D. Hassinger, M.E. Schwab and S.B. Kater, *Sci* **259** (1993) 80-83.
[91] J.E. Vaughan, C.K. Hendrikson and J.A. Grieshaber, *J. Cell Biol.* **60** (1974) 664-672.
[92] M. Berry, P. McConnell and J. Sievers, Dendritic growth and the control of neuronal form, in *Current topic in developmental Biology*, edited by R.K. Hunt (Academic Press, London, 1980).
[93] N.K. Wessels and R.P. Nutall, *Exp. Cell Res.* **115** (1978) 111-122.
[94] D. Bray, *J. Cell Sci.* **37** (1979) 391-410.
[95] M.P. Van Veen, P.C. Letourneau and J. Van Pelt, in *Dynamic mechanisms of neuronal outgrowth*, Ph.D. Thesis, Free University, Amsterdam, 1993.
[96] P.C. Letourneau, T.A. Shattuck and A.H. Resslear, *J. Neurosci.* **6** (1986) 1912-1917.
[97] M.P. Van Veen and J. Van Pelt, *J. Theor. Biol.* **159** (1992) 1-24.
[98] M.P. Van Veen and J. Van Pelt, *Bull. Math. Biol.* **55** (1993) 249-273.
[99] L.G. Bergen and G.G. Borisy, *J. Cell Biol.* **84** (1980) 141-150.
[100] D. Soifer and K. Mack, Microtubules in the nervous system, in *Handbook of Neurochemistry*, edited by A. Lajtha (Plenum Press, New-York, 1984), pp. 245-280.

[101] L.A. Amos and W.B. Amos, *Molecules of the cytoskeleton* (MacMillan Molecular Biology Series, MacMillan Education Ltd., London, 1991).

[102] F. Verde, M. Dogterom, M. Stelzer, E. Karsenti and S. Leibler, *J. Cell. Biol.* **118** (1992) 1097-1108.

[103] M. Dogterom and S. Leibler, *Phys. Rev. Lett.* **70** (1993) 1347-1350.

[104] N.K. Pryer, R.A. Walker, V.P. Skeen, B.D. Bourns, M.F. Soboeiro and E.D. Salmon, *J. Cell Biol.* **103** (1992) 965-973.

[105] G. Shaw and D. Bray, *Exp. Cell Res.* **104** (1977) 55-62.

[106] D.W. Cleveland, M.A. Lopata, P. Sherline and M.W. Kirschner, *Cell* **25** (1981) 537-546.

[107] S. Okabe and N. Hirokawa, *J. Cell Biol.* **107** (1988) 651-664.

[108] P.N. Hoffman, M.A. Lopata, D.F. Watson and R.F. Luduena, *J. Cell Biol.* **119** (1992) 595-604.

[109] C. Faivre, C. Legrand and A. Rabié, *Int. J. Dev. Neurosci.* **3** (1985) 559-565.

[110] A. Matus, *Annu. Rev. Neurosci.* **11** (1988) 29-44.

[111] T.J. Dennerl, P. Lamoureux, R.E. Buxbaum and S.R. Heidemann, *J. Cell Biol.* **109** (109) 3073-3083.

[112] P. Lamoureux, J. Zheng, R.E. Buxbaum and S.R. Heidemann, *J. Cell Biol.* **118** (1992) 665-661.

[113] D. Bray, *Dev. Biol.* **102** (1984) 379-389.

[114] J. Zheng, P. Lamoureux, V. Santiago, T. Dennerl, R.E. Buxbaum and S.R. Heidemann, *J. Neurosci.* **11** (1990) 1117-1125.

[115] R.E. Buxbaum and S.R. Heidemann, *J. Theor. Biol.* **134** (1988) 379-390.

[116] D. Bray, *Trends Neurosci.* **10** (1987) 431-434.

[117] D.J. Golberg and D.W. Burgmeister, *Trends Neurosci.* **12** (1989) 503-506.

[118] J. Van Pelt and R.W.H. Verwer, *Bull. Math. Biol.* **48** (1986) 197-211.

[119] J. Van Pelt, R.W.H. Verwer and H.B.M. Uylings, *J. Neurosci. Meth.* **18** (1986) 153-165.

[120] R.W.H. Verwer, J. Van Pelt and H.B.M. Uylings, An introduction to topological analysis of neurons, in *Quantitative Methods in Neuroanatomy*, edited by F. Steward (Wiley, London, 1992).

[121] H.B.M. Uylings, K. Kuypers, M.C. Diamond and W.A.M. Veltman, *Exp. Neurol.* **62** (1978) 658-677.

[122] C.S. Cohan and S.B. Kater, *Sci* **232** (1986) 1638-1640.

[123] R.D. Fields, E.A. Neatle and P.G. Nelson, *J. Neurosci.* **10** (1990) 2950-2964.

[124] P.G. Haydon, D.P. McCobb and S.B. Kater, *J. Neurobiol.* **18** (1987) 197-215.

[125] M.P. Mattson and S.B. Kater, *Brain Res.* **478** (1989) 337-348.

[126] V. Redher and S.B. Kater, *J. Neurosci.* **12** (1992) 3175-3186.

[127] H. Parnas and L. Segel, *J. Theor.* **84** (1980) 3-29.

[128] W.J. Wadman and J.A. Connor, *Neurosci.* **48** (1992) 293-305.

[129] S.B. Kater, M.P. Mattson, C. Cohan and J.A. Connor, *Trends Neurosci.* **11** (1988) 315-321.

[130] D.J. Golberg, D.W. Burmeister and R.J. Rivas, in *The nerve growth cone*, edited by P.C. Letourneau, S.B. Kater and E.R. Macagno (Raven Press, New-York, 1994), pp. 79-95.
[131] P. Dustin, *Microtubules* (Springer Verlag, Berlin, 1978).
[132] J. Diaz-Nido, R. Armas-Portela, A. Martinez, M. Rocha and J. Avila, in *The nerve growth cone*, edited by P.C. Letourneau, S.B. Kater and E.R. Macagno (Raven Press, New-York, 1994) pp. 39-53.
[133] D. Purves, J.T. Voyvodic, L. Magrassi and H. Yawo, *Sci* **238** (1987) 1122-1126.
[134] N.A. O'Rourke and S.E. Fraser, *Neuron* **5** (1990) 159-171.
[135] H.G.E. Hentschel, D. Samuels and A. Fine, *Physica A* **254** (1998) 46-61.
[136] H.G.E. Hentschel and A. Fine, *Proc. R. Soc. London* **263** (1996) 1-8.
[137] D. Samuels, H.G.E. Hentschel and A. Fine, *Phil. Trans. R. London* **351** (1996) 1147-1156.
[138] W.S. McCulloch and W.H. Pitts, *Bull. Math. Biol.* **5** (1943) 115-143.
[139] J.A. Anderson and E. Rosenfeld, *Neurocomputing* (Foundations of Research, Cambridge, Mass., MIT Press, 1988).
[140] J.A. Anderson and E. Rosenfeld, *Neurocomputing II* (Directions of Research, Cambridge, Mass., MIT Press, 1990).
[141] C. Koch and I. Segev, *Methods in neuronal modeling*, From synapses to networks (Cambridge, Mass. The MIT Press, 1989) p. 524.
[142] R. Durbin, C. Mial and G. Mitchson, *The computing neuron* (Wokingham, Addison Wesley, 1990).
[143] J.J.B. Jack, D. Noble and R.W. Tsien, *Electric current flow in excitable cells* (Oxford University Press, 1983).
[144] W. Rall, Cable theory for dendritic neurons, in *Methods in neuronal modeling*, edited by C. Koch and I. Segev (MIT Press, Cambridge, Mass., 1989) pp. 9-62.
[145] I. Segev, J.W. Fleshman, and R.E. Burke, Compartmental models of complex neurons, in *Methods in neuronal modeling*, edited by C. Koch and I. Segev (MIT Press, Cambridge, Mass., 1989) pp. 63-96.
[146] A. Zador, C. Koch and R.K. Brown, *PNAS* **87** (1990) 6718-6721.
[147] S.W. Jaslowe, *Neurosci.* **47** (1992) 495-519.
[148] B.W. Mel, *Neural. Comp.* **4** (1992) 502-516.
[149] B.W. Mel, The Clusteron: toward a simple abstraction for a complex neuron, in *Advances in neural information processing systems*, edited by J. Moody, S. Hanson and R. Lippmann (Morgan Kaufmann, 1992) pp. 35-42.
[150] B.W. Mel, *J. Neurophysiol.* **70** (1993) 1086-1101.
[151] I. Segev, *Trends Neurosci.* **15** (1992) 414-421.
[152] A. Baginskas, A. Gutman and G. Svirskis, *Neurosci.* **53** (1993) 595-603.
[153] W. Rall and I. Segev, Functional possibilities for synapses on dendrites and dendritic spines, in *Synaptic function*, edited by G.M. Edelman, W.E. Gall and W.M. Cowan (John Wiley and Sons, New-York, 1987) pp. 637-698.

[154] G.M. Sheperd and R.K. Brayton, *Neurosci.* **21** (1990) 6718-6721.
[155] T.J. Sejnowski, Open questions about computation in cerebral cortex, in *Psychological and Biological models*, edited by J.L. McCelland and D.E. Rumelhart (MIT Press, Cambridge, Mass., 1986) pp. 372-389.
[156] T.J. Sejnowski, C. Koch and P.S. Churchland, *Comp. Neurosci.* **241** (1988) 1299-1306.
[157] H. Bras, S. Korogod, P. Gogan and S. Tyc-Dumont, *Europ. J. Neurosci.* **5** (1993) 1485-1493.
[158] S. Korogod, *Biol. Cybern.* **74** (1996) 85-93.
[159] S. Korogod, H. Bras, V.N. Sarana, P. Gogan and S. Tyc-Dumont, *Europ. J. Neurosci.* **6** (1994) 1517-1527.
[160] G. Horcholle-Bossavit, S. Korogod, P. Gogan and S. Tyc-Dumont, *Neurophysiol. Kiev.* **29** (1997) 62-75.
[161] L.J. De Felice and A. Issac, *J. Stat. Phys.* **70** (1992) 339-354.
[162] P. Gogan, S. Tyc-Dumont, I. Schmiedel-Jakob and Y. Chitti, *Biophys. J.* **69** (1995) 299-310.
[163] H. Bras, P. Gogan and S. Tyc-Dumont, *Neurosci.* **22** (1987) 947-970.

CONTRIBUTED PAPER 4

Chemical Waves and Dendrites Navigation during Self-Wiring of Neural Nets

R. Segev and E. Ben-Jacob

1. INTRODUCTION

In mammals, the precise structure of the neural circuits in the brain cannot be stored genetically, since the DNA lacks sufficient memory for such amounts of information. In humans for example, the DNA consists of about 10^8 bases and there are approximately 10^{15} synaptic connections. On the other hand, the network development is not entirely random. It appears that the degree of randomness in the synaptic connections is large on the micro scales (<1 mm) but on the macro scale (>2 cm) the organization is almost entirely deterministic [1]. A fundamental question is, then, what principles determine the network structure, and in particular what is the role of randomness *vs.* stored genetic information of its generation process? Furthermore, since the development is not entirely deterministic, what is the effect of the environment, *i.e.* external stimulus, on the network organization? Naturally, when we discuss the principles that organize the network, we assume that more than one structure of connections is possible. Hence we face the question of characterizing the different possible networks, and their effect on the network functioning.

A major obstacle in unraveling the fundamental principles of adaptive self-wiring is the complexity of the real brain. Hence, much effort is devoted to *in vitro* experimental studies of much simpler 2D systems [2,3]. The theoretical studies presented here are for such 2D systems.

Our model is composed of two main generic features:

1. a navigation strategy for the micro level, *i.e.* length scale below 1 mm or about $10^3 - 10^4$ cells;
2. a mechanism for adaptive self-organization on the meso (about 1 mm – 1 cm) and macro scale (above 1 cm).

The contemporary view is that the brain structure is essentially deterministic on a large scale but probabilistic on a small scale [1]. As a consequence the network has no optimal structure. But we believe that neural networks can be constructed in an optimal way, and this optimal way is derived from the biological mechanisms that construct the network. In any approach to the construction of the network there must be a precise strategy by which the neurons find each other to establish synaptic connections. At the beginning of the growth process the neurite has to migrate from it's own cell soma. The neurite migrates to the area in which it is supposed to form a synaptic connection. In this area there are many neurons, each of which is a possible target cell for the neurite. When the neurite approaches one of the possible target cells, with which it will finally form a synaptic connection, it has to be attracted to that cell soma.

The navigation strategy we propose is based on sophisticated means of chemical signaling for communication and regulation, including repulsive and attractive chemotaxis (chemotaxis is a movement based on a gradient of a chemical agent). For reasons presented in Section 4 we propose that the additional mechanism for self-wiring on the meso and macro scales is based on chemical waves. This proposal is inspired by the $cAMP$ waves employed during aggregation of *Dictyostelium discoiedum* amoeba [4, 5] and is also motivated by the fact that chemical waves (including of $cAMP$) play a role during the brain activity.

2. NEURITE AS AMOEBA WITH A TAIL

Neurites are equipped for navigation with a unit called growth cone [6, 7]. It is known that the growth cone is capable of measuring concentration and concentration gradients of chemicals [3, 8], including the gradients of repulsive and attractive chemotactic agents.

The movement of the growth cones appears to be a non-uniform random walk [9, 10]. The typical growth rate is of the order of micron/minute [11]. In the presence of a chemotactic agent, the movement is biased towards (attractive chemotactic) or away (repulsive chemotactic) from the chemical gradient. The movement is reminiscent of that of amoebae *Dictyostelium discoiedum* [4] in the presence of chemotactic materials. For that reason Lumins refers to the growth cone and its neurite as "amoeba with a tail" [12].

In modeling the neurite navigation we were inspired by the bions model used in the study of amoebae aggregation [5] and by the communicating walker model used in the study of bacterial colonies [13–15]. In the model the growth cones are represented by walkers which perform off-lattice non-uniform random (biased) walk [16]. The chemical dynamics (*e.g.* chemotactic agents, triggering field) are described by continuous reaction-diffusion equations solved on a tridiagonal lattice with a lattice constant $a_0 = 10\ \mu$m. Each of the soma is represented by a stationary (not moving) unit occupying one lattice cell. The neurite are simply defined as the trajectory performed by the growth cone.

We assume that each of the soma cells continuously emits a repulsive agent. When a neurite first sprouts it is mainly affected by the repulsive agent and moves away from its "mother" soma cell. It then continues to move on a trajectory which maximizes the distances from the surrounding soma cells (see [17, 18] for more details).

When a neurite reaches a specific length determined by its soma cell *via* the adjustment of the internal energy of the neurite, it does two things:

1. it switches its chemotactic sensitivity from sensitivity to the repulsive agent to sensitivity to the attractive;
2. it emits a quantum of a triggering material (which satisfies an equation similar to Eq. (1) of [17]). Soma cells in the neighborhood respond by emitting a quantum of attractive agent if they sense an above threshold concentration of the triggering material. As a result, the growth cone moves towards the soma cell with the strongest attractive response (typically, the one closest to the growth cone. The above features are included in the model, as we show in detail in [16–18].

3. A NEED FOR ADDITIONAL MECHANISM

So far we have described a navigation strategy for the microlevel. This strategy on its own is limited in the variety and complexity of the wiring patterns it can produce (see Fig. 1c). In these figure we see that the patterns have simple homogeneous structure of connectivity. Hence the genetic effect on the wiring patterns produced by this strategy has to be limited. At the same time it is known that the brain has a specific organization on the macro-scale (centers of activity over 1 cm) which is genetically determined. In general it is not known what happens on the meso-scale between the apparent randomness of the micro-scale and the deterministic organization of the macro-scale [1]. We believe (unlike Ref. [1]) that the micro-structure is not entirely random nor the macro-structure is fully deterministic. But more important, we expect a gradual transition from the meso-structure to the macro-structure with partial deterministic organization on the meso-scale. If so, there should be an additional mechanism (to the navigation strategy) to enable meso and macro scales self-organization.

4. THE EXCITABLE MEDIA MECHANISM

It is well established that chemical waves, which affect the electrical activity of neurons, can propagate through out the cortex [19–22]. This brings to mind that such chemical waves can regulate the wiring process over the meso and macro scales.

A classical example of self-organization mediated *via* chemical waves is the aggregation of the slime mold *Dictyostelium amoebae* during starvation.

These microbes collectively form an excitable media which produces spiral waves of $cAMP$. The movement of the individual organism is biased by the local gradient of the $cAMP$ (chemotaxis response). The result is a regulated movement towards the center of the colony along well defined stream lines [4,5].

The embryonic media is composed of two main cell types: neurons and glia. The glia cells are the majority of the cell population in the brain (90% of the brains cells are glia) [7]. It is widely accepted that the glia give the structural support to the brain and they play the main part in the neurons nutrition. Inspired by the aggregation of slime mold amoebae we propose an additional role to the glia: they act as an excitable medium and form chemical waves of at least one additional agent referred to as Glia Chemical Agent (GCA) needed for self-wiring on the meso and macro scale. The GCA (which can also be $cAMP$) plays a similar role to that of the $cAMP$ during amoebae aggregation. It acts as an additional chemotactic agent for the growth cone movement.

In order to turn the glia cells into an excitable field in the brain tissue we place randomly additional stationary units (representing the glia) on the 2D hexagonal lattice with a density ρ_G. Typically the 2D density ρ_G should be $0.05 - 0.2$ in numbers. Each glia unit can be in one of three possible states:

1. rest state;
2. excitable state, and
3. refractory state.

Then, at rest, the unit measures the concentration of the chemotaxis at its location, at each time step. Once the concentration is above a threshold level, the unit becomes excited (enter state 2) and starts to emit a quantum of GCA during τ_1 time units. Then it enters the refractory state and it is immune to further excitation. After another τ_2 time units the state progresses to rest state.

Numerical experiment of these system reveals [5] two basic patterns of waves that propagate in the medium. In Figure 2a we show a spiral wave that propagates in the medium in response to appropriate initial conditions, and in Figure 2b we show concentric waves that propagate in response to a periodic pacemaker at the center of the grid.

The measured velocity of $cAMP$ waves is about 300 μm/min [4] and the measured velocity of the waves in the cortex is about 3000 μm/min [22]. The velocity of the waves in simulation is $10a_0$ per 1 time step which corresponds to 10^{-2} cm/s. This is in agreement with the wave velocity measured in experiments (~0.06 cm/s).

We further assume that above a threshold level of the GCA concentration it acts as an additional chemotactic agent (either repulsive or attractive) on the movement of the walkers (growth cones). In Figures 1c–e we demonstrate the GCA effect in a system of 100 cells. The soma are restricted to emit only one neurite each in order to make the wiring pattern more transparent to the eye. In Figure 1c there are no chemical waves and indeed the wiring pattern "looks"

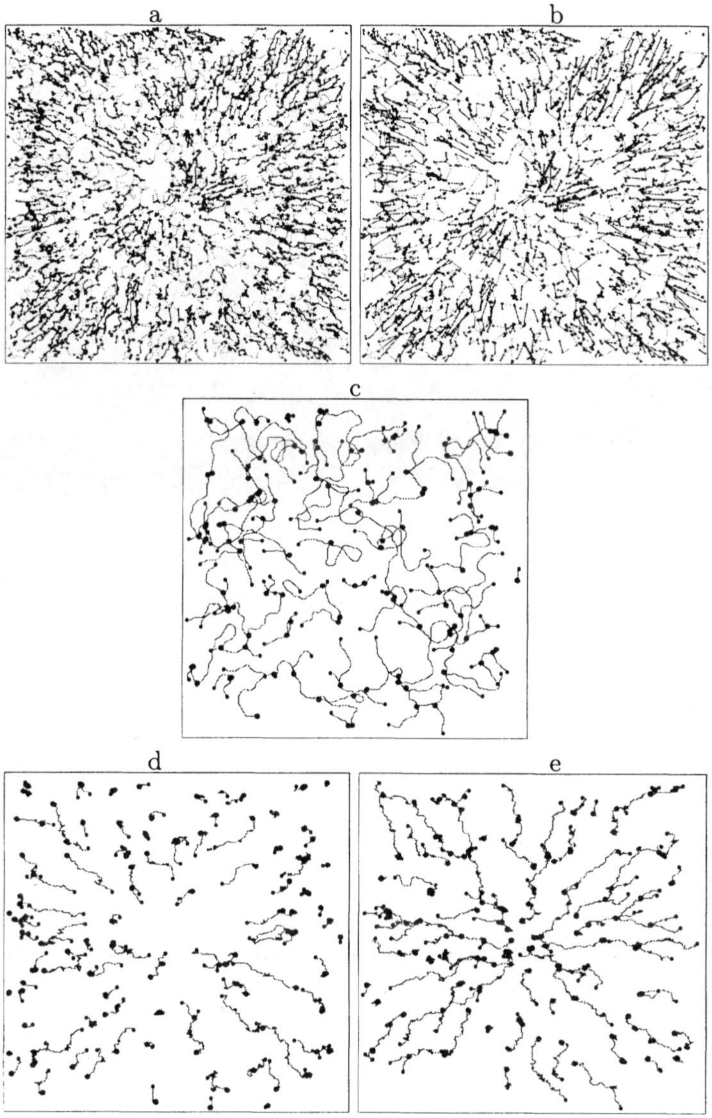

Fig. 1. — a) The effect of spiral wave on the network structure. In the simulation we assumed the GCA is an attractive agent. The spiral wave propagates in response to appropriate initial conditions. b) The same network as in Figure 1a but now the final position of the growth cones is linked by a straight line in order to make the connections pattern more conspicuous. c–f) In each network there are 100 cells which are restricted to emit only one neurite in order to make the connection patterns more conspicuous. c) There is no wave source at the center. The resulting network "looks" random. d) There is a pacemaker at the center of the grid. The GCA is assumed to be a repulsive agent. As a result the neurites migrate out from the center. e) Again a pacemaker at the center but now the GCA is assumed to be an attractive agent.

a b

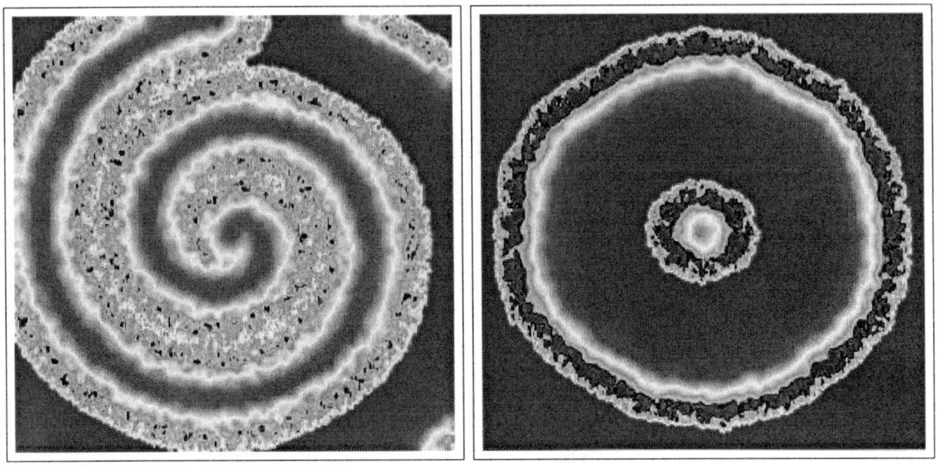

Fig. 2. — Waves propagating in the media for $\rho_B = 0.25$, $\tau_1 = .8$, $\tau_2 = 5$, $D = 10$, and $C_T = .001$. a) Spiral wave propagating in a 2D excitable media on a 200×200 grid. b) Target waves propagating in response to a pacemaker at the center of the grid (again 200×200).

random. Next we placed a pacemaker at the center to produce concentric waves. In Figure 1d we took the GCA to act as an attractive chemotaxis agent, and in Figure 1e to act as a repulsive one.

In Figure 1a we demonstrate the effect of a spiral wave. In this case the system is composed of 1250 cells, each soma emits two neurites and repulsive response is taken. In order to make the pattern of the network more transparent to the eye we line up neurites, *i.e.* we draw a line between the cell soma of and the walkers position (Fig. 1b).

5. CONCLUSIONS

We have presented a navigation strategy for micro-level network organization and an excitable media based wiring mechanism for the meso and macro levels organization. These mechanisms leads to the formation of neural networks with different structures, which can be genetically dependent. Our results raise two fundamental issues:

1. one needs to develop characterization methods (beyond number of connections per neurons) to distinguish the various possible networks;
2. the relations between the network organization and its computational properties and efficiency.

The next step of the endeavor presented here is to include the effect of the network electrical activity on its self-wiring. Once some initial synaptic connection are formed, the network begins its electrical activity. It is natural to expect that from this point onwards the wiring dynamics depends on the electrical activity.

To conclude, we are still far from understanding the emergence of a functioning brain from a collection of neurons. Yet we believe our studies provide a useful first step towards this goal.

Acknowledgements

The research was supported in part by grant No. BSF 92-00051 from the Israel, USA binational foundation, by grant No. 593-95 from the Israeli Academy of Science, the Siegl Prize for Research, Sackler institute and The Adams Super Center.

REFERENCES

[1] M. Abeles, *Corticonics* (Cambridge University Press, 1991).
[2] B. Dwir, G. De Stasio, A. Sadeghi, D. Mercanti, M. Teresa Ciotti, G. Margaritondo and E. Kapon, *Directional growth of neurons on ploy-l-lysine patters, preprint* (1996).
[3] M. Tessier-Lavigne and C.S. Goodman, *Sci* **274** (1966) 1123.
[4] P.C. Newell, *J. Gen.; Microbiol.* **104** (1978) 1.
[5] D. Kessler and H. Levine, *Phys. Rev. E* **48** (1993) 4801.
[6] H.G.E. Hentschel and A. van Ooyen, *Proc. R. Soc. Lond. B* **266** (1999) 2231–2238.
[7] I.B. Levitan and L.K. Kaczmarek, *The Neuron* (Oxford University Press, 1991).
[8] M. Tessier-Lavigne and M. Placzek, *TINS* **14** (1991) 303.
[9] E. Braun (1997) (personal communication).
[10] H.M. Buettner, R.N Pittman and J.K. Ivins, *Develop. Biol.* **163** (1994) 407.
[11] R.W. Gundersen and J.N. Barrett, *Sci* **206** (1979) 1079.
[12] R. Lumins (1997) (personal communication).
[13] E. Ben-Jacob, *Nat* **373** (1995) 566.
[14] E. Ben-Jacob, O. Schochet, A. Tenenbaum, I. Cohen, A. Czirok and T. Vicsek, *Nat* **368** (1994) 46.
[15] E. Ben-Jacob, *Contemp. Phys.* **38** (1997) 205.
[16] R. Segev and E. Ben-Jacob, *Phys. Lett. A* **237** (1998) 307–313.
[17] R. Segev and E. Ben-Jacob, *J. Complex Syst.* **1** (1998) 67–78.
[18] R. Segev and E. Ben-Jacob, *Neural Netw.* (2000) (in press).
[19] J.D. Murray, *Mathematical Biology* (Springer-Verlag, 1989).
[20] V.M Fernandes de Lima *et al., Brain Res.* **614** (1993) 45.
[21] H. Martins-Ferrera and R.J. Do Carmo, *Can. J. Phys. Pharma.* **65** (1987) 1092.
[22] N.A. Grorelova and J. Bures, *J. Neurobiol.* **14** (1983) 353.

CHAPTER VI

The Mouse Embryonic Lung: A Biological Example of Branching Morphogenesis

S. Bellusci, A. Mailleux, D. Ndiaye and J.-P. Thiery

1. INTRODUCTION

The diversity of ramified structures existing in the animal and vegetal worlds has long been a topic of fascination for scientists. So far very few models have been developed to study the molecular basis of the branching process. In mammals, this process takes place during the development of many organs such as the mammary gland, lung, pancreas, kidney, etc. In most cases, ontogenesis starts from a simple structure, usually an epithelial tube surrounded by fibroblastic cells. As development proceeds, buds form to give rise to a ramified, functional structure. In the last decade, significant discoveries have been made to unravel the molecular mechanisms underlying the ramification process. We have used the mouse embryonic lung as a model system to elucidate the molecular basis of branching morphogenesis. The aim of this review is to give to the physicists working on the dynamics of arborescent structures some insight into the mechanism of bud formation. In particular, we will present the role of key signaling molecules in the process of bud formation during lung development.

2. EMBRYONIC LUNG DEVELOPMENT

The respiratory system in mouse is composed of two parts, the trachea and the lung. They both develop at 9.5 days post coïtum (dpc) from the foregut, which at that time is essentially a tube of epithelial cells surrounded by fibroblastic cells [1] (Fig. 1A). At that stage, lung development starts with an outpocketing of the endoderm of the ventral foregut, termed the laryngotracheal groove, which rapidly gives rise to the trachea and the two main bronchi.

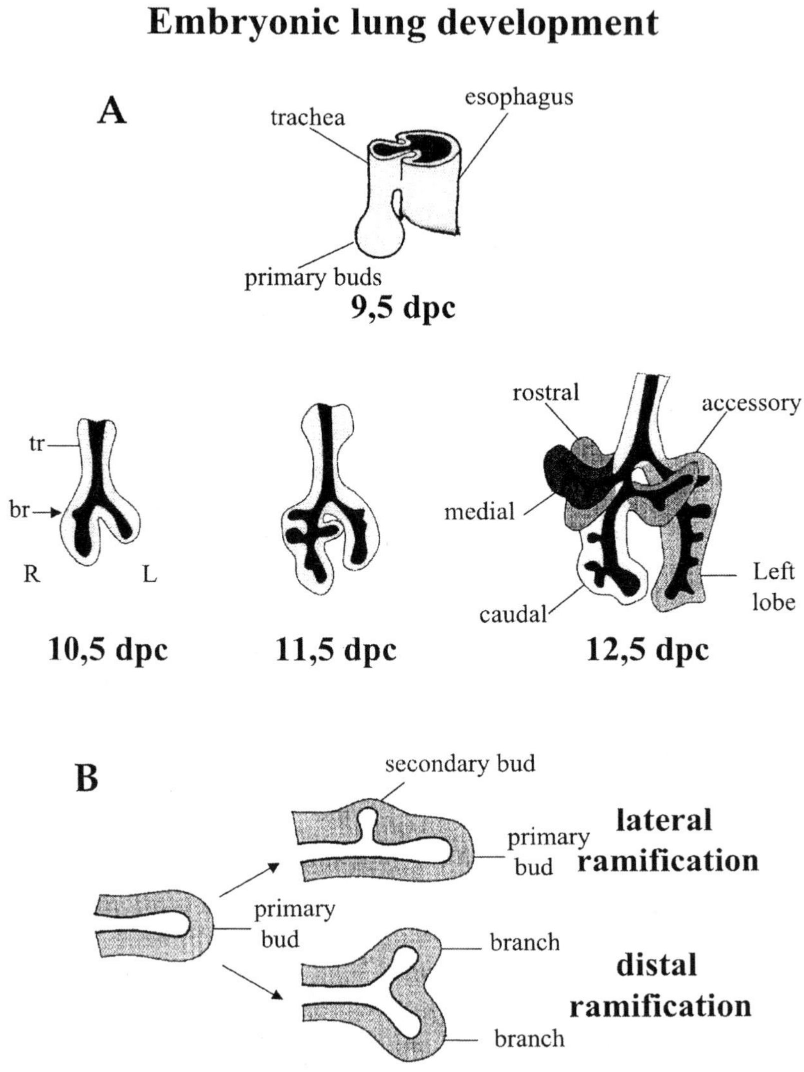

Fig. 1. — Lung branching morphogenesis. A) Embryonic lung development in mouse. Lung development starts at 9.5 dpc with the emergence of two primary buds from the ventral foregut (only one is showed in the picture). At 10.5 dpc, the lung is composed of a trachea (tr) and two bronchi (br). Note the presence of a secondary bud on the right bronchus (red arrow). From 11.5 to 12.5 dpc, the lung undergoes an important process of ramification to give rise to four lobes on the right (called rostral, medial, accessory and caudal lobes) and one lobe on the left. B) Ramifications from a primary bud can occur either laterally (secondary bud formation) or distally (branch formation).

From this point onward, lung morphogenesis involves the repeated cycles of bud initiation, elongation and growth arrest, a stereotypic pattern of branching and budding which can be followed precisely up to 13.5 dpc (see Fig. 1A). The overall result of this cycle process is the elaboration of a tree-like structure, the reproducibility of which, at least in the early phases of branching, is an important feature of lung development in many species. The right and left primary buds grow and sprout several secondary buds in a precise spatiotemporal sequence. This budding process is regulated differently in the right bronchus compared to the left bronchus. This difference leads to the formation an asymmetric structure observable very early in lung morphogenesis.

3. BUD FORMATION IS A MULTISTEP PROCESS

Given the complexity of the branching pattern, it is important to define the words primary bud, secondary bud and branches (Fig. 1B). A primary bud is defined as a knob-like cluster of cells that undergo morphological changes and proliferate to move away from the surface of a preexisting structure. Conversely, a secondary bud develops on a stalk derived from another bud, while branches arise by bifurcation of a terminal bud (see Fig. 1). As indicated above, bud formation involves several phases. The first is initiation, in which are established the boundaries of a bud primordium and the signaling center within it. The second phase, proximal-distal outgrowth, is best characterized by cell proliferation. During this phase, soluble factors such as Sonic hedgehog or Fibroblast Growth Factors, allow the proliferation of the mesenchymal or epithelial tissues respectively. Finally, growth arrest takes place. This phase is due in part to a combination of programmed cell death, expression of growth inhibitors and inhibition of distal mitogen-secreting centers.

Surprisingly enough, we are just beginning to understand how these different steps are controlled. In this review we will focus only on the first two phases of bud formation. Although equally important, the regulation of growth arrest will not be discussed here in detail.

4. INTERACTIONS BETWEEN EPITHELIAL AND MESENCHYMAL COMPARTMENTS ARE ESSENTIAL FOR THE BRANCHING PROCESS

In vertebrates, budding involves dynamic and reciprocal interactions between two cell populations: one fibroblastic and the other epithelial. Evidence of signaling between these two compartments came initially through tissue transplantation experiments. Transplantation studies carried out by Alescio and Cassini in 1962 first showed that the mesenchyme around the distal lung buds would induce ectopic buds when grafted next to the tracheal epithelium which normally never branches [2]. The ectopic buds express genes characteristic of distal endoderm [3]. Conversely, another study showed that the mesenchyme around the trachea would inhibit bud formation when grafted next to the distal epithelium [4]. This observation has received further experimental support since such a graft induces the expression of genes normally transcribed in the trachea [5]. Interposing a filter between the distal mesenchyme and the proximal epithelium does not prevent the induction of ectopic buds in such transplants, indicating that the distal mesenchyme produces soluble factors able to act either on the endoderm or on the basal layer to change cell behavior [6]. Recently, several genes expressed at high level either in the distal epithelium or in the distal mesenchyme have been identified [7]. In the sections which follow we will attempt to explain how these genes might work together to regulate branching morphogenesis in the mouse lung. It is to be hoped that the information provided by the mouse system will lead to a broader understanding of branching morphogenesis in a diversity of organisms.

5. SEVERAL CLASSES OF GENES ARE EXPRESSED IN THE EMBRYONIC LUNG

Given the powerful genetic tools available for studying the fruit fly *Drosophila melanogaster*, most of the genes that are relevant for lung branching morphogenesis were initially discovered in this organism. Mutations in these genes affect the dorso-ventral axis, segment polarity or development of internal organs such as the respiratory system. In most cases vertebrate homologues of these *Drosophila* genes have been characterized and the secreted proteins encoded by these genes have been assigned to four major molecular families: the Transforming Growth Factor beta superfamily (which include the Bone Morphogenetic proteins), the Fibroblast growth Factors, the Wnts and the Hedgehogs. The genes outlined above can be further classified, albeit arbitrarily, on the basis of the spatio-temporal pattern of their expression in the mouse embryonic lung (Fig. 2). The Class I genes (includes among others *Bone Morphogenetic Protein 4* (*Bmp4*), *Sonic hedgehog (Shh)* and *Wnt7b*) are expressed essentially in the distal endoderm. Conversely, Class II genes are expressed in the distal mesenchyme located around the epithelial buds. *Patched* (a Sonic hedgehog receptor), *Wnt2* and *Fgf10* are typical examples of Class II genes. Class III genes have a much broader expression profile, and the transcripts for these genes are found throughout all the mesenchyme and/or epithelium. Two of the three genes of the *Gli* family (see below) belong to this Class.

These highly localized patterns of gene expression suggest that the distal buds are active signaling centers all throughout the branching morphogenesis of the lung. It would appear that factors secreted by the epithelium act on the mesenchyme and *vice versa* to control locally cell proliferation, apoptosis and cell differentiation, the various forces that sculpt the developing lung. To investigate these hypothesized interactions between the epithelial and

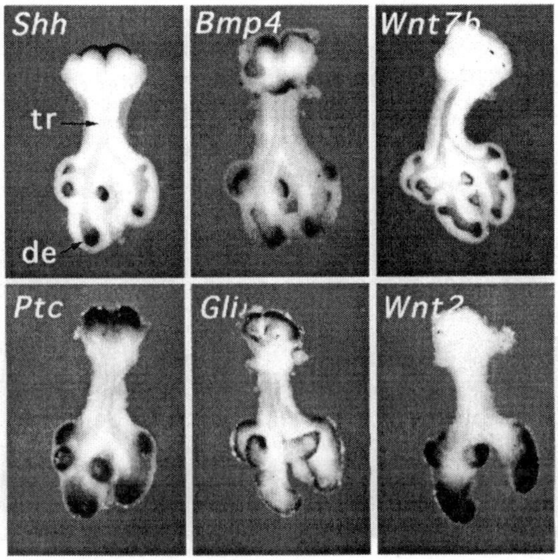

Fig. 2. — Examples of gene expression in the 11.5 dpc lung. Class I genes such as *Sonic hedgehog*, *Bone Morphogenetic Protein 4* and *Wnt7b* are expressed in the epithelium. Note the high level of gene expression in the distal epithelium (de) compared to that seen in the trachea (tr). Class II genes such as *Patched*, *Gli* and *Wnt2* are expressed in the distal mesenchyme. Whole mount *in situ* hybridization using digoxigenin-labeled antisense RNA probes.

mesenchymal tissues during lung morphogenesis, we have used a variety of approaches. These include classical tissue transplantation experiments, *in vitro* studies with purified proteins and genetic modifications (gene overexpression using a lung specific promoter and gene inactivation by homologous recombination in embryonic stem cells). Taken together the results paint a picture which supports the hypothesis offered above.

6. BONE MORPHOGENETIC PROTEIN 4 INHIBITS PROLIFERATION OF THE EPITHELIAL COMPARTMENT

Bmp4, the first Class I gene cited above, encodes a secreted protein belonging to the TGFβ superfamily. In the embryonic lung *Bmp4* is expressed at high level in the distal epithelium but it is also detected at a lower level in the adjacent mesenchyme [7]. This expression pattern suggests a possible role for *Bmp4* in lung morphogenesis. Classical genetics would normally offer the perfect means to test this hypothesis. Unfortunately, embryos homozygous for a mutation in the *Bmp4* gene die around 7.5 dpc [8]. So the classical genetic approach cannot be used in this case to study the role of this gene during lung development.

Since direct inactivation of *Bmp4* in the lung is not currently possible, we have attempted to study the role of this gene by overexpressing it.

We began by generating transgenic embryos overexpressing *Bmp4* under the control of the *human Surfactant Protein C* (*SP-C*) promoter [7], an approach used many times before with considerable success to overexpress specifically a gene in the distal lung epithelium [9]. This promoter allows basal transgene expression from 9.5 dpc up to 15.5 dpc at which time the expression level increases drastically in correlation with epithelial cell differentiation. The results are striking, transgenic lungs overexpressing *Bmp4* are smaller than the normal counterparts taken at the same stage, the decrease in size being associated both with a lower proliferation rate in the epithelium and a partial inhibition of type II pneumocyte differentiation throughout the lung [7]. These results indicate that BMP4 acts both as a negative regulator of epithelial cell proliferation and differentiation in this system.

7. SONIC HEDGEHOG INDUCES MESENCHYMAL CELL PROLIFERATION

Shh is another Class I gene of developmental significance in a number of systems. It encodes a secreted protein which typically exercises its effect on a neighboring population of responding cells. In the embryonic lung, *Shh* is expressed throughout all the lung epithelium but stronger distally than proximally. *Patched*, a gene encoding a receptor for the Sonic hedgehog protein, is expressed mainly in the mesenchyme, suggesting that this signaling molecule elicits its effect mostly in this compartment [10]. To investigate *Shh* function during lung development, we generated embryos overexpressing *Shh* in the distal epithelium. Transgenic embryos die at birth of a respiratory failure [10]. Analyses of the lungs indicate an increase in mesenchymal cell proliferation associated with normal epithelial differentiation. In addition, the blood vessels were not directly apposed to the alveolar spaces as in normal lungs but rather surrounded by multiple layers of cells. This defect in capillary invasion may explain the respiratory failure at birth. What are the genes controlled by Sonic hedgehog in the lung? As had already been demonstrated in *Drosophila*, we have shown that the transcription of *Patched* is up-regulated in *Shh* transgenic lungs [10]. In Drosophila, Hedgehog induces *Patched* transcription *via* Cubitus interruptus (Ci), a zinc finger transcription factor [11]. Three vertebrate homologues of *ci* called *Gli1*, *Gli2* and *Gli3* have been identified and all are expressed in the normal lung mesenchyme [12]. In lungs overexpressing *Shh*, expression of *Gli1* but not *Gli2* nor *Gli3* is up-regulated [10]. Recently inactivation of *Shh* and *Gli2*/*Gli3* in mouse has been reported. In embryos homozygous for a mutation in *Shh*, œsophagus and trachea are fused, and the respiratory epithelium is severely underdeveloped [13,14]. The double mutants *Gli2-/-*/*Gli3+/-* yield a phenotype similar to that seen with *Shh-/-* [15]. In double mutants

Gli2-/-/Gli3-/- the lung fails to form as in *Fgf10-/-* embryos [15]. These results indicate that *Gli* genes play an important role during lung development.

In *Drosophila*, Hedgehog is able to induce the expression of *decapentaplegic* (a *Bmp2/Bmp4* homologue) [11]. In the vertebrate lung, that simple relationship between Sonic hedgehog and Bmps is not conserved [10]. *Bmp4* expression is not increased in lungs overexpressing *Shh*, and *Shh* expression is not changed in lungs overexpressing *Bmp4*. These results indicate that the Sonic hedgehog and BMP4 signaling pathways act independently during lung ontogenesis. By contrast, our work on the role of FGF10 during lung morphogenesis shows that Sonic hedgehog controls *Fgf10* expression [16]. In conclusion, SHH regulates the state of proliferation of the mesenchymal tissue and controls specific gene expression as *Ptc*, *Gli* and *Fgf10*.

8. FIBROBLAST GROWTH FACTOR 10 IS ASSOCIATED WITH DIRECTIONAL GROWTH OF THE EPITHELIAL BUDS

As already summarized above, Alescio and Cassini's experiment in 1962 established that one or more factors expressed in the distal mesenchyme were able to induce ectopic bud formation from the trachea. This original result has more recently expanded through the use of a mutated form of the FGF receptor 2 IIIb which blocks FGF signaling during lung development. Expression of this mutated receptor in the distal lung epithelium (using the *SP-C* promoter) severely inhibited lung development in a very specific way [17]. The trachea and the two main bronchi were present and absolutely normal, but secondary buds and branches are absent. The transgenic mice die at birth, a result indicating that one or more members of the FGF family are essential for normal branching and thus survival. An obvious next step was to determine which FGF was important *in vivo* to control the branching process. To date, the FGF family comprises 19 members. Of those FGF1, FGF7 and FGF10 are expressed at high level in the developing lung [16]. FGF receptor 2-IIIb binds all three of these FGF members with a high affinity making them likely candidates for the regulation role in bud formation *in vivo*. Expression pattern analysis revealed that *Fgf10*, but not *Fgf1* nor *Fgf7*, is expressed in the lung mesenchyme at 11.5 dpc, a stage by which branching has already occurred. *Fgf1* and *Fgf7* are expressed only later in lung development (13.5 and 14 dpc respectively) [16]. *Fgf10* is expressed dynamically in the mesenchyme adjacent to the distal buds from the earliest stages of lung development. The temporal and spatial pattern of gene expression outlined above suggests that *Fgf10* plays an important role in directional outgrowth and possibly induction, of epithelial buds [16] (Fig. 3). That hypothesis has been tested on isolated distal lung endoderm cultured in Matrigel (a commercially available mixture of extracellular matrix protein). In the absence of growth factors, the lung endoderm fails to proliferate and dies. In the presence of FGF10, multiple primary buds are observed after as little as 24 hours of culture. After 48 hours of culture, branches are observed

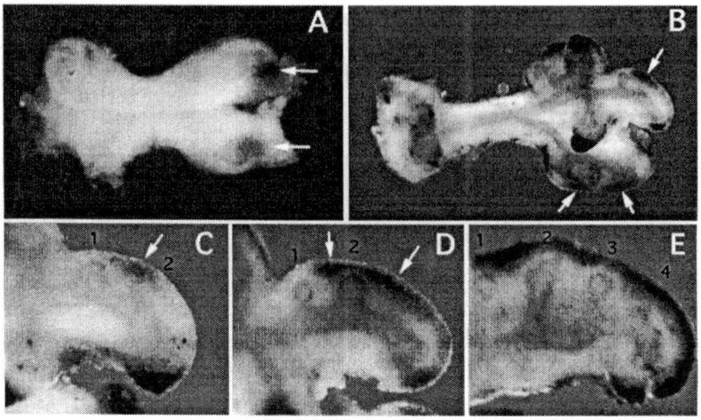

Fig. 3. — Dynamic pattern of expression of *Fgf10* during lung morphogenesis. A) *Fgf10* is expressed as early as 10.5 dpc in the distal mesenchyme of the two bronchi (white arrow). B) At 11.5 dpc, *Fgf10* is expressed in specific regions of the mesenchyme. This pattern of expression is associated with bud formation (the white arrows indicate areas where *Fgf10* is weakly expressed). C–E) *Fgf10* expression in the right caudal lobe between 11.5 dpc and 12.5 dpc (lungs were taken every 8 hours). The expression pattern suggests a role in the directional growth of the buds (see text). The numbers indicate the positions of the secondary buds forming laterally.

(see Fig. 4). Bud formation is associated with increased cell proliferation in the epithelium. Definitive evidence for *Fgf10*'s key role in the branching process comes from gene inactivation experiments. Embryos homozygous for a mutation in the *Fgf10* gene die at birth completely lacking limbs and possessing a severely altered respiratory system [18, 19]. Bronchi formation is completely abolished; the trachea is the only structure to form.

The pattern of *Fgf10* expression suggests that positive and negative regulators produced by the endoderm are able to act on the mesenchyme to regulate levels of *Fgf10* expression. So far the positive regulators are still unknown. Surprisingly, Sonic hedgehog which acts as a positive regulator of *Fgf* expression in many developmental systems has been shown to be a negative regulator of *Fgf10* expression. Several lines of evidence are supporting this conclusion:

(1) in transgenic lungs overexpressing *Shh*, *Fgf10* is down regulated [7];

(2) in lungs where *Shh* has been inactivated, *Fgf10* is clearly up regulated [14] and

(3) *in vitro* experiments with primary cultures of lung mesenchyme show that *Sonic hedgehog* elicits a decrease in *Fgf10* expression [20].

In conclusion, FGF10 is so far the only factor which has been associated directly with bud formation: localized sources of FGF10 in the mesenchyme regulate epithelial cell proliferation and bud outgrowth.

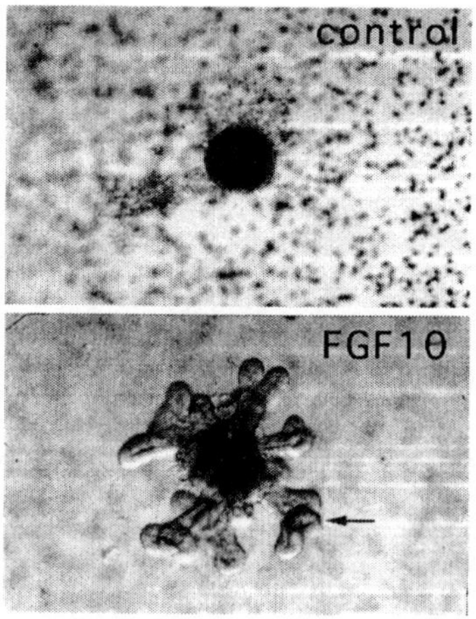

Fig. 4. — Activity of FGF10 in Matrigel cultures of mesenchyme-free endoderm. A) Endoderm grown for 48 hours in control medium, without FGFs. Most cells have died, but a few remain as a dense aggregate. B) Culture with 250 ng/ml of FGF10. Endoderm has grown and formed primary buds and branches (black arrow).

9. TWO VERTEBRATE HOMOLOGS OF THE DROSOPHILA GENE SPROUTY ARE EXPRESSED IN THE DEVELOPING LUNG

As previously mentioned, the powerful genetic tools available for studying the fruit fly *Drosophila melanogaster* make this model organism relevant to identify new genes regulating branching morphogenesis. The complex network of tubules forming the *Drosophila* respiratory system arises after gastrulation through a highly reproducible process of branching morphogenesis. The process begins with 10 pairs of tracheal placodes located on each side of the developing embryo. Each placode, initially formed from 50 to 80 epithelial cells, gives rise to 6 primary buds composed of 3 to 20 cells each. Each of these primary buds grows toward a specific target location. In contrast with the budding process occurring in the vertebrate lung, formation of the primary buds in *Drosophila* involves only cell migration (no cell proliferation). Later in development, a second round of branching takes place at the terminal part of each bud. The most distal cell sends out cytoplasmic extensions within which a lumen forms to accommodate a continuous network of tubules. These extensions correspond to the distal branches seen in mouse lung development. Formation of both primary buds and distal branches is dependent upon the expression of *branchless* and *breathless*, two *Drosophila* genes encoding a Fgf-like secreted molecule and its associated tyrosine kinase receptor, respectively [21, 22]. Breathless is expressed homogeneously throughout all the tracheal cells, whereas Branchless is expressed by a restricted number of cells surrounding the placode at precise locations at which primary buds will grow [21]. It is clear now that Branchless is involved in directional growth of the primary buds and subsequent branches, and that FGF10 is directly related to Branchless from a functional point of view. Recently, *sprouty*, a new mutant affecting tracheal placode morphogenesis has been characterized [23]. In this mutant, additional branches form from the primary buds. Genetic and molecular analysis show that Sprouty is involved in a negative feedback loop to regulate Branchless activity. Sprouty is a 63 kDa membrane-bound protein with a cysteine-rich domain which is conserved between vertebrates and invertebrates. Sprouty contains a putative signal peptide domain for secretion, but in spite of this, the protein is typically found associated with the plasma membrane. This result has been confirmed by a recent work by Casci *et al.*. They show that Sprouty is an intracellular protein which is bound to the cytoplasmic face of the membrane and which acts downstream of tyrosine kinase receptors by inhibiting the Ras pathway [24]. Thus, Sprouty appears to be a general inhibitor of growth factors signaling through tyrosine kinase receptors. Others and we have recently screened the Expression Sequence Tag databases of GenBank and found four mouse homologues of *Drosophila sprouty* [25–27]. In the developing lung, *Sprouty4* is expressed exclusively in the mesenchyme [25] whereas *Sprouty2* is found only in the epithelium, at least up to 13.5 dpc [26]. SPROUTY proteins could thus be involved in regulating growth factor activity in either or both the epithelial and the mesenchymal tissues. For example a recent study has shown that inactivation of *Sprouty2* in the embryonic lung using antisense oligonucleotides leads to an 80% increase in bud number (compared to control lungs or lung treated with sense oligonucleotides) [26]. Although the mechanism of action of *Sprouty2* remains to be elucidated, an obvious initial hypothesis is that this increase in bud number reflects an increase in FGF10 activity.

10. TOWARD AN INTEGRATED MODEL OF BUDDING

As outlined earlier, different growth factors expressed either in the epithelium or mesenchyme are regulating the branching process. These factors are inducing cell proliferation, apoptosis and cell differentiation. The goal of this section is to integrate seemingly disparate data concerning the regulation of the branching process into a coherent overview. As mentioned above, the complete cycle of budding consists of three phases: bud initiation, bud outgrowth and growth arrest (Fig. 5). The molecular mechanisms controlling the initiation phase are still unclear and will certainly be the focus of future studies. In particular it will be extremely important to identify the first genes expressed in the region where the bud will form. These early genes are likely to regulate the expression of the genes that we have already identified.

During the growth phase, FGF10 is so far the key molecule, and in the future it will be of vital importance to identify positive regulators allowing such precise

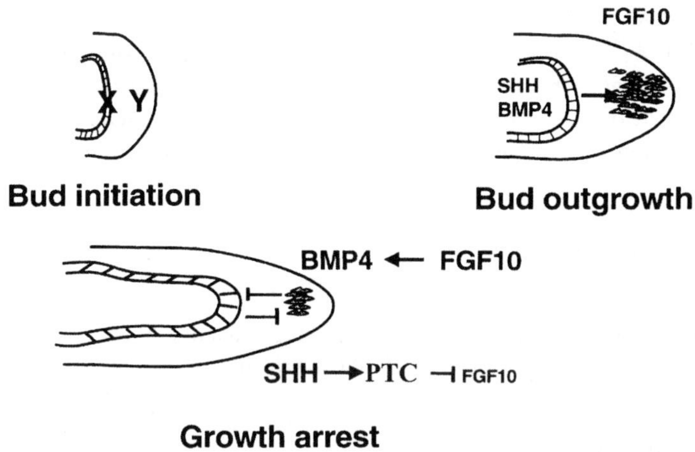

Fig. 5. — Integrated model of budding. Three phases are observed (1) bud initiation; (2) bud outgrowth and (3) growth arrest. During the initiation phase, unknown factors expressed in the epithelium (called X) and in the mesenchyme (called Y) are involved in the decision of the endoderm to form a bud. During the bud outgrowth phase, FGF10 in the distal mesenchyme acts on the endoderm to promote its proliferation. Concomitantly, SHH and BMP4 expression is building up in the endoderm. Factors responsible to positively regulate *Fgf10* and *Shh* expression have not yet been identified. By contrast, FGF10 induces *Bmp4* expression in the endoderm. During the growth arrest phase, high levels of BMP4 inhibit epithelial cell proliferation. In addition, SHH negatively regulates *Fgf10* expression which in turn leads to growth arrest of the epithelial bud.

spatio-temporal expression. FGF10 expressed in the distal mesenchyme probably acts on the epithelium through FGFR2-IIIb to promote cell proliferation. FGF10 could also modify cell-cell and cell-extracellular matrix interactions to facilitate or promote cell migration [28]. On the other hand, SHH secreted by the distal epithelium induces cell proliferation in the mesenchymal compartment through PTC and GLI. This process would allow directional growth of the buds toward the localized domain of *Fgf10* expression until a negative feedback loop blocks bud outgrowth. We still don't know what are the positive regulators of *Shh* and *Fgf10* expression. For example, factors like the WNTs (*Wnt7b* is expressed in the epithelium [7]) could regulate *Fgf10* expression.

During the growth arrest phase, Sonic hedgehog in the epithelium negatively regulates *Fgf10* expression in the mesenchyme (Fig. 5). The precise nature of this inhibition is unclear, and more experiments will have to be done to determine if the inhibition is directly promoted by SHH and at what threshold level the effect is elicited. In parallel, FGF10 expressed in the mesenchyme upregulates *Bmp4* expression in the epithelium, which in turn inhibits epithelial cell proliferation [20]. This model, built essentially on signaling molecules, is obviously incomplete. Other molecules such as the extracellular matrix proteins, the integrins and the metalloproteases certainly play important roles in the bud outgrowth and growth arrest phases. In the future, we will take into consideration discoveries concerning signaling pathways, cell-cell and cell-extracellular matrix interactions to propose a more realistic model of the lung branching process.

Acknowledgements

We are very thankful to Prof. Bill Thomas for critical review of this manuscript and to Dominique Morineau for the photographs.

REFERENCES

[1] A.A.W. Ten-Have Opbroek, *Exp. Lung Res.* **17** (1991) 11.
[2] T. Alescio and A. Cassini, *J. Exp. Zool.* **150** (1962) 83.
[3] J.M. Shannon, *Dev. Biol.* **166** (1994) 600.
[4] N.K. Wessels, *J. Exp. Zool.* **150** (1962) 83.
[5] J.M. Shannon, L.D. Nielson, S.A. Gebb and S.H. Randell, *Dev. Dyn.* **212** (1998) 482.
[6] T.V. Taderera, *Dev. Biol.* **16** (1967) 489.
[7] S. Bellusci, R. Henderson, G. Winnier, T. Oikawa and B.L. Hogan, *Development* **122** (1996) 1693.
[8] G. Winnier, M. Blessing, P.A. Labosky and B.L.M. Hogan, *Genes Dev.* **9** (1995) 2105.
[9] S.E. Wert, S.W. Glasser, T.R. Korfhagen and J.A. Whitsett, *Dev. Biol.* **156** (1993) 426.
[10] S. Bellusci, Y. Furuta, M.G. Rush, R. Henderson, G. Winnier and B.L. Hogan, *Development* **124** (1997) 53.
[11] P.W. Ingham, *Curr. Op. Genet. Dev.* **5** (1995) 492.
[12] J.C. Grindley, S. Bellusci, D. Perkins and B.L. Hogan, *Dev. Biol.* **188** (1997) 337.
[13] C.V. Pepicelli, P.M. Lewis and A.P. McMahon, *Current Biol.* **8** (1999) 1083.
[14] Y. Litingtung, L. Lei, H. Westphal and C. Chiang, *Nat. Genet.* **20** (1998) 58.
[15] J. Motoyama, J. Liu, R. Mo, Q. Ding, M. Post and C.C. Hui, *Nat. Genet.* **20** (1998) 54.
[16] S. Bellusci, J. Grindley, H. Emoto, N. Itoh and B.L. Hogan, *Development* **124** (1997) 4867.
[17] K. Peters, S. Werner, X. Liao, S. Wert, J. Whitsett and L. Williams, *Embo. J.* **13** (1994) 3296.
[18] H. Min, D.M. Danilenko, S.A. Scully, B. Bolon, B.D. Ring, J.E. Tarpley, M. DeRose and W.S. Simonet, *Genes Dev.* **12** (1998) 3156.

[19] K. Sekine, H. Ohuchi, M. Fujiwara, M. Yamasaki, T. Yoshizawa, T. Sato, D. Yagishita, D. Matsui, Y. Koga, N. Itoh and S. Kato, *Nat. Genet.* **21** (1999) 78.
[20] D. Lebeche, S. Malpel and W. Cardoso, *Mech. Dev.* **86** (1999) 125.
[21] D. Sutherland, C. Samakovlis and M.A. Krasnow, *Cell* **87** (1996) 1091.
[22] L. Glazer and B.-Z. Shilo, *Genes Dev.* **5** (1991) 697.
[23] N. Hacohen, S. Kramer, D. Sutherland, Y. Hiromi and D. Krasnow, *Cell* **92** (1998) 253.
[24] T. Casci, J. Vinos and M. Freeman, *Cell* **96** (1999) 655.
[25] A.A. de Maximy, Y. Nakatake, S. Moncada, N. Itoh, J.P. Thiery and S. Bellusci, *Mech. Dev.* **81** (1999) 217.
[26] J.D. Tefft, M. Lee, S. Smith, M. Leinwand, J. Zhao, P.J. Bringas, D.L. Crowe and D. Warburton, *Current Biol.* **9** (1999) 219.
[27] G. Minowada, L.A. Jarvis, C.L. Chi, A. Neubuser, X. Sun, N. Hacohen, M.A. Krasnow and G.R. Martin, *Development* **126** (1999) 4465.
[28] B.L.M. Hogan, *Cell* **96** (1999) 225.

CHAPTER VII

Branched Structures, Acinus Morphology and Optimal Design of Mammalian Lungs

B. Sapoval, M. Filoche and E.R. Weibel

1. INTRODUCTION

The purpose of this lecture is to demonstrate that for good efficiency of the diffusive transfer of oxygen to blood, the unit transfer system, namely the lung acinus, should not be too large. This result is a general consequence of diffusive screening effects. The constraint of maximal efficiency then imposes that the lung has to be divided into a large number of small efficients units. As the lung is space-filling, the air access to these units has to be a branched geometry.

There are essentially two steps in the mammalian pathway for oxygen transport from the air to the blood [1]. The initial step occurs in the lung and is the transport of oxygen in the airways to the pulmonary gas exchanger. The second step is the transfer of oxygen across the tissue membrane that separates air and blood to the red blood cells where it is absorbed on the hemoglobin. An important characteristic of this system is that the gas exchanger forms a very large surface between air and blood, about 130 m^2 in humans, and that this surface is connected to the outside air by a system of branched airways that have their origin in the trachea. Of approximately 23 generations of airways, the first 14 are purely conducting pipes whereas the walls of the most seven peripheral generations are densely occupied by gas exchange units in the form of tightly packed side chambers called alveoli. These most peripheral airways form the acinus, a unit of gas exchange tissue connected to the airway system by an airway of generation 17, on average. This will be explained below.

In the proximal airways, oxygen is transported by mass flow of air due to ventilation, which is about 7 l min^{-1} at rest and rises up to 68 l min^{-1} in exercise condition. The diameter $d(Z)$ of these airways falls gradually with each generation Z of dichotomous branching according to $d(Z) = d(0).2^{-Z/3}$.

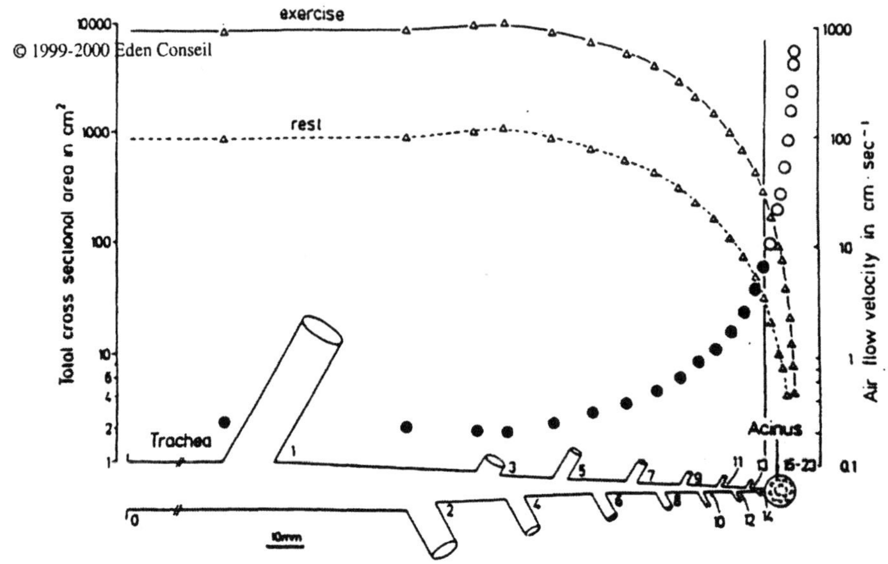

Fig. 1. — Simplified representation of the bronchial tree in the human lung. Only one between two of the airways is represented at each bifurcation of order Z. Also shown in the figure is the progressive increase of the total cross-section of the airways (circles) at each generation and the corresponding decrease in the airflow velocity at rest and under exercise (triangles). The data comes from experimental and morphometric studies of the human lung [1,2].

Since the number of branches doubles with each generation, the total airway cross-section increases so that the air flow velocity U falls as air moves inwards (Fig. 1) until the speed of diffusion of oxygen becomes higher than the air flow velocity [1,2]. In the most peripheral airways of the acinus, oxygen diffuses into quasi-static air in the gas exchange region. Diffusion is driven by an oxygen gradient because it is finally dissolved in the surface of the gas exchanger tissue. It diffuses across the tissue membrane to be absorbed by the red blood cells that move through the capillaries in a steady flow driven by circulation of the blood. Hence the blood functions as an oxygen sink.

The transition from mass air-flow to diffusion occurs in the most peripheral conducting airways and in those airways that lead into the acinus. Below this transitional region, we are in a purely diffusive regime. Oxygen enters the acinus across a narrow cross-section (about 7×10^{-2} cm^2) and then interacts with a complex irregular gas exchange surface of about 5 cm^2 formed by the complex of alveoli around the acinar airways. In a steady state circumstance, the oxygen concentration obeys the Laplace equation. This system then belongs to a class of systems in which the transport towards an irregular surface is due to a field derived from a Laplacian potential [3]. These are common in many natural or technical processes. For instance, in the design of high current batteries, one uses porous electrodes: the current in the electrolyte is proportional to an electric field that is derived from the Laplacian electric potential. Other systems are plant roots or villi in the human intestine [4]. Most often, these systems work in steady state conditions, but the case of the electrochemical exchange of charge across rough or porous electrodes is special in the sense that it can be studied by applying an a.c. voltage to an irregular electrode and monitoring the a.c. response. Oxygen partial pressure difference between inspired oxygen and capillaries is equivalent to such an a.c. voltage. Frequency dependent transport can easily be studied experimentally by impedance spectroscopy [3]. These last studies constitute the experimental support of the considerations that are used below. In the case of a membrane, neutral diffusing species, here O_2 molecules, are brought to the membrane surface by diffusion currents due to concentration gradients. In a battery, the ions are transported by the electric field which is the gradient of the electric potential. The transport across the membrane and the absorption by blood plays the same role as the redox reaction on the battery electrode.

In the context of electrochemistry, the concept of screening is classical. It describes the fact that, in an irregular electrode, the geometrical protuberances receive a much larger current than the deeper regions which are said to be partially or totally "screened". Screening here is due to the fact that a molecule starting on a diffusion source and diffusing randomly through the gas has a higher probability to hit the membrane on the most accessible regions (entrance or protuberances) of the system. If it is absorbed at the very first hit, because the membrane permeability is large, only a fraction of the geometry will be reached. In the case of the acinus, the existence of the deeper regions would then be of no use because diffusing O_2 molecules would be absorbed *before* ever reaching these regions. The role of screening is qualitatively described in Figure 2. The geometry of the exchange surface then plays an essential role in the screening efficiency.

On the other hand, if the permeability is small, linked to finite oxygen solubility and diffusivity in the membrane medium, the molecule will not be absorbed at first hit but after many collisions with the membrane. In that situation, the majority of the gaseous molecules have a fair chance to reach the deeper regions as indicated schematically in Figure 2b. The effect of screening then depends on the membrane permeability. Conversely, the global flux across a membrane with small intrinsic permeability will increase if the total surface is increased by geometrical irregularity. This is the case of the lung. The necessity for irregular or fractal surfaces with large areas has been already suggested as one possible reason for the existence of fractal morphology in living systems [5]. A preliminary approach indicating that such morphology could be considered as optimized for the mammalian acinus was given in [6].

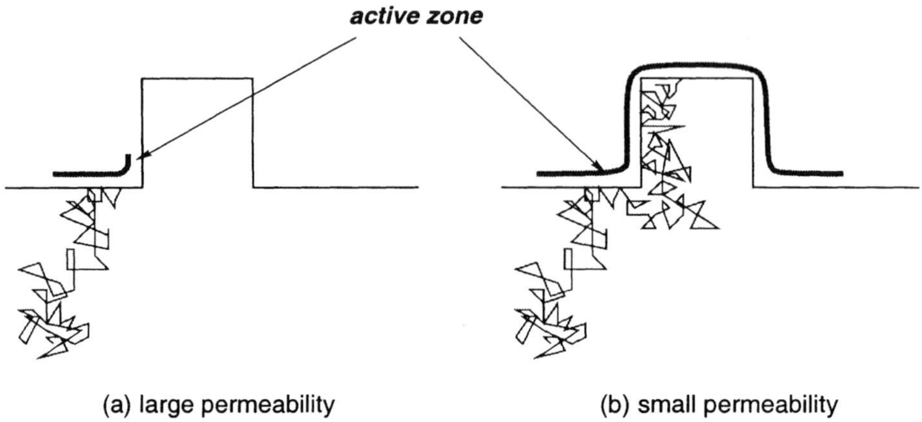

Fig. 2. — Schematic representation of the screening effect: particles (or random walkers) diffusing from the source have a large probability to hit the membrane on its salient features and a small probability to reach the deeper regions of an irregular membrane by escaping collisions with these features. If the permeability is infinite, the absorbing region are the salient features of the geometry (left). On the opposite, if the permeability of the membrane is small, the particle is not absorbed at first hit. It bounces again and again and has a larger probability to explore the deeper regions of the structure (right). The size of the active zone for absorption then increases when the permeability decreases.

The role of screening was already proposed as a qualitative hypothesis by Weibel [1] who suggested that, because of progressive oxygen absorption along the acinar pathway, the mean partial pressure (or concentration) over the entire pathway would decrease.

The relative influence of diffusivity, permeability and geometry on the flux then becomes a fundamental question. This is what we discuss quantitatively here. The problem has been considered in many details, theoretically, experimentally and numerically. A complete list of references up to 1995 is given in reference [3]. Fractal geometry has played an important role by offering simple models of strongly irregular systems [7,8]. However the results used here apply in general to arbitrary irregular geometry with minor modification. They are based on both theoretical arguments, numerical simulations and experiments which exemplify the "robust" character of the considerations that are used below.

In this course, we first recall briefly the essential results from the diffusion towards and across an irregular or fractal membrane. A definition of an optimal diffusion exchanger is proposed. It expresses that the optimum system is that which permits the maximum oxygen flux in a given volume, given the values of alveolae size and of the physical transport parameters diffusivity and permeability, which govern the phenomenon. This optimization criterion is then compared with the geometry of pulmonary acini in several animal species. We find that in normal breathing regime, at rest, the conductance of the system is diminished due to the existence of screening.

In the following, the ability of a system to transfer oxygen is characterized by a diffusion "conductance". The word conductance, or inverse resistance, is a classical word to describe the properties of electrodes. Here, the conductance is the flux per unit partial oxygen concentration (or in an equivalent manner for unit oxygen partial pressure). So each time the word conductance is used, the reader can read "flux for a unit concentration at the entrance of the system under consideration".

2. STRUCTURE AND GEOMETRY OF PULMONARY ACINUS

The pulmonary acinus is defined as the gas exchange region of the airway system. Here, the blood is brought so close to the air that oxygen can be transferred from air to blood by diffusion across the thin tissue barrier. At the same time, CO_2 is discharged from blood into air.

The acinus is formed on the last few generations of the airway tree in that the smooth wall of conducting airways becomes transformed in scalloped surfaces. This corresponds first to a reduction in tissue mass, with only a thin membrane wrapping the blood capillaries [1]. Second, this is where the airway surface is transformed into a complex of small cuplike alveoli which become densely packed so that the acinus has the appearance of a foam-like structure with each of the "foam bubbles" open to an airway branch.

The airways continue to branch by dichotomy into the acinus for 6–9 generations counting from the appearance of the first alveolus. The airways are blind ending so that the acinus is a compact bounded structure of defined size as shown in Figure 3 [9]. It thus appears as the unit of the gas exchanger of the lung, of which there are 30 000 in the human lung. One acinus contains about 10 000 alveoli. As the intraacinar airways branch, their diameter and length are gradually reduced by about 6 – 10% with each generation. This slight reduction will be ignored in the following because it does not influence significantly the results.

The essential feature of acinar airways is that their wall is completely lined with the gas exchange surface. This surface is increased, in the human lung, by a factor of 10 above the smooth outline of the duct because of the formation of alveoli [10]. The effect of this feature is that air, as it enters into the acinar airways, is exposed to the reduced partial pressure p_{O_2} near the alveolar surface caused by O_2 extraction by the blood flowing past the exchange surface.

The basic structure of the acinus is the same in all mammalian lungs; it has been studied in some detail for the mouse, rat, rabbit, and human lungs [9,10]. In the smaller species, the most peripheral 6 generations of airways are densely

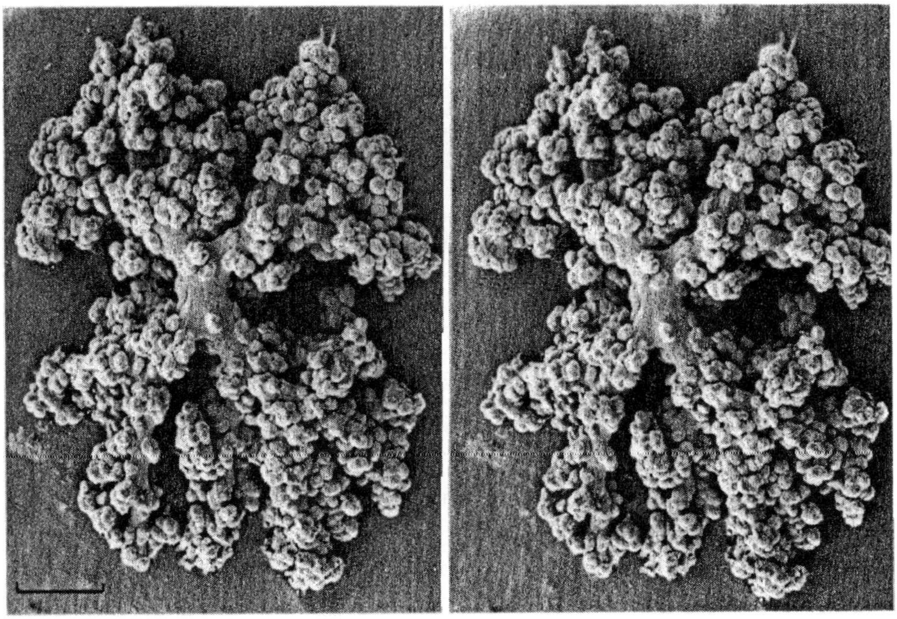

Fig. 3. — Stereo pair of of scanning electron micrographs of a slightly stretched acinus from a rat lung, to show arrangement of alveolar complexes around the alveolar ducts. Scale marker 500 μm. (From Rodriguez *et al.* [9].)

covered with alveoli (Fig. 3). In large lungs, such as the human, the acini appear very large and the acinar airways comprise about 9 generations. But in the first three generations, the airway wall carries only a few scattered alveoli. In these lungs, the structural equivalent to the acinus in smaller lungs is what has been called the "1/8 subacinus" in which all airways are densely covered with gas exchange tissue [10]. The size of the acinus is a function of body size.

3. PHYSICAL PARAMETERS DETERMINING THE ACINUS EFFICIENCY

In this paper, we consider that the air flow velocity drops abruptly to zero at some stage of the respiratory tree. Figure 1 shows, on the example of the human airway tree, that air flow velocity drops rapidly as the region of the acinus is approached because the total airway cross-section increases steeply [11]. From there and below, we deal then with a static diffusion problem. We use this static version of the plateau regime.

In such a situation, the net diffusion transfer is determined by only three factors: the geometry of the system, the diffusivity D of oxygen and the permeability W of the membrane. The linear passive permeability is found from the net flux across a flat membrane of area A separating a zone with concentration c at the surface and zero concentration on the other side where the flux can be written as (cAW). The membrane here is supposed to be passive and we neglect back diffusion coming from the other side of the membrane where the concentration is supposed to be zero; this is justified by the fact that the oxygen-loaded red blood cells are steadily carried away by blood flow.

For our purpose, the membrane can be considered as a thin layer of water of thickness τ. The effective permeability of the barrier is determined by two consecutive processes. First, the dissolution of oxygen in water at the gas membrane surface and second the diffusion of oxygen across this thin layer. The solubility β of oxygen in water is equal to 1.41 nmol cm^{-3} torr^{-1}. Working here in terms of absolute concentrations, this corresponds to a numerical value of 2.4×10^{-2} molecule of oxygen per unit volume of water for 1 molecule of gaseous oxygen per unit volume. The diffusion coefficient of oxygen in the membrane tissue is close to the one in water and plasma. Calling it $D_{\mathrm{O_2,water}}$, the permeability of the membrane can be written

$$W = \beta \, \frac{D_{\mathrm{O_2,water}}}{\tau}. \tag{1}$$

One should also take into account the probability P of capture of oxygen by red blood cells. This is the quantity which corresponds to the factor $\theta_{\mathrm{O_2}}$ called the binding rate of oxygen by blood. We assume here that $P = 1$. From the values of these coefficients, one can define a length Λ which is the ratio

$$\Lambda = \frac{D}{W} = \left(\frac{D_{\mathrm{O_2,air}}}{\beta D_{\mathrm{O_2,water}}} \right) \tau. \tag{2}$$

This length plays a dominant role in this problem [3]. For a two-dimensional system ($d = 2$), it is the length of the perimeter of a region where screening effects can be neglected. For a three-dimensional system ($d = 3$), it is the length of the perimeter of a region in a planar cut where screening effects can be neglected. For this reason, the length Λ can be named the "unscreened perimeter length". The crucial fact here is that, as shown below, this length is comparable to the perimeters of real acini. This corresponds to partial screening.

The barrier thickness varies slightly between different mammalian species as discussed in detail in reference [1]. A good order of magnitude is $\tau \approx 10^{-4}$ cm. Using the known value for the diffusivity of oxygen in air ($D_{\mathrm{O_2,air}} = 0.2$ cm^2 s^{-1}) and using $D_{\mathrm{O_2,water}} = 33 \times 10^{-6}$ cm^2 s^{-1} (at 37 °C), one obtains

$$\Lambda(\mathrm{cm}) = 2.53 \times 10^5 \ \tau \ (\mathrm{cm}). \tag{3}$$

The numerical value of Λ for mammals is then of order $\Lambda \approx 10$ cm. Note that in the case of fishes, which breathe water, solubility and diffusivity are the same in both fluid and $\Lambda = \tau$.

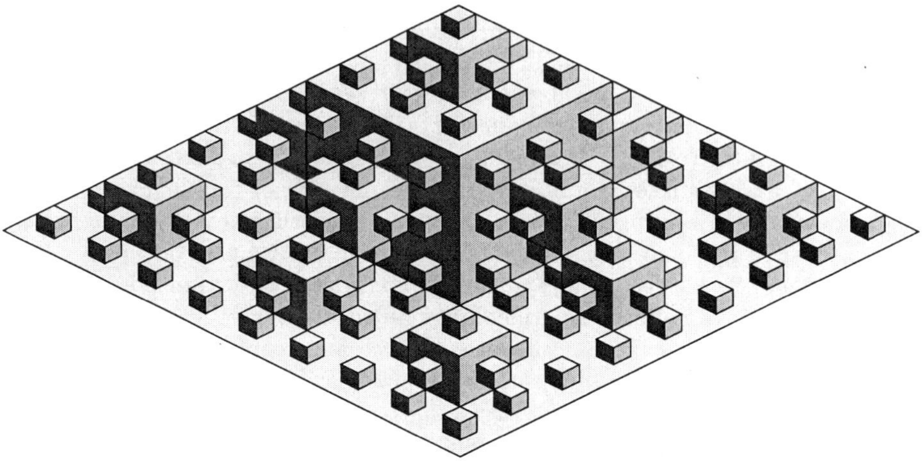

Fig. 4. — Schematic representation of a deterministic fractal membrane system in $d = 3$. Here the fractal dimension is $D_\mathrm{f} = \mathrm{Log}13/\mathrm{Log}3$. The membrane has a smaller scale ℓ below which it can be considered as smooth. The size of the system is L. We assume that the ventilation maintains a constant concentration c_0 at a diffusion source situated near the entrance of the system.

To introduce the parameters of the geometry, a self-similar surface such as that shown in Figure 4 will be considered. This figure, which is not representative of an acinus, serves only to define the quantities that we need to use. The smaller flat element is a square of side ℓ, the smaller cutoff. The size or diameter of the entire system, here the side of the cubic basis is called L. The other quantity which characterizes such a geometry is the fractal dimension D_f. For a fractal surface, the total area A_T is given by the relation [7]

$$A_\mathrm{T} = \ell^2 \left(\frac{L}{\ell}\right)^{D_\mathrm{f}}. \tag{4}$$

In the case of the acinus, ℓ would correspond to the typical diameter of an alveolus. The simplified diffusion system that we consider has a diffusion source at the bottom of the figure where the concentration of diffusing particles is constant in time. In the case where the diffusion source is near the membrane, the system presents a macroscopic diffusion conductance Y_M defined by

$$\Phi = Y_\mathrm{M}\, c_0 \tag{5}$$

where Φ is the flux and c_0 is the concentration at the source. The admittance is then the flux per unit pressure, or concentration of oxygen.

To understand the role of Λ, it is useful to consider diffusion on a 2D square lattice with spacing a. On such a lattice, the values of D and W are $D = a^2\omega/4$ and $W = a\sigma\omega/4$. The quantity ω is the jump probability per unit time and σ is the absorption probability at each molecular collision with the membrane. If, at each collision with the membrane, the diffusing molecule has a probability σ to be absorbed, it will be, on average, absorbed after $1/\sigma$ collisions. By bouncing again and again before absorption, it travels deeper into the gaseous space. During that random walk, the molecule explores the membrane in an approximately uniform manner because the number of steps in the gas is much larger than the number of sites touched on the membrane. On average, when the molecule has explored $1/\sigma$ sites or, equivalently, a length $a/\sigma = D/W = \Lambda$, the walk is terminated by the effective permeation across the membrane. In 2D, the length Λ characterizes then the average *perimeter* length of a zone in which the molecule will be really absorbed. It plays then a dominant role to characterize the overall efficiency of screening phenomena and the net diffusion flux.

Another way to understand the role of Λ is to compare directly in 3D the conductance to reach with the conductance to cross a surface of area A_{T} and of size or diameter L_{A}. Here the "diameter" of an irregular surface is the diameter of the smaller sphere that surrounds the surface. The conductance to reach the surface is of order DL_{A} whereas the conductance to cross is of order WA_{T}. They are equal when $D/W = \Lambda = A_{\mathrm{T}}/L_{\mathrm{A}}$. The length $A_{\mathrm{T}}/L_{\mathrm{A}}$, ratio of the 3D surface area by its diameter, is the average total length (or perimeter) of a random plane section of the surface. This section is a curve, possibly disconnected, and $A_{\mathrm{T}}/L_{\mathrm{A}}$ is its total length.

For a given system geometry, when the membrane permeability is small enough, the concentration or partial pressure of oxygen is constant everywhere in the volume. This corresponds to a classical regime in which the surface works uniformly. In this low permeability regime, the membrane conductance is simply proportional to the total area of the membrane $\ell^2\ (L/\ell)^{D_{\mathrm{f}}}$ [3] and

$$Y_{\mathrm{M}} = \ell^2 \left(\frac{L}{\ell}\right)^{D_{\mathrm{f}}} W. \tag{6}$$

For higher values of the permeability, one enters a regime where only a part of the total surface is used for transfer due to screening. In the case of a fractal membrane the conductance is within good approximation [3, 12]

$$Y_{\mathrm{M}} = L^2 \left(\frac{D}{\ell}\right)^{\frac{D_{\mathrm{f}}-2}{D_{\mathrm{f}}-1}} W^{\frac{1}{D_{\mathrm{f}}-1}}. \tag{7}$$

Note that in both regimes the conductance is larger than the conductance WL^2 that would present a flat membrane. The crossover between the two regimes is obtained when equation (6) = equation (7) or

$$L = \ell \left(\frac{\Lambda}{\ell}\right)^{\frac{1}{D_{\mathrm{f}}-1}} \quad \text{or} \quad \Lambda = \ell \left(\frac{L}{\ell}\right)^{D_{\mathrm{f}}-1}. \tag{8}$$

It is a mathematical result that the cut by a plane of a fractal surface of dimension D_f is a fractal line with dimension $D_f - 1$ [13]. The quantity $\ell\ (L/\ell)^{D_f-1}$ is then the total length of the cut by a plane of our initial surface. The crossover then occurs when the length of a cut of one acinus membrane by a plane is equal to Λ. From the point of view of optimization, note that in the weak permeability regime, the entire membrane is working whereas in the high permeability regime, the active surface is only a fraction of the total surface [3, 14]. The conductance however being an increasing function of W, the partially working membrane presents a conductance larger than in the low permeability case.

For given values of ℓ, L and Λ, the conductances given by equations (6) and (7) increase with the fractal dimension. This fact corresponds to the intuitive idea that, if a large exchange surface is needed, a space filling surface with $D_f = 3$ will be the more efficient. We will consider that this approximation is valid. The use of a unique value for ℓ within the acinus is permitted because the various feature sizes keep the same order of magnitude [10]. In particular, a cut of the acinus presents an irregular, and homogeneous, foam like pattern.

Using $D_f = 3$, the weak permeability conductance will be

$$Y_M = \frac{L}{\ell}\ L^2\ W \tag{9}$$

and the high permeability conductance is

$$Y_M \approx \left(\frac{D}{W\ell}\right)^{\frac{1}{2}} L^2\ W. \tag{10}$$

The crossover now occurs for a size $L_c \approx (\Lambda\ell)^{1/2}$ or equivalently for $\Lambda \approx L_c^2/\ell$. One can also consider the area of the active surface as a function of Λ and compare it to the total surface available for exchange [14, 15]. This will be pursued in the next section for a more realistic geometry.

4. A BETTER GEOMETRICAL MODEL: THE "HILBERT" ACINUS

The real acinus geometry is markedly different from the type of geometry which is schematized in Figure 4. The main difference is that the entrances of real acini are narrow as compared to the depth of the structure. In the language that is used in [15], the acini should be considered as deeply porous structures. In that case it is known that the crossover between the low and large permeability regimes occurs for Λ somewhat larger than L^2/ℓ or a size L somewhat smaller than $(\Lambda\ell)^{1/2}$.

For this reason, we have numerically studied the conductance in $d = 2$ of what could be called an "Hilbert acinus". The geometry of an "Hilbert acinus" is suggested by the structure of the Hilbert curve, one of the first examples of a space filling curve in $d = 2$ [16]. Hilbert acini are shown in Figure 5 for the

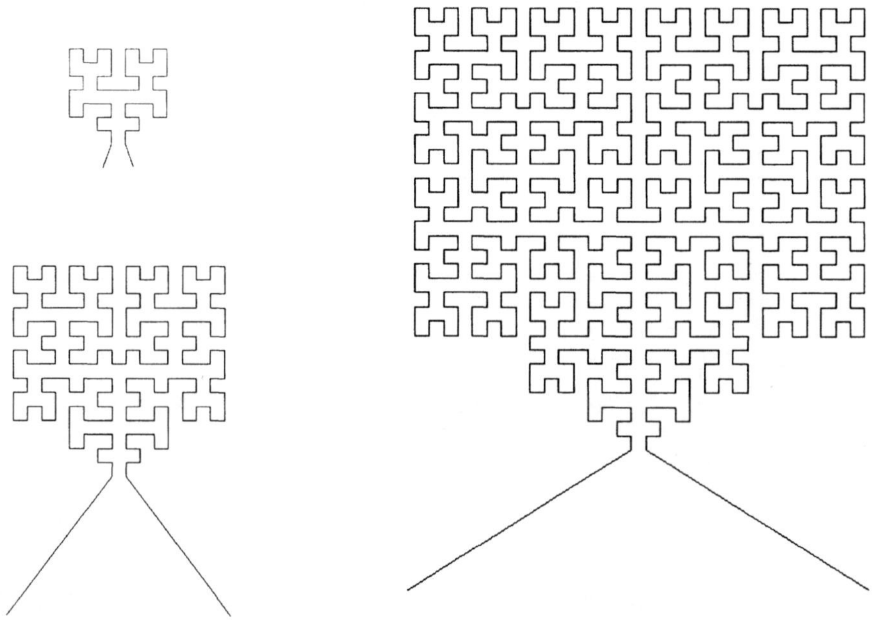

Fig. 5. — "Hilbert" acini of generations 2, 3 and 4. The perimeters are respectively $53\ \ell$, $213\ \ell$ and $853\ \ell$.

second, third, and fourth generation of fractality. Their respective perimeter length is $L_{\mathrm{p},2} = 53\,\ell$, $L_{\mathrm{p},3} = 213\,\ell$, and $L_{\mathrm{p},4} = 853\,\ell$. This structure is the main part of the Hilbert curve. The Hilbert acinus exhibits some of the topological properties of real acini. It is a branched structure with narrow access that gives access to larger and larger surface. The dependence on Λ of the conductance of these structures has been computed using the finite element method [17].

For a given value of Λ, the conductance can be written in the form $Y(\Lambda) = W\ L_{\mathrm{eq}}(\Lambda)$ where the length $L_{\mathrm{eq}}(\Lambda)$ would be the perimeter of a system which would present the same conductance if no screening would exist (at infinite diffusivity). For each acinus, one can then define for each value of the unscreened perimeter, a screening efficiency E by

$$E = \frac{L_{\mathrm{eq}}(\Lambda)}{L_{\mathrm{p}}}. \tag{11}$$

This ratio measures the fraction of the perimeter which really works *as if the diffusion was infinitely fast.* This quantity is plotted in Figure 6.

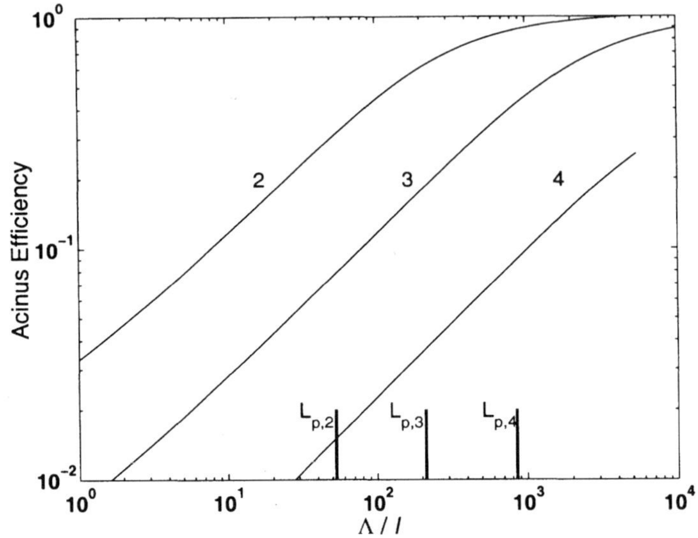

Fig. 6. — Screening efficiency of Hilbert acini of generations 2, 3, and 4.

One observes that, for Λ equal to the perimeter L_{p}, the relative conductance surface is found to be around 20% for the generation 3 and 10% for generation 4. This indicates that screening may have a strong effect on the effective flux.

One could object to the use of the Hilbert geometry as a model for the real acini, which are 3D structures. In principle, screening effects are less important in 3D. However, the real acinus is hierarchical and, more important, the gas phase presents *no loops*. For this reason, the real acinus does not deviate significantly from the Hilbert acinus model.

5. SMALLER IS BETTER: AN OPTIMIZED ACINUS

It has been indicated above that the flux is an increasing function of the fractal dimension. Space filling acini, close to those described in the introduction, then exhibit optimal morphology. But, besides that criterion, their exists a direct influence of the sizes of the individual acini compared to Λ. This criterion is determined by the value of the length Λ. One can guess that if the organ size has a size such that it works partially because of screening effects, there is a misuse of space. The value of Λ gives a natural scale to compare the size of the active surface to the size of the organ itself. This is discussed now in a quantitative manner using the expressions of the conductance obtained for the open fractal.

In both regimes described above, the conductance increases with the size L. But, if one thinks of optimized exchangers, one should take into account that the exchange must take place in a finite volume. Here, the available space is the thoracic volume. In that sense, the *best* exchanger should really be characterized by an *optimized conductance per unit volume* or specific conductance $Y_{\text{spec.}} = Y_{\text{M}}/L^3$. The specific conductances take respectively the values W/ℓ and $(D/W\ell)^{1/2}\, L^{-1}\, W$ in the low and high permeability regime.

A surprising result is obtained: for given D and W, in the high permeability regime where screening plays a role, the specific conductance *decreases* with the size of the exchanger. This is indicated schematically in Figure 7. The conclusion is that the best exchanger should have a size smaller, or on the order of, the crossover size $(\Lambda\ell)^{1/2}$. This seems to be an interesting result because it shows, on general grounds, that if diffusion to a membrane with finite permeability plays a role in the respiratory system, this system has to be made of accessible but independent sacs, therefore *has to be* some sort of *branched system* in order to achieve maximum specific conductance and then maximum global flux. A too large system, with $L \gg (\Lambda\ell)^{1/2}$ would present a poor specific conductance.

The above discussion introduces a direct link between two different concepts: the fact that the conductance per unit volume is maximized and the fact that there is no screening and that the membrane works uniformly. This fundamental result, namely, *the individual acini should not be too large* is illustrated schematically by comparing the net flux across an acinus of the fourth generation and four acini of the third generation occupying approximately the same volume. These systems are shown in Figure 8. Due to screening, the large acinus works poorly for the realistic values of Λ/ℓ which are about 10^3 for the mammals that we study as shown below.

In summary, for the system to be optimized, its size should not be too large. If, however, it is significantly smaller, the specific conductance which has the value W/ℓ remains constant. In other words, the above criteria is robust in the sense that a possible range of sizes corresponds to a maximum specific conductance. But, although it is sufficient that the size is smaller than the critical size, the system should not be too small because it has to be fed by an hydrodynamic flux which itself must be submitted to the weakest possible resistance. The hydrodynamic resistance increasing like the inverse 4th power of the airway diameter, the entrance should not be too small. On the other hand, there is no gain in efficiency in further decreasing the size of the acini as far as the acini are already small enough to be optimized.

At this stage, it should be emphasized that fractal geometry is used here only for the sake of writing algebraic expressions for the quantities of interest or for offering suitable geometrical models. The ideas behind are more general and can be applied whatever the geometry and also whatever the microscopic details which make, for instance, the membrane inhomogeneous in thickness. In $d = 2$ especially, it has been shown that these ideas apply to any irregular geometry, fractal or not and whatever the structure of the membrane [15, 18]. This is

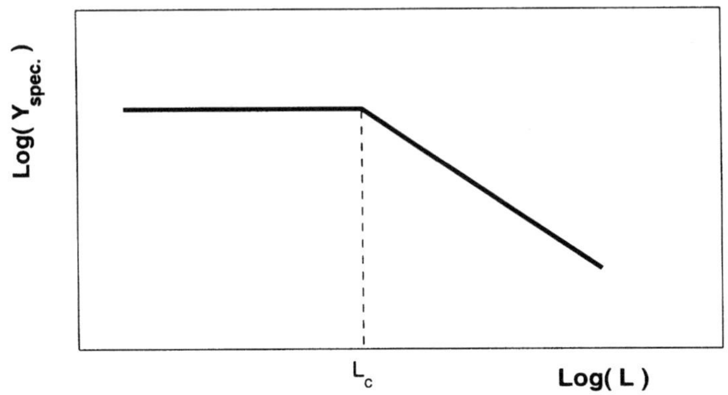

Fig. 7. — Schematic behavior of the specific conductance of the membrane as a function of the size of the fractal membrane for given unscreened length and smaller cutoff ℓ. The optimum for the acinus for which $D_f = 3$ should be found below the crossover, located around $L_c = (\Lambda\ell)^{1/2}$. The acinus then should not be too large to be efficient. For the Hilbert acinus, the cross-over for a critical size slightly smaller than $(\Lambda\ell)^{1/2}$.

why the above considerations seem to apply to the acinus of several animals, as discussed below, although their real geometry is not a simple fractal [10,20].

6. COMPARISON WITH ANATOMICAL DATA

To give a quantitative comparison of the above arguments with mammalian lung morphometry, we use a simplified model of the acinus geometry. We consider the acinus as a dense arrangement of cubic alveoli of side ℓ up to the diameter size L. In this simplified picture, ℓ is of the order of the diameter of an alveolus while L is of the order of the macroscopic diameter of one acinus. The length of the perimeter of a cut is then of the order of $\ell\ (L/\ell)^2$.

Following these guidelines, one can make a comparison for several animals: the mouse, the rat, the rabbit and the human lung using the data of references [9–11]. In these papers, the quantities of interest which have been determined experimentally are the acinus volume V_a together with the alveolar surface per acinus S_a. In terms of the above quantities, $V_a \approx L^3$ and if we think of cubic alveoli with side ℓ with 5 active faces (at least one face must be open to give access), $S_a \approx 5\ (L/\ell)^3\ \ell^2 \approx 5L^3/\ell$. The quantity ℓ can then be found as equal to $5V_a/S_a$. If our simplified model geometry is correct, the length $\ell \approx 5V_a/S_a$ which represents the smaller cutoff of the structure should be found of the same order of magnitude as the typical alveolar diameter. The numerical values are given in Table I. The values of ℓ found from the cubic alveoli model

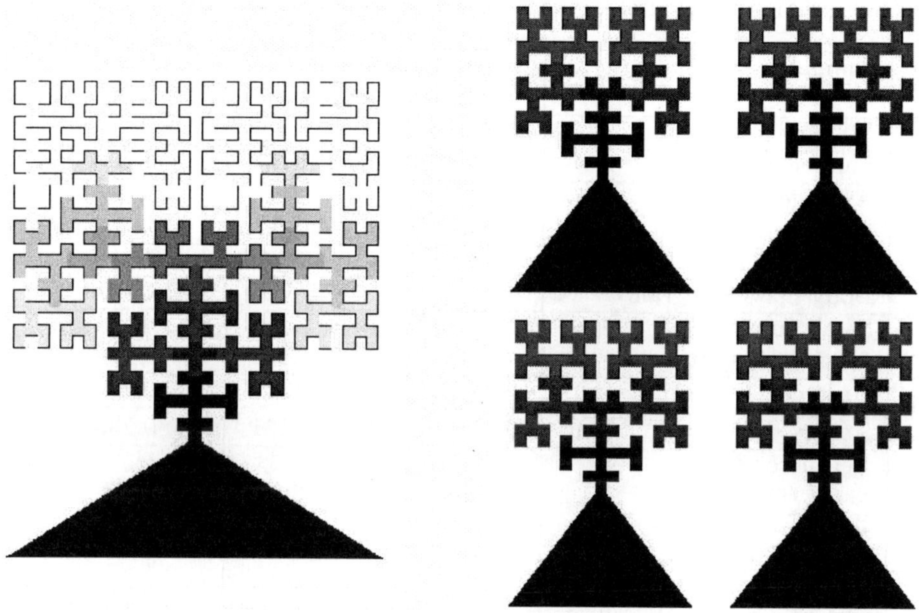

Fig. 8. — Illustration of the principle "smaller is better" for the acinus. The representation shows a change of color at each equiconcentration value smaller than the previous one by a factor 2 (stripe representation of isobars in log scale). In the same global volume, here surface, 4 acini of generation 3 present a total conductance larger than one acini of generation 4 for $\Lambda/\ell = 200$. Most of the generation 4 acinus perimeter does not participate to the exchange while the concentration is more uniform in generation 3 acini. For much larger values of the unscreened perimeter, the two systems would behave equivalently.

are a little smaller but have the same order of magnitude as the sizes of the sacs directly observed in the pictures of references [9–11]. This indicates that the arrangement of "cubic alveoli" is a reasonable approximation of the real geometry. Note that, in the same model geometry, the diameter L of the acini is of order $V_{\rm a}^{1/3}$.

Several remarks can be formulated on the data given in Table I. First, the acini perimeters and Λ are both of the same order, a few centimeters. Second, the values of the relative perimeter $L_{\rm p}/\ell$ would correspond to "Hilbert acini" of generation between 3 and 4. For the values of the unscreened perimeters (Λ/ℓ) found in Table I, of order 1000, this would correspond to a screening efficiency between 15% and 60%.

These parameters correspond to a breathing regime at rest. One can then observe that, in this situation, the screening effects reduce the fraction of the

Table I. — Geometrical parameters for mammals. The data given in bold are anatomical data (from Weibel *et al.*).

Animal	Mouse	Rat	Rabbit	Human 1/8 subacinus
Volume V_a (10^{-3} cm^3)	**0.41**	**1.70**	**3.40**	**23.4**
Surface S_a (cm^2)	**0.42**	**1.40**	**1.65**	**8.63**
Average membrane thickness τ (10^{-4} cm)	**0.30**	**0.40**	**0.50**	**0.62**
Lung volume V_L (cm^3)	**1.49**	**13.1**	**79.2**	**4341**
Body mass M (gram)	**30**	**500**	**3 000**	**70 000**
Small cutoff $\ell = 5V_a/S_a$ (μm)	49	61	103	136
Unscreened length Λ (cm)	7.25	10	12.5	15.5
Acinus size $L = V_a^{1/3}$ (cm)	0.074	0.121	0.151	0.286
Perimeter $L_p \approx \ell(L/\ell)^2$ (cm)	1.12	2.43	2.21	6.01
Relative perimeter L_p/ℓ	228	380	215	448
Relative unscreened length Λ/ℓ	1480	1665	1220	1150

acinus membrane which is really used for oxygen transfer. This reduction factor can reach one order of magnitude. On the contrary, in the exercise condition, the increase of the air velocity displaces the transition zone between convection and diffusion deeper in the structure. In our model, this would correspond to a change from generation 4 to generation 3 or 2 [21]. As shown in Figure 6, this would induce a drastic increase in the overall lung efficiency, as observed experimentally. This may also explain why, in case of emphysema, patients have a spontaneous tendency to breathe more rapidly.

7. CONCLUSIONS

In summary, screening effects play an important role in the physiology of oxygen transfer in the acini of mammalian lungs. The competition between the speed to reach the membrane (measured by the oxygen diffusivity in air) and the speed to cross the membrane (measured by the permeability) can be characterized by a length Λ, equal to the ratio of these transport coefficients. This length approximately determines the perimeter of the acinus above which screening effects reduce the efficiency of the system.

Based on this approximation, it has been found that too large acini would be poorly efficient. In the mammalian lung, Λ is of order of 10 cm. Therefore, in order to be efficient, the human lung has to be divided into many small diffusive cells. Furthermore, a space filling surface, described by a fractal dimension $D_f \approx 3$, is an optimized geometry. These conclusions are general and apply for any structure. In particular, the best specific conductance is of order W/ℓ, which is the ratio of the membrane permeability to the diameter of the alveolar sacs. This suggests that the metabolic rate, proportional to the total oxygen flux, should be simply proportional to the volume of the lung multiplied by this specific conductance [21].

In order to obtain a more quantitative discussion of the screening effects in the acinus, a new geometrical model, called the Hilbert acinus, has been proposed and studied numerically. In this geometry, for realistic values of the physical and geometrical parameters, screening effects are present at rest and that the active surface is smaller than the total available membrane surface. From the figures, it corresponds to an efficiency comprised between 10% and 60%. The study of the model also indicates that, in the exercise situation, the diffusion moves deeper into the structure, increasing strongly the overall transfer efficiency.

This study suggests that the complexity of the lung geometry might be the result of optimization under physical constraints of the respiratory function. How this optimization relates to Darwinian evolution by natural selection remains an open question.

The Laboratoire de Physique de la Matière Condensée is Unité Mixte de Recherches du CNRS, UMR No. 7643 of CNRS.

REFERENCES

[1] E.R. Weibel, *The Pathway for Oxygen* (Harvard University Press, Cambridge, 1984).

[2] E.R. Weibel, *Morphometry of the Human Lung* (Springer-Verlag, 1963).

[3] B. Sapoval, *Transport across Irregular Interfaces: Fractal electrodes, membranes and catalysts*, in "Fractals and Disordered Systems" 2nd., edited by A. Bunde and S. Havlin (Springer-Verlag, 1996) pp. 232- 261. A popularization version can be found, in French, *Pour la Science* **198** (1994) 30-37.

[4] G.H. Bell, D. Emslie-Smith and C.R. Paterson, *Physiology and Biochemistry* (9th Ed. Churchill Livinstone: Edimburgh, 1976).

[5] B. Sapoval, *Fractals* (Aditech, Paris, 1990).

[6] B. Sapoval, *Fractals in Biology and Medicine*, edited by T.F. Nonnenmacher, G.A. Losa and E.R. Weibel (Birkhäuser-Verlag, Ble 1994) pp. 241-249.

[7] B. Mandelbrot, *The Fractal Geometry of Nature* (W.H. Freeman and Co, San Francisco, 1982).

[8] B. Sapoval, *Universalités et Fractales* (Flammarion, Paris, 1997).

[9] M. Rodriguez, S. Bur, A. Favre and E.R. Weibel, *Amer. J. Anat.* **155** (1987) 143.

[10] B. Haefeli-Bleuer and E.R. Weibel, *Anatom. Rec.* **220** (1988) 401.
[11] E.R. Weibel, *Respiratory Physiology, an Analytical Approach*, edited by H.K. Chang and M. Paiva (M. Dekker, Inc., 1989) p. 1.
[12] E. Chassaing and B. Sapoval, *J. Electrochem. Soc.* **141** (1994) 2711-2715.
[13] K. Falconer, *Fractal Geometry: Mathematical Foundations and Applications* (John Wiley and Son Ltd., New York, 1990).
[14] R. Gutfraind and B. Sapoval, *Fractals in Biology and Medicine*, edited by T.F. Nonnenmacher, G.A. Losa and E.R. Weibel (Birkhäuser-Verlag, Bâle 1994) pp. 251-265 and *J. Phys. I France* **3** (1993) 1801-1818.
[15] B. Sapoval, M. Filoche, K. Karamanos and R. Brizzi, *Eur. Phys. J. B* **9** (1999) 739.
[16] see, for instance, H.-O. Peitgen, H. Jürgens and D. Saupe, *Chaos and Fractals* (Springer-Verlag, New York, 1992) p. 388.
[17] M. Bernadou *et al.*, Collaboration Modulef: "Modulef : une bibliothèque modulaire d'éléments finis", INRIA 1985 (France).
[18] M. Filoche and B. Sapoval, *Eur. Phys. J. B* **9** (1999) 755.
[19] M. Filoche and B. Sapoval, *J. Phys. I France* **7** (1997) 1487-1498.
[20] E.R. Weibel, *Fractals in Biology and Medicine*, edited by T.F. Nonnenmacher, G.A. Losa and E.R. Weibel (Birkhäuser-Verlag, Bâle, 1994).
[21] B. Sapoval *et al.* (to be published).

CONTRIBUTED PAPER 5

Quantitative Studies of Branching Morphogenesis in the Developing Kidney

I.S. Harper, L.A. Cullen and J.F. Bertram

1. INTRODUCTION

Branching morphogenesis is a key event in the development of many epithelial organs, including the lung, prostate gland, mammary gland, salivary gland and kidney. The branching pattern within each organ is generally considered to be highly organized, and certainly is important in establishing the characteristic architecture of the adult organ. Clinically, the pattern of branching morphogenesis in the developing permanent kidney (metanephros) is believed to be a major determinant of the total number of nephrons (filtering units of the kidney) in adult life. After successive generations of branching, the normal human kidney can contain between 300 000 and over 1 million nephrons [1]. Reduced nephron number has been associated with the development of essential hypertension [2], chronic renal failure [4] and the long-term success of renal allografts [3]. Ureteric duct branching morphogenesis therefore is not only a critical event in kidney development, but also possibly a determinant of adult kidney function and disease.

1.1. Epithelial – Mesenchymal Interaction

Kidney development is characterized by a series of reciprocal inductive interactions between the epithelial ureteric bud and the metanephric mesenchyme. Metanephric development is initiated when the epithelial bud emerges from the Wolffian duct and grows caudally into the adjacent mesenchyme. The mesenchymal cells induce growth and branching of the ureteric bud. Through repeated patterns of division, the branches of the ureteric bud ultimately give rise to the collecting duct system of the kidney and the ureter. Simultaneously, the widened tips (ampullae) of the branching tubules induce the surrounding mesenchyme firstly to survive and proliferate, and secondly to undergo conversion from mesenchyme to epithelium. These epithelial cells differentiate into nephrons, ultimately connecting to the collecting duct system. Our understanding of the genetic and molecular regulation of kidney development has been described in a number of recent reviews by Horster *et al.* [9], Burrow [5], Clark and Bertram [6], and Kuure *et al.* [10].

1.2. Pattern of Ureteric Branching Morphogenesis in Humans

Microdissection of human embryos by Osathanondh and Potter [12,13] demonstrated the characteristic pattern of branching morphogenesis in the human metanephros. Other detailed reports have been provided by Oliver [11] and more recently Al-Awqati and Goldberg [1]. The first branch emerges as a bud from the Wolffian duct. The ampulla of this bud divides producing two new ampullae, each responsible for the formation of a new tubule.

Osathanondh and Potter [12, 13] described the pattern of branching in four periods. Period one involves active branching of collecting tubules with the advancement of nephrons. The bud enters the surrounding metanephric mesenchyme and branches dichotomously. Each tip continues to advance to the periphery where it eventually divides, producing two daughter branches. Dilatation of these first generations results in formation of the renal pelvis and calyces. The next generations become the collecting tubules. After the ureteric bud has produced 3–5 generations of branches, nephron induction commences. Each tip now induces the mesenchyme to epithelialise forming a nephron; this nephron eventually forms a connection to the advancing tip. As the ampulla divides producing two new ampullae, the nephron is carried with one ampulla while the other induces a new nephron. However, there is recent data to suggest that there is also a "subtle control mechanism" that can add new branches at sites other than the tips. Internodal branches developing off much older branches possibly result in the optimal arrangement of the final ureteric tree (reviewed in Davies and Davey [7], http://cpmcnet.columbia.edu/dept/genetics/kidney/movies.html).

Period two begins when the ampulla can no longer divide, but can now induce nephrons when already having a nephron attached. Instead of individually attaching to collecting tubules, nephrons become arranged in arcades.

Period three begins when the ampullae advance beyond the point of arcade attachment and subsequently ampullae seldom divide, nephrons individually attach along the entire length of the duct. Eventually, the ampulla ceases function and disappears. Terminal tubules are the 15th generation of the original pelvic ducts.

Period four occurs just prior to birth and commences at the time the ampulla disappears. This period involves only interstitial growth and differentiation. The absence of an ampulla means further terminal branching of collecting tubules and induction of nephrons is impossible, and so all nephrons are formed before birth.

1.3. *In vitro* Models of Branching Morphogenesis

The molecular regulation of kidney development and ureteric branching morphogenesis has been extensively studied *in vitro*. Whole rat and mouse metanephroi are able to grow and differentiate *in vitro* much as they do *in vivo*. This is a very powerful, and the most widely used *in vitro* model. Both branching of the ureteric epithelium, and mesenchyme to epithelium conversion forming nephrons, occur *in vitro*. Exogenous growth factors and other molecules can be added to the culture medium in order to examine the roles of these molecules in renal development.

Despite the power of metanephric organ culture as a model for defining the roles of specific molecules in kidney development, relatively little has been learnt about the *pattern* of growth of the ureteric tree. We have been developing a method for examining in three dimensions the branching morphogenesis of the metanephric kidney. Quantitative and objective 3-dimensional measurements of parameters such as number of branches; number of branch generations; length of specific branches and of the whole tree; angles between branches; and rate of growth (before and after branching) will provide accurate indices of growth and branching.

2. A QUANTITATIVE APPROACH TO 3D BRANCHING

Whole mouse (Balb c) kidneys were obtained at embryonic day 12 (E12) and cultured for up to 48 hours. These whole metanephroi were then stained with a fluorescent antibody (Calbindin-D28K, Sigma-Aldrich Pty. Ltd., Castle Hill, Australia) to specifically stain the branching ureteric duct (Fig. 1a).

Metanephroi were then optically sectioned with a 10x lens on a Leica TCS NT confocal microscope. Since the lens has a measured axial resolution (FWHM) of 10 μm, sections were taken at intervals of $4-5$ μm to satisfy the requirements for Nyquest sampling. The 2D optical slices of the trees were 8 bit grey level images, typically 512×512 pixels, and between $20 - 50$ images were required to sample a metanephros. Each metanephros was then segmented (Fig. 1b) and volume- or surface-rendered using Voxblast (version 1.4, Vaytek, Iowa) in order to reconstruct the 3D volume for display and analysis (Figs. 1e-g). This software interpolates sections as necessary to fulfill a requirement for cubic voxels, and also provides facilities for slice tracing (used to demarcate individual branches of the ureteric tree) and linear measurements in 3D space.

Our initial attempt to generate and measure the 3D skeleton has been in part manual and in part automatic, and somewhat laborious. It is now described. Using the Slice Tracing feature in Voxblast, voxels of specific intensity (for example, white or 255 on the 8-bit grayscale) were drawn in the midpoint of

Table I. — Measurements of ureteric branches of an E12 mouse metanephros cultured for 2 days. Values for mean length are mean ± standard deviation.

Generation	No. of Branches	Total length (μm)	Mean length (μm)
1	1	356	356
2	2	274	137 ± 14
3	4	700	175 ± 51
4	8	1376	172 ± 50
5	6	941	157 ± 69

each branch in each section. In order to facilitate, and to some extent validate the selection of the midpoint in the individual sections, a duplicate stack of the same sections was binarised and then thinned to a single pixel skeleton (Fig. 1c) in a batch process using ScionImage (Scion Corporation, Maryland, USA). The resultant stack of 2D skeletons (Fig. 1d) could be scrolled through and inspected using ScionImage alongside the ureteric tree stack being analyzed in Voxblast. Importantly, this also facilitated the identification of branch points. The simultaneous inspection of corresponding ureter and skeleton sections was essential for determination of branch points as well as confirming the position of the rest of the skeleton. Once the approximate skeleton had been drawn, the distance between consecutive hand-drawn voxels (depicting the skeleton) was easily measured and summed using Voxblast's measurement feature. These results are shown in Table I and Figure 1h.

3. COMMENTS ON THE METHODOLOGY

We have developed the first method for measuring in 3 dimensions the structure of the ureteric tree in the developing permanent kidney (metanephros). The method involves: (1) whole mount immunostaining of the ureteric tree; (2) optical sectioning of the whole tree using confocal microscopy; (3) segmentation and volume rendering of the tree; (4) manual skeletonization of tree profiles in the optical sections; (5) overlaying of skeletons from 2D sections; (6) manual construction/identification of a final 3D skeleton; and (7) computerized 3D measurement of the length of the whole skeleton and individual branches. While this method represents an advance on methods for measuring ureteric branching from a 2-dimensional perspective, the procedure clearly needs further refinement. At present, the skeletonization of the tree is performed manually, and therefore must necessarily involve a significant degree of subjectivity. The overlaying of section skeletons to subjectively generate a final 3-dimensional skeleton is carried out for validation purposes only, and the 3D skeleton manually constructed by painting voxels in a 3D software programme such as Voxblast. Ideally for any quantitative method for large scale use, the skeletonizing procedure should be done accurately, objectively and

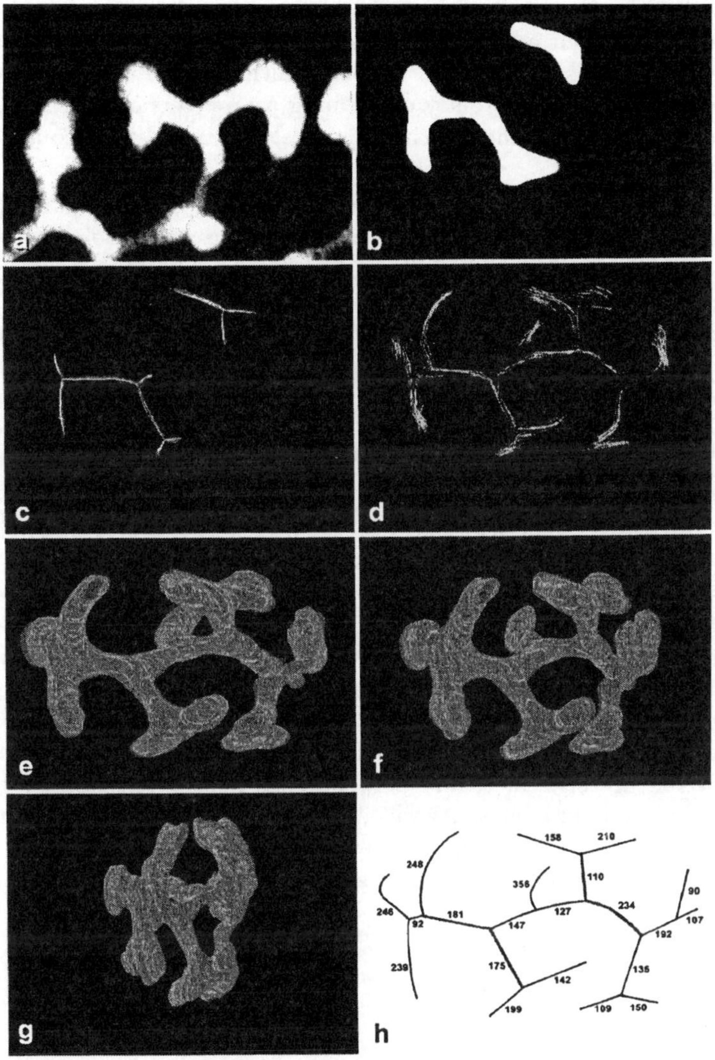

Fig. 1. — The process of serial imaging, volume rendering and manual production of the skeleton for subsequent measurement is shown. a) A typical confocal section of a fluorescently-labeled 12 day-old metanephros. The subsequent segmentation and skeletonization is shown for a different metanephros. The ureteric region in each section was segmented by thresholding and binarisation (b). Individual sections were also skeletonized (c), producing a stack of skeletons (d) which was then used for inspection and comparison with the surface rendered tree (e). In f) and g) the surface rendered tree is rotated 30° and 60° to illustrate its three-dimensional structure. By scrolling through the tree, each branch segment could be traced and measured, and a final mathematical tree constructed (h, units in μm). The first branch (trunk) of the tree is in the centre and is 356 μm long.

automatically. Furthermore, the method at present does not readily provide information on branch angles. Volume and surface area can be determined from volume/surface rendering in Voxblast, although we do not present such data here. We are thus currently developing a new version of the method to automatically determine and measure the skeleton in 3D to account for these deficiencies.

Image analysis and processing has previously been used to measure the length of ureteric branches. Gilbert *et al.* [8] analyzed fluorescein-labeled ureteric trees from E14 rat metanephroi. The kidneys were optically sectioned every 15 μm, and the sections were segmented, smoothed and thresholded prior to skeletonization. The skeletonization was performed automatically. These skeletons were compared to the original images. The "mean skeleton" and measurements of the total length of the ureteric tree were then calculated in 2-dimensions. The number of branch points and analysis of their location was also performed on the binary skeleton (Gilbert *et al.* [8]). Despite the final data being in two dimensions and not taking account of the fact that even in culture the ureteric tree grows in three dimensions, this contribution by Gilbert *et al.* [8] is important because it demonstrated for the first time how confocal microscopy can be used in conjunction with automatic image analysis to quantify ureteric branching and growth.

Two other groups have since analyzed the branching pattern of the ureteric bud. Qiao *et al.* [14] produced 3-dimensional reconstructions of the ureteric tree, also from a set of fluorescein-labeled 10 μm optical sections. However, no quantitation was performed. Van Adelsberg [15] also analyzed the branching pattern of the ureteric duct, and paid particular attention to the symmetry of the dichotomous branches. Again, image analysis was used in combination with confocal sections of immunostained whole mounts. The technique assesses the relative proportions of symmetric and asymmetric branch points.

4. CONCLUDING REMARKS

Until now the structure of the developing ureteric tree has mostly been described in purely qualitative terms. A few quantitative studies have been performed, but none have taken into consideration the true three-dimensional structure of the tree. The approach described here provides the first length measurements of ureteric branches that take account of the true 3-dimensional structure of the tree. Clearly, the technique requires improvements, both in terms of objectivity, robustness, and automation. An accurate and automated technique for measuring the ureteric tree in three dimensions is essential for achieving a new level of understanding of the pattern of ureteric growth and branching within the developing kidney. Ideally, such a technique could not only be used to assess static trees (*e.g.* after fixation), but also dynamic ureteric trees that are continuing to grow.

Acknowledgements

Our research on kidney development is supported by an Australian Research Council Small Grant. Luise Cullen is supported by a Biomedical Research Scholarship from the Australian Kidney Foundation.

REFERENCES

[1] Q. Al-Awqati and M.R. Goldberg, *Kidney Int.* **54** (1998) 1832-1842.
[2] B.M. Brenner, D.L. Garcia and S. Anderson, *Am. J. Hypertens.* **1** (1988) 355-347.
[3] B.M. Brenner and E.L. Milford, *Am. J. Kidney Dis.* **21** (1993) 66-72.
[4] B.M. Brenner and G.M. Chertow, *Am J. Kidney Dis.* **23** (1994) 171-175.
[5] C. Burrow, *Pediatr. Nephrol.* **14** (2000) 240-253.
[6] A.T. Clark and J.F. Bertram, *Curr. Opin. Nephrol. Hypertens.* **9** (2000) 247-251.
[7] J.A. Davies and M.G. Davey, *Pediatr. Nephrol.* **13** (1999) 535-541.
[8] T. Gilbert, C. Cibert, E. Moreau, G. Geraud and C. Merlet-Benichou, *Kidney Int.* **50** (1996) 783-795.
[9] M.F. Horster, G.S. Braun and S.M. Huber, *Physiol. Rev.* **79** (1999) 1157-1191.
[10] S. Kuure, R. Vuolteenaho and S. Vainio, *Mech. Dev.* **92** (2000) 31-45.
[11] J. Oliver, *Nephrons and kidneys; a quantitative study of development and evolutionary mammalian renal architectonics* (New York, Hoeber Medical Division, Harper & Row, 1968).
[12] V. Osathanondth and E.L. Potter, *Arch. Pathol.* **76** (1963a) 277-289.
[13] V. Osathanondth and E.L. Potter, *Arch. Pathol.* **76** (1963b) 290-302.
[14] J. Qiao, H. Sakurai and S.K. Nigam, *Dev. Biol.* **96** (1999) 7330-7335.
[15] J. van Adelsberg, *Dev. Genet.* **24** (1999) 299-308.

CONTRIBUTED PAPER 6

Morphogenic Responses of Mammary Epithelial Cells Grown in Biological Semi-Solid Substrates

V. Fafeur, A. Delannoy-Courdent, J.C. Bout and B. Vandenbunder

1. INTRODUCTION

Morphogenetic movements of epithelial cells are critical to normal and pathological development of tissues and organs. Characteristically, epithelial cells are tightly connected through several junctional structures and this cohesiveness allows continuous lining in the body. During normal development, epithelial cells can modulate their cohesiveness to acquire diverse morphological forms for their specific functions, such as the thyroid follicles, the kidney tubules or the complex branching and alveolar structures found in the lung, salivary and mammary glands. These supracellular structures are not only determined by the genome of the cells, but are also governed by local molecular signals, such as growth factors, by adhesive interactions of epithelial cells with themselves or other cell types (cell-cell adhesion) or extra-cellular matrix substrates (cell-matrix adhesion) and by their function in the organism (for reviews see Bissel *et al.* [1] and Gumbiner [2]). The understanding of deregulated adhesive properties of epithelial cells is of major importance during tumor development, since more than 80% of the human tumors are epithelial in origin. It is assumed that the development of malignant tumors, in particular the transition from benign lesions to invasive, metastatic tumors, is characterized by the ability of a cancerous epithelial cell to overcome cell-cell adhesion and cell-substrate adhesion, and to invade surrounding tissue. Tumoral and metastatic development can therefore be described as the result of deregulated growth and adhesiveness, and increased invasiveness of epithelial cells.

The molecular and cellular mechanisms that determine how various epithelial acquire their specific morphologies are far to be understood. In this study, we wished to investigate the morphogenic ability of epithelial cell lines cultured in semi-solid matrix substrates, which allow formation of three-dimensional supracellular structures.

2. EXPERIMENTS AND METHODS

The conditions of culture are indicated in Figure 1. Briefly, cells were embedded and cultured in a gel layer of matrix proteins: collagen I or Matrigel (a commercial preparation of the basement membranes, which are thin extracellular matrices underlying cells, its major components are: laminin, collagen IV, heparan sulfate proteoglycans, entactin and nidogen). In addition, we added or not to the culture medium a growth factor, scatter factor/hepatocyte growth factor (SF/HGF), which is known to stimulate proliferation, motility and morphogenesis of epithelial cells. The two mammary cell lines, NME and NMF cells (NME: epithelioid, NMF: fibroblastic), are epithelial subpopulations isolated

⟹

Fig. 1. — Morphogenic capacities of two mammary epithelial cell lines grown in biological semi-solid substrates: **2D growth on plastic**: NME (A) and NMF (B) cells (2 500 cells/ml) were seeded on plastic into 24-wells and were cultured for four days. Whereas NME cells grow in tightly packed islets, with a typical epitheloid morphology, NMF cell grow as fusiform individual cells, with an elongated fibroblastic morphology. **2D growth on Matrigel**: NME (C) and NMF cells (D) (20 000 cells/ml) were seeded on a layer of Matrigel into 24-wells and were cultured for two days. Whereas NME cells grow as aggregated structures (compact group of cells), NMF cells form a network structure. **3D growth in Matrigel**: NME (E) and NMF (F) cells were mixed with Matrigel (8.2 mg/ml) (500 cells/300 μl into 24-wells) and were cultured for ten days. Both NME and NMF cells grow as compact structures. It is worth noting that NMF cells display cellular extensions during the first days of culture (see F), but with time these visible cellular extensions are lost and the cells group each other to form compact structures. **3D growth in Matrigel**: the conditions of culture are identical to E and F, except that SF/HGF was added in the growth media. Both NME (G) AND NMF (H) cell respond to the morphogenic effect of SF/HGF, with different characteristics. Budding of a group of cells occurs on the surface of the aggregated structures of NME cells, while individual cells protrude and expands as branches from NMF aggregates. **3D growth in collagen**: NME (I) and NMF (J) cells were mixed with collagen I (2 mg/ml) (500 cells/300 μl into 24-wells) and were cultured for ten days. NME cells grow as tightly connected structures, while NMF cells grow as scattered structures. **3D growth in Matrigel/collagen**: NME (K) and NMF (L) cells were mixed with a mixture of collagen I and Matrigel (8.2 mg/ml) (500 cells/300 μl into 24-wells) and were cultured for ten days. NME cells started by forming cell clusters and then form elongated and branched structures of similar size. NMF cells give rise to elongated and branched structures, with trunk and ramifications. Bar = 100 μm.

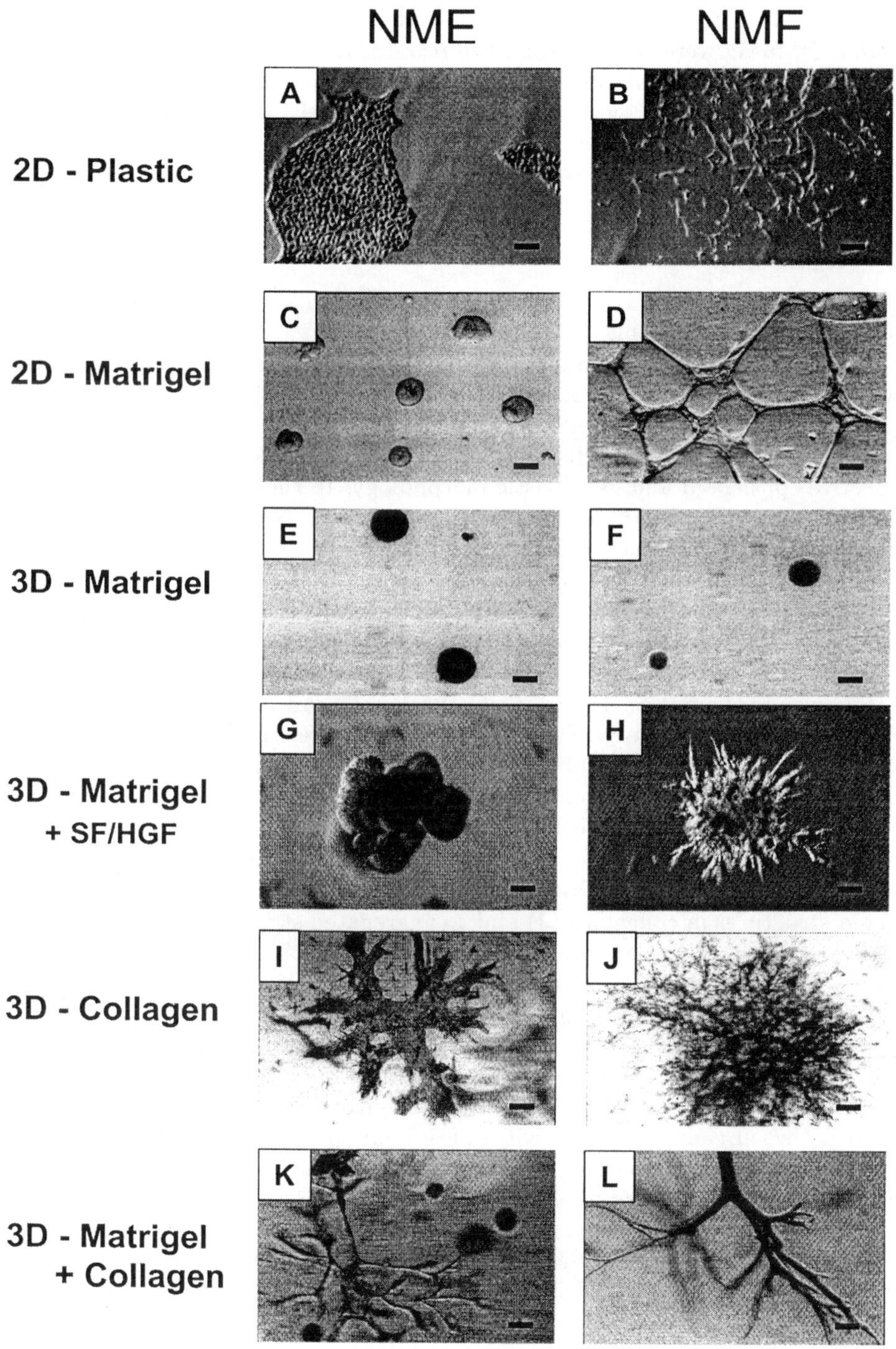
NME
NMF
2D - Plastic
2D - Matrigel
3D - Matrigel
3D - Matrigel
+ SF/HGF
3D - Collagen
3D - Matrigel
+ Collagen
A
B
C
D
E
F
G
H
I
J
K
L

from the NMuMg epithelial cell line issued from a normal mouse mammary gland. NMuMg were subcloned on the basis of their morphotype and allowed to characterize factors influencing invasiveness of epithelial cells [3]. In particular, NMF cells are more invasive than NME cells, in agreement with the proposal that transition from an epitheloid to a fibroblastic morphotype marks invasiveness in clinical and experimental cancer.

3. RESULTS

The conclusions from this study are the following.

1 - Epithelial cells elaborate supracellular structures in matrix substrates which are not evidenced, nor easily expected from their behavior in conventional culture on a plastic substrate. When cultured on plastic, NME grow as cell islets (A) and NMF cells as isolated cells (B); this agrees with their respective epithelioid and fibroblastic morphology. In these conditions, it can be demonstrated that NMF cells are more motile than NME cells and that they have lost some epithelial characteristics, including loss of cell-cell association during proliferation (A), loss of expression of the E-cadherin adhesion molecule and absence of responsiveness to the SF/HGF growth factor (data not shown). In contrast, using 2- or 3-dimensional (2D, 3D) culture in a matrix substrate, we show that the cells are still competent to associate and to elaborate specific morphogenic structures. For example, NMF cells can form network structures (D), aggregated structures (F), or branched structures (H, L), and they respond to the morphogenic effect of SF/HGF (compare F and H). The variety of these morphogenic responses demonstrates that these matrix substrates provide a biological support to favor cell-cell, as well as cell-matrix interactions.

2 - The same epithelial cell line can display distinct morphogenic properties according to the matrix substrate. For example, NMF cells give rise to scattered structures in collagen I (J) and to aggregated structures in Matrigel (F). By varying either the protein concentration or the protein composition of the substrate, we found that both parameters contribute to the shape of these structures (NME cells, compare E, I, K; NMF cells compare F, J, L). In particular, individual NMF cells are more elongated in collagen 1 (J) than in Matrigel (F), indicating that they elaborate more and/or stronger cell-matrix interactions in collagen 1 than in Matrigel. The fact that some morphogenic changes are already detectable within few hours (see 3) also indicate that most of these interactions are elaborated from readily available cell-matrix adhesion molecules. The corresponding cell-matrix adhesion molecule expressed by these cells, including members of the integrin family, remains to be identified.

3 - The same matrix substrate can allow distinct epithelial morphogenic properties according to the genotype of the cells. For example, when grown on matrigel the NME cells form aggregated structures (C), while NMF cells form a network structure (D). We documented the progressive elaboration of these structures within 24 h, following the initial seeding of individual cells.

Within less than 8 h, NME cells move and group each other, while NMF cells both move and group each other and protrude visible cellular extensions, which were subsequently stabilized when one cell contacts another one, this process allowed the formation of aligned cells in a network structure. In addition, by increasing the initial number of cells, the final number of structures (in aggregates or as a network) was not increased in parallel, rather the size of these structures increased (data not shown). This indicates that cells attracted each other within areas of Matrigel of similar size. We further analyzed whether different cell-cell interactions are displayed according to the cell type within the same substrate, using samples depicted in K and L. The results can be summarized as follows: a) using electron microscopy, we found that NMF cells are more loosely connected to each other than NME cells, nonetheless they still associate through conventional cellular junctions, b) by measuring expression of cadherin cell-cell adhesion molecules, we found that NME cells express E-cadherin but not NMF cells, which rather express another cadherin, which remains to be identified.

4 - Each cell type has a specific morphogenic program, which can be stimulated by a growth factor. Indeed, SF/HGF has distinct morphogenic effects on NME and NMF cells: it induces budding of NME cells (compare E and G) and branching of NMF cells (compare F and H). It can be postulated that the cellular mechanisms for budding and branching are distinct: for budding a group of cells are expanding together from the initial aggregate (E, G), while for branching one single cell protrudes and expands from the initial aggregate, and is secondarily followed by other cells (F, H). It should be stressed that the known scattering effect of SF/HGF cannot be evidenced when NMF cells are grown on plastic, since these cells are spontaneously scattered (B). By finding a branching effect of SF/HGF using cell cultures in the presence of a matrix substrate (H), we contradict the conclusion that invasive cells are unresponsive to the action of a growth factor. It is unknown whether SF/HGF induces these different morphogenic responses in NME and NMF cells through distinct signal transduction mechanisms.

4. SUMMARY

In summary, these manipulations document several novel and intriguing results, which allow biologists to formulate new questions and/or hypothesis. In particular, we found that highly motile, invasive and cancerous cells (NMF) are capable to form associated and branched structures and can respond to growth factor, whereas most studies emphasize the demonstration that invasive cancerous cells have an anarchic growth and unregulated adhesion properties. In the future, experiments will be aimed to characterize the molecules discriminating between the morphogenic capacities of NME and NMF cells.

REFERENCES

[1] Bissell, Weaver V.M., Lelievre S.A., Wang F., Petersen O.W. and Schmeichel K.L., *Cancer Res.* **1** (1999) 1757-1763.
[2] Gumbiner, *Cell* **84** (1996) 345-357.
[3] Van den Broecke, Vleminckx K., De Bruyne G., Van Hoorde L., Vakaet L., Van Roy F. and Mareel M., *Clin. Exp. Metastasis* **14** (1996) 282-296.

CHAPTER VIII

Vascular Development: Design Principles and Morphometric Analysis of a Branched Vascular Tree

F. le Noble, W. Hacking, D. Slaaf and H. Struijker-Boudier

1. GENERAL INTRODUCTION

The formation of new blood vessels is an important process in embryonic development and in (patho) physiological conditions like cyclic vessel growth in the endometrium, wound healing, diabetic retinopathy, and tumor growth. Lack of blood vessel growth is a key event in pathologies like essential hypertension, peptic ulcers, and bowel atresia. At present clinical trials are undertaken with anti-angiogenic drugs to interfere with tumor development [2, 3, 7, 14, 19]. Pro-angiogenic therapies, or "therapeutic angiogenesis" are employed in the treatment of ischemia related events such as coronary artery disease ("infarction") and peripheral artery occlusive diseases [11, 20, 56, 59, 67]. This therapy aims at stimulation of collateral artery formation to the ischemic region thereby restoring flow and oxygen delivery.

Although the therapeutic application of agents influencing angiogenesis is promising, it is still subject of critical analysis. Systemic administration of growth factors for example also influences cardiovascular hemodynamics. The efficacy of collateral vessel formation depends mainly on the local hemodynamic consequences of the intervention. Modeling studies integrating hemodynamics, metabolic status and vascular growth may provide a valuable tool in understanding the basic principles and help to identify the site and time-course at which the (anti)angiogenic drugs can be applied within a network. This chapter outlines some of the basic characteristics of structure and function of the vascular system. It will deal with a description of growth processes during embryogenesis and (patho) physiological adaptation. Modeling studies with regard to hemodynamics and metabolic demand will be presented and finally, some parameters used to analyze vascular structures will be discussed.

2. THE CIRCULATION

In order to construct a realistic computer model enabling simulation of vascular development it is necessary to understand the principal design in relation to the function of the vascular tree. This paragraph outlines some of the basic characteristic of the morphology and the hemodynamics of the cardiovascular system.

In higher vertebrates, blood is pumped through a closed fluid circuit, the vascular system. Based on functional morphology, the vascular system can be divided in the arterial (delivering blood to peripheral tissues), the capillary (exchange vessels at the tissue level), and the venular system (returning blood to the heart). The arterial system can be divided in three compartments based on function and vessel dimensions: the conduit vessels, the resistance sized vessels, and the microcirculation. The perfusion pressure in the vascular system is generated by the beating heart.

Blood pressure within the arterial system varies during the heart cycle. Because the conduit vessels are very elastic, they function as temporal buffer for blood during the ejection phase of the heart (the windkessel function). Effectively this reduces the afterload of the heart and high pressure elevations. The static pressure within the arterial system, *i.e.* the mean arterial pressure, depends on the cardiac output and the peripheral vascular resistance. Cardiac output is a function of heart rate and stroke volume. Peripheral vascular resistance is influenced by blood viscosity, vessel diameter, and vessel length according to Poiseuille's law. Cardiac output (CO) and mean arterial pressure (MAP) can be measured directly. Peripheral resistance can be calculated from the ratio of MAP and CO.

The resistance sized vessels and the microcirculation contribute to the peripheral vascular resistance. For a single vessel, resistance can be calculated according to Poiseuille's law (Tab. I). Vessels are part of a vascular network. The vascular architecture, the interconnection of single vessels (parallel coupling of vessels, arcades) also contributes to total vascular resistance. Since lumen diameter stands to the power of 4 in Poiseuille's law, it is considered to be an important determinant of resistance. Diameter control can be exerted by either acute or long-term adaptations.

Acute control includes nervous mechanisms, myogenic response (diameter change in response to pressure change), flow dependent dilation and vasomotion. Long-term regulation of lumen diameter can be divided into a structural and a functional component. Structural changes include changes in vascular wall mass (smooth muscle hypertrophy/hyperplasia) and structure (radial growth of lumen). Long-term functional changes include alterations in the regulation of vascular tone for example by changes in sensitivity for circulating

Table I. — Definitions of resistance, shear stress, and circumferential wall stress.

1. Resistance for a vessel segment according to Poiseuille's law
$R = \frac{8L\eta}{\pi r^4}$ L, vessel length; η, viscosity; r, radius. Unit: Pa m^{-3} s
2. Shear Stress for a vessel segment
$\tau = \frac{4\eta Q}{\pi r^3}$ Q, flow; η, viscosity; r, radius. Unit: N/m^2
3. Circumferential Wall Stress for a vessel segment (Laplace)
$\sigma = \frac{Pr}{W}$ P, pressure; r, radius; W, wall thickness. Unit: N/m^2

vasoactive hormones or alterations in nervous innervation. Changes in network geometry, due to the formation or regression of vessels, are also regarded as a long-term control mechanisms.

The peripheral circulation performs several functions such as meeting the metabolic demand of tissues and maintaining central hydrostatic pressure through modulation of vascular diameter, hence, peripheral resistance. The largest pressure drop takes place along the arterioles of the microcirculation. Arterioles are therefore considered to play an important role in hemodynamic homeostasis. For the cells in the tissues, the arterioles of the microcirculation are important since they control flow distribution, and thereby, oxygen and nutrient distribution.

Tight regulation is necessary to keep flow to the tissue within safe boundaries and allow proper exchange of oxygen and nutrients according to Starling's law. However, high rises in pressure in the proximal part of the tree have to be avoided since they may result in vascular damage. Although resistance is the subject of many investigations, resistance itself can not be sensed by the vasculature. It is a calculated parameter that determines flow to the target tissue given a certain pressure drop along a vascular segment of network. Changes in resistance that occur during fetal development or disease should be related to their hemodynamic determinants. If perfusion of an organ remains constant, delivery of nutrients and oxygen, and removal of metabolites seem to prevail. The resulting change in pressure may be just the consequence.

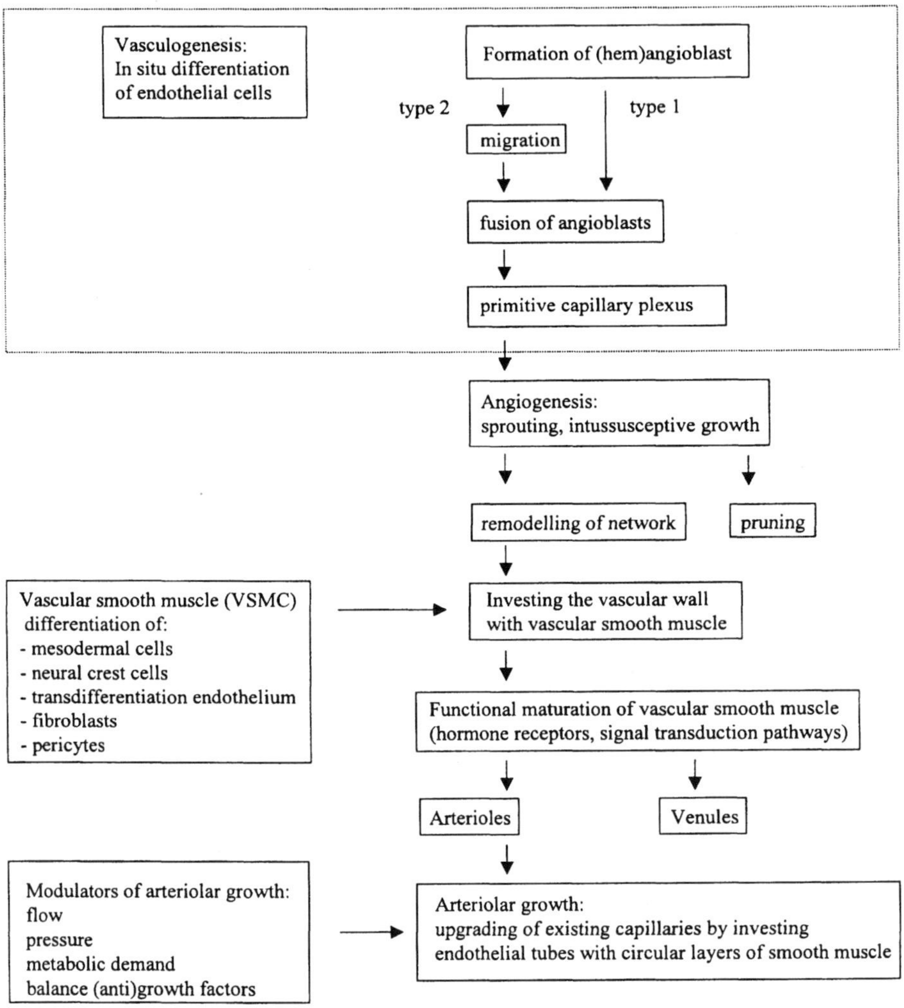

Fig. 1. — Schematic representation of the processes involved in vascular development.

3. FORMATION OF THE VASCULAR TREE

During embryonic development, endothelial precursor cells can be detected along with primitive blood cells, (hem)angioblasts [10, 47]. They are located in discrete zones referred to as blood islands. Subsequent fusion of blood islands is the first evidence for a primitive vasculature. The formation of these earliest blood vessels is called vasculogenesis (see Fig. 1). Morphogenetically, vasculogenesis can be divided in vasculogenesis types 1 and 2 [49]. Type 1 refers to angioblasts that arise in the place were the vessels are formed, for example the

a)
b)

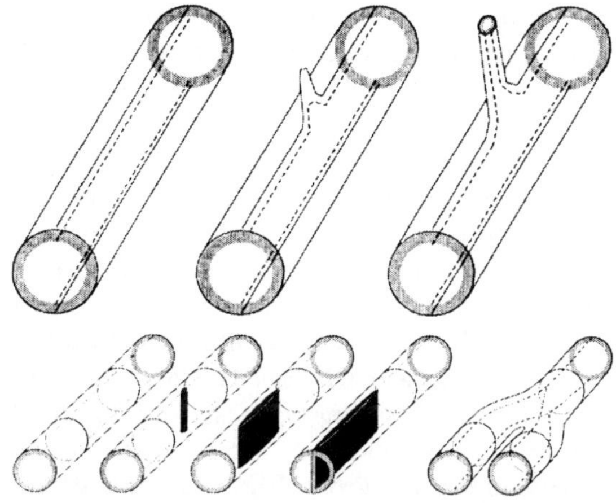

Fig. 2. — a) Sprouting angiogenesis. b) Non-sprouting angiogenesis, intussusceptive growth.

dorsal aorta, whereas type 2 refers to angioblasts that migrate to the site of vessel formation, for example the posterior cardinal veins. The first primitive blood vessels do not carry flow yet.

Angiogenesis is defined as the formation of new blood vessels as outgrowths of pre-existing vessels [49]. Morphologically, angiogenesis can be divided in sprouting angiogenesis and non-sprouting angiogenesis (Fig. 2). Sprouting angiogenesis (Fig. 2a) involves degradation of the basement membrane surrounding the endothelial layer, migration of the endothelial cell towards the angiogenic stimulus, proliferation of the endothelial cell, lumen formation and fusion with nascent tubes.

Non-sprouting angiogenesis or intussusceptive growth, is a process of splitting ("like a zipper") of an existing vessel by transcapillary pilars of extracellular matrix (Fig. 2b) [16, 24, 57]. Furthermore, non-sprouting angiogenesis may occur as the result of proliferation of endothelial cells inside the vessel lumen and the subsequent fusion and splitting of the capillaries [48].

Vasculogenesis has been found to be operative in the vascularization of endodermal organs. Other organs (brain, kidney) may depend on angiogenesis. Growth of new vessels into developing organs can result from invaginations of primitive vessels into the organ. However, the developing organs usually possess endothelial precursor cells, or even small capillary networks (without flow). It has been shown that outgrowth of the vascular system may be partly due to incorporation of these pre-existing endothelial cells located in the organ into the vascular wall or network of already existing and expanding vessels.

In summary (Fig. 1), endothelial cells used for vessel formation may arise from *in situ* differentiation in the organ, from sprouting of an existing vessel into the organ, or originate from a plexus elsewhere and after migration contribute to the development of the organ vasculature. Hence, heterogeneity in the behavior of adult endothelial cells along the vascular tree may be related to the embryonic origin [10]. This aspect should be considered in the construction of computer models simulating embryonic angiogenesis.

The newly formed vascular plexus is rapidly remodeled into a structure resembling the mature branching pattern. This is called pruning. Both vasculogenesis and angiogenesis result in an excess of endothelial tubes. In general the redundant endothelial cells and capillary tubes are lost. The redundant endothelial cells may be used for reassemblence into growing or expanding vessels. The endothelial cells may also de-differentiate into other cell types, or rapidly disappear. It is not clear whether the latter is dependent on apoptosis.

Although a lot is known about angiogenesis and differentiation of endothelial cells, surprisingly little is known about the formation of arterioles and venules (Fig. 1). Arterioles and venules grow (Fig. 3) by investing capillary tubes with circular layers of vascular smooth muscle cells [32, 63]. During embryonic development, vascular smooth muscle cells seem to originate from the splanchnic mesoderm and from neural crest cells [6]. Vascular smooth muscle cells may also be recruited from local mesenchyme surrounding the developing blood vessel. Recent experimental evidence indicates that a small portion of endothelial cells of the primitive aorta may transdifferentiate into vascular smooth muscle cells and contribute to the formation of the media [71].

In adult tissues, it has been shown that at the transition of a terminal arteriole and a capillary, vascular smooth muscle cells, which can originate from the arteriole, migrate towards the capillary tube, and subsequently surround the capillary tube and form a circular layer [50, 52]. At the cellular level this may involve proliferation of an existing vascular smooth muscle cell and subsequent migration of the daughter vascular smooth muscle cell through the extracellular matrix towards the capillary (Fig. 3). The mechanisms regulating the direction of migration through the extracellular matrix and the remodeling around the capillary tubes are not known. Also pericytes have been implicated to play a role in this process [61, 62].

Recent experimental evidence suggests that in adult tissues the origin of contractile cells involved in arteriolization may also be located outside the vascular tree. Fibroblasts can be recruited from the interstitium surrounding the vascular network and incorporated in the arteriolar tree and differentiate into cells with a contractile phenotype [52, 63].

The factors controlling the three-dimensional architecture of a vascular bed characteristic for a given organ are poorly understood. Homeobox genes [15], tissue oxygen concentration, extracellular matrix structure [70], ion currents may be important in the regulation of organ development and vascular architecture. The architecture is also dependent on local remodeling, apoptosis (programmed cell death), flow and pressure. However, the ontogeneity of organ characteristic features of endothelium, vascular smooth muscle cells, and vascular architecture remain speculative. It is tempting to suggest that they

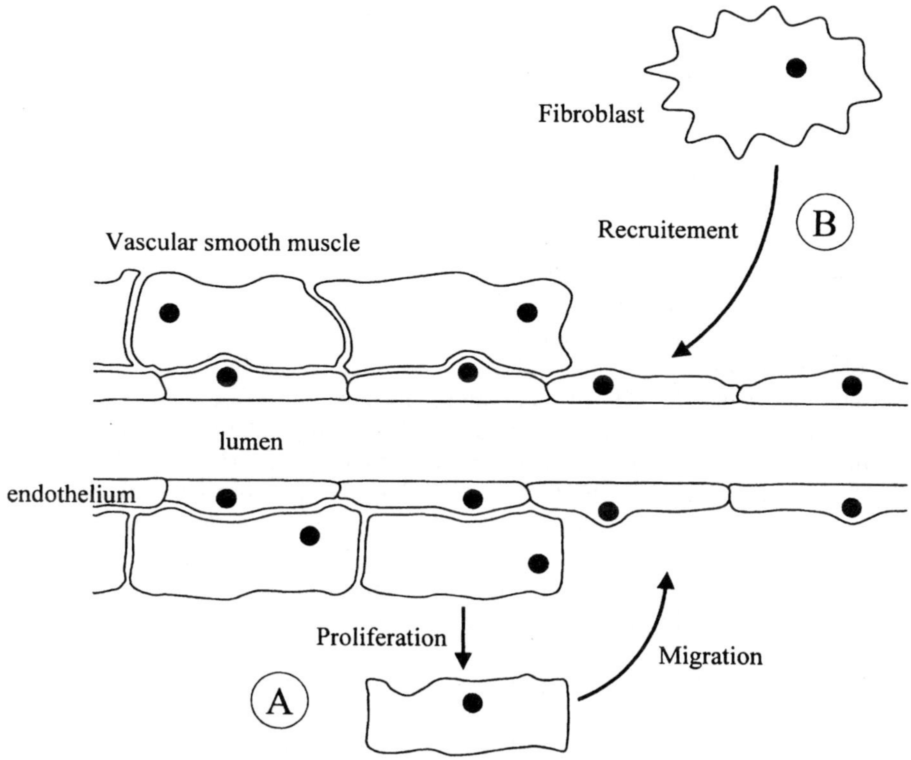

Fig. 3. — Schematic representation of processes involved in growth of arterioles. Note that the vascular smooth muscle cells may be recruited from the terminal arteriole A), or from a location outside the vasculature B).

are related to the embryonic origin and modulated by the local environment, hemodynamics and metabolism.

4. GROWTH FACTORS

The introduction of molecular biology has led to the identification of a range of chemical and mechanical mediators involved in the design of the vascular system. It is now well established that angiogenesis depends on a delicate balance between angiogenic and anti-angiogenic molecules. In the case of tumor growth, this balance has shifted towards "uncontrolled" pro-angiogenesis [27], whereas essential hypertension has been linked to lack of proper angiogenesis [40]. Mechanical factors can influence growth by altering the expression and/or secretion of growth factors and related genes, or influence the sensitivity for growth factors.

It is beyond the scope of this chapter to review all the mediators and molecular mechanisms. The reader is referred to some excellent reviews with regard to this topic [4, 9, 28, 31, 54]. The most widely studied factors with regard to the previously described growth processes will be briefly discussed.

4.1. Chemical Mediators

Basic fibroblast growth factor (bFGF) is a potent angiogenic factor both *in vitro* and *in vivo*. It has been postulated that bFGF is important in the initial induction of hemangioblasts from the primitive mesoderm [18].

Vascular endothelial growth factor (VEGF) is the key regulator of embryonic vasculogenesis and angiogenesis [17]. It can be secreted by both mesenchymal and stromal cells. VEGF can stimulate both proliferation and migration of endothelial cells. The *in vivo* effects of VEGF during embryogenesis depend on the expression of the VEGF receptors, VEGF-R1 (of flt-1) and VEGF-R2 (or flk-1). Absence of the VEGF-R2 receptor is lethal during early embryonic development. The vascular phenotype is characterized by a defect in the formation of both angioblasts and hemangioblasts. The experimental data suggest that the VEGF-R2 receptor is important for the induction of the first endothelial cell differentiation. In knock-out models of the VEGF-R1 receptor, angioblasts are present but endothelial cells are not remodeled into blood vessels. It is postulate that the VEGF-R1 receptor is crucial for the formation of endothelial tube structures. Secretion of VEGF can be influenced by local oxygen tension; a reduction in oxygen availability (hypoxia) stimulates the release of VEGF. This is related to the presence of a HiF-1 (hypoxia inducible factor-1) binding site in the promotor region of the VEGF gene. HiF-1 is a protein and a mediator of transcriptional control of the VEGF gene. The HiF-1α subunit becomes more stable at low oxygen levels, hence, during hypoxia more functional HiF-1 is available and VEGF will be transcribed.

The endothelial cell receptor tyrosine kinases Tie-1 and Tie-2 play an important role in the establishment of capillary integrity and remodeling into large caliber vessels, respectively. Mice lacking the Tie-1 gene display leaky capillaries, extravasation of erythrocytes, and a generalized oedema. The branching pattern of the large arteries is not affected. Although Tie-1 may not be involved in the process of vasculogenesis and angiogenesis, it is crucial for the structural integrity of (early) microvascular endothelium and control of fluid exchange. The natural ligands for the Tie-1 receptor are not known yet.

In the Tie-2 knock-out mouse, the absence of the Tie-2 receptor resulted in the absence of angiogenic sprouts, a ruptured and disorganized aorta and improper remodeling of small (capillary) vessels into large caliber vessels. It is postulated that Tie-2 plays an important role in vessel maturation, organization of a branched network and remodeling of small into large vessels. The naturally occurring activating ligand for the Tie-2 receptor is angiopoietin-1, the natural antagonist angiopoietin-2. A deficiency in angiopoietin-1 results in a similar vascular phenotype as the Tie-2 knock-out. Detailed ultrastructural analysis revealed impairment of the formation of tissue-folds (see non-sprouting angiogenesis), lack of recruitment of pericytes/vascular smooth muscle cells, and disorganized collagen. This suggests a role for angiopoietin-1 in endothelial-extracellular matrix interaction.

Using a molecular biological approach it has been postulated that from the onset of early angiogenesis, there is already a genetic difference between endothelial cells used to build an arteriole or a vein [71]. This distinction was revealed by the expression of the ligand ephrin-B2 on arterial cells, whose cognate receptor ephrin-B4 was expressed on venous cells. It has been hypothesized that within a layer or plexus of endothelial cells, the endothelial cell expressing the ephrin-B2 ligand interacts with the endothelial cell expressing the ephrin-B4 receptor, resulting in remodeling into two tubes, an arteriole and a venule. Hence, besides the difference in hemodynamics, oxygenation, and structural anatomy, there is a genetic difference as well. This suggests that there are already arterial and venous angioblasts, and perhaps arterial and venous specificity in angiogenesis and vascular maturation, although the initiators are not yet known.

The formation and growth of arterioles and venules involves construction of a vascular wall with a media consisting of vascular smooth muscle cells [8, 32]. Platelet Derived Growth Factor (PDGF) plays an important role in the recruitment of mesenchymal precursor cells (both pericytes and vascular smooth muscle cells) in the vascular wall of developing arterioles and venules. It can be secreted by endothelial cells. Especially the PDGF-β receptor has been implicated to play a role in the recruitment of vascular smooth muscle cells and pericytes in the small vessels, whereas the PDGF-α receptor is more associated with large caliber vessels. Beside recruitment, PDGF has been associated with the regulation of proliferation of vascular smooth muscle cells. The effects of PDGF are counteracted by Transforming Growth Factor-β (TGF-β). TGF-β may inhibit growth and migration of endothelial cells and stimulate the formation of extracellular matrix proteins and stimulate maturation and differentiation of vascular smooth muscle. TGF-β is secreted by the endothelial cells after attachment of a mural cell with the endothelium.

4.2. Mechanical Mediators

Vessels are subjected to mechanical forces [70], distension due to the blood pressure, and a frictional force due to the flow of blood. The detailed mechanisms of mechanotransduction is outlined in the chapter by Levi, and are reviewed in detail by Davies [12], Skalak [62] and Ingber [33].

In short: an increase in pressure can stimulate the expression of the proto-oncogenes c-fos and c-jun, modulate the balance between PDGF and TGF-β, activate the renin-angiotensin system and, as a consequence, induce smooth muscle cell growth. Changes in flow can alter the endothelial release of nitric oxide, which in turn can influence vascular diameter but also regulate smooth muscle cell growth. Furthermore nitric oxide can influence the activity of matrix metalloproteinases, which is important with respect to remodeling of the vascular wall. The transduction of mechanical stresses involves activation of ion channels at the level of the cell membrane, conformational changes of integrins, and activation of focal adhesion kinases. This in turn activates intracellular MAP kinases which ultimately results in gene expression and protein synthesis. The intracellular orchestration of the mechanotransduction is determined by the architectural organization of the cytoskeleton. Recently Ingber [33] postulated the tensegrity model which describes how mechanical energy may be converted into chemical energy at the molecular level, and explains how stress induced conformational changes of the cytoskeleton can change the intracellular distribution and function of signaling molecules.

In summary, we have identified growth factors and receptors that are involved in the induction of endothelial cells, organization of capillary integrity and network structures, remodeling in large caliber vessels and finally the recruitment of vascular smooth muscle cells (Fig. 1). Induction of cell differentiation and the early formation of the vascular network depends on a delicate balance of growth factors. This balance can be altered by environmental factors (tissue pO_2) and the perfusion status of the vasculature. The genesis of specific arteriolar and venular characteristics, phenotypical changes of vascular smooth muscle during maturation [39], as well as organ specific features remains to be elucidated.

Understanding the regulation of the vascular branching pattern is another challenge [44]. Although we can describe growth processes morphologically, the control of three-dimensional organization of vascular structures during embryonic development remains a mystery. It should be realized that the structures are seldom random. Recent experimental evidence obtained from the fruit-fly (Drosophila melanogaster) and the mouse lung (see chapter by Bellusci) indicates the existence of genes directing the branching pattern [5, 13]. The genes trachealess/branchless, pointed and FGF-10 are implicated to play a role in the initial budding and the subsequent formation of sprouts at the growing tips, whereas sprouty and Sonic hedgehog are implicated to function as feedback inhibitors. The molecular machinery determining patterning of the vascular tree remains to be investigated, but by analogy with the genes trachealess and branchless, HiF-1 (hypoxia inducible factor-1) may be an important candidate [60]. Finally, what determines the expression pattern of the patterning genes, and how is this patterning information encoded in the genome?

5. HEMODYNAMICS, METABOLIC DEMAND AND VASCULAR NETWORK ADAPTATION

It is recognized that the mechanisms controlling vascular development under physiological conditions include a) metabolic demand [1, 65] b) wall shear stress [26] and c) circumferential wall stress adaptation [31, 53, 56, 62]. The final architecture might be the result of these processes striving for an optimal value.

Vascular adaptations as to meet metabolic demand, depend on tissue oxygen tension and cellular oxygen consumption. Thus another origin of the growth stimulus resides within the tissue. Adaptation to wall shear stress and circumferential wall stress are the result of the perfusion status of the vessels. Therefore, perfusion controls growth from within the existing vascular tree, although the initial change may be induced by tissue factors.

The first studies indicating a role for hemodynamic factors in the control of vascular development were already performed more than a century ago by Thoma [68]. Looking at the microcirculation of the chick embryo vitelline membrane (yolk-sac vasculature), Thoma observed that vessels with a high flow widen, whereas vessels with a low flow regressed or disappeared. Although he was unable to measure flow and pressure, he was the first to postulate that hemodynamics can influence vascular development. About 30 years later Murray [46] proposed that vascular networks adapt their geometry on the basis of the principle of minimalization of work required to maintain adequate blood flow. He proposed that lumen radius is proportional to the cube root of the volume flow rate. Based on careful analysis of the vascular morphology, the relationship seems to hold in many cases (especially the larger vessels), and similar relations were even found in other fluid transport systems as well [36].

It is currently known that vessels can adapt their dimensions to match changes in pressure and flow according to shear stress and circumferential wall stress adaptation [45, 53]. Shear stress is the force exerted on the blood vessel due to shear of the flowing blood along the vascular wall (Tab. I). Circumferential wall stress is related to the force distending the blood vessel, the intra-luminal pressure (Tab. I). To evaluate the effects of shear stress and circumferential wall stress adaptation on vascular development, a distinction has to be made between acute and long-term effects and between effects on a single vessel and the vascular network. Furthermore, the effects may be dependent on the developmental stage, *i.e.*, embryonic *versus* adult, of the vascular bed examined.

5.1. Flow

For a single vessel, adaptation to shear stress implies that an acute reduction in flow is counteracted by a lumen reduction, whereas an increment in flow is counteracted by a lumen increase. Since shear stresses seem to be rather constant along the vascular tree, it has been suggested that vessels strive to adapt around an optimal shear stress value [34, 37, 73]. Increments in tissue flow, for example during exercise, cause an acute increase in arteriolar diameter. Chronically, it can result in a structural radial increase in lumen diameter. Although the acute response involves the local release by the endothelium of vasodilator substances like nitric oxide and prostacyclin, the nature of the switch from an acute functional to a long-term structural adaptation is not known. Long-term increments in flow also stimulate the growth of arterioles and capillaries.

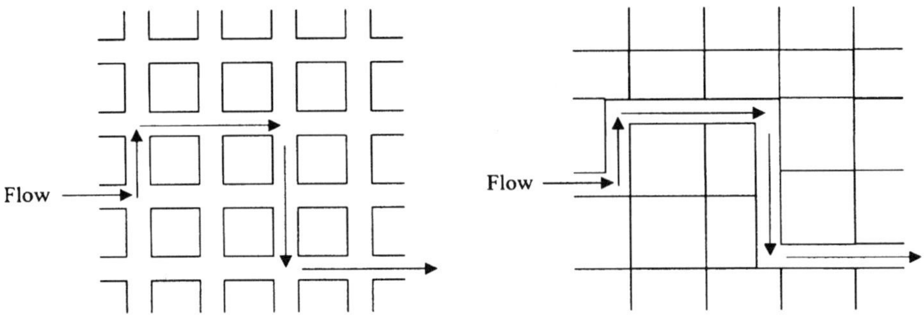

Fig. 4. — Effects of flow on remodeling of a network.

Reductions in flow have been shown to cause regression and apoptosis in both arterioles and capillaries.

Computer modeling studies have shown that adaptation to shear stress alone is not sufficient to explain the maintenance of a stable vascular network [26]. Vessels preferential for carrying flow were predicted to increase lumen diameter, whereas vessels perpendicular to the preferential flow direction, were predicted to regress (Fig. 4). Eventually, this process leads to regression of almost all arterioles, venules and capillaries, and finally to the formation of one large A-V (arterio-venous) shunt. Since *in vivo* this is not a general phenomenon, adaptation to shear stress alone can not explain the formation of a stable vascular network. Thus, although shear stress may be regulated within a narrow range, it is not the primary regulated variable.

5.2. Pressure

Vessels also adapt to changes in intra-luminal pressure according to wall stress adaptation (Tab. I). Wall stress adaptation suggests that changes in vessel diameter of wall thickness may compensate for changes in pressure. Experimental evidence in both humans and experimental animal models show a close correlation between the level of blood pressure and wall thickness. In pathological conditions like hypertension ("high blood pressure") prolonged increases in pressure can result in increased media thickness and/or remodeling of vascular smooth muscle cells around a more narrow lumen [45].

Pressure can also influence microvascular network architecture. In both primary and secondary forms of hypertension, the increase in pressure is accompanied by a structural reduction in the number of small arterioles and capillaries, a phenomon called rarefaction [40]. It has been postulated that given the higher perfusion pressure, both the structural lumen reduction and the regression of vessels are adaptions to normalize tissue flow. Hence, not pressure, but normalizing flow to the tissues prevails.

Computer modeling studies have shown that the increases in pressure can indeed explain rarefaction based on circumferential wall stress adaptation [51]. In short, vessel diameter and network boundary pressures were assigned using experimental data of a vascular network. Subsequently, flow (Qij) through each segment was calculated. Subsequently, pressure was increased by (a realistic) 35%. Arterioles were allowed to autoregulate their diameter (an assumption which may not be accurate in early embryonic networks), which was adjusted iteratively until Qij was restored. In this situation, wall stresses in the terminal arterioles generally decreased. A stress-growth rule was applied, stress above a threshold (8×10^4 dynes/cm^2) causes growth, whereas stress below the threshold (5×10^4 dynes/cm^2) causes regression. The model predicted a regression of about 22% of the terminal arterioles in response to the pressure increase which is in agreement with experimental *in vivo* data. Based on the same rules a reduction in pressure was predicted to result in arteriolar growth. In case of absence of autoregulation as occurring at the capillary level, increases in pressure were predicted to induce the upgrading of back to back connecting capillaries into arterioles and resulting in pronounced arcade (arteriolar-arteriolar connections) formation. Increasing capillary wall thickness as alternative to normalize wall stress. In conclusion, modeling according to circumferential wall stress alone can explain both regression and formation of vessel. The model itself, however, is not sufficient to explain the (embryonic) genesis, and maintenance of a realistic vascular structure.

Both the shear and the wall stress adaptation models assume the presence of an autoregulatory response in an already existing tree. This response depends on the ability to regulate tone, hence vascular diameter. This involves modulation of the contractile status of the vascular smooth muscle cells. In the adult vascular smooth muscle cells, there are hormone receptors which are coupled to a signal transduction pathway and ion channels, which upon activation modulate the actin-myosin interaction, and thereby influence the contractile status. During embryogenesis, vascular smooth muscle cells may be recruited during early development and be morphologically present, however, the establishment of a regulatory function (which depends on the ability to synthesize vasoactive hormones, on the presence of the hormone receptors, ion channels, a proper signal transduction pathway coupled to the contractile machinery etc.) may take until the last part of fetal gestation [39] or even until after birth. Therefore, if adaptation to changes in the perfusion status of the vasculature also hold during embryogenesis and fetal development, influencing the structure of the vascular tree, through modulation of conductance (both structural changes in lumen diameter as well as structure of the network architecture), may represent the only means for vessels to normalize shear and wall stress.

5.3. Metabolic Demand

The main function of the capillary network is to provide an exchange organ for oxygen and nutrients, and removal of metabolites. To ensure metabolic homeostasis, the capillary network would have to meet organ requirements. Numerous experimental studies both in humans and animals support the idea that the capillary network adapts as to meet metabolic requirements. The higher the metabolic rate, the more capillaries. An increase in metabolic rate (exercise) is associated with an increase in capillarization, and, depending on the time-course, an increase in number and size of arterioles. Reductions in metabolic rate (for example during immobilization of skeletal muscle) on the other hand results in regression of capillaries. The mechanisms behind the metabolic control of vascular growth are complex and reviewed in detail by Adair [1] and Ferrara [17]. In short: at the biochemical level, an increase in (aerobic) metabolism induces an increased oxygen consumption. A biochemical state in which oxygen demand is not met by oxygen supply is called hypoxia. At the cellular level hypoxia is associated with tissue release of vasodilator substances like adenosine that a) can increase arteriolar lumen diameter and attract flow to the hypoxic regions, and b) stimulate angiogenesis and arteriogenesis. Furthermore, hypoxia has been associated with the activation of hypoxia "sensitive" genes like HiF-1 (hypoxia inducible factor-1) that in turn can activate the transcription and release of vascular endothelial growth factor (VEGF) which in turn results in a stimulation of vascular growth.

Pathological tumor growth is also associated with adaptation to metabolic demand. It is now well established that tumor growth is dependent on the formation of a supportive vascular network around the growing tumor. The fast growing tumor cells, which need a lot of oxygen and nutrients, create a hypoxic core due to their excessive growth and high metabolic demand. At the molecular level this is accompanied by activation of HiF-1 [17, 60], a.o.VEGF secretion and the subsequent induction of angiogenesis from nearby vessels. The new vessel network usually is not sufficient to meet the metabolic demand of the growing tumor resulting in a necrotic tumor core. Cell death results in the additional release of intra-cellular growth factors.

Physiological and pathological adaptation to metabolic demand may share a common mechanism or molecular pathway, the difference is found in the resulting architecture of the vascular networks. Tumor associated networks can be described as having a chaotic appearance, with pronounced dilation of feeding arterioles and tortuousity of capillaries [41, 42]. At the capillary level, the basement membrane does not have to be continuous and gaps between endothelial cells can be present. The origin of this pattern and the hemodynamic consequences (important for drug delivery to the tumor) are subject for investigation.

With regard to oxygen delivery and metabolic demand, an interesting model has been proposed by Gottlieb [22, 23]. The model is based on three simple parameters:

1. diffusion distance (d) from the center of a cell to the nearest point in the existing vasculature;

2. critical diffusion distance for oxygen (D);
3. tissue expansion coefficient.

At the start of the simulation, the distance of each (fictive!) cell in the tissue to the nearest point in vasculature is calculated. If $d <= D$ than the cell in the tissue is within a safe distance from the vasculature, and therefore not hypoxic. If $d > D$, than the cell in the tissue is relatively hypoxic. As a result the cells sends an angiogenic signal to the vasculature. The model now assumes that only the endothelial cell receiving the highest concentration of angiogenic stimulus (the nearest endothelial cell) generates a sprout with a length (l) towards the cell. This new sprout reduces the diffusion distance from the cell to vasculature, if now $d <= D$ the cell is not hypoxic anymore and stops generating angiogenic substances. Subsequently, the tissue expands, the number of cells increases, and distances are again calculated. To include arcades, the models assumes that if two sprouts are very near to each other (anastomoses boundary), they fuse.

Modeling vascular trees based on local oxygen availability proposes an explanation for the creation of a vascular tree during fetal development. Interestingly, based on simple rules with regard to oxygen diffusion, that are furthermore biologically feasible (HiF-1 and VEGF expression), it is possible to create realistic looking organ specific arterial architectures. It is tempting to suggest that the various vascular patterns found in different tissues are determined by local metabolic demand. However, there are some shortcomings to the model. First, a growing endothelial sprout does not carry flow yet, hence may not contribute to the oxygen delivery. Second, the released angiogenic substance is diluted in the tissue volume. If diffusion is allowed in all directions, the concentration declines rapidly according to $[\text{angio}]\times(1/r^3)$, in which [angio] is the initial concentration, and r the distance to the cell. Based on hormone kinetics and receptor affinity, either a lot of hormone must be produced, or existing endothelial tubes have to be very near. Furthermore, a diffusable substance theoretically reaches all neighboring endothelial cells, but only the nearest endothelial cell is allowed to grow. This raises a question with regard to coordination. Assume this is indeed the case *in vivo*, how do endothelial cells communicate in order to allow only the growth of one sprout? Since this is difficult to conceive physiologically, biophysicists made an addition to the model based on ion flow. If we assume flux of ions from the cell in the tissue towards the vasculature, with the electrical resistance in the tissue being regulated by the tissue pO_2 (low $pO_2 -> $ low resistance), ions will only flow along the pathway with the smallest resistance. Hence, since there is only one pathway with the lowest resistance (or preferential flow channel), the ion flow may direct and coordinate the growth of the new sprout.

The models presented here are based on physiological principles. They offer an interesting first order approach to understand formation and maintenance of the vascular tree. Models may also be constructed based on physical concepts [30], power optimization principles [66] or even stochastic processes [55]. One model is presented in detail by Fleury and Schwartz (Contributed Paper 7). Interesting examples (by Chaplain and Anderson) can also be found on internet: http://www.mcs.dundee.ac.uk:8080/~sanderso/research.html

6. ANALYSIS OF VASCULAR STRUCTURE

A vascular network consists of a multitude of vessel segments with a variety of diameters which are usually connected in a complex manner. Using computer models, growth and remodeling of a vascular network may be simulated. Results are often called "realistic", however, how can we reduce "the general impression of size and shape" to a verifiable and measurable parameter? The standard morphometric parameters used to quantify the vascular anatomy may be divided into dimensional parameters, a.o. diameter, length or volume, and topological parameters which describe the actual branching pattern.

Two of the most commonly used parameters to describe the distribution of microvessels in a tissue are microvascular density and microvascular density index [58, 65]. These parameters can be obtained by applying quantitative stereology (counting intersections with a grid), and are a function of both vessel length and diameter. They describe the space or volume occupied by the vascular tree in a given tissue or surface area. In muscle preparations, the capillary to fiber ratio and capillary surface area are often used to describe microvascular changes with respect to changes in metabolic behavior.

To construct a mathematical network model, we need to know how the vessels are interconnected, the topological structure. The network topology indicates the interconnection of vessels regardless of their lengths, diameters, or other dimensional measures. To describe the vascular branching pattern, topological ordering rules can be applied (see Fig. 5). A vascular tree can be ordered from proximal to distal, centrifugal or generation ordering (Fig. 5a) [21, 43], or from distal to proximal (Fig. 5b: Strahler ordering) [35, 64, 69, 72]. Subsequently, topological measures like mean centrifugal order, bifurcation ratio, and asymmetry can be calculated. The mean centrifugal order is defined as the centrifugal order averaged over all segments of the tree. When comparing trees, it should be taken into account that this parameter is tree-size dependent.

The bifurcation ratio is defined as: $\mathrm{Rb}(s) = (n.$ of vessel of Strahler order $s-1)/(n.$ of vessel of Strahler order $s)$ [29].

For a node with two distal subtrees attached, having $N1$ and $N2$ terminal vessels asymmetry is defined as $\mathrm{ASYM} = (N2-N1)/(N2+N1-2)$ (see Fig. 6).

Besides just describing the tree, the calculated parameter should preferentially have large discriminatory properties, ultimately one value allowing only one structure. The efficiency (E) of a parameter to differentiate between two trees can be calculated as the ratio of M (the number of different values obtained) and T (the number of different trees), $E = M/T$ $(0 < E < 1)$ [25].

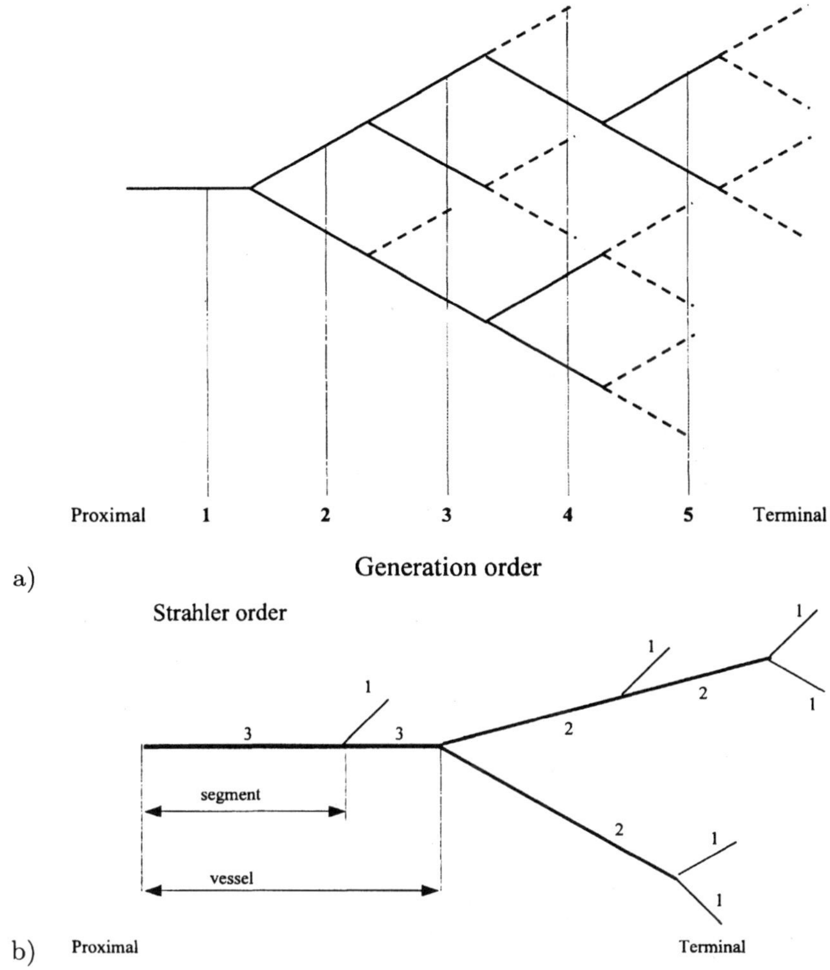

Fig. 5. — a) Ordering of a vascular tree, generation analysis. The continuous line represents the branched arteriolar tree, the dotted lines represent the capillaries. b) Ordering of a vascular tree, Strahler analysis. The continuous line represents the branched arteriolar tree.

Two trees are topologically different if one of them can not be transformed into the other by rotation of subtrees.

Using a set of all topologically different trees with 31 fixed terminal segments (generating 10 905 different trees) the efficiency of asymmetry, mean centrifugal order and branch-node ratio were 0.398, 007 and 0.001, respectively [25]. Hence, the parameters *per se* have low discriminatory properties. The combination of asymmetry and mean centrifugal order resulted in an efficiency of 0.779 and may provide a valuable tool [25].

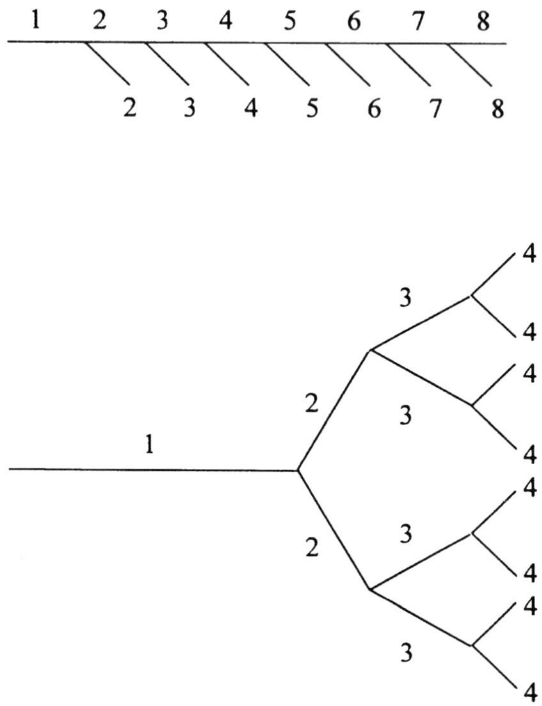

Fig. 6. — Vascular (a)symmetry. Note the difference in generation distribution between the asymmetric (top) and the complete symmetric tree (bottom).

The relative cumulative generation order distribution can be used to describe growth of a developing tree (see Fig. 7). Theoretically, in a pure dichotomous vascular tree, new segments can be added (Fig. 8) to the existing tree according to terminal growth (at the most distal part of the tree) or segmental growth (along existing segments). Theoretically, growing according to terminal growth (TG) and segmental growth (SG) can create all possible vascular (dichotomous) patterns provided the right strategy is applied. Vascular trees may be described by a stratified combination of both processes. Experimentally obtained topological order distributions can be fitted with the distributions of SG, TG or a combination with the distribution being proportional to the contribution of either process (see Fig. 7). This technique is useful to describe the site of action of angiogenic substances.

Conversely, the vascular structure was generated according to a certain growth principle, hence, the given structure contains information about the processes that contributed to its design. If the vasculature results from the equilibrium of growth control mechanisms, each with its specific contribution to the structure, it is possible to analyze, given a certain structure, when and in which quantity

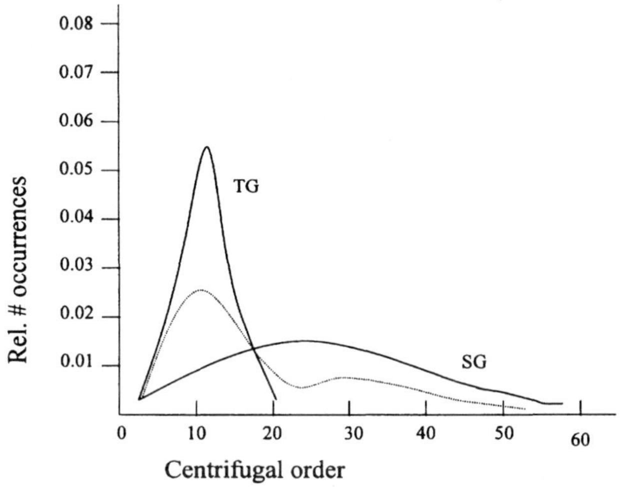

Fig. 7. — Relative cumulative centrifugal order distribution for terminal growth (TG) and segmental growth (SG). Note the differences in mean value, and the distribution about the mean. The dotted line represents the summated curve for 50% TG and 50% SG.

each mechanism governed control. By strategically applying topological rules (for example a time dependent contribution of terminal and segmental growth) to a tree, it is possible to obtain a retrospective insight in its morphogenesis. The challenge remains to integrate the available knowledge on the genetic and hemodynamic regulation of the branching pattern, similar to the concept of the branching embryonic lung, with the growth governing rules of a model. Ideally, the parameter settings at which the model can generate trees resembling the *in vivo* observed branching pattern, can be linked to a specific gene-product or indicate the balance of growth factors and growth inhibitors.

In many tissues, the branching pattern is not dichotomous but exhibits arteriolar-arteriolar connections or arcades. Analysis of arcading networks is difficult. To characterize arcading networks a combination of a) number of arcade loops in the network, b) branch-node ratio with the network and, c) the number of vessel segments in an arcade loop can be used [38]. The branch-node ratio in the network is determined by counting the total number of branch-nodes in the arcade network divided by the number of perimeter branch-nodes in the network. If the number of branch-nodes (P) and number of arcades are known (A), the number of segments can be calculated: $S = P + A - 1$. In a pharmacological experiment looking at drugs influencing arteriolar arcade formation, we found that the branch-node ratio was very sensitive in discriminating between the treatment groups. The number of segments in an arcade loop is reported rather constant (about 5) in different networks of different species. It is

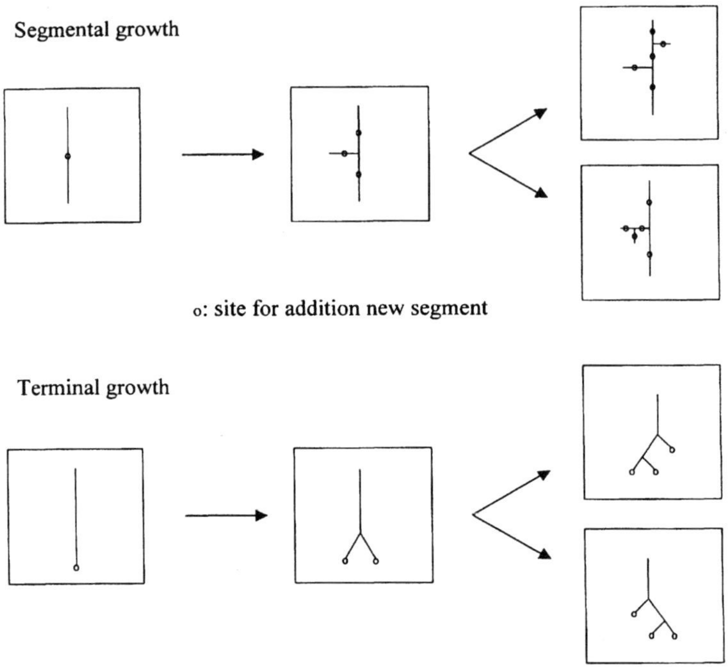

Fig. 8. — Modeling Segmental (top) and Terminal (bottom) growth. Segmental growth allows addition of new segments along an existing segment; terminal growth allows addition only at terminal segments.

postulated that arcading networks can result from arteriolization of back to back connecting capillaries in a terminal arteriole. The stimulus can be an increase in circumferential wall stress in these capillaries (the recuited vascular smooth muscle cells increase wall thickness thereby restoring wall stress) or the release of specific growth factors for vascular smooth muscle like angiotensin-II (see chapter by Levi) [38].

Functional morphometry is classifying vessels according to their function in distributing the flow over the organ. For example, at rest, a certain distribution of flow occurs. When vasodilated a different partition of flow can be observed. The topology should be such that a separation in different classes allows for maximal recognition of the different functions. In certain preparations like the cremaster muscle, vessels can be identified by their location in the network relative to their branching order and function. In this case, the large feed arteriole is called the first order A1 vessel. All side-branches are called second order, or A2 vessel. Lateral off-shoots of second order vessel with smaller diameters are classified as third order arterioles, A3. All side-branches of A3 vessels with distinctly smaller diameters were classified as fourth order vessels,

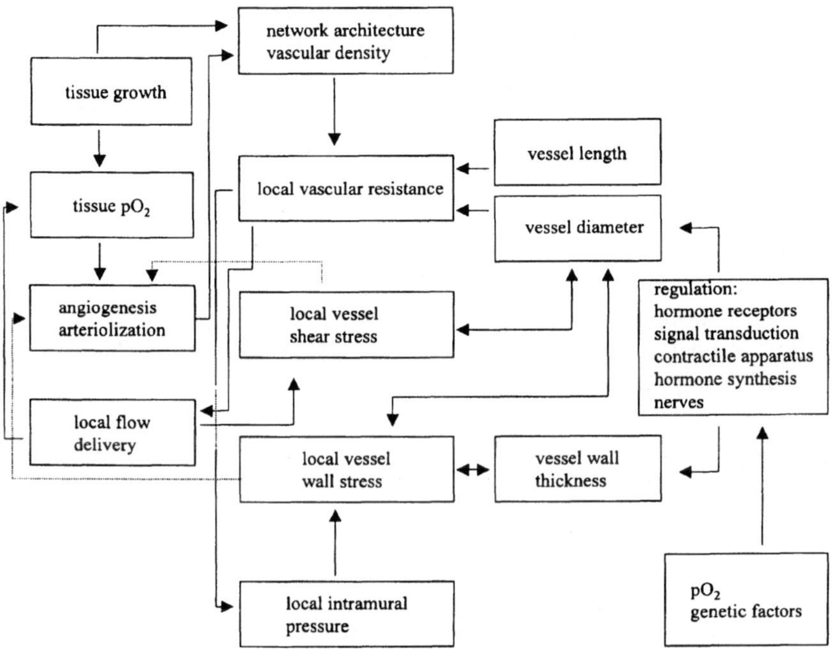

Fig. 9. — Schematic representation of complex interaction between hemodynamics, tissue oxygenation, diameter regulation, and vascular network adaptation.

A4. In function, the vessel types differ in vascular tone and in distribution of $\alpha 1$ and $\alpha 2$ adrenoceptors. Changes in architecture can be represented as a ratio, number of A4 per A3 vessel, A2 per A1 before and after complete vasodilation.

7. CONCLUSION

Vascular networks can no longer be regarded as static structures. Genetic, hemodynamic and metabolic factors provide a complex interplay of mechanisms to model and remodel the architecture of vascular networks continuously. The complex nature (see Fig. 9) stresses the need to develop (computer) models enabling a detailed view on the contribution of each of the individual growth governing forces (for example by generating computer knock-out models). Since abnormalities in vascular development and architecture are now believed to be important pathogenic mechanisms in diseases such as hypertension and tumor growth, this knowledge may provide the basis for new treatment strategies. For the purpose of designing a realistic model both knowledge of biological and physical principles should be integrated and related to the function of the structure. For example, the migration of angioblasts in the yolk-sac may be represented by "random walkers" whereas the primitive fluid movements in the primitive embryo may be described by "viscous fingering". Finding a common language, linking biological processes with the physical counterpart, can result in an exciting new area of research and allow a more rational design of pharmaceutical drugs aimed at manipulating the growth and design of the vascular tree.

Acknowledgements

The authors would like to thank J. Willems for his technical support and Prof. Dr. Axel Pries (Free University, Berlin) for his stimulating discussions on network analysis.

REFERENCES

[1] T.H. Adair, W.J. Gay and J. Montani, *Am. J. Physiol.* **259**; *Regulatory Integrative Comp. Physiol.* **28** (1990) R393-R404.
[2] C. Amant, L. Berthou and K. Walsh, *Drugs* **58** (1999) 33-36.
[3] C.T. Baillie, M.C. Winslet and N.J. Bradley, *Br. J. Cancer* **72** (1995) 257-267.
[4] L. Beck, P.A. D'Amore, *FASEB J.* **11** (1997) 365-373.
[5] S. Bellusci, J. Grindley, H. Emoto, N. Itoh and B.L. Hogan, *Development* **124** (1997) 4867-4878.
[6] M. Bergwerff, M.E. Verberne, M.C. Deruiter, R.E. Poelmann and A.C. Gittenberger-De Groot, *Circ. Res.* **82** (1998) 221-231.
[7] V. Brower, *Nat-Biotechnol.* **17** (1999) 963-968.
[8] I. Buschmann and W. Schaper, *N.I.P.S.* **14** (1999) 121-125.
[9] P. Carmeliet and D. Collen, *TCM* **7** (1997) 271-281.
[10] J.D. Coffin, J. Harrison, S. Schwartz and R. Heimark, *Dev. Biol.* **148** (1991) 51-62.
[11] T. Couffinhal, M. Silver, L.P. Zheng, M. Kearney, B. Witzenbichler and J.M. Isner, *Am. J. Pathol.* **152** (1998) 1667-1679.
[12] P.F. Davies, *Phys. Rev.* **75** (1995) 519-560.
[13] A.A. De Maximy, Y. Nakatake, S. Moncada, N. Itoh, J.P. Thiery and S. Bellusci, *Mech. Dev.* **81** (1999) 213-216.
[14] J. Denekamp, *British J. Radiol.* **66** (1993) 181-196.
[15] E.M. De Robertis, G. Oliver and C.V.E. Wright, *Sci Am.* (1990) 26-32.
[16] V. Djonov, M. Schmid, S.A. Tschanz and P.H. Burri, *Circ. Res.* **86** (2000) 286-292.
[17] N. Ferrara and T. Davis-Smyth, *Endocrine Rev.* **18** (1997) 4-25.
[18] I. Flamme and W. Risau, *Development* **116** (1992) 435-439.
[19] J. Folkman, *Antiangiogenic gene therapy*, in Proceedings of the National Acadamy of Sciences USA 95 (1998) pp. 9064-9066.
[20] J. Folkman, *Circ.* **97** (1998) 1108-1110.

[21] P. Gaehtgens, K. Ley and A.R. Pries, *Topological approach to the analysis of microvessel structure and hematocrit distribution*, in Microvascular networks: experimental and theoretical studies, edited by Popel and Johnson (Basel: Karger, 1985) pp. 52-60.
[22] M.E. Gottlieb, *Vascular Networks: fractal anatomies from non-linear physiologies*, in Proceedings 13th annual conference IEEE (Orlando, 1991).
[23] M.E. Gottlieb, *The VT model: 1 deterministic model of angiogenesis and biofractals based on physiological rules*, in Proceedings IEEE, 17th annual northeast bioengineering Conf. (1991) pp. 38-39.
[24] J.P. Van Groningen, A.C.G. Wenink and L.H.M. Tester, *Anat. Embryol.* **184** (1991) 65-70.
[25] W. Hacking, *The interaction between vascular structure and blood flow distribution* Ph.D. Thesis, University of Amsterdam, Department of Biophysics (1995).
[26] W.J.G. Hacking, E. Vanbavel and J.A.E. Spaan, *Am. J. Physiol.* **270** (1996) H364-H375.
[27] Z.C. Han and Y. Liu, *Int. J. Hematol.* **70** (1999) 68-82.
[28] D. Hanahan, *Sci* **277** (1997) 48-50.
[29] R.E. Horton, *Geol. Soc. Am. Bull.* **56** (1945) 275-370.
[30] A.G. Hudetz, P *Microvasc. Res.* **45** (1993) 1-10.
[31] O. Hudlicka, *Microcirculation* **5** (1998) 7-23.
[32] J.E. Hungerford and C.D. Little, *J. Vasc. Res.* **36** (1999) 2-27.
[33] D.E. Ingber, *Ann. Rev. Physiol.* **59** (1997) 575-599.
[34] A. Kamiya, R. Bukhari and T. Togawa, *Bull. Math. Biol.* **46** (1984) 127-137.
[35] G.S. Kassab, C.A. Rider, N.J. Tang and Y.B. Fung, *Am. J. Physiol.* **265** (1993) H350-H365.
[36] M. LaBarbera, *Sci* **249** (1990) 992-1000.
[37] B.L. Langille and F. O'Donnell, *Sci* **231** (1986) 405-407.
[38] F.A.C. Le Noble, L.C.G.A. Kessels-Van Wylick, W.J.G. Hacking, D.W. Slaaf, M.G.A. Oude Egbrink and H.A.J. Struijker-Boudier, *J. Vasc. Res.* **33** (1996) 480-488.
[39] F.A.C. Le Noble, K. Ruijtenbeek, S. Gommers, J.G. De Mey and C.E. Blanco, *Am. J. Physiol.* **278** (2000) H1261-H1268.
[40] F.A.C. Le Noble, F.R.M. Stassen, W.J.G. Hacking and H.A.J. Struijker Boudier, *J. Hypertens.* **16** (1998) 1563-1572.
[41] J.R. Less, M.C. Posner, T.C. Skalak, N. Wolmark and R.K. Jain, *Microcirculation* **4** (1997) 25-33.
[42] J.R. Less, T.C. Skalak, E.M. Sevick and R.K. Jain, *Cancer Res.* **51** (1991) 265-273.
[43] K. Ley, A.R. Pries and P. Gaehtgens, *Microvasc. Res.* **32** (1986) 315-332.
[44] R.J. Metzger and M.A. Krasnow, *Sci* **284** (1999) 1635-1639.
[45] M.J. Mulvany and C. Aalkjaer, *Phys. Rev.* **70** (1990) 921-961.

[46] C.D. Murray, *The physiological principle of minimum work. The vascular system and the cost of blood volume*, in Proceedings of the National Acadamy of Sciences USA 12 (1926) pp. 207-214.
[47] D.M. Noden, *Am. Rev. Respir. Dis.* **140** (1989) 1097-1103.
[48] S. Patan, B. Haenni and P.H. Burri, *Microvasc. Res.* **51** (1996) 80-98.
[49] T.J. Poole and J.D. Coffin, *Morphogenetic mechanisms in avian vascular development*, in Issues in Biomedicine: The development of the vascular system, edited by R.N. Feinberg, G.K. Sherer and R. Auerbach (Basel: Karger, 1991) Vol. 14, pp. 25-36.
[50] R.J. Price, G.K. Owens and T.C. Skalak, *Circ. Res.* **75** (1994) 520-527.
[51] R.J. Price and T.C. Skalak, *Microvasc. Res.* **47** (1994) 188-202.
[52] R.J. Price and T.C. Skalak, *Microcirculation* **5** (1998) 39-47.
[53] A.R. Pries, T.W. Secomb and P. Gaehtgens, *Circ. Res.* **77** (1995) 1017-1023.
[54] W. Risau, *Nat* **386** (1997) 671-674.
[55] K. Sandau and H. Kurz, *J. Microscopy* **175** (1994) 205-213.
[56] W. Schaper and W.D. Ito, *Circ. Res.* **79** (1996) 911-919.
[57] P. chlatter, M.F. Konig, L.M. Karlsson and P.H. Burri, *Microvasc. Res.* **54** (1997) 65-73.
[58] G.W. Schmid-Schoenbein, B.W. Zweifach and S. Kovalcheck, *Microvasc. Res.* **14** (1977) 303-317.
[59] F.W. Sellke and M. Simons, *Drugs* **58** (1999) 391-396.
[60] G.L. Semenza, *Biochem. Pharmacol.* **59** (2000) 47-53.
[61] D. Shepro and N. Morel, *FASEB J.* **7** (1993) 1031-1038.
[62] T.C. Skalak and R.J. Price, *Microcirculation* **3** (1996) 143-165.
[63] T.C. Skalak, R.J. Price and P.J. Zeller, *Microcirculation* **5** (1998) 91-94.
[64] A.N. Strahler, *Bull. Geol. Soc. Am.* **63** (1952) 1117-1142.
[65] D.M. Strick, R.L. Waycaster, J.P. Montani, W.J. Gay and T.H. Adair, *Am. J. Physiol.* **260** (1991) H1385-H1389.
[66] L.A. Taber, *Biophysic. J.* **74** (1998) 109-114.
[67] S. Takeshita, L.P. Zheng, E. Brogi, M. Kearney, L.Q. Pu, S. Bunting, N. Ferrara, J.F. Symes and J.M. Isner, *J. Clin. Invest.* **93** (1994) 662-670.
[68] R. Thoma, *Untersuchungen ber die Histogenesis und Histomechanik des Gefass systems (German)* (Stuttgart: Enke, 1893).
[69] E. Vanbavel and J.A.E. Spaan, *Circ. Res.* **71** (1992) 1200-1212.
[70] R.B. Vernon and H.E. Sage, *Am. J. Pathology* **147** (1995) 873-883.
[71] H.U. Wang, Z.F. Chen and D.J. Anderson, *Cell* **93** (1998) 741-753.
[72] E. Weibel, *Morphometry of the human lung* (Springer, Berlin, 1963).
[73] M. Zamir, *J. Gen. Physiol.* **69** (1977) 449-461.

CHAPTER IX

Mechanics of the Large Artery Vascular Wall

B. Levy and A. Tedgui

1. INTRODUCTION

The aorta and large arteries are generally thought of as conduit vessels whose main function is to provide a conduit for blood flow to reach the peripheral tissues. However, because the pressure and flow curves are not a simple ratio, it has long been recognized that the cardiovascular system functions in more complex fashion than merely a simple resistance to blood flow. Blood pressure is highest at the beginning of the systemic circulation; the decrease of blood pressure is not linear with vessel diameter or distance in the vascular tree. Blood pressure decrease ranges from 30 to 40% of the aortic pressure in vessels down from 250 to 50 μm in diameter [1–3] while most of the pressure drop occurs in the terminal arterioles with diameters smaller than 100 μm and which branch into numerous small capillaries. The site of the largest pressure drop may differ between tissues; however, in vessels smaller than 60 μm, no correlation has been found between the central arterial pressure and microvascular pressure which suggests that perfusion pressure is being controlled in these blood vessels and those with lower diameter [4].

Arterial blood flow is actually determined by blood pressure as well as by impedance of the vascular system. The latter is dependent on static and dynamic pressure-volume relations of blood vessels. Thus, any realistic representation of vascular properties must account for the visco-elastic properties of the vasculature [5].

2. PARAMETERS IN VESSEL WALL MECHANICS

2.1. Forces and Stresses

Forces in the wall of a cylindrical tube can be oriented in three directions: circumferential, radial and longitudinal (Fig. 1). Since the blood vessel may be

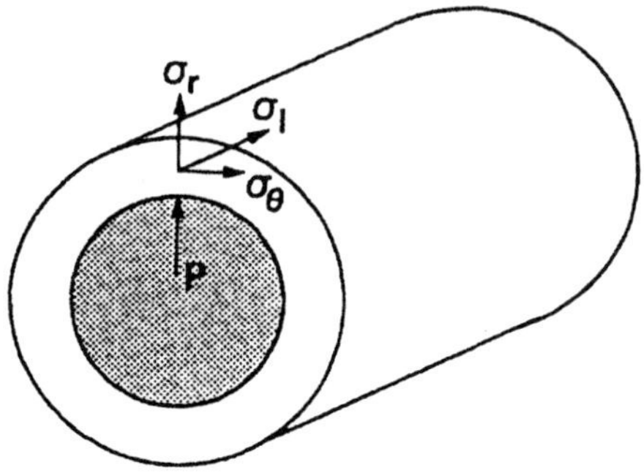

Fig. 1. — Pressurization of blood vessels results in stress development in three different directions: tangential (σ_θ), radial (σ_r), and longitudinal (σ_l).

considered to be incompressible [6,7], dimension changes in one direction will result in dimension changes in at least one other direction. Moreover, arteries are anisotropic, *i.e.* stiffness is dependent on direction. It has been shown that vessel length increases when a vessel is inflated, but that this amount depends on the vessel type. Most of our knowledge about the anisotropy of arteries has been derived from studies on aortas [8] and carotid arteries [6]. In carotid arteries, longitudinal stiffness is about half of the tangential stiffness [6,9].

The "transverse anisotropy" was related to vascular tethering and to the distending pressure: arteries shorten as soon as they are released from their attachments in the body. Actually, most of the longitudinal distension of an arterial segment is related to the transmural pressure. When a segment of artery is submitted, *in vitro*, to increases in transmural pressure, its length increases by 10 to 20% for pressure varying from 0 to 50 mmHg (Fig. 2).

The longitudinal distension is less marked in older individuals, from 40% at age 12 to 12% or less after age 60 [10]. Recent results confirmed these results in rats' carotid arteries evidencing that aging alters arterial wall anisotropy [11].

The circumferential force developed in a tube is quantified by the tension T, which is calculated according to Laplace's law [12]:

$$T = P\,r \quad [\mathrm{N\ m^{-1}}]$$

where P [N m^{-2}] is the transmural pressure and r [m] the internal radius.

Stress (σ) refers to forces developed per unit area and is calculated as:

$$\sigma = T/h \quad [\mathrm{N\ m^{-2}}]$$

where h [m] is the wall thickness.

The wall thickness and tension however depend on the transmural pressure *i.e.* on the circumferential strain of the artery. In linear mechanics, the relation between strain (ε), the relative increase in length (%), and stress (σ) is linear

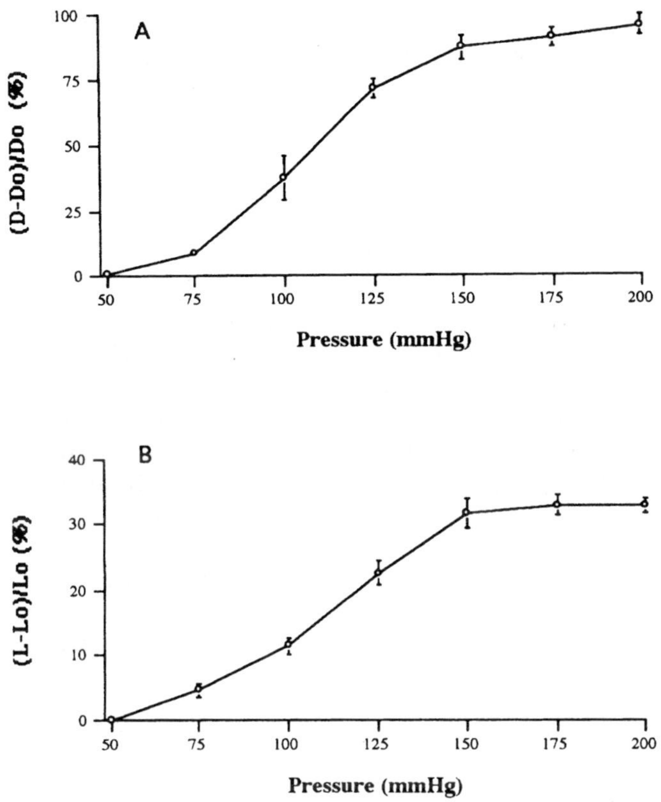

Fig. 2. — Changes (mean $\pm$ SEM) in diameter A) and length B) relative to their values at $P = 50$ mmHg in normotensive Wistar Kyoto rats.

in a wide range and can be expressed by the elastic modulus E, also called Young's modulus.

$$E = \sigma/\varepsilon \text{ [N m}^{-2}\text{]}.$$

In this case, E remains constant. Biological materials, however, do not possess a constant elastic modulus, and E is a function of strain. Yet, for each small increase in strain, an elastic modulus can be calculated, which holds only for one strain level. This elastic modulus is called incremental elastic modulus, Einc, and is thus the differential of the stress-strain relation. Einc is often expressed in the C.G.S unit dyne cm^2, which equals 10^{-1} N m^2.

2.2. Compliance

Though incremental elastic modulus is a useful parameter to describe mechanical properties of a tissue, in blood vessels it is also important to know the relation between pressure changes and volume changes. The compliance of a chamber is a quantitative description of its overall wall properties; compliance C is defined as the change in volume due to a change in pressure, that is:

$$C = \Delta V/\Delta P \quad [\mathrm{m}^5\ \mathrm{N}^{-1} \text{ or ml/mmHg}]$$

where ΔV is the change in volume [m^3, ml, or μl], and P, the change in transmural pressure [N m^{-2} or mmHg].

Compliance is determined not only by the mechanical properties of the wall but also by the geometry and size of the vessel. Moreover, because the pressure-volume relations of most biological systems are non-linear, compliance is a pressure-dependent quantity. Compliance should be clearly distinguished from a related quantity, capacity. Capacity refers to the amount of volume that a chamber can contain at any pressure; capacity depends on compliance and on the volume already contained in the chamber at zero transmural pressure. This introduces another quantity important and particularly difficult to measure, the unstressed volume, which is the volume contained in the vessel when the pressure is negligible.

In the study of isolated vessels and when using *in vivo* non invasive ultrasonic measurements of vessel diameters, it is convenient to calculate the compliance per unit vessel length, or cross sectional compliance Ccs, which corresponds with changes in lumen cross-sectional area:

$$Ccs = \Delta CSA/\Delta P \quad [\mathrm{m}^4\ \mathrm{N}^{-1} \text{ or } \mathrm{mm}^2\ \mathrm{mmHg}^{-1}]$$

where CSA is the cross sectional area.

Finally, the compliance can be normalized to the vessel volume at a given transmural pressure to yield the distensibility D.

$$D = C/V_{\mathrm{P}} \quad [\mathrm{m}^2\ \mathrm{N}^{-1}\ \mathrm{mmHg}^{-1}].$$

3. DETERMINATION OF MECHANICAL PROPERTIES OF BLOOD VESSELS

The determination of the mechanical properties of blood vessels requires simultaneous measurement of blood vessel dimension (or volume) and internal pressure. Several experimental methods have been used to assess these parameters *in vitro, in situ* and *in vivo*.

3.1. *In vitro* Measurements

Since the work of Mulvany and Halpern [13], the *in vitro* determination of the stress-strain relationship of vessel rings is the most commonly used experimental method to assess the mechanical properties. Arterial rings, as small as $150 - 200\ \mu$m diameter, are mounted by inserting two thin threads through the vascular lumen and attaching these threads to support plates at both sides of the vessel. One plate is connected to a force transducer and the second one to a displacement transducer; the vessel ring is immersed in an organ chamber containing a physiological solution. Various drugs can be added in the bathing solution to assess the vessel ring mechanical properties under different pharmacological conditions. The isometric myograph technique is commonly used and allows the measurement of the wall tension- ring circumference relationship under control, active (phenylephrine), and passive (sodium nitroprusside, or potassium cyanide poisoning).

Cylindrical segments of blood vessels can be cannulated, pressurized and perfused at controlled flow rates. This technique has been developed for use in large [14] and small vessels [15] and is widely used since the Halpern's system is commercially available. This method provides a more physiological loading of the vessel and the possibility for more precise pharmacological interventions. Halpern and Kelly [16] have summarized the differences between the isometric myograph and the cannulated vessel techniques. In cannulated vessels (and not in isometric myograph)

1) diameter is allowed to change;
2) shape is cylindrical;
3) a transmural pressure exists across the wall and thus the filtration rate through the vessel wall can be measured in nearly physiological conditions;
4) physiological axial length can be conserved;
5) pharmacological agents can be perfused and/or superfused.

The relation between the mechanical variables stress and strain and their distribution in the vessel wall depends on the experimental method used. In a comparative study, Cox [17] demonstrated that stress at equal levels of strain was higher in rings than in cylindrical segments from identical vessels. A comparison of dimensions from ring and cylindrical segments also showed significant differences between actual and calculated vessel radius. Furthermore, likely because of the presence of specific and non specific peptidases in the endothelial layer, the dose-effect relationship of constriction induced by peptide agents is quite different when the peptide is applied intraluminally or extraluminally [18]. Finally, the ring technique is relatively easy to perform and give valuable information concerning the physiology, the reactivity and the pharmacological properties of large arteries and resistance vessels. However, for precise *in vitro* measurement of the stress-strain relationship, it is recommended to use the method of pressurized and perfused cannulated vessels.

We developed an *in situ* experimental model allowing us to simply and accurately measure the static compliance of the carotid artery [19–21]. A segment of common carotid artery (22 – 25 mm length) free of collateral branches is isolated *in vivo* and *in situ*. A computer-controlled servo-system allows us to record the steady-state volume pressure relationship and therefore to evaluate the static mechanical properties of the isolated segment of common carotid artery. These measurements are performed under controlled conditions in the presence of a physiological vasomotor tone and after total abolition of the local smooth muscle tone by incubation with potassium cyanide. It is thus possible to compare the mechanical properties of the studied artery under active and passive smooth muscle conditions.

The advantages of such a model are the following:

1) to preserve the integrity of the endothelium in the isolated segment of artery. At the end of the experiment after excising and washing, the absence of fixation of Evan's blue dye by the wall indicated that the endothelial surface remained unaltered;

2) to keep the dissected, unexposed and non excised segment of carotid artery to its physiological length and in its normal fluid environment;

3) to avoid the collapse of thc vessel. The assessment of only static and not dynamic mechanical properties of the studied segment of artery remains the main limitation of such experimental model.

Using ultrasonic echographic measurements of vessel diameter, we developed an *in vitro* model to study the reactivity and static and dynamic mechanical behavior of the rat carotid artery under controlled flow and pressure conditions [14]. The carotid artery is cannulated *in vivo*, before it is excised, and kept during the whole experiment at its original longitudinal stress and dimension. Ultrasonic echo-tracking techniques allow for the measurement of the internal diameter with an accuracy of less than 10 μm. The *in vitro* isolated artery can then be studied under controlled pressure and flow conditions. The isolated segment of artery can be submitted to steady or oscillatory pressure and flow.

Finally, the diameter-pressure static relationship of smaller muscular arteries can be assessed *in situ* by using optical measurement of diameter. Videomicroscopy is used to measure diameters of mesenteric arteries (500 to 100 μm) submitted to imposed levels of pressure or flow [22–24]. The limitation of such an experimental model does not depend on the dimension measurement but rather on the accuracy of pressure measurements. More sophisticated techniques of pressure measurement such as the servo-null technique [25] must be used to obtain reliable simultaneous measurements of phasic pressure in small arteries.

Figure 3 shows typical recordings of pressure-diameter recordings under basal condition, under "active" conditions (maximal vasoconstriction), and after poisoning of the smooth muscle cell with potassium cyanide (passive conditions) in a mesenteric artery. Using the previously described formulas and definitions, the compliance, distensibility, and elastic modulus can be derived from the pressure-diameter relationships. This experimental model also permits investigation of the flow-induced dilation of the isolated segment of vessel and the effect of the endothelium on the arterial mechanical properties.

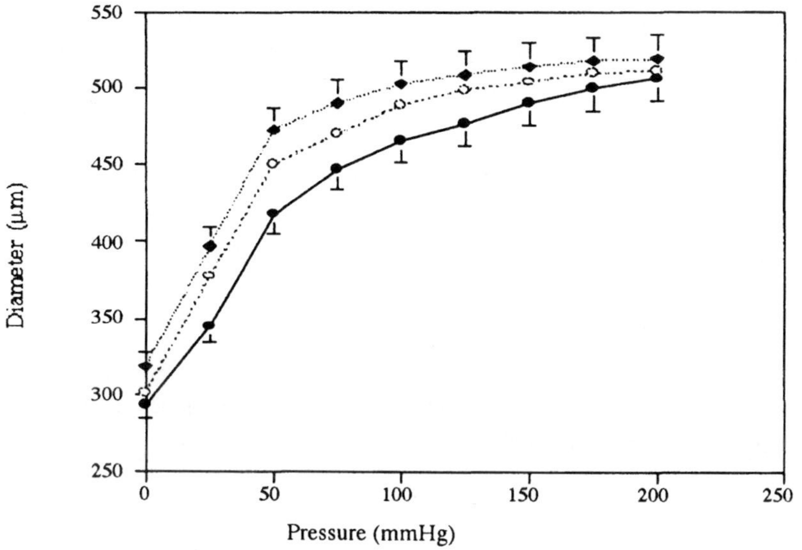

Fig. 3. — Pressure diameter relationship in a rat mesenteric artery under control conditions (open circles), after activation by phenylephrine (closed circles), abolition of the smooth muscle tone (diamonds).

3.2. *In vivo* Measurements

New ultrasonic technologies have been developed allowing us to precisely measure, both *in vitro* and *in vivo*, instantaneous arterial diameter [26]. The probe consists of a 10 MHz strongly focused piezoelectric transducer operated in the pulse-echo mode. A stereotaxic arm permits motion of the transducer in x, y and z coordinates with micrometric steps to place the probe perpendicularly to the arterial axis in its largest cross-sectional dimension. The transducer is positioned so that its focal zone is located in the center of the artery; thus, the back scattered echoes from both the anterior and posterior walls can be easily visualized. A typical radio frequency signal is then displayed on a computer monitor interfaced to the transducer system evidencing signals from the anterior and the posterior wall. These signals are only visible as the ultrasound beam crosses the axis of the vessel and are characterized by a first high-amplitude signal followed by a relatively silent acoustic period and then a second high-amplitude signal. The sample rate of the system is 5000 Hz and its resolution is close to 2 μm [27]. The highly non-linear characteristics of blood vessels require the study of dynamic elastic properties to be performed while applying small pressure perturbations superimposed on a constant mean level of pressure or strain. Under these conditions, linearization of the mechanical properties is allowed and modeling facilitated. By using ultrasonic devices, several authors reported differences in static and dynamic arterial mechanical properties evidencing smaller dynamic than static distensibility and, therefore, suggesting a significant viscosity of the arterial wall under *in vivo* conditions [28, 29]. It was constantly shown that the wall stiffness increased with frequency. At frequencies of 1 Hz, the increase is already high, but between 2 and 10 Hz, marked increases could be measured. Bergel [30] found a ratio of dynamic to static incremental elastic modulus at 1 Hz to range from 1.1 to 1.7 depending on the vessel type; the viscous component of the visco-elastic properties increases toward the peripheral arteries.

3.2.1. Pulse Wave Velocity

The older way of estimating the distensibility of the arterial system consists of measuring the arterial pulse wave velocity. The Moens-Korteweg equation relates pulse wave velocity (PWV) and elastic modulus E

$$PWV^2 = Eh/2\rho R \quad [\mathrm{m\ s^{-1}}]$$

where h is the wall thickness [m], R the vessel internal radius [m], r the specific mass of blood [kg m^{-3}].

Assuming a non compressible vascular wall and small strains, PWV is equal to:

$$PWV^2 = V\Delta P/\rho\Delta V.$$

The measurement of the pulse wave velocity in clinical practice is a simple and reliable method to assess global mechanical properties of the aorta and large elastic arteries [31].

3.2.2. Vascular Impedance

Mechanical properties of the vascular bed can be determined *in vivo* as well. From aortic input impedance measurements, systemic arterial compliance can be determined [32]. However, the value of the calculated compliance strongly depends on the mathematical model for the arterial tree.

Vascular impedance expresses the relation of the force acting in the bloodstream to the resulting motion of blood. This relationship is a function of the physical properties of blood and the blood vessels. Vascular impedances are always ratios of pressure to flow; three different types have been defined:

longitudinal impedance: $Z_\mathrm{L} = (-\mathrm{d}P/\mathrm{d}x)/Q$;

input impedance: $Z_x = P/Q$;

transverse impedance: $Z_w = P/(-\mathrm{d}Q/\mathrm{d}x)$.

P is the blood pressure and Q is the volumetric rate of flow in the longitudinal (x) direction.

Impedance spectrum is in complex numbers, defined by pressure and flow or their derivatives. The impedance equations apply only to sinusoidal signals which can be derived from natural pulse waves *in vivo* by methods of frequency analysis. An impedance must be expressed as a spectrum of its complex values over a range of frequencies.

3.2.2.1. Longitudinal Impedance (Z_L) is the ratio of the pressure *gradient* to flow and is thus the pulsatile analogue of vascular resistance. The use of the pressure *gradient* makes this impedance a property of a vascular segment of unit length, depending of the local properties of the vessel. Therefore, Z_L is usually expressed in dyne sec/cm^5 per cm vessel length.

3.2.2.2. Input Impedance (Z_x) is the ratio of pressure to flow at a particular vascular cross section. It is thus a property of a particular site in the circulation, which is regarded as the *input* to all the vascular tree distal to that point; it is determined by the properties of the distal bed as well as local conditions. The input impedance at the end of a vessel is often referred to as its terminal impedance. If the distal bed is no more than a continuation of the local vessel, with exactly the same physical properties, the input impedance would be an expression of those uniform properties, called the characteristic impedance (Z_c). Z_x and Z_c would thus be identical in a uniform tube terminated by a matching impedance. This condition does not occur in the circulation because the consecutive segments of the vascular tree differ in their dimension and their elasticity, and hence in their characteristic impedance. This mismatching generates reflected waves which influence the shape of the blood pressure and flow waves. Thus, the observed pressure and flow at a given point of arterial system are a mixture of incident and reflected waves. The relation of Z_c to the properties of large vessels can be approximated as:

$$Z_c = \rho PWV/A$$

where A is the vessel cross sectional area.

Aortic input impedance in the ascending aorta has been measured by many investigators, in man as well as in the dog and in the rat. A typical spectrum in the dog is shown in Figure 4.

The input impedance falls from 0 to 2 Hz and fluctuates with frequency thereafter. The modulus at zero frequency (mean pressure/mean flow) is 10 to 20 times greater than the rest of the frequency spectrum. The phase of the input impedance is negative at low frequencies, signifying that flow leads pressure. The phase angle approaches zero as frequency rises and becomes positive or turns downward again between 4 and 8 Hz. Characteristic impedance calculated by averaging the observed input impedances at frequencies from 2 to 12 Hz are ranging between 5 and 10% of the modulus at zero frequency (equal to the total peripheral systemic resistance).

Significant changes of arterial input impedance can occur in the course of normal physiological adjustments of the circulation or in response to vasoactive drugs. Basically, three mechanisms may change the arterial impedance:

- constriction or dilatation of arterioles modifies the peripheral reflection factors and thus alters the size of the frequency-dependent oscillations in

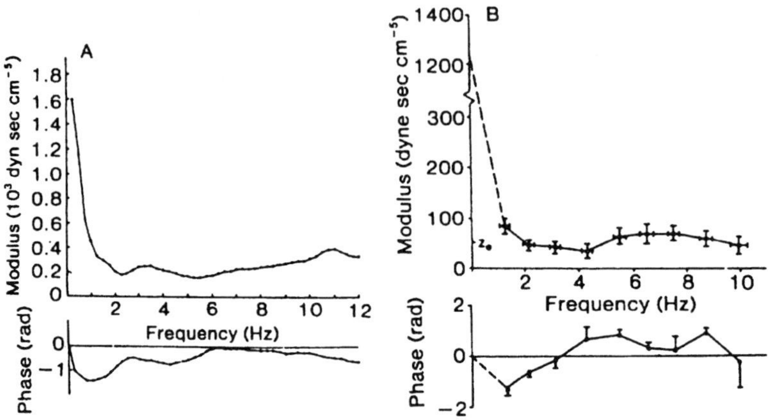

Fig. 4. — Aortic input impedance in the ascending aorta in the dog (see text for details about these impedance diagrams).

the input impedance spectrum. Constriction enlarges the reflection factors whereas arteriolar dilation reduces reflection. The input resistance, or impedance at zero frequency, is more greatly altered by arteriolar vasomotion;

- an increase in elastic modulus increases phase velocity and thus shifts the Z_x minima to higher frequencies. The smooth muscle activated in this case is that of the artery concerned (*e.g.* the aorta), rather than the peripheral vessels. The displacement of important reflection site to position nearer the heart has the same effect;
- any alteration of cross sectional area of the vessel changes its characteristic impedance. According to the equation $Z_c = \rho PWV/A$, the characteristic impedance, Z_c, increases as the radius becomes smaller and *vice versa.* This response interacts with the previous one (variations in elastic modulus) because Z_c is directly proportional to phase velocity. If wave velocity and radius both change exactly in the right proportions, the characteristic impedance remains unaltered. This occurs in many experimental conditions with opposite changes induced by vasoactive agents on the vessel diameter and its elastic modulus.

REFERENCES

[1] R.W. Gore, *Circ. Res.* **34** (1974) 581-591.
[2] B.W. Zweifach, *Circ. Res.* **34** (1974) 843-857.
[3] M.J. Mulvany and C. Aalkjaer, *Physiol. Rev.* **70** (1990) 921-961.
[4] R.W. Gore and H.G., *Fed. Proc.* **34** (1975) 2031-2037.

[5] O. Frank, *Z. Biol.* **85** (1926) 91-130.
[6] T.E. Carew, R.N. Vaishnav and D.J. Patel, *Circ. Res.* **23** (1968) 61-68.
[7] P.B. Dobrin and A.A. Rovick, *Am. J. Physiol.* **217** (1969) 1644-1652.
[8] D.J. Patel, J.S. Janicki and T.E. Carew, *Circ Res.* **25** (1969) 765-779.
[9] O. Lichtenstein, M.E. Safar, P. Poitevin and B.I. Levy, *Hypertension* **26** (1995) 15-19.
[10] W.R. Milnor, *Hemodynamics* (William and Wilkins Ed., Baltimore, 1989) pp. 71-73.
[11] M.A. Gaballa, C.T. Jacob, T.E. Raya, J. Liu, B. Simon and S. Goldman, *Hypertension* (1998) (in press).
[12] C.G. Caro, T.J. Pedley, R.C. Schroter and W.A. Seed, *The mechanics of the circulation* (Oxford University Press, Oxford, 1978).
[13] M.J. Mulvany and W. Halpern, *Nat* **260** (1976) 617-619.
[14] L. Caputo, A. Tedgui and B.I. Levy, *Circulation Res.* **77** (1995) 303-309.
[15] B.R. Duling and R.J. Rivers, *Isolation, cannulation and perfusion of microvessels*, edited by C.H. Baker and W.L. Nastuk, Microcirculatory Technology (Academic Press, Orlando, 1986) pp. 265-280.
[16] W. Halpern and M. Kelly, *Blood Vessels* **28** (1991) 245-251.
[17] R.H. Cox, *Am. J. Physiol. (Heart Circ. Physiol. 13)* **244** (1983) H298-H303.
[18] B.J. Fallon, N. Stephens, J.R. Tulip and A.M. Haegerty, *Am. J. Physiol.* **268** (1995) H670-H678.
[19] B.I. Levy, J.B. Michel, J.L. Salzmann, M. Azizi, P. Poitevin, M.E. Safar and J.P. Camilleri, *Circ. Res.* **63** (1988) 227-239.
[20] B.I. Levy, J. Benessiano, P. Poitevin and M.E. Safar, *Circ. Res.* **66** (1990) 321-328.
[21] B.I. Levy, M. Duriez, M. Phillipe, P. Poitevin and J.B. Michel, *Circulation* **90** (1994) 3024-3033.
[22] H.Y. Qiu, D. Henrion and B.I. Levy, *Hypertension* **24** (1994) 317-321.
[23] H.Y. Qiu, D. Henrion and B.I. Levy, *Hypertension* **24** (1994) 474-479.
[24] H.Y. Qiu and B. Valtier, *J. Pharmacol. Toxicol. Method* **33** (1995) 159-170.
[25] W.M. Chilian, C.L. Eastham and M.L. Marcus, *Am. J. Physiol.* **25** (1986) H779-H788.
[26] A.P.G. Hoeks, P.J. Brands, F.A.M. Smeets and R.S. Reneman, *Ultrasound Med. Biol.* **16** (1990) 121-128.
[27] Y. Tardy, D. Hayoz, J.P. Mignot, P. Richard, H.R. Brunner and J.J. Meister, *J. Hypertens. Suppl.* **10** (1992) S105-S109.
[28] P. Boutouyrie, Y. Bezie, P. Lacolley, P. Challande, P. Chamiot-Clerc, A. Benetos, J.F. de la Faverie, M. Safar and S. Laurent, *Arterioscler. Thromb. Vasc. Biol.* **17** (1997) 1346-1355.
[29] O. Lichtenstein, M.E. Safar, E., Mathieu, P. Poitevin and B.I. Levy, *Hypertension* **32** (1998) 346-351.
[30] D.H. Bergel, *J. Physiol. (London)* **156** (1961) 458-469.
[31] R. Asmar, A. Benetos, J. Topouchian, P. Laurent, B. Pannier, A.M. Brisac, R. Target and B.I. Levy, *Hypertension* **26** (1995) 485-490.

CONTRIBUTED PAPER 7

A Link Between Dendritic Growth and Remodeling of Blood Vessels

V. Fleury and L. Schwartz

1. INTRODUCTION

As explained by B. Lévy and F. le Noble in this book, the hydrodynamics of blood vessels ("hemodynamics") is rather complex and both formation and function of vessels are very dependent on the actual mechanics of the vascular wall. Especially, the endothelium cells (see previous chapters) are able to perform feedback actions towards mechanical forces. These capabilities are rooted in mechanoreceptors which have not yet been clearly identified. Unfortunately, mechanics is often a difficult matter [1] and one hardly sorts out the contributions of strain, stress, pressure gradients etc. to the morphogenesis and function of vascular systems. The mathematical description which is involved in mechanical problems ("tensorial formalism") is difficult, and it sounds at first glance impossible to give a simple model of vessel growth which would aim at describing the morphogenesis of vascular trees from first principles at the scale of, say, an entire organ. One would think it necessary to model 3D vessels with some detail, and then perform actual growth of these vessels under the action of both chemical and mechanical factors.

We shall however give in the following a simple model for the concomitant growth of arterial and venal trees. This model does not incorporate the detail of vessel structure, but it is nevertheless able to give insight into the self-organization of 2D and 3D vasculatures. This model will be based on simple concepts from out-of-equilibrium physics, which have been put forward initially in the context of crystal growth, and which are now being extended to other areas. Especially, and most interestingly, these models are currently being translated to problems in geology, where formation of river networks is also dependent on hydrodynamics (see Meakin's chapter in this book).

Much to the surprise of participants to the Les Houches School dedicated to branched systems, normal blood vessels and normal river networks seem to share many characteristics. Furthermore, in some abnormal instances, braided rivers may form, which are quite reminiscent of abnormal vasculatures which form in pathological situations. Therefore, one may expect general features of vasculature formation to be independent of the detail of the cellular behavior (what is known as "generic" features in physics, or as "universality classes"). From the biological point of view, it is clear that the detail of the vasculature structure at the macro-scale cannot be entirely coded inside the genome, because such a scheme would require an enormous amount of data. Then, models of branched growth might be exportable from one scientific field to another.

2. DESCRIPTION OF THE MODEL

As explained by le Noble, one essential step in vessel formation is "pruning", *i.e.*, the maturation of an array of small capillaries which progressively transforms itself into the branching vasculature. We shall not be interested, here, in how the capillary array (the so called *plexus*) is actually formed [2], but only with the dynamics of the maturation process [3].

If we consider the capillary bed, prior to maturation, as a discretized lattice of small tubes in which blood flows, and if we introduce the sensitivity of the endothelial cells to shear stress only, it is easy to construct a physical model of vascular tree morphogenesis in the following way.

In a capillary tube (and in a crude approximation) the pressure drop between two vertices (two capillary crossings) is related to the flow V in the middle of the tube that links the two vertices by the Poiseuille law:

$$(R^2/\eta)\mathrm{grad}P = -V, \tag{1}$$

where R is the radius of the tube, and η the viscosity of the blood (there should also be a prefactor for the geometry of the vessel section). If we consider, now, the entire lattice of the capillary bed, the pressure at each vertex is obtained, in the simplest hypothesis, by solving the conservation equation for the fluid flow (incompressibility), which, in such an array of tubes, is given by:

$$\Delta P = 0. \tag{2}$$

We next implement the growth of vessels by the following simple rule: wherever the shear is large, the local capillary segment is progressively replaced by a segment of vessel. The justification of such a growth rule can be found in a long list of work [4–6] (and references therein), and also in the previous chapters. The rationale is that nature has provided endothelium cells with a mechanism for vessel enlargement, in regions where the drag is too high. This sensitivity of the cells is purely local. Cells sense the mechanics of the flow, and react to it locally (no emission of chemotactants). Of course, considering only the effect of shear, and not the absolute value of the pressure is a simplification that will need to be modified in future work. The progressive change of the capillary diameter amounts to introducing a speed v of growth of the branched pattern which is proportional to the shear stress (v *is not* the fluid speed V). We take as simplifying assumption that the wider blood vessels are so much wider than the capillaries, that the pressure drop across them is extremely small (stated otherwise, the vein or artery cross section is much larger than the cross section of the capillaries).

The shear stress σ at the very surface of one small tube is proportional to the gradient of the fluid speed taken at that surface, so it is proportional to the fluid speed V in the middle of the tube, it is also proportional to viscosity (the higher the viscosity, the higher the shear), and to the inverse of the radius of the tube (the larger the tube, the smaller the shear). In the end, the shear stress σ, the fluid speed in the center of the tube and the pressure gradient are related by:

$$\sigma = 2(\eta/R)V = (R/4)\text{grad}P. \tag{3}$$

As a consequence, the direction in which the pattern of large vessels is more likely to elongate is the direction of high gradient of pressure, and the growth speed of the vascular interface through the capillary bed will be proportional, in this simple model, to the gradient of P taken at the border of the tree of large vessels (the so-called harmonic measure in physics):

$$v = -k\text{grad}P \tag{4}$$

where k is a biophysical, normalizing parameter and v is the local, instantaneous, "growth speed" of the vessel tree (again, not the fluid speed V), this "growth speed" varies from place to place. It is known that transfer of fluid shear stress forces to the cell occurs first at the luminal cell surface. Plasma membrane molecules are therefore candidate mechanotransducers. This question has been extensively reviewed in references [4,5]. The parameter k in our model represents very crudely the time scale of the sequence of events (membrane transduction)-(cell reaction) until a large vessel is completed. Whatsoever, this time scale is much larger than the time scale of the establishment of the pressure drop itself. Indeed, the time scale for physical change of the vessel is of the order of tens of minutes to hours, while the time scale for the establishment of the pressure pattern is of the order of seconds. In the end, the model, in this simplifying version, is identical to the so-called Diffusion Limited Aggregation (DLA) [7], model of growth, in its Dielectric Breakdown version [8], which is well known to give fractal branching patterns.

The dielectric breakdown phenomenon is the propagation of an electrostatic discharge in an insulator (say, lightning in air). It is well known to give branched patterns. In a crude analogy, the branching vasculature is a spark that propagates on a net.

The DLA model is discussed by several authors in this book, especially by Meakin, in the context of river networks. The application of DLA to geology

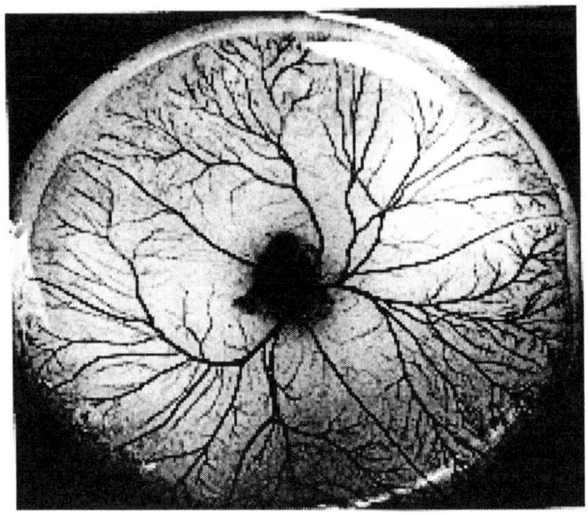

Fig. 1. — The chick embryo at an early developmental stage, showing the vasculature in the yolk sac.

assumes only that the flow in the stream erodes progressively the banks of the river which enlarges. Hence the flow increases, and the river widens even more etc. ("snow-ball" effect). A branched pattern self-organizes spontaneously.

We should insist on two facts. First that in the description of vessel maturation from a capillary *plexus*, the "growth" of the vascular tree is not actually a growth, but more a replacement or a selection process. The segments of vessels in the branching arterioles or venules are present in the first place, in the form of capillaries. Some of the latter were selected to enlarge and elongate, some others have faded away, and vanished, and some others remain capillaries, but the arteries and veins do not actually grow at tips. One never observes an artery or a vein with a dead end, that grows towards another vessel end. This is why the connection of arteries to veins is "automatic", and does not require a specific search of arteries towards veins and *vice versa.* (This is at variance with existing models, like Meinhardt's model [9–11] which require a rather redundant signaling of arteries towards veins and *vice versa.*) The morphogenic information is carried by the flow, which makes the morphogenesis process quite economical.

3. 2D NUMERICAL SIMULATION

The DLA process forms self-organized branched patterns, and it is easy to perform such a growth in a situation which evokes a biologically relevant case [3]. We took as simple example the *area vasculosa* of the chick embryo

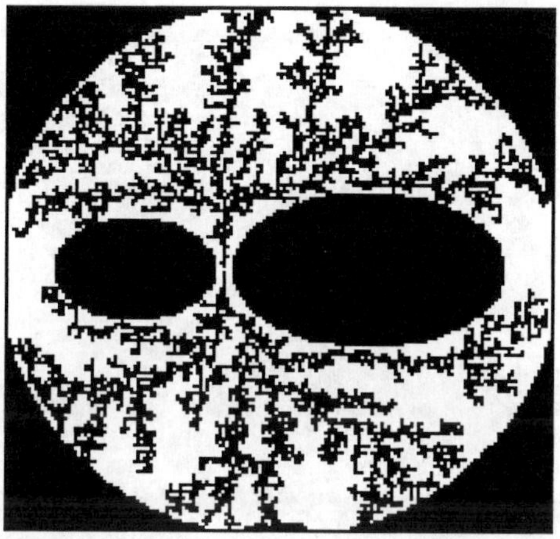

Fig. 2. — DLA growth in a geometry which is as close as possible to the avian embryo.

(Figs. 1 and 2). In this instance, we have a round vessel "far away" on the yolk sac, and a mass of capillaries inside a flat circular area (the *area vasculosa* itself). The embryo occupies a globular region at one pole of the yolk sac, which we model by two ovals (say the head and body). Rudiments of arteries are put at right angles between the head and body, which represent crudely the heart structure at an early developmental stage (this cross geometry is anatomical).

The result of such a simulation is a branching pattern which is not deterministic, but which has some deterministic flavor to it, because the shape of the yolk sac and of the growing embryo determines the overall distribution of vessels.

In his chapter, Meakin strongly warns that one should not be satisfied with a mere resemblance between an experimental branched network and a simulated one. However, this is especially true when the assumptions of the model are not well established. In the case of vessel growth, as modeled here, the endothelium reaction to shear is a well known fact.

4. A MODEL OF 3D GROWTH, AND ITS NUMERICAL SIMULATION

Unsatisfied with a mere resemblance between the vasculatures and DLA patterns, we may try a step further. We propose a simple extension of the DLA model, which will enable us to simulate the maturation of a capillary *plexus* inside a 3D organ [12].

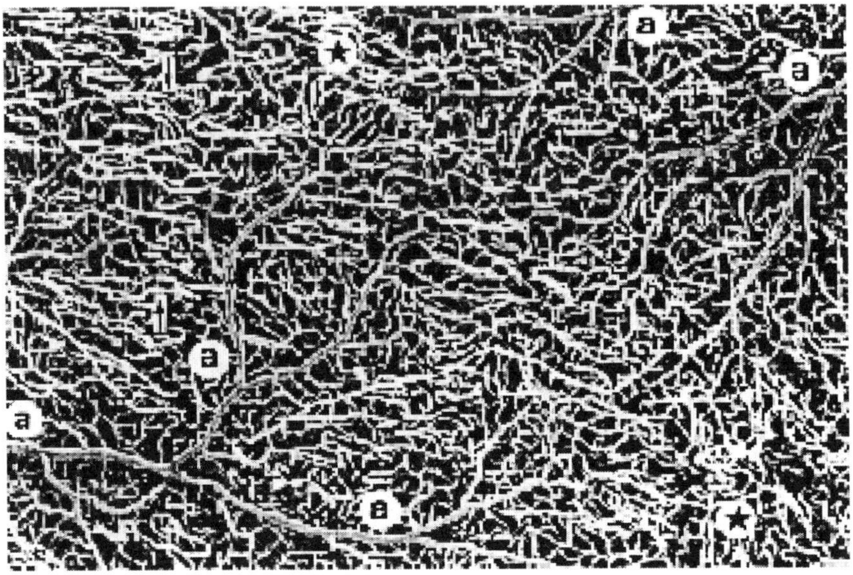

Fig. 3. — An araldite cast of a 3D *plexus* in the choroid *vasculosa* of the pig (courtesy of Simoens [13]). In this region, and at that developmental stage, only a capillary *plexus* is visible (except for a few vessels which start forming).

The "DLA by shear stress" mechanism which we have described in the previous section is able to explain the formation of an arterial bed in a flat two-dimensional primary plexus. But it is still insufficient to explain the pattern of vessels in an organ, say, the ocular *vasculosa*, because in this instance, there exist in the same zone long venules *and* arterioles that intertwin, and a capillary bed further down (Figs. 3 and 4).

Venules and arterioles interdigitate and irrigate the *same* region in the capillary bed. This fact is impossible to explain with the basic "river network" model described above, when *two* vascular trees, *two* 3D rivers must be considered.

This is why we need an additional essential fact, which is indeed observed in real vasculogenesis: capillary regression [13]. As the vessels penetrate into the capillary bed, the capillaries that are connected to them regress and disappear. This does not come as a surprise, since there is almost no flow in them. Eventually, vessels connect to the capillary bed by the tips, and not by the sides of the tubes.

Now we show that this feature: regression of capillaries wherever the shear is too low, is sufficient to construct a model of 3-D vasculature maturation of an entire organ as follows. Starting from an arterial rudiment, a first tree forms in a 3-D *plexus*, that grows following the shear stress in a DLA fashion. Capillary regression disconnects the tree from the capillaries, except at tips.

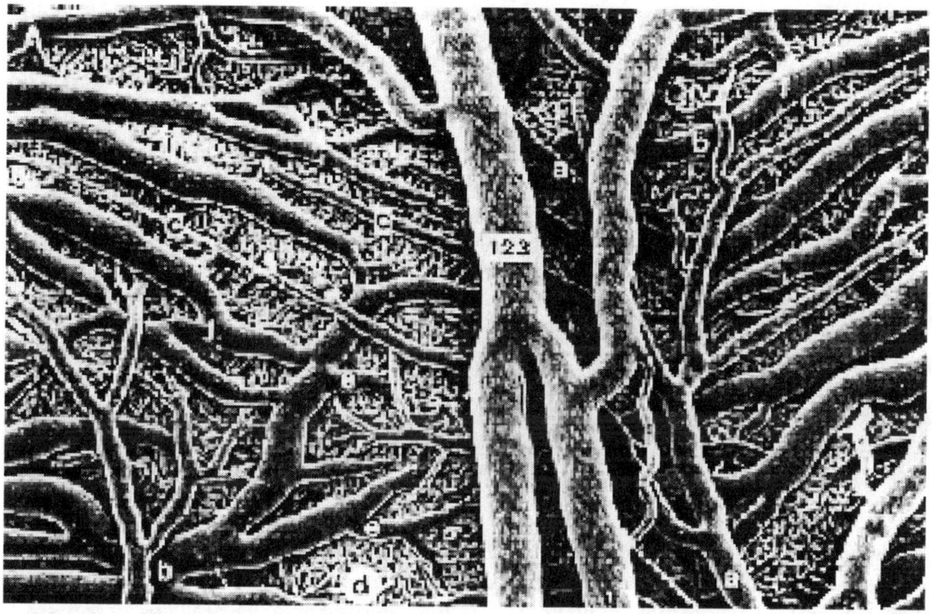

Fig. 4. — An araldite cast of a 3D vasculature in the choroid *vasculosa* of the pig (courtesy of Simoens [13]). Now, we see a very complex pattern of vessels which is obviously 3D, and which contains such situations as veins overlapping arteries, veins overlapping veins etc.

The remaining *plexus* is almost identical to the primary *plexus*, except for links that have been cut off. From the topological point of view, there is little difference with the primary *plexus*, because capillaries are markedly more numerous than vessels (compare Figs. 3 and 4).

Now, this capillary *plexus* serves for the growth of a second tree, the venal tree. Starting at some point from a venule rudiment, and progressing in the same way, a second tree will grow across the same 3-D *plexus*, ignoring the presence of the (disconnected) arterial tree, and ending in the capillaries which do not transform into large vessels. Eventually, the capillaries disconnect from the main veins also, except at tips. The final aspect of the vasculature decomposes into two regions: 1) the capillaries that did not transform nor disappear, where exchange occurs; 2) the region of vascular trees, from which capillaries have disappeared. This region shows a complex interlace of arterioles and venules, with no apparent logic, since arterioles and venules have ignored each other during growth. One will observe venules overlapping venules, venules overlapping arterioles etc. But considered independently, both trees will branch in an analogous fashion, and grow towards the capillary bed. There, they interdigitate, since both arterioles and venules are formed from a subset of tubes that, in the first place, where percolating. Despite the apparent randomness, the venules

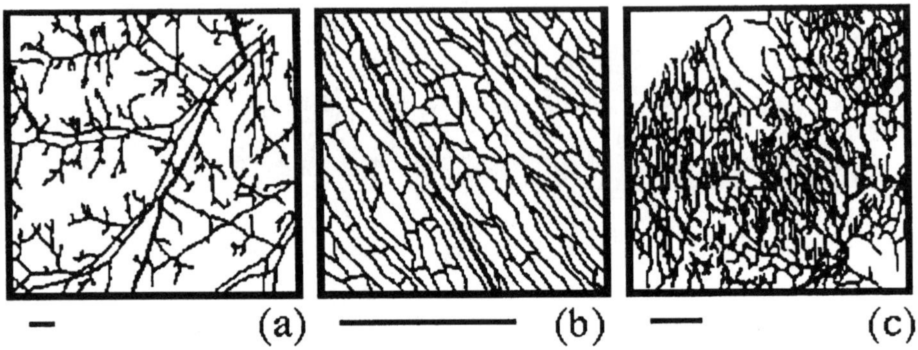

Fig. 5. — Skeletonized images of vascular networks, as found in existing literature. a) Normal subcutaneous arteriovenous network; b) normal subcutaneous capillary network; c) tumor network [14].

will statistically grow towards arterioles causing a statistical "pairing" because the source term of P, which drives the growth of the venules, is at the tips of the arterioles, where blood flows out. This "veins towards arteries" tropism is scale invariant. The pairing will be all the more evident as the maturation of both types of vessels is close in time. We have implemented numerically the process described in the previous paragraph. The result (see color plates 5.A and 5.B) shows all the features evoked above. In the end, the process reveals a remarkable economy of means in that:

1) it uses only local biomechanical information, in the form of the local shear;
2) since it uses the shear flow as long range information, it constructs the vasculature in the regions of actual flow;
3) it uses the same capillary bed for arteries and veins;
4) it preserves the vital demands during construction (compatible with ontogeny), and it can and also be evolved easily (compatibility with phylogeny), since progressive improvement of the feedback loop "shear-sensing-O-capillary transformation" will progressively form improved vasculatures.

5. COMMENT ON EXISTING WORK

The fractal analysis of vasculatures has already been investigated by several authors [14]. The DLA model itself has even been put forward as a good model of vascular growth. However, an important clarification must be made here. On purely theoretical grounds, diffusion of molecules, such as growth factors, can induce the growth of a DLA tree. In almost all studies so far, it has been argued that the DLA aspect of a normal vasculature was linked to diffusion limited growth in a field of biochemicals concentration.

Figure 5 shows a typical example of patterns analyzed in an article which compares a normal situation to a cancerous one. One sees in the left image a typical normal vascular tree. However, it is methodologically erroneous to compare the left image to the middle image (a normal capillary bed), because the left image is a pruned capillary bed. The normal vasculature in Figure 5a was initially, before maturation, a capillary bed like the one in Figure 5b, and the set of vessels shown in the left image never actually "grew", although they seem to have dead ends. The differences in diffusion kinetics as supposedly inferred from a difference between images like Figures 5a-c is much more likely a difference in shear stress sensing, *i.e.* a difference in the behavior of mechanoreceptors. What the cancerous vasculature shows is a badly connected capillary mesh, which does not prune normally. In an image like Figure 5a, the true medium in which the vessels "grew" has actually disappeared (it was the capillary plexus, which regressed when the larger vessels formed, note that the scales in the normal vasculature and in the capillary bed are totally different). The Laplacian field inducing the DLA-like pattern was in fact restricted to the inside of those tubes, which are no longer there. Moreover, even if one insisted on invoking transport of morphogens, it is very likely that such a transport would be much more rapid by the inside of the pattern, where there is flow, than by the surrounding tissue. Hence, the true medium for a morphogen field would still be the lumen of the vessels.

However, in an image like Figure 5a, there still exist, in fact, many vessels comparable to the ones in Figure 5b. They connect the tips of the arterioles to the tips of the venules: these capillaries have disappeared in the resolution of the image.

6. REASONABLE MODIFICATIONS OF THE MODEL

We acknowledge that the model presented here is an oversimplified picture of vessel maturation. Several features in the model could be modified to get a more accurate and realistic morphogenesis.

First of all, the vessels are not divided into capillaries, on the one hand, and large vessels, on the other hand. There exists a continuum of sizes, corresponding to varying degrees of hydrodynamic resistance. This could be incorporated in a refined version of this model (see an interesting attempt by Honda [15], who tried a more complex model).

Second, the time scale of the vessel enlargement and the density of capillaries in the plexus could be dependent on the cell metabolism, especially, on hypoxia. Then, the pattern would self-organize in such a way that it would spontaneously drain the hypoxic regions (this amounts to turn k into a function of oxygen concentration, in Eq. (4)). Such a feature could also be incorporated into the model, although not very easily, because it requires solving an advection-diffusion equation in the network, for the gas concentration.

Thirdly, the shear stress, as invoked in the models above, is only the scalar component of the shear at the very surface of the cell wall. We neglected the tensorial aspects. In fact, we have only considered the effect of the flow, inside the tube, on the tube itself. But a flow passing through such a tube will also provoke a shear on the neighboring tubes, because the tissue is elastic, and mechanical shear is transmitted across the tissue to the surrounding regions. This will imply a recirculation of vessel growth, and formation of loops, as emphasized by Couder in his chapter, because the stress matrix has non-diagonal components (applying a force in one direction induces a deformation in the perpendicular one). This is at variance with geological models, in which tensorial mechanics is not present at the scale of the river width. A simple model could be implemented in which shear on one lattice bond might deform other bonds in the neighborhood. Especially, the presence of two vessels with opposite flows increases the shear on both vessels. Hence, the statistical pairing of arteries and veins is increased by taking into account the contribution of shear inside one vessel, to the shear in the neighboring one.

Another essential feature which is not incorporated in the present model is elongation of vessels in the tangential direction. Indeed, we have only implemented the enlargement of the diameter under a strong flow. However, even large vessels keep on elongating, under the action of the internal flow. Hence, in a simulation like a DLA simulation, an elongation of existing branches should be added to the actual growth at tips.

Finally, we should add, at least *en passant*, a remark which relates the model to brain formation. In the model presented here, a massive excess of vessel segments exists at first, which is progressively remodeled into a tree. Many useless capillaries regress and eventually disappear. That tree provides sufficient drainage of the tissue, and is more economical in terms of metabolic maintenance of itself. A similar situation exists in synaptic morphogenesis. It appears that, in many instances, neural development starts by formation of random connections which are in massive excess, and which progressively "prune" into the final synaptic pattern, whose structure is more clearly and economically ramified (see the chapter by Ternaux). In this instance, the flux is a flux of neurotransmitters, but the pruning agent is not clearly identified (nerve impulse). Should the mechanics and hydrodynamics of vesicle transport in neural pseudo-pods ("processes") play also a role remains an open question.

In general terms, growth of branched systems by maturation processes might have been selected by nature because they allow the transport system to carry out its task prior to any evolution (the primary plexus or the primary synaptic mesh work, although they might not be optimal), and to progressively transform it into a better shape. Hence, this sort of [mesh + pruning] process provides both a starting point and a "fitting" mechanism (in the genetic sense). We may then add to the warning by Paul Meakin, that one should be very cautious with models of branched growth in biology which do not provide a reasonable phylogeny, or to models which require the branched organ or structure to be optimal right away.

REFERENCES

[1] Y.C. Fung, Biomechanics Motion Flow Stress and Growth (Springer-Verlag, New-York, 1990).

[2] W. Risau, Mechanisms of angiogenesis, *Nat* **386** (1987) 671-674.

[3] V. Fleury and L. Schwartz, Diffusion limited aggregation from shear stress as a simple model of vasculogenesis, *Fractals* **7** (1999) 33-39.

[4] C.F. Dewey Jr., S.R. Bussolari, M.A. Gimbrone and P.F. Davies, The dynamic response of vascular endothelial cells to fluid shear stress, *J. Biomech. Eng.* **103** (1981) 177-188.

[5] P.F. Davies, Flow mediated endothelial mechanotransduction, *Physiolog. Rev.* **75** (1995) 519-551.

[6] A. Tedgui, S. Lehoux and B. Levy, Mechanical Factors and Vascular Biology, Biology of the Arterial Wall, edited by B. Levy and A. Tedgui (Kluwer Academic Press, Norwell, 1999), pp. 72-100.

[7] T.A. Witten and L.M. Sander, *Phys. Rev. Lett.* **47** (1981) 1400-1404.

[8] L. Niemeyer, L. Pietronero and A.H.J. Wiesmann, Fractal Dimension of Dielectric Breakdown, *Phys. Rev. Lett.* **52** (1984) 1033-1036.

[9] H. Meinhardt, Morphogenesis of lines and nets, *Diff.* **6** (1976) 117-123.

[10] A.J. Koch and H. Meinhardt, Biological pattern formation: from basic mechanisms to complex structures, *Int. J. Bif. Chaos* **7** (1997) 1-26.

[11] V. Fleury, Branching Morphogenesis in a Class of Reaction-Diffusion Models, *Phys. Rev. E* **61** (2000) 4156-4160.

[12] V. Fleury and L. Schwartz, Modelisation of 3D vasculature by interlaced DLA, *Fractals* **10** (2000) (to appear).

[13] P. Simoens, Morphology study of the vasculature in the orbit and eyeball of the pig, Thesis to the faculty of Veterinary Medicine (State University of Ghent, Ghent, 1985).

[14] Y. Gazit, D.A. Berk, M. Leunig, L.T. Baxter and R.K. Jain, Scale invariant behavior and vascular network formation in normal and tumor tissue, *Phys. Rev. Lett.* **75** (1995) 2428-2431.

[15] H. Honda and K. Yoshizato, Formation of the branching pattern of blood vessels in the wall of the avian yolk sac studied by a computer simulation, *Develop. Growth. Differ.* **39** (1997) 581-589.

CHAPTER X

Adaptive Branching During Colonial Development of Lubricating Bacteria

I. Golding, I. Cohen, I.G. Ron and E. Ben-Jacob

1. INTRODUCTION

Many bacterial strains exhibit beautiful branching patterns during colonial development. Much effort is devoted to the search for basic principles of self-organization (growth, communication, regulation and control) on the cellular and multi-cellular levels. Our approach is to use the successful conceptual framework for branching growth patterns in non-living systems as a tool to understand their significantly more complex biological counterparts.

Here we focus on three examples of bacterial branching patterns: 1) basic branching exhibited by *P. dendritiformis* var. *dendron.* 2) Chiral branching exhibited by *P. dendritiformis* var. *chiralis.* 3) Vortex branching exhibited by *P. vortex.* All the above strains are lubricating bacteria. The first two strains are swimming bacteria and the third strain are swarming bacteria. Characteristic patterns are shown in Figure 1.

Drawing on the analogy with diffusive patterning in non-living systems [1–5], we can state that complex patterns are expected. The cellular reproduction rate that determines the growth rate of the colony is limited by the level of nutrients available for the cells. The latter is limited by the diffusion of nutrients towards the colony (for low nutrient substrate). Hence colony growth under certain conditions should be similar to diffusion limited growth in non-living systems as mentioned above [1–3]. The study of diffusive patterning in non-living systems teaches us that the diffusion field drives the system towards decorated (on many length scales) irregular fractal shapes [6–10]. Indeed, bacterial colonies can develop patterns reminiscent of those observed during growth in non-living systems [11–20]. But, this is certainly not the end of the story. In fact, the colonies exhibit far richer behavior. This is ultimately a reflection

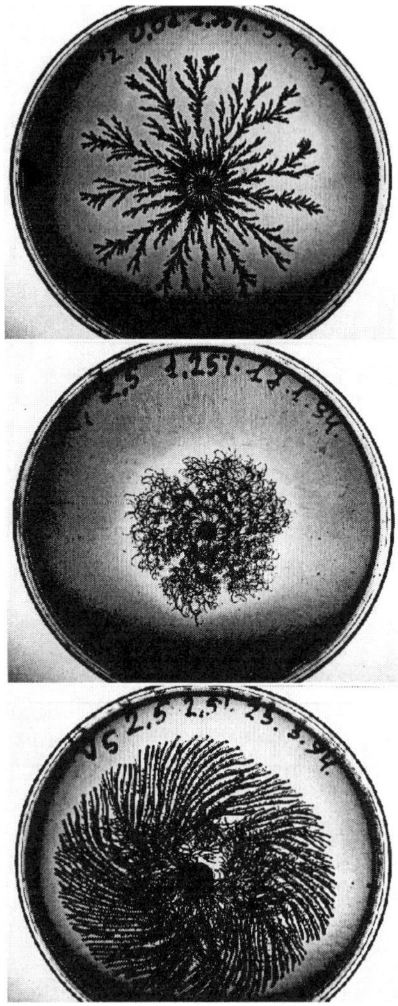

a b c

Fig. 1. — The three characteristic branching patterns discussed in this chapter: a) simple branching exhibited by the $\mathcal{T}$ morphotype grown at 0.01 g/l peptone and 1.75% agar concentration. b) Chiral branching pattern exhibited by the $\mathcal{C}$ morphotype grown at 2.5 g/l peptone and 1.25% agar concentration. c) Vortex branching exhibited by the $\mathcal{V}$ morphotype grown at 2.5 g/l peptone and 2.5% agar concentration.

of the additional levels of complexity involved [11, 14, 21–26]. The building blocks of the colonies are themselves living systems, each having its own autonomous self-interest and internal degrees of freedom. Yet, efficient adaptation of the colony to adverse growth conditions requires self-organization on all levels which can only be achieved *via* cooperative behavior of these individual cells. Thus, pattern formation at the colony level may be viewed as the outcome of a dynamical interplay between the micro-level (the individual cell) and the macro-level (the colony) [11, 12, 14]. For this interplay to work, the effects of changes at the micro-level must make themselves felt at the macro-level. This is why the notion of a singular perturbation, discovered to be a key for understanding pattern selection in non-living systems, will be of even more importance here.

We present experimental observations together with results of numerical simulations. We seek to elicit, from the experimental observations and the biological knowledge, the generic features and basic principles needed to explain the biological behavior and to include these features in the model. We will demonstrate that such modeling, with close comparison to experimental observations, can be used as a research tool to reveal new understanding of the biological systems.

Generic modeling is not about using a sophisticated mathematical description to dress pre-existing understanding of complex biological behavior. Rather, it means a cooperative approach, using existing biological knowledge together with mathematical tools and synergetic point of view for complex systems to reach a new understanding (which is reflected in the constructed model) of the observed complex phenomena.

The generic models to date can be grouped into two main categories:

1. Discrete models such as the Communicating Walkers models of Ben-Jacob *et al.* [21,23,25] and the Bions model of Kessler and Levine [27,28]. In this approach, the microorganisms (bacteria in the first model and amoebae in second) are represented by discrete, random walking entities (walkers and bions, respectively) which can consume nutrients, reproduce, perform random or biased movement, and produce or respond to chemicals. The time evolution of the chemicals is described by reaction-diffusion equations;

2. Continuous or reaction-diffusion models [29, 30]. In these models the microorganisms are represented *via* their 2D density, and a reaction-diffusion equation of this density describes their time evolution. This equation is coupled to the other reaction-diffusion equations for the chemical fields. In the context of branching growth, this idea has been pursued recently by Mimura and Matsushita *et al.* [31, 32], Kawasaki *et al.* [33], Kitsunezaki [34] and Kozlovsky *et al.* [35]. A summary and critique of this approach can be found in [36] and [37].

Here we use both the discrete and continuous models. Section 2 is devoted to the basic branching patterns. Chiral branching is described in Section 3. One of the key communication features is chemotaxis, which is described in Section 4. Section 5 describes the vortex branching. In Section 6 we present the important phenomenon of bursts of sectors during branching growth. Section 7 deals with our experiments with antibiotics, and Section 8 is the concluding one.

2. BASIC BRANCHING PATTERNS

2.1. Observations and Biological Background

For the experiments described in this paper we use three groups of bacterial strains, all of which we initially isolated from plates containing *Bacillus subtilis* [11, 14, 38, 39]. The strains were subsequently identified as belonging to the genera Paenibacillus. The strains display characteristic and distinct colonial morphologies and thus are characterized as three different morphotypes. We define morphotype as a group of bacteria exhibiting typical colonial patterns. Two strains belong to the same morphotype if under similar growth conditions they develop colonies of the same morphology for a range of growth conditions. Colonial growth patterns of a morphotype are inheritable and can be generated following inoculation with a single cell [2, 40] (note that this definition does not exclude the possibility that bacteria from different species will belong to the same morphotype, or that a specific species can belong to two different morphotypes depending on conditions).

The basic branching patterns described in this section belong to the $\mathcal{T}$ morphotype. Morphologies and studies of $\mathcal{T}$ morphotype are presented in [11, 24]. A representative strain from $\mathcal{T}$ morphotype was identified in [38] as *Paenibacillus dendritiformis* strain T168. Colonial patterns of the $\mathcal{T}$ morphotype on semi-solid agar are characterized by branched patterns with the branching growing in tip-splitting manner.

Morphologies and studies of chiral branching patterns, the $\mathcal{C}$ morphotype (for the *P. dendritiformis*) are presented in [14, 23, 41]. When grown on semi-solid agar, colonies of bacteria from the $\mathcal{C}$ morphotype typically have thin branches, all twisted with the same handedness. New branches are side-branches of existing branches (Sect. 3.2).

Morphologies and studies of $\mathcal{V}$ morphotype are presented in [13, 14, 21, 42]. A representative strain from $\mathcal{V}$ morphotype was identified in [39] as *P. vortex* strain V168. When grown on hard agar, these swarming bacteria organize themselves into vortices. Colonial patterns of these bacteria are branchy, where each branch is led by a droplet of cells. Each leading droplet consists of up to million cells that rotate around a common center (hence the term vortex).

Based on our experience, strict protocol has to be followed for reproducible results, as the colonial development is very sensitive to small changes in the growth conditions. We followed a variation of the protocol described in [43]. The bacteria are grown on agar substrate in standard plastic petri dishes with a diameter of 88 mm. The medium is prepared by mixing in one liter of H_2O varying amounts of agar (Bacto-Agar; Difco Laboratories), varying amounts of peptone (Bacto-Peptone; Difco Laboratories), 5 g of NaCl and 5 g of K_2HPO_4. The amounts of peptone range from 0.1 g (poor nutrient) to 10 g (rich nutrient). The amounts of agar range from 5 g/l (0.5%, very soft agar) to 25 g/l (2.5%, hard agar). Agar concentrations of more than 0.8% render the substrate semi-solid at room temperature and prevent the bacteria from swimming inside it.

The mixture is autoclaved at 121 °C and 1.1. At for 45 min and then allowed to cool at room temperature. The mixture is poured into petri dishes, 12.5*cc* into each plate (exact and uniform agar thickness is important for reproducible patterns). The plates are allowed to stand closed (but not sealed) at room temperature for 2 – 3 days, until the weight of the plates is decreased by 1 g per plate due to water evaporation.

For growth in the presence of an antibiotic, the drug is added after the agar is taken from the autoclave and cooled below 40 °C.

In most of our experiments the bacteria are inoculated at the center of the plate. We use 5 μl droplets containing $10^4 - 10^5$ cells.

2.1.1. *Food Consumption, Reproduction and Starvation*

P. dendritiformis and *P. vortex*, like most bacteria, reproduce by fission of the cell into two daughter cells which are practically identical to the mother cell. The crucial step in the cell division is the replication of the genetic material and its sharing between the daughter cells. Hasty replication of DNA might lead to many errors – most organisms limit the rate of replication to about 1000 bases per second. Thus reproduction must take at least a minimal time τ_R. This reproduction time τ_R is about 25 min in Paenibacillus.

For reproduction, as well as for movement and other metabolic processes, bacteria and all other organisms need an influx of energy. Any organism which does not get its energy directly from sunlight (by photosynthesis) needs an external supply of food. In the patterning experiments the bacteria eat nutrient from the agar. As long as there is enough nutrient and no significant amount of toxic materials, food is consumed (for cell replication and internal processes) at the maximal rate Ω_c. To estimate Ω_c we assume that a bacterium needs to consume an amount of food C_R of about 3×10^{-12} g. It is 3 times its weight – one quanta for doubling body mass, one quanta used for movement and all other metabolic processes during the reproduction time τ_R, and one quanta is for the reduced entropy of making an organized cell out of food. Hence Ω_c is about 2 fg/s (1 fg $= 10^{-15}$ gram).

If nutrient is deficient for a long enough period of time, the bacterial cells may enter a special stationary state – a state of a spore – which enables them to survive much longer without food. The bacterial cells employ very complex mechanisms tailored to the process of sporulation. They stop normal activity – like movement – and use all their internal reserves to metamorphose from an active volatile cell to a sedentary durable "seed". While the spores themselves do not emit any chemicals (as they have no metabolism), the pre-spores (sporulating cells, see Fig. 9) do not move and emit a very wide range of waste materials, some of which are unique to the sporulating cell [44–46]. These emitted chemicals might be used by other cells as a signal carrying information about the conditions at the location of the pre-spores. Ben-Jacob *et al.* [24, 25, 47] suggested that such materials are repelling the bacteria ("repulsive chemotactic signaling") so that they escape a dangerous location.

When bacteria are grown in a petri dish, nutrients are usually provided by adding peptone, a mixture including all the amino acids and sugars as source of carbon. Bacteria which are not defective in synthesis of any amino acid can grow also on a minimal agar in which a single source of carbon and no amino acids are provided. Such growth might seem to be easier to model as the growth is limited by the diffusion of a single chemical. However, during growth on minimal agar there is usually a higher rate of waste product accumulation, introducing other complications into the model. Moreover many of our strains are auxotrophic *i.e.* defective in synthesis of some amino acids and they need an external supply. Providing the bacteria with these amino acids and only a single carbon source poses the question as to what is the limiting factor in the growth of the bacteria. For all those reasons we prefer to use peptone as a nutrient source.

2.1.2. Bacterial Movement

In the course of evolution, bacteria have developed ingenious ways of moving on surfaces [48]. Motility on surfaces is often associated with swarming or gliding [49–54], but $\mathcal{C}$ and $\mathcal{T}$ morphotype swim in a thin layer of liquid on top of the agar [25, 48, 55]. Swimming is a solitary movement done in liquid. A swimming bacterium moves in nearly straight runs, interrupted by short periods of tumbling. The tumbling event is a random rotation in one location. The direction of the next run is dictated by the final orientation of the tumbling bacterium.

A swimming bacterium propagates itself by rotating a bundle of flagella [56, 57]. Each flagellum is a helical protein filament which is hooked to a molecular engine transversing the bacterial membrane. The engines can rotate the flagella clockwise or counterclockwise. When the bacterium turns most of them counterclockwise, the flagella form an aligned bundle and push the bacterium forward. When they turn clockwise, the flagella disjoin and the bacterium tumbles.

In our experiments, the bacteria do not swim in a pre-supplied liquid, but in a layer of fluid on the surface of the agar. The bacterial cells move individually and at random in the same manner as flagellated bacteria move in wet mounts (*i.e.*, nearly straight runs separated by brief tumbling). Swimming takes place only in sufficiently thick surface fluid. Microscope observations reveal no organized flow-field pattern.

Based on microscope observations of movement and electron microscope observations of flagella we identify the movement of $\mathcal{C}$ and $\mathcal{T}$ morphotype as swimming. Cells tumble about every $\tau_{\mathrm{T}} \approx 1-5$ s depending on external conditions. The speed v of the bacterium between tumbling events is very sensitive to conditions such as the liquid viscosity, temperature and pH level. Typically, it is of the order of $1 - 10$ μm/s.

Swimming can be approximated by a random walk with variable step size [58]. At low bacterial densities the random walk can be described by a diffusion equation with a diffusion coefficient $D_{\mathrm{b}} \equiv v^2\,\tau_{\mathrm{T}} = 10^{-8} - 10^{-5}$ cm^2/s. Low bacterial densities mean that the mean free path between bacterial collisions l_c is longer than the tumbling length $l_T \equiv v\tau_T$, thus collisions between the bacteria can be neglected. The mean free path (or collision length) is

$$l_c \propto \begin{cases} \rho^{-\frac{1}{3}} & \text{in 3 dimensions} \\ \sigma^{-\frac{1}{2}} & \text{in 2 dimensions} \end{cases} \tag{1}$$

where ρ is the 3D bacterial density and σ is the 2D density – the projection of ρ on the surface.

At high densities ($l_c < l_T$), the collisions cannot be neglected. In attempting to approximate the dynamics in those conditions, one may want to consider the time of straight motion to be l_c/v instead of τ_T. Hence D_b depends on the bacterial density to yield

$$D_b \propto \begin{cases} v\rho^{-\frac{1}{3}} & \text{in 3D} \\ v\sigma^{-\frac{1}{2}} & \text{in 2D.} \end{cases} \tag{2}$$

This approximation is valid under the assumptions that a collision event is identical to a tumbling event (abrupt uncorrelated change in direction of motion), that a tumbling event is independent of the collisions, and that the speed between such events is not affected by their frequency.

The assumption that a collision event is like a tumbling event poses many problems. Even if the bacteria do not activate a special response to collision it is unrealistic to assume that collisions are elastic, or that the flagella adopt immediately the new orientation which changes during collisions. Thus it is reasonable to assume strong correlation between the cell's orientation before collision and the cell's orientation after collision. In addition, the orientation after the collision should be biased according to the average direction of motion of the surrounding bacteria, as they carry the liquid with them. The important parameter is not the collision length l_c but re-orientation time τ_r. The re-orientation time is the time it takes a bacterium to loose memory of its initial orientation, *i.e.* the time span on which the final orientation has effectively no correlation with the initial orientation. At low densities the re-orientation time τ_r is equal to the tumbling time τ_T. As the density rises and the collisions become more frequent, τ_r decreases. τ_r defines the densities above which the constant diffusion coefficient $D_b \equiv v^2\tau_T$ is not a good approximation. It is quite possible that these densities are high enough so as to make the velocity and even the type of motion dependent on bacterial density, making relation (2) irrelevant. In any case, high cellular densities do mean an effective decrease in the diffusion coefficient related to the bacterial movement.

When swimming in an unstirred liquid, very low cellular densities also effect the movement. The bacteria secrete various materials into the media and some of them, *e.g.* enzymes and other polymers, significantly change the physical properties of the liquid making it more suitable for bacterial swimming. The secretion of these materials depends on cellular density, thus at not-too-high densities the speed of swimming rises with the cellular density. Hence the diffusion coefficient related to the bacterial movement should be a non-monotonic function of the bacterial density. Moreover, the specific functional form might depend on the specific bacterial strain.

In other conditions there is a similar but more pronounced effect. On semi-solid surface the bacteria cannot swim at all inside the agar and they have to produce their own layer of liquid to swim in. To produce such fluid the bacteria secrete lubricant (wetting agents). Other bacterial species produce known extracellular lubricants (such as surfactants, see [59–62] and references therein, or the extracellular slime produced by *Proteus mirabilis* [63]). There are various materials (various cyclic lipopeptides have been identified) which can draw water from the agar. The composition and properties of the lubricant of *P. dendritiformis* is not known, but we will assume that higher concentration of lubricant is needed to extract water from a drier agar, and that the lubricant is slowly absorbed into the agar (or decomposes). It is possible that the mechanism by which *P. dendritiformis* produces the layer of liquid is different from what we postulate here, but as long as the liquid is self-produced and facilitates the bacterial movement (as revealed by the observations) the bio-chemical details of the liquid do not change the validity of the models we present here. A single bacterium on the agar surface cannot produce enough fluid to swim in, thus the bacteria cannot break out of the fluid layer and the branches of a $\mathcal{T}$ or $\mathcal{C}$ colony can be defined by this fluid. Whenever bacteria enter the shallower parts of the layer, at the edge of the branch, they become sluggish, indicating that the depth of the layer effects the bacterial movement. It can be argued that in such cases the bacterial speed is related to the bacterial density by a power law (at least in low densities). Not only is the diffusion coefficient related to the bacterial movement a non-monotonic function of the bacterial density (as in a liquid agar), but it also vanishes for extremely low densities. In this case it is clear that the specific functional form depends on the specific bacterial strain (*B. subtilis*, for example, cannot move at all under such conditions).

2.2. Experimental Observations: Branching Growth of Bacterial Colonies

Some additional examples of the patterns exhibited by colonies of the $\mathcal{T}$ morphotype are shown in Figures 2, 3 and 4. For intermediate agar concentrations (about 1.5% – 1.5 g in 100 ml), at very high peptone levels (above 10 g/l) the patterns are compact (Fig. 3a). At somewhat lower but still high peptone levels (about 5 – 10 g/l) the patterns exhibit quite pronounced radial symmetry and may be characterized as dense fingers (Fig. 3b), each finger being much wider than the distance between fingers. For intermediate peptone levels, branching patterns with lower fractal dimension (reminiscent of electro-chemical deposition) are observed (Fig. 3c). The patterns are "bushy", with branch width smaller than the distance between branches. As the peptone level is lowered,

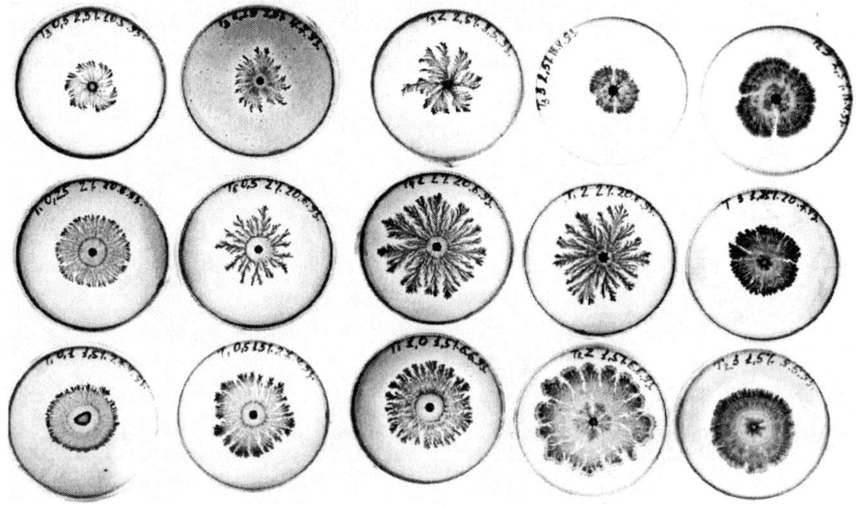

Fig. 2. — Patterns exhibited by the $\mathcal{T}$ morphotype as function of peptone level (increasing from left to right) and agar concentration (1.5% bottom row, 2% middle row, 2.5% top row).

the patterns become more ramified and fractal–like. Surprisingly, at even lower peptone levels (below 0.25 g/l for 2% agar concentration) the colonies revert to organized structures: fine branches forming a well defined global envelope. We characterize these patterns as fine radial branches (Fig. 3d). For extremely low peptone levels (below 0.1 g/l), the colonies lose the fine radial structure and again exhibit fractal patterns (Fig. 4). For high agar concentration the branches are very thin (Fig. 4b).

At high agar concentration and very high peptone levels the colonies display a structure of concentric rings (Fig. 5). At high agar concentrations the branches also exhibit a global twist with the same handedness (weak chirality), as shown in Figure 6. Similar observations during growth of other bacterial strains have been reported by Matsuyama *et al.* [19,20]. We refer to such growth patterns as having weak chirality, as opposed to the strong chirality exhibited by the $\mathcal{C}$ morphotype.

A closer look at an individual branch (Fig. 7) reveals the phenomenon of density variations within the branches. These 3-dimensional structures arise from accumulation of cells in layers. The aggregates can form spots and ridges which are either scattered randomly, ordered in rows, or organized in a leaf vein-like structure. The aggregates are not frozen; the cells in them are motile and the aggregates are dynamically maintained.

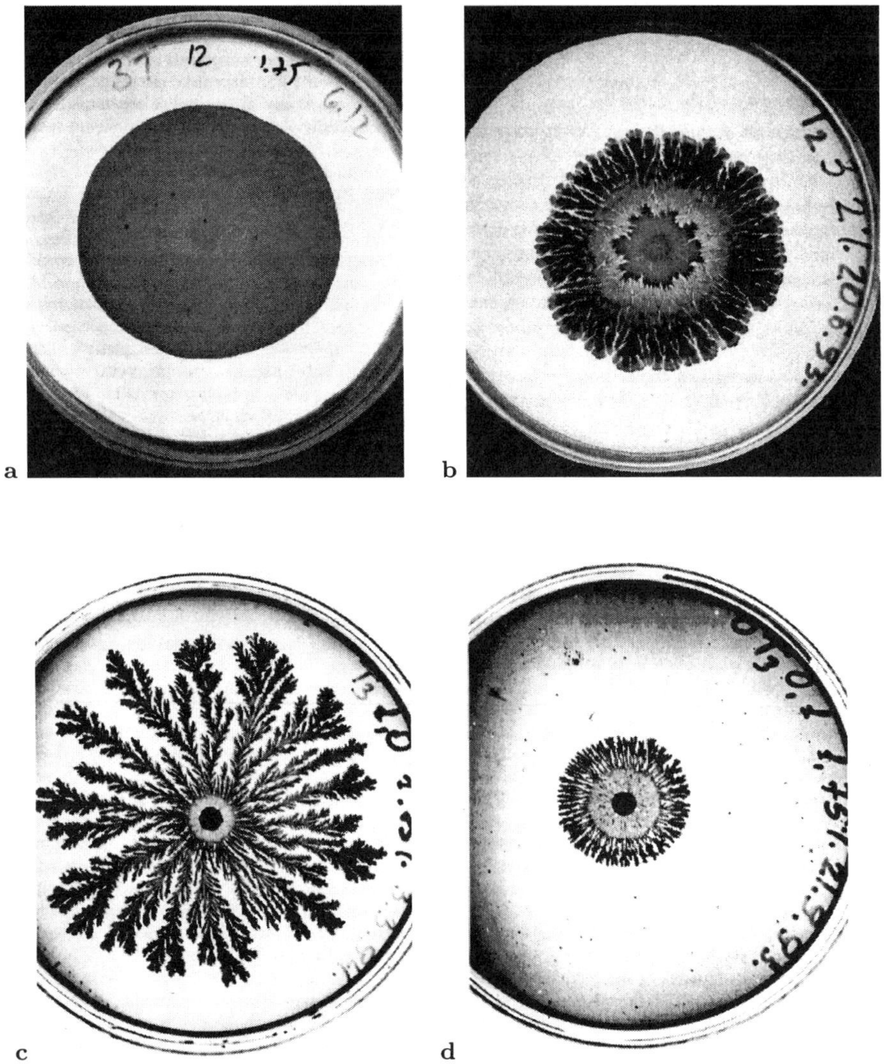

Fig. 3. — Examples of typical patterns of $\mathcal{T}$ morphotype for intermediate agar concentration. a) Compact growth for 12 g/l peptone level and 1.75% agar concentration. b) Dense fingers for 3 g/l peptone and 2% agar. c) Branching fractal pattern for 1 g/l peptone and 1.75% agar. d) A pattern of fine radial branches for 0.1 g/l peptone and 1.75% agar.

Under the microscope, cells are seen to perform a random-walk-like movement in a fluid. The boundary of the fluid defines a local boundary for the branch (Fig. 8). Whenever the cells are active, the boundary propagates slowly

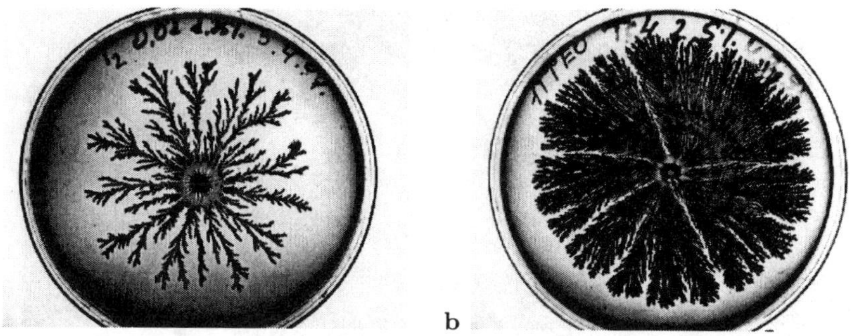

a b

Fig. 4. — Colonial patterns of $\mathcal{T}$ morphotype. a) Fractal pattern for 0.01 g/l peptone level and 1.75% agar concentration. b) Dense branching pattern for 4 g/l peptone and 2.5% agar. Note that the branches are much thinner than those in Figure 3b, *i.e.* the branches are thinner for higher higher agar concentrations.

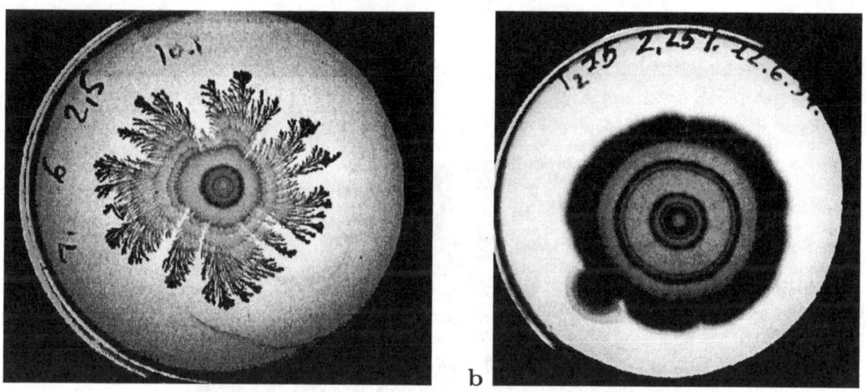

a b

Fig. 5. — a) Pattern of concentric rings superimposed on a branched colony for 2.5 g/l peptone level and 2.5% agar concentration. b) Concentric rings in a compact growth for 15 g/l peptone level and 2.25% agar concentration.

as a result of the cellular movement pushing the envelope forward and production of additional wetting fluid. Electron microscope observations reveal that these bacteria have flagella for swimming.

The observations reveal also that the cells are active at the outer parts of the colony, while closer to the center the cells are stationary and some of them sporulate (form spores) (Fig. 9). It is known that certain bacteria respond to adverse growth conditions by entering a spore stage until more favorable growth conditions return. Such spores are metabolically inert and exhibit a marked resistance to the lethal effects of heat, drying, freezing, deleterious chemicals, and radiation.

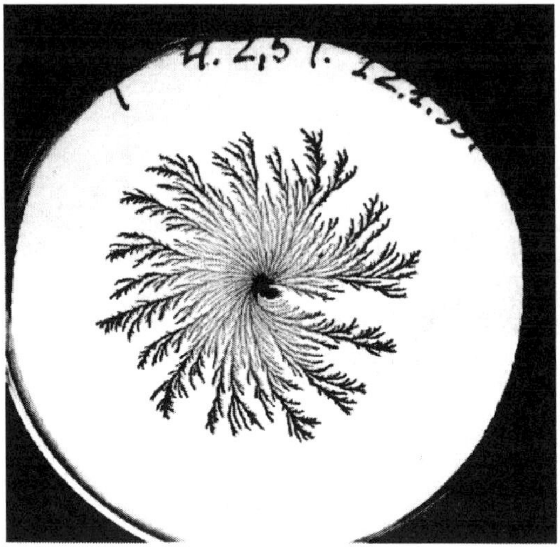

Fig. 6. — Weak chirality exhibited by the $\mathcal{T}$ morphotype during growth on 4 g/l peptone and 2.5% agar concentration.

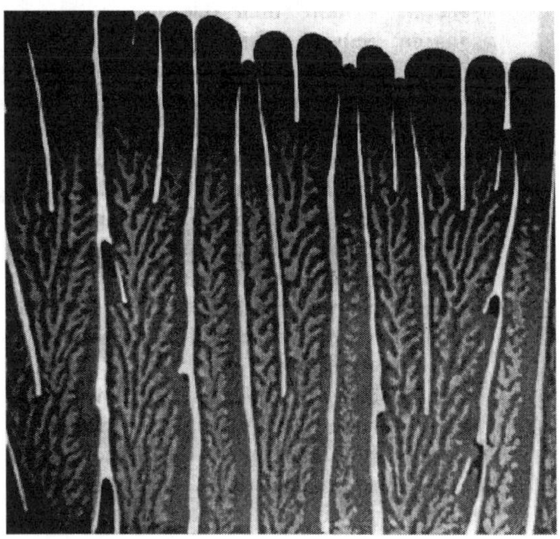

Fig. 7. — Density variations within branches of a colony of $\mathcal{T}$ morphotype. Optical microscope observations $\times 50$.

At very low agar concentrations (below 0.5%) the bacteria swim inside the agar and not on its surface. Between 0.5% and 1% agar concentration some of the bacteria move on the surface and some inside the agar.

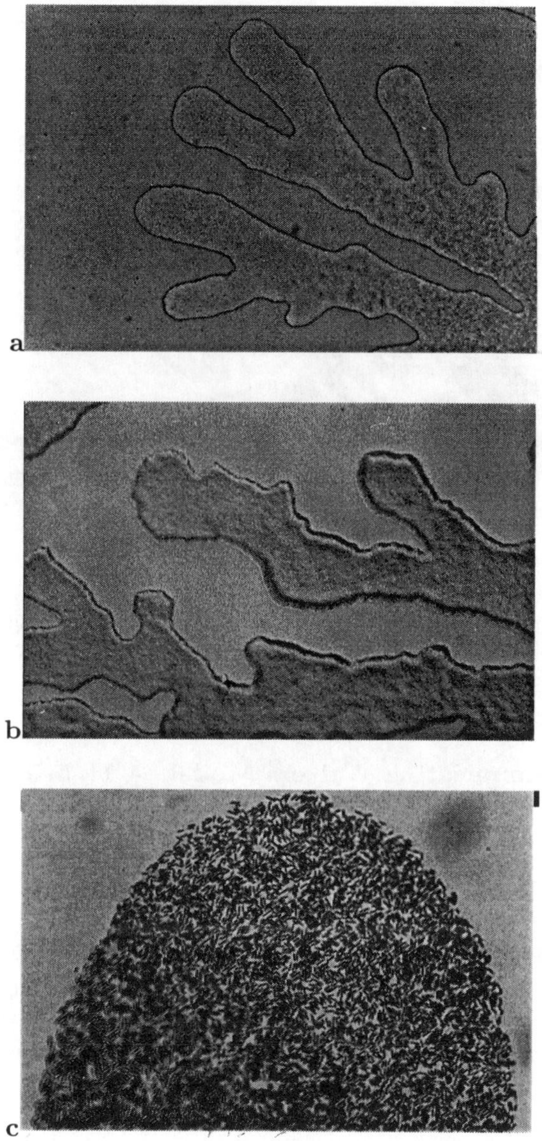

Fig. 8. — Closer look on branches of a colony. a) $\times 50$ magnification shows the sharp boundaries of the branches. The width of the boundary is in the order of micron. b) Nomarsky (polarized light) microscopy shows the hight of the branches and their envelope. What is actually seen is the layer of lubrication fluid, not the bacteria. If the colony is left to dry, the same observation technique reveals a layer of bacteria – thiner layer with more rugged surface. c) $\times 500$ magnification shows the bacteria inside a branch. Each bar is a single bacterium. There are no bacteria outside the branch.

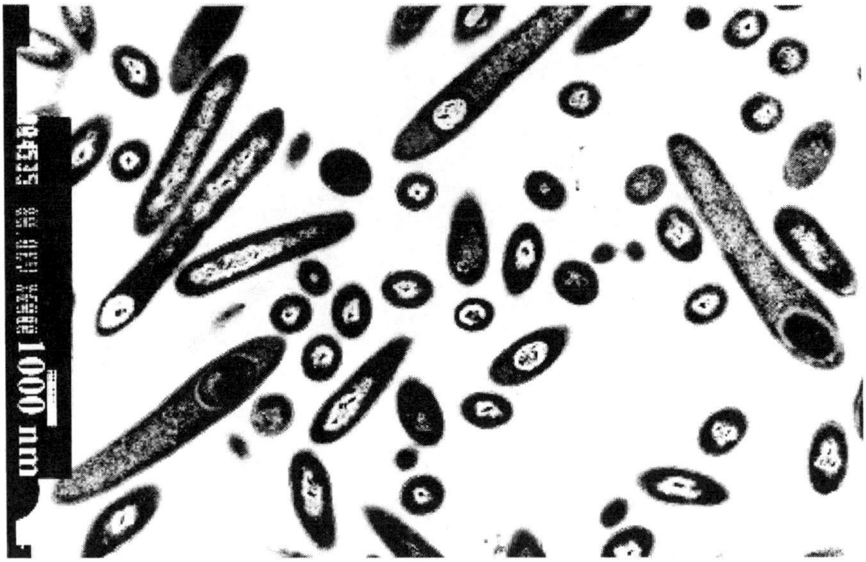

Fig. 9. — Electron microscope observation of $\mathcal{T}$ bacteria. Round or oval shapes with a bright center are spores. Elongated shapes are living cells. The cells engulfing oval shapes are pre-spores.

2.3. The Communicating Walkers Model: A Hybrid Model

The Communicating Walkers model [25] was inspired by the diffusion-transition scheme used to study solidification from supersaturated solutions [64–66]. The former is a hybridization of the "continuous" and "atomistic" approaches used in the study of non-living systems. The diffusion of the chemicals is handled by solving a continuous diffusion equation (including sources and sinks) on a tridiagonal lattice with a lattice constant a_0. The bacterial cells are represented by walkers allowing a more detailed description. In a typical experiment there are $10^9 - 10^{10}$ cells in a petri-dish at the end of the growth. Hence it is impractical to incorporate into the model each and every cell. Instead, each of the walkers represents about $10^4 - 10^5$ cells so that we work with $10^4 - 10^6$ walkers in one numerical "experiment".

The walkers perform an off-lattice random walk on a plane within an envelope representing the boundary of the wetting fluid. This envelope is defined on the same triangular lattice where the diffusion equations are solved. To incorporate the swimming of the bacteria into the model, at each time step each of the active walkers (motile and metabolizing, as described below) moves a step of size $d < a_0$ at a random angle Θ. Starting from location $\vec{r_i}$, it attempts to move to a new location $\vec{r_i'}$ given by:

$$\vec{r_i'} = \vec{r_i} + d(\cos\Theta; \sin\Theta). \tag{3}$$

If $\vec{r_i'}$ is outside the envelope, the walker does not move. A counter on the segment of the envelope which would have been crossed by the movement $\vec{r_i} \to \vec{r_i'}$ is increased by one. When the segment counter reaches a specified number of hits N_c, the envelope propagates one lattice step and an additional lattice cell is added to the colony. This requirement of N_c hits represent the colony propagation through wetting of unoccupied areas by the bacteria. Note that N_c is related to the agar dryness, as more wetting fluid must be produced (more "collisions" are needed) to push the envelope on a harder substrate.

Motivated by the presence of a maximal growth rate of the bacteria even for optimal conditions, each walker in the model consumes food at a constant rate Ω_c if sufficient food is available. We represent the metabolic state of the i-th walker by an "internal energy" E_i. The rate of change of the internal energy is given by

$$\frac{\mathrm{d}E_i}{\mathrm{d}t} = \kappa C_{\text{consumed}} - \frac{E_m}{\tau_{\mathrm{R}}}, \tag{4}$$

where κ is a conversion factor from food to internal energy ($\kappa \cong 5 \times 10^3$ cal/g) and E_m represent the total energy loss for all processes over the reproduction time τ_{R}, excluding energy loss for cell division. C_{consumed} is $C_{\text{consumed}} \equiv \min(\Omega_{\mathrm{C}}, \Omega'_{\mathrm{C}})$, where Ω'_{C} is the maximal rate of food consumption as limited by the locally available food [67]. When sufficient food is available, E_i increases until it reaches a threshold energy. Upon reaching this threshold, the walker divides into two. When a walker is starved for long interval of time, E_i drops to zero and the walker "freezes". This "freezing" represents entering a pre-spore state (starting the process of sporulation, see Sect. 4.1.2).

We represent the diffusion of nutrients by solving the diffusion equation for a single agent whose concentration is denoted by $n(\vec{r}, t)$:

$$\frac{\partial n}{\partial t} = D_n \nabla^2 C - bC_{\text{consumed}}, \tag{5}$$

where the last term includes the consumption of food by the walkers (b is their density). The equation is solved on the tridiagonal lattice. The simulations are started with inoculum of walkers at the center and a uniform distribution of the nutrient.

Results of numerical simulations of the model are shown in Figure 10. As in the case of real bacterial colonies, the patterns are compact at high nutrient levels and become fractal with decreasing food level. For a given nutrient level, the patterns are more ramified as the agar concentration increases. The results shown in Figure 10 do capture some features of the experimentally observed patterns. However, at this stage the model does not account for some critical features, such as the ability of the bacteria to develop organized patterns at very low nutrient levels. Ben-Jacob *et al.* [2, 47, 68, 69] suggested that chemotactic signaling must be included in the model to produce these features (see Sect. 4).

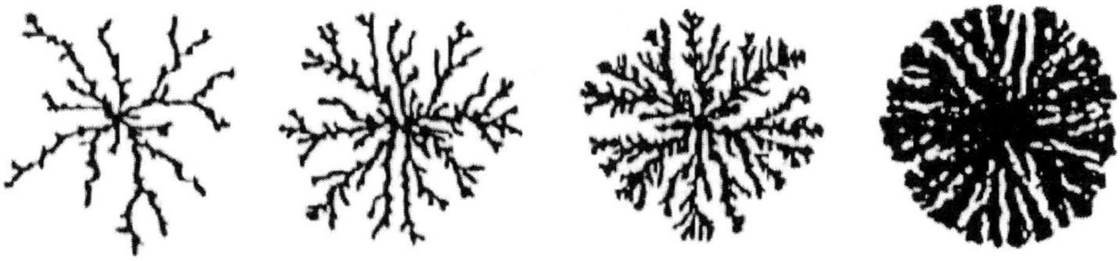

Fig. 10. — Colonial pattern of the Communicating Walkers model. Here $N_c = 20$ and n_0 is 6, 8, 10 and 30 from left to right respectively.

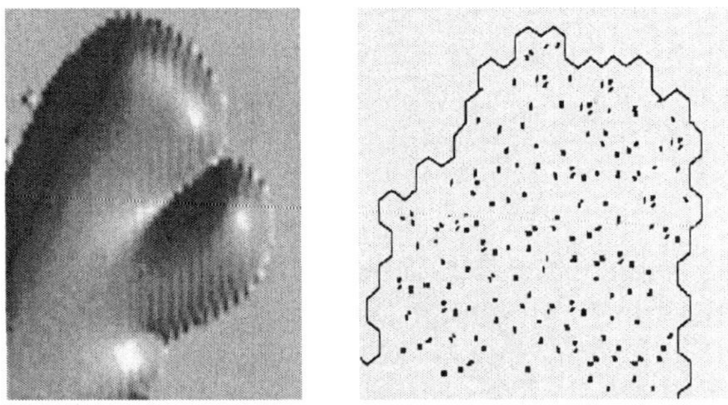

Fig. 11. — Closer look on simulated colonies. On the right: a tip of a branch in the Communicating Walkers model. The boundary of the branch and walkers can be seen. On the left: lubrication at a tip of a branch in the Lubricating Bacteria model.

2.4. The Lubricating Bacteria Model

The Lubricating Bacteria model is a reaction-diffusion model for the bacterial colonies [37]. This model includes four coupled fields. One field describes the bacterial density $b(\vec{x}, t)$, the second describe the height of lubrication layer in which the bacteria swim $l(\vec{x}, t)$, third field describes the nutrients $n(\vec{x}, t)$ and the fourth field is the stationary bacteria that "freeze" and begin to sporulate $s(\vec{x}, t)$ (see Sect. 4.1.2).

The dynamics of the bacterial field b consists of two parts; a diffusion term which is coupled to the lubrication field and a reaction part which contains terms for reproduction and death. Following the same arguments presented for the Communicating Walkers model, we get a reaction term of the form $(\kappa b \min(\Omega_C, n) - E_m b/\tau_R)$. Assuming that nutrient is always the factor limiting the bacterial growth we get, upon rescaling, the growth term $bn - \mu b$ (μ constant).

We now turn to the bacterial movement. In a uniform layer of liquid, bacterial swimming is a random walk with variable step length and can be approximated by diffusion. The layer of lubrication is not uniform, and its height affects the bacterial movement. An increase in the amount of lubrication decreases the friction between the bacteria and the agar surface. The term "friction" is used here in a very loose manner to represent the total effect of any force or process that slows down the bacteria. As the bacterial motion is over-damped, the local speed of the bacteria is proportional to the self-generated propulsion force divided by the friction. It can be shown that variation of the speed leads to variation of the diffusion coefficient, with the diffusion coefficient proportional to the speed to the power of two. We assume that the friction is inversely related to the local lubrication height through some power law: friction$\sim l^{\gamma}$ and $\gamma < 0$. The bacterial flux is:

$$\vec{J}_{\mathrm{b}} = -D_{\mathrm{b}} l^{-2\gamma} \nabla b. \tag{6}$$

The lubrication field l is the local height of the lubrication fluid on the agar surface. Its dynamics is given by:

$$\frac{\partial l}{\partial t} = -\nabla \cdot \vec{J}_{\mathrm{l}} + \Gamma b n (l_{\mathrm{max}} - l) - \lambda l \tag{7}$$

where $\vec{J}_{\mathrm{l}}$ is the fluid flux (to be discussed), Γ is the production rate and λ is the absorption rate of the fluid by the agar. λ is inversely related to the agar dryness.

The fluid production is assumed to depend on the bacterial density. As the production of lubrication probably demands substantial energy, it also depends on the nutrient's level. We assume the simplest case where the production depends linearly on the concentrations of both the bacteria and the nutrients.

The lubrication fluid flows by diffusion and by convection caused by bacterial motion. A simple description of the convection is that as each bacterium moves, it drags along with it the fluid surrounding it.

$$\vec{J}_{\mathrm{l}} = -D_{\mathrm{l}} l^{\eta} \nabla l + j \vec{J}_{\mathrm{b}} \tag{8}$$

where D_{l} is a lubrication diffusion constant, $\vec{J}_{\mathrm{b}}$ is the bacterial flux and j is the amount of fluid dragged by each bacterium. The diffusion term of the fluid depends on the height of the fluid to the power $\eta > 0$ (the nonlinearity in the diffusion of the lubrication, a very complex fluid, is motivated by hydrodynamics of simple fluids). The nonlinearity causes the fluid to have a sharp boundary at the front of the colony, as is observed in the bacterial colonies (Fig. 11).

The complete model for the bacterial colony is:

$$\begin{aligned}
\frac{\partial b}{\partial t} &= D_{\mathrm{b}} \nabla \cdot (l^{-2\gamma} \nabla b) + bn - \mu b \\
\frac{\partial n}{\partial t} &= D_n \nabla^2 n - bn \\
\frac{\partial l}{\partial t} &= \nabla \cdot (D_{\mathrm{l}} l^{\eta} \nabla l + j D_{\mathrm{b}} l^{-2\gamma} \nabla b) + \Gamma b n (l_{\mathrm{max}} - l) - \lambda l \\
\frac{\partial s}{\partial t} &= \mu b.
\end{aligned} \tag{9}$$

Fig. 12. — Growth patterns of the Lubricating Bacteria model, for different values of initial nutrient level n_0. The apparent (though weak) 6-fold anisotropy is due to the underlying tridiagonal lattice.

The second term in the equation for b represents the reproduction of the bacteria. The reproduction depends on the local amount of nutrient and it reduces this amount. The third term in the equation for b represents the process of bacterial "freezing". For the initial condition, we set n to have uniform distribution of level n_0, b to have compact support at the center, and the other fields to be zero everywhere.

Preliminary results show that the model can reproduce branching patterns, similar to the bacterial colonies (Fig. 12). At low values of absorption rate, the model exhibits dense fingers. At higher absorption rates the model exhibits finer branches. We also obtain finer branches if we change other parameters that effectively decrease the amount of lubrication. We can relate these conditions to high agar concentration.

We can now check the effect of bacterial discreteness on the observed colonial patterns. Following Kessler and Levine [70], we introduce the discreteness of the system into the continuous model by repressing the growth term at low bacterial densities ("half a bacterium cannot reproduce"). The growth term is multiplied by a Heaviside step function $\Theta(b - \beta)$, where β is the threshold density for growth. In Figure 13 we show the effect of various values of β on the pattern. High cutoff values make the model more sensitive to the implicit anisotropy of the underlying tridiagonal lattice used in the simulation. The result is dendritic growth with marked 6-fold symmetry of the pattern. Increased values of cutoff also decrease the maximal values of b reached in the simulations (and the total area occupied by the colony).

The reason for the pattern turning dendritic is as follows: the difference between tip-splitting growth and dendritic growth is the relative strength of the effect of anisotropy and an effective surface tension [1]. In the Lubricating Bacteria model there is no explicit anisotropy and no explicit surface tension. The implicit anisotropy is related to the underlying lattice, and the effective surface tension is related to the width of the front. The cutoff prevents the

Fig. 13. — The effect of a cutoff on the growth patterns in the Lubricating Bacteria model. Aside from the cutoff, the conditions are the same as in the middle pattern of Figure 5, where the maximal value of b was about 0.025. The values of the cutoff β are, from left to right, 10^{-6}, 10^{-5} and 3×10^{-5}. The 6-fold symmetry is due to anisotropy of the underlying lattice which is enhanced by the cutoff.

growth at the outer parts of the front, thus making it thinner, reduces the effective surface tension and enables the implicit anisotropy to express itself.

We stress that it is possible to find a range of parameters in which the growth patterns resemble the bacterial patterns, in spite a high value of cutoff. Yet the cutoff does not improve the model in any sense, it introduces an additional parameter, and it slows the numerical simulation. We believe that the well-defined boundary makes the cutoff (as a representation of the bacterial discreteness) unnecessary.

2.5. The Non-Linear Diffusion Model

It is possible to introduce a simplified model, where the fluid field is not included, and is replaced by a density-dependent diffusion coefficient for the bacteria $D_b \sim b^k$ [71, 72]. Such a term can be justified by a few assumptions about the dynamics at low bacterial and low lubrication density:

- the production of lubricant is proportional to the bacterial density to the power $\alpha > 0$ ($\alpha = 1$ in the previous mode);
- there is a sink in the equation for the time evolution of the lubrication field, *e.g.* absorption of the lubricant into the agar. This sink is proportional to the lubrication density to the power $\beta > 0$ ($\beta = 1$ in the previous mode);
- over the bacterial length scale, the two processes above are much faster than the diffusion process, so the lubrication density is proportional to the bacterial density to the power of β/α;
- the friction is proportional to the lubrication density to the power $\gamma < 0$.

Given the above assumptions, the lubrication field can be removed from the dynamics and be replaced by a density dependent diffusion coefficient. This coefficient is proportional to the bacterial density to the power $k \equiv -2\gamma\beta/\alpha > 0$

Fig. 14. — Growth patterns of the Kitsunezaki model, for different values of initial nutrient level n_0. Parameters are: $D_0 = 0.1$, $k = 1$, $\mu = 0.15$. The apparent 6-fold symmetry is due to the underlying tridiagonal lattice.

A model of this type is offered by Kitsunezaki [34]:

$$\frac{\partial b}{\partial t} = \nabla(D_0 b^k \nabla b) + nb - \mu b \tag{10}$$

$$\frac{\partial n}{\partial t} = \nabla^2 n - bn \tag{11}$$

$$\frac{\partial s}{\partial t} = \mu b. \tag{12}$$

For $k > 0$ the 1D model gives rise to a front "wall", with compact support (*i.e.* $b = 0$ outside a finite domain). For $k > 1$ this wall has an infinite slope. The model exhibits branching patterns for suitable parameter values and initial conditions, as depicted in Figure 14. Increasing the initial nutrient level makes the colonies more dense, similarly to what happens in the other models.

As in the Lubricating Bacteria model, adding the "Kessler and Levine correction" to the model, *i.e.* making the growth term disappear for $b < \beta$, does not seem to make the patterns "better", or closer to the experimental observations (Fig. 15). The apparent increased sensitivity to the implicit anisotropy results from the narrowed front, which decreases the effective surface tension.

3. CHIRAL BRANCHING PATTERNS

3.1. Observations

Chiral asymmetry (first discovered by Louis Pasteur) exists in a whole range of scales, from subatomic particles through human beings to galaxies, and seems to have played an important role in the evolution of living systems [73, 74]. Bacteria display various chiral properties. Mendelson *et al.* [75–78] showed that long cells of *B. subtilis* can grow in helices, in which the cells form long strings that twist around each other. They have shown also that the chiral

Fig. 15. — Growth patterns of the Kitsunezaki model, with a cutoff correction. Cutoff value $\beta = 0.1$, all other parameters as in Figure 14, right pattern. The apparent 6-fold symmetry is due to the underlying tridiagonal lattice.

characteristics affect the structure of the colony. Ben-Jacob *et al.* [11, 14, 23] have found yet another chiral property – the strong chirality exhibited by the $\mathcal{C}$ morphotype. Here, the flagella handedness acts as a microscopic perturbation which is amplified by the diffusive instability, leading to the observed macroscopic chirality. This appears to be analogous to the manner in which crystalline anisotropy leads to the observed symmetry of snowflakes [1]; more about this later.

3.1.1. *A Closer Look at the Patterns*

$\mathcal{C}$ morphotype exhibits a wealth of different patterns according to the growth conditions (Fig. 16). As for $\mathcal{T}$ morphotype, the patterns are generally compact at high peptone levels and become ramified (fractal) at low peptone levels. At very high peptone levels and high agar concentration, $\mathcal{C}$ morphotype conceals its chiral nature and exhibits branching growth similar to that of $\mathcal{T}$ morphotype.

Below 0.5% agar concentration the $\mathcal{C}$ morphotype exhibits compact growth with density variations. These patterns are almost indistinguishable from those developed by the $\mathcal{T}$ morphotype. In the range of 0.4%–0.6% agar concentration the $\mathcal{C}$ morphotype exhibits its most complex patterns (Fig. 17). Surprisingly, these patterns are composed of chiral branches of both left and right handedness. Microscopic observations reveal that part of the growth is on top of the agar surface while in other parts the growth is in the agar. Our model of the chiral growth explains that indeed growth on top of the surface and in the agar should lead to opposite handedness.

Optical microscope observations indicate that during growth of strong chirality the cells move within a well defined envelope. The cells are long relative to those of $\mathcal{T}$ morphotype, and the movement appears correlated in orientation (Fig. 18). Each branch tip maintains its shape, and at the same time the tips keep twisting with specific handedness while propagating. Electron microscope observations do not reveal any chiral structure on the cellular membrane [24].

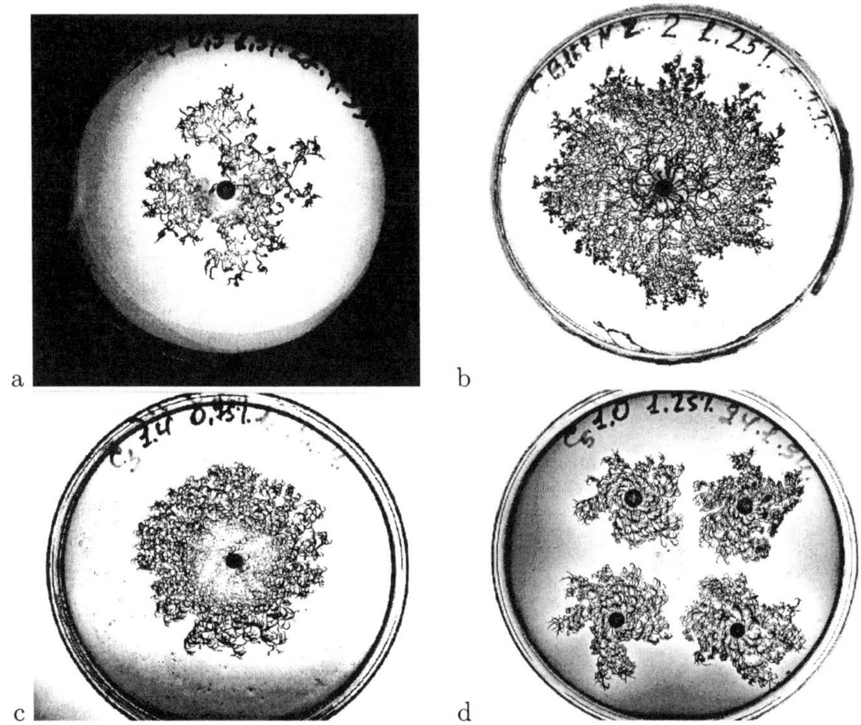

Fig. 16. — Patterns exhibited by the $\mathcal{C}$ morphotype for different growth conditions. a) Thin disordered twisted branches at 0.5 g/l peptone level and 1.5% agar concentration. b) Thin branches, all twisted with the same handedness. at 2 g/l peptone level and 1.25% agar concentration. c) Pattern similar to (b) but on softer agar: 1.4 g/l peptone level and 0.75% agar concentration. d) Four inocula on the same plate, conditions of 1 g/l peptone level and 1.25% agar concentration.

3.2. The Communicating Spinors Model

The Communicating Spinors Model was developed to explain the chirality of the $\mathcal{C}$ morphotype colonies. Our purpose is to show that the flagella handedness, while acting as a singular perturbation, leads to the observed chirality. It does so in the same manner in which crystalline anisotropy leads to the observed symmetry of snowflakes [1].

It is known [57, 79, 80] that flagella have specific handedness. Ben-Jacob *et al.* [23] proposed that the latter is the origin of the observed chirality. In a fluid (which is the state in most experimental setups), as the flagella unfold, the cell tumbles and ends up at a new random angle relative to the original one. The situation changes for quasi 2D motion – motion in a "lubrication" layer thinner then the cellular length. We assume that in this case, of rotation in a

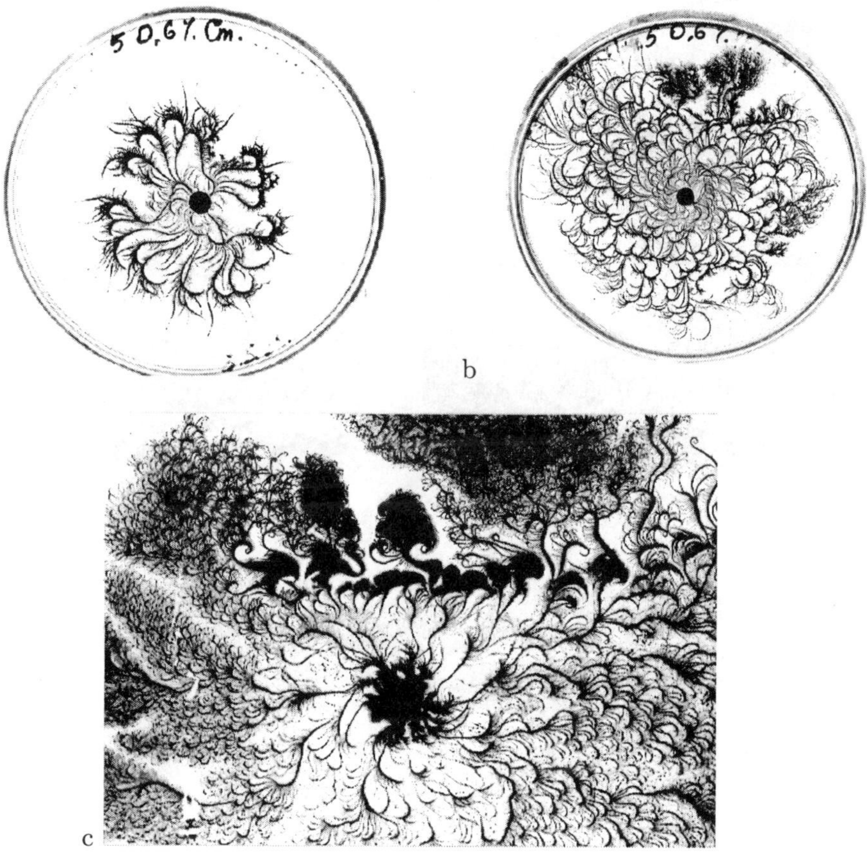

a b c

Fig. 17. — In agar soft enough for the bacteria to swim in it, the branches lose the one-side handedness they have on harder agar. The two colonies of a) and b) are of 5 g/l peptone level and 0.6% agar concentration. The two patterns are of two stains of the C morphotype, strains whose patterns are indistinguishable on harder agar. c) Closer look ($\times 10$ magnification) on a colony grown at 8 g/l peptone level and 0.6% agar concentration.

plane, the tumbling has a well defined handedness of rotation. Such handedness requires, along with the chirality of the flagella, the cells' ability to distinguish up from down. The growth in an upside- down petri- dish shows the same chirality. Therefore, we think that the determination of up *vs.* down is done either *via* the vertical gradient of the nutrient concentration or *via* the vertical gradient of signaling materials inside the substrate or *via* the friction of the cells with the surface of the agar. The latter is the most probable alternative; soft enough agar enables the bacteria to swim *below* the surface of the agar which leads to many changes in the patterns, including reversing the bias of the branches.

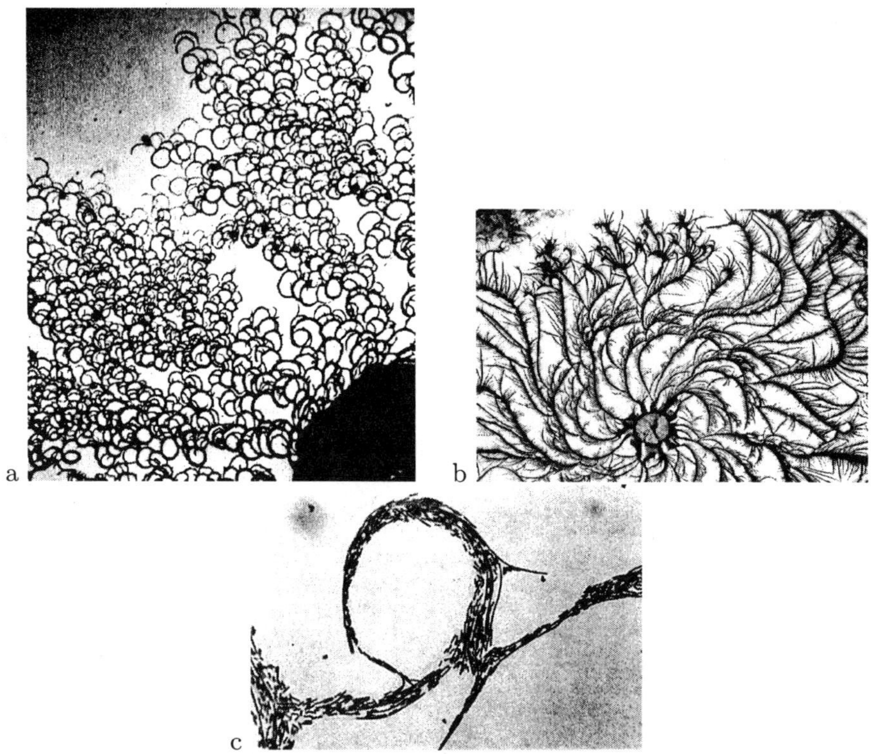

Fig. 18. — Optical microscope observations of branches of $\mathcal{C}$ morphotype colony. a) $\times 20$ magnification of a colony at 1.6 g/l peptone level and 0.75% agar concentration, the anti-clockwise twist of the thin branches is apparent. The curvature of the branches is almost constant throughout the growth. b) $\times 10$ magnification of a colony at 4 g/l peptone level and 0.6% agar concentration at branches are not thin, but have a feathery structure. The curvature of the branches varies, but it seems that at any given stage of growth the curvature is similar in all branches. That is, the curvature is a function of colonial growth. c) $\times 500$ magnification of a colony at 1.6 g/l peptone level and 0.75% agar concentration. Each line is a bacterium. the bacteria are long ($5 - 50$ μm) and mostly ordered.

To cause the chirality observed on semi-solid agar, the rotation of tumbling must be, on average, less than 90° and relative to a specific direction. Co-alignment (orientational interaction) limits the rotation. We further assume that the rotation is relative to the local mean orientation of the surrounding cells.

To test the above idea, we included the additional assumed features in the Communicating Walkers model [25], changing it to a "Communicating Spinors" model (as the particles in the new model have an orientation and move in quasi-1D random walk). The Communicating Walkers model [25] was inspired by the diffusion-transition scheme used to study solidification from supersaturated solutions [64–66]. The former is a hybridization of the "continuous" and "atomistic" approaches used in the study of non-living systems. Ben-Jacob *et al.* have presented in the past a version of the Communicating Spinors model for the chiral growth [23]. The model we present here is closely related to a previous model of the chiral growth, but it differs in two features. The first is the orientation field (see below), which was discontinuous piecewise constant and in this model it is continuous piecewise linear. The second difference is the definition of a single run of a spinor (the stretch between two tumbling events), which was defined as one run per one time unit (*i.e.* each step is a run) and now is defined as variable number of steps in the same direction.

The representation of bacteria as spinors allows for a close relation to the bacterial properties. The bacterial cells are represented by spinors allowing a more detailed description. At the end of the growth of a typical experiment there are $10^8 - 10^9$ bacterial cells in the petri- dish. Thus it is impractical to incorporate into the model each and every cell. Instead, each of the spinors represents about $10 - 1000$ cells, so that we work with $10^4 - 10^6$ spinors in one numerical "experiment".

Each spinor has a position $\vec{r}_i$, direction θ_i (an angle) and a metabolic state ("internal energy") E_i. The spinors perform an off-lattice constrained random walk on a plane within an envelope representing the boundary of the wetting fluid. This envelope is defined on the same tridiagonal lattice where the diffusion equations are solved. To incorporate the swimming of the bacteria into the model, at each time step each of the active spinors (motile and metabolizing, as described below) recalculate its direction θ'_i and moves a step of size d to this direction.

The direction in which each spinor moves is determined in two steps; first the spinor decides whether it should continue the current run, that is to continue in the same direction $\theta'_i = \theta_i$. In the basic version of the model (see Sect. 4.3 for extension of the model) the decision is random with a specific probability p to continue the run. The resulting runs have geometric distribution of lengths, with mean run length of d/p. Once a spinor decides to change direction, the new direction θ'_i is derived from the spinor's previous direction by

$$\theta'_i = P(\theta_i, \Phi(\vec{r}_i)) + Ch + \xi + \omega \tag{13}$$

Ch and ξ represent the new features of rotation due to tumbling. Ch is a fixed part of the rotation and ξ is a stochastic part, chosen uniformly from an interval $[-\eta, \eta]$ (η constant). ω is an orientation term that takes, with equal probabilities, one of the values 0 (forward direction) or π (backward direction). This orientation term gives the spinors their name, as it make their re-orientation invariant to forward or backward direction. $\Phi(\vec{r}_i)$ is the local mean orientation in the neighborhood of $\vec{r}_i$. P is a projection function that represents the orientational interaction which acts on each spinor to orient θ_i

along the direction $\Phi(\vec{r}_i)$. P is defined by

$$P(\alpha, \beta) = \alpha + (\beta - \alpha). \qquad (14)$$

Once oriented, the spinor advances a step d in the direction θ'_i, and the new location $\vec{r'}_i$ is given by:

$$\vec{r'}_i = \vec{r}_i + d\,(\cos\theta'_i, \sin\theta'_i)\,. \qquad (15)$$

The movement is confined within an envelope which is defined on the tridiagonal lattice. The step is not performed if $\vec{r'}_i$ is outside the envelope. Whenever this is the case, a counter on the appropriate segment of the envelope is increased by one. When a segment counter reaches N_c, the envelope advances one lattice step and a new lattice cell is occupied. Note that the spinor's direction is not reset upon hitting the envelope, thus it might "bang its head" against the envelope time and time again. The requirement of N_c hits represent the colony propagation through wetting of unoccupied areas by the bacteria. Note that N_c is related to the agar dryness, as more wetting fluid must be produced (more "collisions" are needed) to push the envelope on a harder substrate.

Next we specify the mean orientation field Φ. To do so, we assume that each lattice cell (hexagonal unit area) is assigned one value of $\Phi(\vec{r})$, representing the average orientation of the spinors in the local neighborhood of the center of the cell. The value of Φ is set when a new lattice cell is first occupied by the advancement of the envelope, and then remains constant. We set the value of $\Phi(\vec{r})$ to be equal to the average over the orientations of the N_c attempted steps that led to the occupation of the new lattice cell. The value of Φ in any given point inside the colony is found by linear interpolation between the three neighboring centers of cells. Clearly, the model described above is a simplified picture of the bacterial movement. For example, a more realistic model will include an equation describing the time evolution of Φ. However, the simplified model is sufficient to demonstrate the formation of chiral patterns. A more elaborate model will be presented elsewhere [81].

Motivated by the presence of a maximal growth rate of the bacteria even for optimal conditions, each spinor in the model consumes food at a constant rate Ω_c if sufficient food is available. We represent the metabolic state of the i-th spinor by an "internal energy" E_i. The rate of change of the internal energy is given by

$$\frac{\mathrm{d}E_i}{\mathrm{d}t} = \kappa C_{\text{consumed}} - \frac{E_{\text{m}}}{\tau_{\text{R}}}, \qquad (16)$$

where κ is a conversion factor from food to internal energy ($\kappa \cong 5 \times 10^3$ cal/g) and E_{m} represent the total energy loss for all processes over the reproduction time τ_{R}, excluding energy loss for cell division. C_{consumed} is $C_{\text{consumed}} \equiv \min(\Omega_{\text{C}}, \Omega'_{\text{C}})$, where Ω'_{C} is the maximal rate of food consumption as limited by the locally available food (Sect. 2). When sufficient food is available, E_i increases until it reaches a threshold energy. Upon reaching this threshold, the spinor divides into two. When a spinor is starved for long interval of time, E_i

drops to zero and the spinor "freezes". This "freezing" represents entering a pre-spore state (starting the process of sporulation, see Sect. 4.3).

We represent the diffusion of nutrients by solving the diffusion equation for a single agent whose concentration is denoted by $n(\vec{x}, t)$:

$$\frac{\partial n}{\partial t} = D_n \nabla^2 C - bC_{\text{consumed}}, \tag{17}$$

where the last term includes the consumption of food by the spinors (b is their density). The equation is solved on the same tridiagonal lattice on which the envelope is defined. The length constant of the lattice a_0 must be larger than the size of the spinors' step d. The simulations are started with inoculum of spinors at the center and a uniform distribution of the nutrient. Both Φ and the spinors at the inoculum are given uniformly distributed random directions.

Results of the numerical simulations of the model are shown in Figure 19. These results do capture some important features of the observed patterns: the microscopic twist Ch leads to a chiral morphology on the macroscopic level. The growth is *via* stable tips, all of which twist with the same handedness and emit side-branches. The dynamics of the side-branches emission in the time evolution of the model is similar to the observed dynamics.

For large noise strength η the chiral nature of the pattern gives way to a branching pattern (Fig. 20). This provides a plausible explanation for the branching patterns produced by $\mathcal{C}$ morphotype grown on high peptone levels, as the cells are shorter when grown on a rich substrate. The orientation interaction is weaker for shorter cells, hence the noise is stronger.

4. CHEMOTAXIS AND CHEMOTACTIC SIGNALING

4.1. Biological Background – Chemotaxis in Swimming Bacteria

Chemotaxis means changes in the movement of the bacteria in response to a gradient of certain chemical field [58,82–86]. The movement is biased along the gradient either in the gradient direction or in the opposite direction. Usually chemotactic response means a response to an externally produced field, like in the case of chemotaxis towards food. However, the chemotactic response can be also to a field produced directly or indirectly by the bacterial cells. We will refer to this case as chemotactic signaling. The bacteria sense the local concentration R of a chemical by membrane receptors binding the chemical's molecules [82, 84]. It is crucial to note that when estimating gradients of chemicals, the bacterial cells actually measure changes in the receptors' occupancy and not in the concentration itself. When put in continuous equations [37, 87], this indirect measurement translates to measuring the gradient

$$\frac{\partial}{\partial x} \frac{R}{(K+R)} = \frac{K}{(K+R)^2} \frac{\partial R}{\partial x} \tag{18}$$

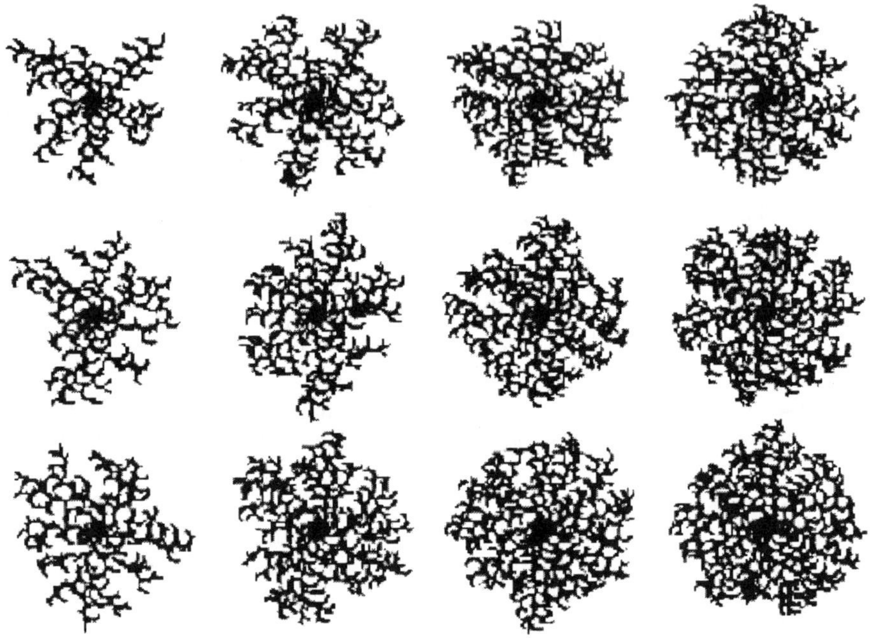

Fig. 19. — A morphology diagram of the Communicating Spinors model for various values of N_c and initial n concentration n_0. $Ch = 6°$, $\eta = 3°$, $d = 0.2$, $p = 0.5$.

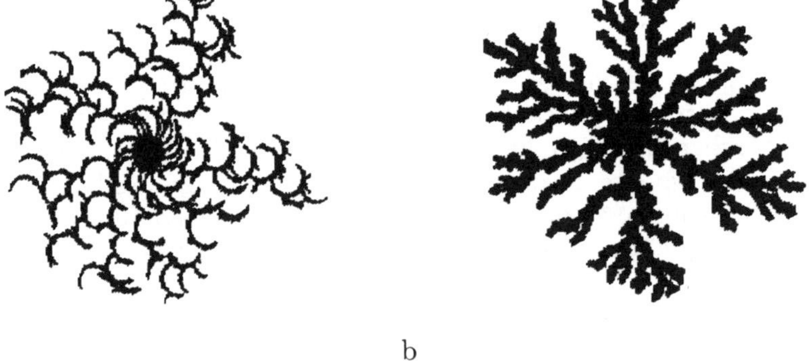

a b

Fig. 20. — When the noise η is increases to $\eta = 180°$ the tumbling of the spinors becomes unrestricted. Their movement becomes like that of the $\mathcal{T}$ bacteria and accordingly the simulated colonial pattern is like that of $\mathcal{T}$ morphotype. On the left $\eta = 3°$, on the right $\eta = 180°$.

where K is a constant whose value depends on the receptors' affinity, the speed in which the bacterium processes the signal from the receptor, etc. This means that the chemical gradient times a factor $K/(K+R)^2$ is measured, and it is known as the "receptor law" [87].

In a continuous model, we incorporate the effect of chemotaxis by introducing a chemotactic flux $\vec{J}_{\text{chem}}$:

$$\vec{J}_{\text{chem}} \equiv \zeta(\sigma)\chi(R)\nabla R \tag{19}$$

$\chi(R)\nabla R$ is the gradient sensed by the bacteria (with $\chi(R)$ having the units of 1 over chemical's concentration). $\chi(R)$ is usually taken to be either constant or the "receptor law". $\zeta(\sigma)$ is the bacterial response to the sensed gradient (having the same units as a diffusion coefficient times the units of the bacterial density σ). It is positive for attractive chemotaxis and negative for repulsive chemotaxis.

Ben-Jacob *et al.* argued [2, 47, 68, 69] that for the colonial adaptive self-organization the bacteria employ three kinds of chemotactic responses, each dominant in different regime of the morphology diagram (the claim was made for $\mathcal{T}$ morphotype, but the same holds for their relatives $\mathcal{C}$ morphotype). One response is the food chemotaxis mentioned above. It is expected to be dominant for only a range of nutrient levels (see the "receptor law" below). The two other kinds of chemotactic responses are chemotactic signaling. One is repulsive chemotactic signaling, a long-range signal. The repelling chemical is secreted by starved bacteria at the inner parts of the colony. The second signal is a short-range attracting signal. The length scale of each signal is determined by the diffusion constant of the chemical agent and the rate of its spontaneous decomposition.

4.1.1. *Amplification of Diffusive Instability Due to Nutrients Chemotaxis*

In non-living systems, more ramified patterns (lower fractal dimension) are observed for lower growth velocity. Based on growth velocity as function of nutrient level and based on growth dynamics, Ben-Jacob *et al.* [25] concluded that in the case of bacterial colonies there is a need for a mechanism that can both increase the growth velocity and maintain, or even decrease, the fractal dimension. They suggested food chemotaxis to be the required mechanism. It provides an outward drift to the cellular movements; thus, it should increase the rate of envelope propagation. At the same time, being a response to an external field it should also amplify the basic diffusion instability of the nutrient field. Hence, it can support faster growth velocity together with a ramified pattern of low fractal dimension.

4.1.2. *Repulsive Chemotactic Signaling*

We focus now on the formation of the fine radial branching patterns at low nutrient levels. From the study of non-living systems, it is known that in the same manner that an external diffusion field leads to the diffusion instability, an internal diffusion field will stabilize the growth. It is natural to assume that some sort of chemotactic agent produces such a field. To regulate the organization of the branches, it must be a long-range signal. To result in radial branches it must be a repulsive chemical produced by bacteria at the inner parts of the colony. The most probable candidates are the bacteria entering a pre-spore stage.

If nutrient is deficient for a long enough time, bacterial cells may enter a special stationary state – a state of a spore – which enables them to survive much longer without food. While the spores themselves do not emit any chemicals (as they have no metabolism), the pre-spores (sporulating cells) do not move and emit a very wide range of waste materials, some of which unique to the sporulating cell. These emitted chemicals might be used by other bacteria as a signal carrying information about the conditions at the location of the pre-spores. Ben-Jacob *et al.* [24,25,47] suggested that such materials are repelling the bacteria ("repulsive chemotactic signaling") as if they escape a dangerous location.

4.1.3. Attractive Chemotactic Signaling

The characteristic length of both types of chemotaxis fields used above – food chemotaxis and repulsive chemotactic signaling – is longer then the typical branch width. The length scale of the 3-dimensional structures (Fig. 7) indicates the existence of a dynamical process with a characteristic length shorter than the branch width. The accumulation of cells in the suggests the existence of a short range self-attracting mechanism. Motivated by work on attractive chemotactic signaling in *Escherichia coli* [22,88–90], such tactics was tested as a means of producing the 3-dimensional structures. Ben-Jacob *et al.* [2,47,68,69] proposed that the transition between fractal-like branched patterns and dense fingering patterns results from an interplay between the three kinds of chemotactic response. As was said above, this interplay and its dependence on the growth parameters is a complex one and needs more study.

4.2. Modeling the Effect of Chemotaxis on Branching Growth

In a continuous model, we incorporate the effect of chemotaxis by introducing a chemotactic flux $\vec{J}_{\text{chem}}$:

$$\vec{J}_{\text{chem}} \equiv \zeta(\sigma)\chi(r)\nabla r \tag{20}$$

$\chi(r)\nabla r$ is the gradient sensed by the cell (with $\chi(r)$ having the units of 1 over chemical's concentration). $\chi(r)$ is usually taken to be either constant or the "receptor law". $\zeta(\sigma)$ is the bacterial response to the sensed gradient (having the same units as a diffusion coefficient). In NLDm the bacterial diffusion is $D_b = D_0 b^k$, and the bacterial response to chemotaxis is $\zeta(b) = \zeta_0 b \left(D_0 b^k\right) = \zeta_0 D_0 b^{k+1}$. ζ_0 is a constant, positive for attractive chemotaxis and negative for repulsive chemotaxis.

Fig. 21. — The effect of chemotaxis on the growth in CWm. Growth conditions and model's parameters are the same in all patterns, excluding parameters related to chemotaxis. The different shades represent different growth times, and indicate the speed of colonial growth. In the center: a typical colony without chemotaxis. On the left: chemotaxis towards food is included in the model. The growth velocity is almost doubled, but the pattern is essentially unchanged by food chemotaxis. On the right: repulsive chemotactic signaling is included in the model. The pattern is of fine radial branches with circular envelope, like the bacterial colony presented in Figure 3d.

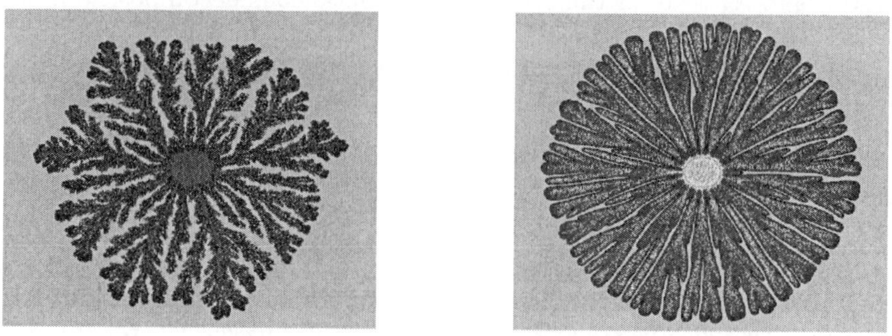

Fig. 22. — Growth patterns of NLDm with food chemotaxis (left) and repulsive chemotactic signaling (right) included. $\chi_{0\mathrm{f}} = 3$, $\chi_{0\mathrm{r}} = 1$, $D_\mathrm{r} = 1$, $\Gamma_\mathrm{r} = 0.25$, $\Omega_\mathrm{r} = 0$, $\Lambda_\mathrm{r} = 0.001$. Other parameters are the same as in Figure 14, middle pattern. The apparent 6-fold symmetry is due to the underlying tridiagonal lattice.

The hypothesis about chemotaxis was tested in CWm and in NLDm. In Figures 21 and 22 it is shown that as expected, the inclusion of food chemotaxis in both models led to a considerable increase of the growth velocity without significant change in the fractal dimension of the pattern.

The equation describing the dynamics of the chemorepellent contains terms for diffusion, production by pre-spores, decomposition by active bacteria and spontaneous decomposition:

$$\frac{\partial r}{\partial t} = D_\mathrm{r}\nabla^2 r + \Gamma_\mathrm{r} s - \Omega_\mathrm{r} b r - \lambda_\mathrm{r} r \tag{21}$$

where D_r is the diffusion coefficient of the chemorepellent, Γ_r is the emission rate of repellent by pre-spores, Ω_r is the decomposition rate of the repellent by active bacteria, and λ_r is the rate of self decomposition of the repellent. In CWm b and s are replaced by active and inactive walkers, respectively.

In Figures 21 and 22 the effect of repulsive chemotactic signaling is also shown. In the presence of repulsive chemotaxis the patterns in both models become much denser with a smooth circular envelope, while the branches are thinner and radially oriented.

Several mechanisms controlling the attractant production were proposed in [22, 88–90]. The simplest of which is a constant production of attractive material a by the active walkers, making the time evolution of a similar to that of r.

$$\frac{\partial a}{\partial t} = D_a \nabla^2 a + \Gamma_a b - \Omega_a a b - \lambda_a r \tag{22}$$

where b is the local density of active bacteria. The second term in the RHS represents the production of attractant. Unlike the repellent production it is done by the active bacteria. The values of the parameters D_a (diffusion coefficient), Γ_a (production rate of the attractant), Ω_a (rate of attractant segregation by bacteria) and λ_a (rate of the attractant's spontaneous decomposition) should be such that the effective length scale of the attraction will be no more than a branch's width. An example of numerical simulations when chemoattractant is included is shown in Figure 23. Clearly it does lead to the formation of 3D structures.

4.3. Modeling the Effect of Chemotaxis on Chiral Branching

The colonial patterns of $\mathcal{C}$ morphotype (*e.g.* Fig. 19) are rarely as ordered as the simulated patterns of the Communicating Spinors model. For example, the branches of the observed colonies usually have varying curvature. In the simulations of $\mathcal{C}$ morphotype shown in Figure 19 all the branches have a uniform curvature. One of the reasons for this difference is the simplifications taken during the model's development. A more elaborate model that we will present [81] will be a better description of the colony. However, some of the observed features can be explained in the context of the Communicating Spinors model. In some of the observed patterns (Fig. 17b), the curvature of the branches has a distinct relation to the branch's radial orientation (the orientation relative to the radial direction): the curvature is smaller when the branch is in the radial orientation and larger when the branch is orthogonal to that orientation. This brings to mind the radial organization of branches in the $\mathcal{T}$ morphotype, and indeed we were able to explain the chiral pattern with the aid of the same concept – repulsive chemotaxis.

Chemotaxis was introduced in previous versions of the Communicating Walkers model by varying, according to the chemical's gradient, the probability of moving in different directions [2, 25]. Modulating the directional probability is not the way bacteria implement chemotaxis – they modulate the

Fig. 23. — A structure of aggregates within a branch of a simulated colony with attractive signaling chemotaxis. The contours are equi-density contours of the walkers' "field".

length of runs. However, the growth of $\mathcal{T}$ morphotype is insensitive to the details of the movement. Modulating the directional probability is as good an implementation of chemotaxis as many other implementations (it was chosen for computational convenience). The pattern of the $\mathcal{C}$ morphotype is based on amplification of microscopic effects (singular perturbation) such as the left bias in the bacterial tumbling. Small differences in the microscopic dynamics of chemotaxis might affect the global pattern. Indeed we found that modulating the directional probability yield unrealistic results in the simulations of $\mathcal{C}$ morphotype. We had to resort to the bacterial implementation of chemotaxis – modulating the length of runs according to the chemical's gradients.

When modulating the length of runs of walkers or spinors one must be careful not to change the particles' speed. Such change is not observed in experiments [83,91] and it has far reaching effects on the dynamics. Changing the particles' speed is like changing the diffusion coefficient of the bacterial density field, a change that can have undesirable effects on the pattern.

a b

Fig. 24. — The effect of repulsive chemotactic signaling on the Communicating Spinors model. a) Without chemotaxis. b) With repulsive chemotaxis. The Spinors are repelled from the inner parts of the colony. The resulting curvature of the branches is reduces when they are in the radial direction. In spite the reversed handedness, the pattern resemble Figure 17b.

Modulating the length of spinors' runs without changing their speed can be done by modulating the number of steps that compose a single run (that was our motivation for dividing the runs into steps). Since the mean number of steps in a run is determined by the reorientation probability p, chemotaxis should modulate this probability. For chemotaxis, the probability of changing direction by the i-spinor in one time step is (for a repellent R):

$$p^* = p + \chi(R)\partial_{\theta_i} R \tag{23}$$

where R is measured at the spinor position $\vec{r}_i$, $\chi(R)$ is the same as in the continuous model (either constant or the "receptor law") and ∂_{θ_i} is directional derivative in the spinor's direction θ_i. p^* is truncated to within the range $[0, 1]$ as it is a probability. The length of the resulting runs will depend on the runs' direction, where a spinor moving up the gradient of the repellent will have shorter *mean* run length than the same spinor moving down the gradient.

The production and dynamics of the repulsive chemotactic signaling in the Communicating Spinors model is the same as in the Non-Linear Diffusion model, see equation (21) (with s representing the density of spinors that "freezed"). The patterns resulting from including repulsive chemotaxis in the model have indeed branches with variable curvature, as can be seen in Figure 24. The curvature is smaller for branches in the radial direction. Food chemotaxis also varies the branches' curvature, but in a less ordered manner, not similar to the observed bacterial patterns.

Under different growth conditions the C morphotype can produce very different patterns. As mention above, if the agar is soft enough the bacteria can

Fig. 25. — The snake-like branches observed in Figure 17a can be reproduced by the Communicating Spinors model. Ch is a continuous function of the colony's radius (the same function in a, b, and c). Maximal value of Ch is $8°$, minimal value is $-2°$. a) With repulsive chemotactic signaling. b) Without chemotaxis. c) With food chemotaxis. The best resemblance to the observed colony is obtained with repulsive chemotactic signaling.

move inside it. In such case, the bias in the bacterial movement might change or even reverse, and it is manifested in the curvature of the branches. Widely changing curvature of the branches can be seen in Figure 17a. The agar hardness was tuned such that in the beginning of the growth the bacteria could swim inside the agar, but they are forced to swim *on* the agar by the end of the growth due to the marginal water evaporation during the growth. In Figure 25 we demonstrate the models' ability to explain such patterns by changing the spinors' bias Ch during the simulation. Ch is set to be a continuous random function of the colonial size, which is constrained only at the beginning and end of growth to have certain values. The function for Ch is the same in all the images of Figures 25, only that in various types of chemotaxis are used. As can be seen, repulsive chemotactic signaling is needed to explain the observed bacterial patterns.

4.4. Weak Chirality in *P. Dendritiformis*

Colonies of $\mathcal{T}$ morphotype grown on hard substrate (above 2.0% agar concentration) exhibit branching patterns with a global twist with the same handedness, as shown in Figures 6 and 26. Similar observations during growth of other bacterial strains have been reported by Matsuyama *et al.* [19, 20]. We refer to such growth patterns as having weak chirality, as opposed to the strong chirality exhibited by the $\mathcal{C}$ morphotype.

In [23], Ben-Jacob *et al.* proposed that, in the case of $\mathcal{T}$ morphotype, it is the high viscosity of the "lubrication" fluid during growth on a hard surface that replaces the cell-cell co-alignment of the $\mathcal{C}$ morphotype that limit the rotation of tumbling. They further assumed that the rotation should be relative to a specified direction. They used gradient of a chemotaxis signaling field (specifically, the long-range repellent chemotaxis) as a specific direction, rather than the local mean orientation field which is used in the case of $\mathcal{C}$ morphotype. It was shown in [23] that inclusion of the above features in the Communicating

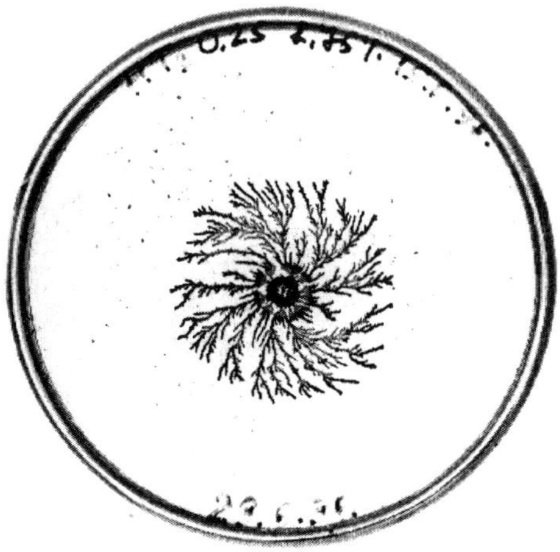

Fig. 26. — Weak chirality (global twist of the branches) exhibited by the $\mathcal{T}$ morphotypefor a peptone level of 0.25 g/l peptone level and agar concentration of 1.75%.

Walkers model indeed leads to a weak chirality which is highly reminiscent of the observed one. The idea above also provides a plausible explanation to the observations of weak chirality by Matsuyama *et al.* [20] in strains defective in production of "lubrication" fluid.

4.4.1. *Weak Chirality – The Non-Linear Diffusion Model*

In the reaction-diffusion model, weak chirality can obtained by modifying the chemotactic mechanism and causing it to twist: We alter the expression for the chemotactic flux $\vec{J}_{\text{chem}}$ (Eq. (19)) so that it is not oriented with the chemical gradient (∇R) anymore. Instead it is oriented with a rotated vector $\hat{\mathbf{R}}(\theta)\nabla R$, where $\hat{\mathbf{R}}(\theta)$ is the two-dimensional rotation operator and θ is the rotation angle. The chemotactic flux is thus written:

$$\vec{J}_{\text{chem}} = \zeta(b)\chi(R)(\hat{\mathbf{R}}(\theta)\nabla R). \tag{24}$$

The effect of rotating the repulsive chemotaxis, as depicted in Figure 27, is to make the pattern chiral, with the degree of chirality determined by the rotation angle θ.

One must note that adding a similar rotation to the food chemotaxis does not have the same effect, because the nutrient gradients do not possess the long-range, radial nature of the chemorepellent gradients.

4.4.2. *Weak Chirality – The Communicating Spinors Model*

As was demonstrated in Section 3.2, the Communicating Spinors model is robust enough to reproduce patterns of $\mathcal{T}$ morphotype, as well as patterns of $\mathcal{C}$

Fig. 27. — Growth patterns of the Non-Linear Diffusion model with a "squinting" repulsive chemotactic signaling, leading to weak chirality. Parameters are as in the previous picture, $\theta = 43^\circ$.

morphotype. Here we use it to model patterns of weak chirality. Two simulated $\mathcal{T}$-like colonies are shown in Figure 28. Figure 28a shows colony with radial branches while Figure 28b shows colony with weak chirality and thinner branches. In the two simulations the spinors have exactly the same response to chemotaxis and the same bias $Ch = 9^\circ$. The two runs differ in the freedom of rotation η; in 28a the spinors can rotate freely ($\eta = 180^\circ$) while in 28b the spinors rotation is somewhat limited ($\eta = 35^\circ$, while for the colony of strong chirality $\eta = 5^\circ$).

It seems that both models – the Non-Linear Diffusion model and the Communicating Spinors model – can capture the essential features of the observed weak chirality. Yet a closer examination reveals that the description of the two models is incompatible. In the Non-Linear Diffusion model the bias from the direction of the gradient is through the chemotaxis process. The spinors, like the bacteria, cannot modulate their runs as a function of the difference between their direction and that of the gradient; they do not know what is the direction of the gradient, only the directional derivative along their path. As was demonstrated in Figure 28, one of the key features for the weak chirality in the spinors model is the correlation in orientation of the spinors (through Φ). In fact in this model the twist of the branches stems from the deviation of the runs' direction from the orientation of neighboring spinors. The twist of the branches is related only indirectly, through the neighbors' orientation, to the chemorepellent's gradient. A continuous model of such process should include information about the mean orientation of the bacterial cells. It should include chemotaxis without rotation, anisotropic diffusion (smaller diffusion

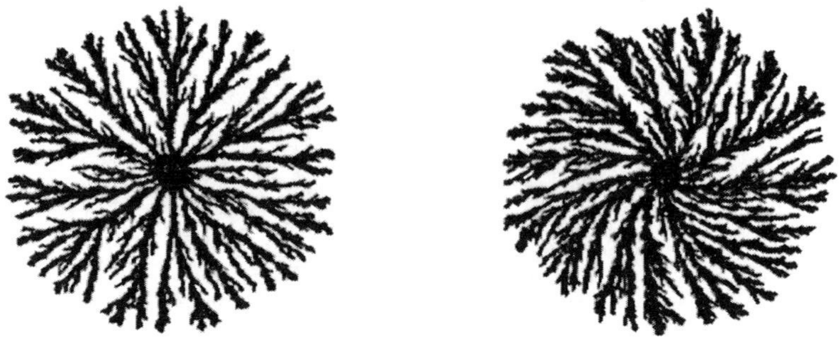

a b

Fig. 28. — Weak chirality of the $\mathcal{T}$ morphotype is modeled by the Communicating Spinors model. Both simulations are with repulsive chemotactic signaling and with bias in the walkers rotation. a) With free rotation ($\eta = 180°$) The pattern is branched, without apparent chirality. expressed. b) With constrained rotation ($\eta = 35°$) weak chirality is expressed.

coefficient in orientations orthogonal to the mean orientation of neighbors) and a rotation on the diffusion operator. Such a model will be presented elsewhere [81].

The discrete spinors model allows for a detailed representation of the bacterial properties. The macroscopic dynamics and resulting patterns, however, are similar in both models – apparently the growth does not amplify the difference in the microscopic dynamics. Thus the "unrealistic" microscopic description of the Non-Linear Diffusion model does not rule it out as an approximation to the growth dynamics of bacterial colonies with weak chirality.

5. BRANCHING PATTERNS OF SWARMING BACTERIA

5.1. Observation of Bacterial Vortices

A wide variety of branching patterns are exhibited by the $\mathcal{V}$ morphotype as the growth conditions are varied. Some representative patterns are shown in Figure 29. Each branch is produced by a leading droplet of cells and emits side branches, each with its own leading droplet (Fig. 30).

Microscopic observations reveals that each leading droplet consists of hundreds to millions of cells that circle a common center (hence the term vortex) at a cell speed of about 10 μm/s. Both the size of a vortex and the speed of the cells can vary according to the growth conditions and the location in the colony, as is shown in Figures 29, 30. Within a given colony, both clockwise and anti-clockwise rotating vortices are observed. The vortices in a colony can

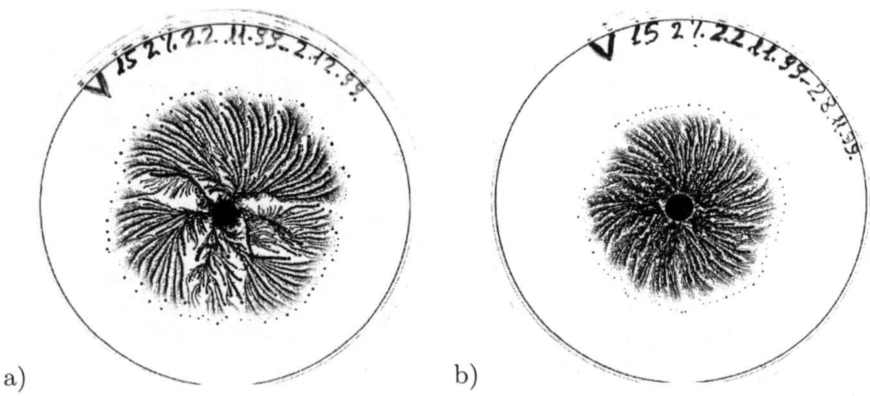

Fig. 29. — Representative patterns observed during colonial development of the $\mathcal{V}$ morphotype. The growth conditions are 15 g/l peptone level and 2% agar concentration.

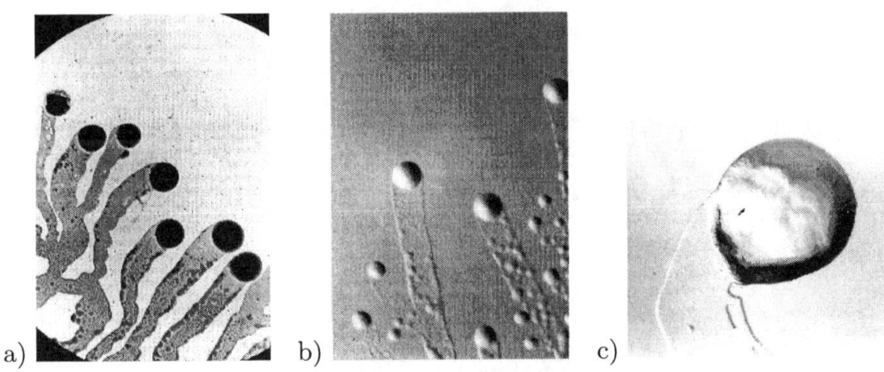

Fig. 30. — Microscope observations of vortices. Magnification is $\times 16$ a, b) and $\times 50$ c). Growth conditions are 4 g/l peptone level and 2.5% agar concentration.

also consist of either a single or multiple layers of cells. We occasionally observed vortices with an empty core, which we refer to as "bagel" shaped. After formation, the number of cells in the vortex increases, the vortex expands and it translocates as a unit. The speed of the vortices is slower than the speed of the individual cells circulating around the vortex center.

Bacterial cells are also contained in the trails left behind the leading vortices (Fig. 30). Some are immobile while others move, swirling with complex dynamics; groups containing a few to thousands of cells move in various directions, changing direction abruptly and sometimes returning to a previous direction. When two such groups pass by each other they can unite into a single group, or they might remain separate despite the close contact. The migrating groups of cells are very reminiscent of the "worm" motion of slime mold or schools of multicellular organisms. The whole intricate dynamics is confined to the trail of the leading vortex, and neither a single cell nor a group of moving cells can pass the boundary of the trail. Only vortices formed in the trails can break out of the trail and create a new branch.

Microscopic observations of the $\mathcal{V}$ morphotype reveal that the bacterial motion is performed in a fluid on the agar surface. We do not observe tumbling motion nor movement forward and backward. Rather, the motion is forward along the cell's long axis, and the cells tend to move in the same direction and speed as the surrounding cells, in a synchronized group. Electron microscope observations show that the bacteria have flagella which indicates the movement to be swarming.

5.2. Modeling the Collective Migration

At the end of growth, a colony contains about $10^9 - 10^{10}$ cells. Hence it is impractical to incorporate each cell into the model. Instead, groups of cells are represented by particles, each representing about $10^2 - 10^3$ cells, such that there are $10^6 - 10^8$ particles (swarmers) in a simulated colony. The i-th swarmer is characterized by its location $\vec{r_i}$, its velocity $\vec{v_i}$ and two internal degrees of freedom ("internal energy" E_i and the amount of cell-associated slime it carries s_i) as discussed below. The equation describing the change in the swarmer's position is simply:

$$\frac{\mathrm{d}\vec{r_i}}{\mathrm{d}t} = \vec{v_i} \; . \tag{25}$$

The description of the velocity change is more intricate:

$$\frac{\mathrm{d}\vec{v_i}}{\mathrm{d}t} = \Gamma \frac{\vec{v_i}}{|\vec{v_i}|} - \nu \vec{v_i} + \mu \left(\langle \vec{v} \rangle_{i,\epsilon} - \vec{v_i} \right) - \chi_\sigma \nabla \sigma + \vec{\xi} \; . \tag{26}$$

The terms on the right hand side of equation (26) are: 1) A propulsion force Γ, which acts in the direction of $\vec{v}_i$. 2) Friction force with the surroundings with a friction coefficient ν. These two terms tend to set the swarmers' speed to be Γ/ν. 3) The velocity-velocity interaction which leads to alignment of the velocity to the neighbors' mean velocity $\langle \vec{v} \rangle_{i,\epsilon}$ averaged over a radius ϵ. 4) A hard-core interaction term, which leads to a repulsion of the swarmer from regions of high density. The local density of the swarmers is denoted by σ. 5) A random force $\vec{\xi}$. We emphasize that this term is included to demonstrate that the dynamics of the model is not sensitive to noise.

In reality the coefficients Γ, μ and ν of equation (26) depend on the local amount of extracellular wetting fluid. In the model this feature is included by the fact that a swarmer can move only if the level of extracellular wetting fluid in its vicinity $w(\vec{r_i})$ is larger than a threshold value W. Each swarmer carries s_i amount of fluid and releases it as a slime if the surface is dry ($w(\vec{r_i}) < W' < W$). When a swarmer encounters a place where w is below W (*i.e.* the colony boundary), its motion is reflected and it deposits fluid at the boundary.

The above features are sufficient to capture the bacterial collective migration. However an additional feature has to be included in order to explain the emergence of vortices.

5.3. Rotational Chemotaxis and Vortex Formation

It has previously been shown [92, 93] in related models, that geometrical constraints of reflecting boundaries can lead to a circular collective motion. This result suggests that a radial inward "force" can lead to vortex formation in bacterial colonies. We first propose, in this regard, that a chemotactic response to self-emitted attractant provides such an inward "force".

We further propose that as a chemotactic mechanism the individual cells weakly vary the propulsion force according to the local concentration of a chemomodulating agent. For high concentrations of a chemoattractant the force is decreased. In a group of cells, such a response creates a gradient of propulsion force. Together with the velocity-velocity interaction, this imposes a torque or local vorticity on the average motion of the cells. Therefore, a swarmer moving at an angle to the chemical gradient, is subjected to a torque which causes the swarmer to twist towards the direction of the local gradient of the chemoattractant. The term describing this attractive rotational chemotactic response is

$$-\chi_a \frac{1}{v_i} \vec{v_i} \times (\vec{v_i} \times \nabla a), \tag{27}$$

where a is the concentration of the attractant and χ_a is a prefactor, which is a measure of the sensitivity to the chemomodulator. This term should be added to the right hand side of equation (26), and Γ should vary in space according to a. The variation of a over time is described by the diffusion equation (28) which includes a source term of emission of chemomodulator by the swarmers at a rate Γ_a and a spontaneous decay term at a rate λ_a:

$$\frac{\partial a}{\partial t} = D_a \nabla^2 a + \Gamma_a \sigma - \lambda_a a \; . \tag{28}$$

5.4. Modeling the Cooperative Organization of Colonies

We have shown that chemomodulation can indeed lead to the formation of stationary vortices (fixed in size and at a fixed location, Fig. 31a), rotating "bagels" (Fig. 31b) and other elements. All these elements are of a length-scale comparable to or smaller than an individual branch of a colony. During

a) b) c)

Fig. 31. — Numerical simulations of the model with the inclusion of attractive rotational chemotaxis (Eqs. (25–28)). The arrows indicate the direction and magnitude of the swarmers' velocity. We started the simulations with randomly scattered swarmers which are initially assumed random velocities (the values of parameters used: $\Gamma = 0.3$, $\nu = 0.1$, $\chi_\sigma = 0.2$, $\xi = 0.2$, $\chi_{\mathrm{A}} = 0.2$, $D_{\mathrm{A}} = 0.1$, $\Gamma_{\mathrm{A}} = 1.0$, $\lambda_{\mathrm{A}} = 0.01$). a) Spontaneous formation of vortices ($\mu = 0.1$). b) Formation of a "Bagel"-like pattern ($\mu = 0.3$). c) Experimentally observed "Bagel" for comparison.

colonial development these elements are organized to form the observed global pattern (Fig. 29).

When modeling colony formation, we must take into account that bacterial cells in a colony do not move in a predetermined space, and their number and state of activity is not conserved. While the colony expands and changes its shape, cells reproduce and in the case of $\mathcal{V}$ morphotype, also sporulate. To provide the means for reproduction, movement and other metabolic processes, the cells consume nutrients from the environment.

As with the Communicating Spinors model, we represent the metabolic state of the i-th swarmer by an "internal energy" E_i (Sect. 3.2). The dynamics of the "internal energy" and of the consumed nutriment n is the same as in Section 3.2. As was proposed in Section 4.2, we assume that starved cells (swarmers for which E_i is zero or below a certain level) emit a diffusive compound at a fixed rate Γ_{r}, and this chemical acts as a chemorepellent modulator. The effect of the repellent on the swarmer movement is represented by a term similar to (27) but with opposite sign. This results from increase in the propulsion force for high concentration of r. After the inclusion of both the chemoattractant and chemorepellent terms, equation (26) becomes:

$$\begin{aligned} \frac{\mathrm{d}\vec{v}_i}{\mathrm{d}t} &= \Gamma\frac{\vec{v}_i}{v_i} - \nu\vec{v}_i - \chi_\sigma\nabla\sigma + \mu\Big(\langle\vec{v}\rangle_{i,\epsilon} - \vec{v}_i\Big) \\ &\quad -\frac{1}{v_i}\vec{v}_i \times (\vec{v}_i \times (\chi_a\nabla a - \chi_r\nabla r)) + \vec{\xi}. \end{aligned} \tag{29}$$

We assume that the repellent material decays slowly, so that its concentration r is almost constant over distances comparable to the typical size of a vortex (long range chemotaxis). This is in contrast with the attractant concentration which is assumed to vary considerably within a vortex (short range chemotaxis).

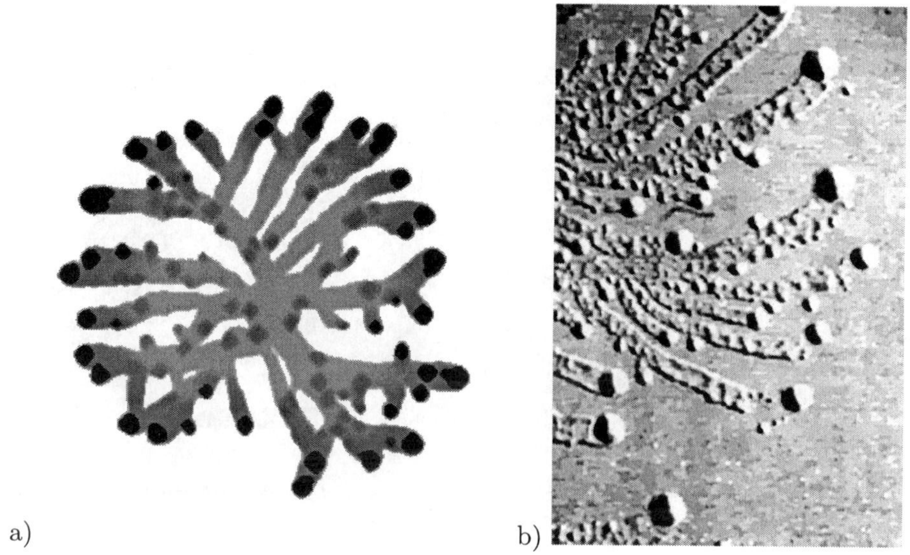

Fig. 32. — a) Simulations of colony development. The gray levels represent swarmers' density. Both repellent and attractant rotational chemotaxis are included, as well as food diffusion, reproduction, sporulation and production of wetting fluid (parameters' value are $\Gamma = 0.3$, $\nu = 0.1$, $\mu = 0.1$, $\chi_\sigma = 0.13$, $\xi = 0.2$, $\chi_{\rm A} = 0.23$, $\chi_{\rm R} = 0.05$, $W = 50.0$, $\kappa = 1.0$, $\Omega_{\rm C} = 0.3$, $\tau_{\rm R} = 33.0$, $E_{\rm m} = 0.01$, $D_{\rm A} = 0.1$, $\Gamma_{\rm A} = 1.0$, $\lambda_{\rm A} = 0.03$, $D_{\rm R} = 0.1$, $\Gamma_{\rm R} = 1.0$, $\lambda_{\rm R} = 0.00001$ and initial concentration of nutrient is 20.0 everywhere). b) Example of observed pattern, for comparison. Magnification is $\times 16$. The growth conditions are 4 g/l; 2.5%.

Thus although the functional form of the repulsion term is similar to the term for attraction, it has a very different effect on the bacterial motion. It affects each vortex as a single unit and provides a mechanism for regulating colonial structure during colonial development. On each vortex, the repulsion acts to push the vortex outward on a curved trajectory. In Figure 32 we show results of numerical simulations where both long range repulsive and short range attractive chemotaxis are included. We also included in the simulations a finite probability per unit time (about $1\%/\tau_{\rm R}$) of the swarmers to become immobile. The immobile swarmers at the trail have similar probability to regain mobility. This was motivated by microscopical observations revealing the presence of cells left behind the advancing vortices.

6. SECTOR FORMATION IN BRANCHING COLONIES

6.1. Observations

There is a well known observed (but rarely studied) phenomenon of bursts of new sectors of mutants during the growth of bacterial colonies (see [94–97],

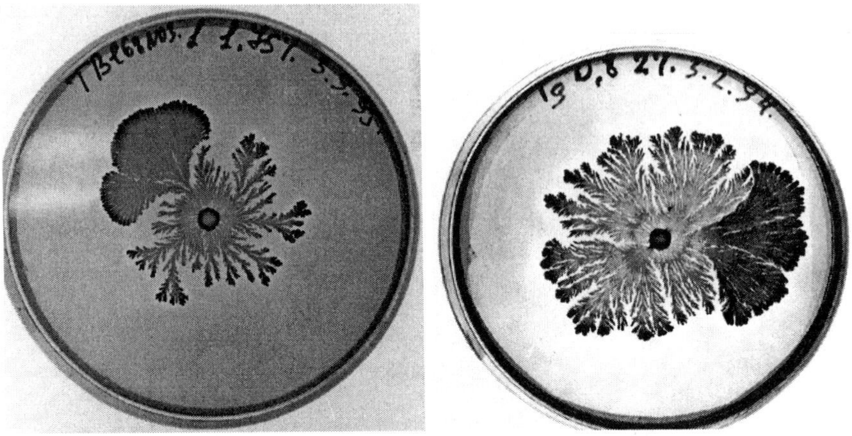

Fig. 33. — Emerging sectors in branching colonies of *P. dendritiformis* var *dendron*, obtained at 1 g/l peptone and 1.75% agar a), 0.8 g/l peptone and 2% agar b).

Ref. [98] contains pictures of emerging sectors in a yeast colony). If the mutants have the same growth dynamics as the "normal", wild-type, bacteria they will usually go unnoticed, unless some property such as coloring distinguishes them. Different coloring may result from different enzymatic activity (natural coloring) or from a different response to a staining process (artificial coloring). In both cases the mutation is not neutral in the strictest sense, but it is neutral as far as the dynamics is concerned. If, however, the mutants have different growth dynamics, a distinguished sector with a different growth pattern might indicate their presence.

In Figure 33 we show observations of bursts of sectors during branching growth of *P. dendritiformis* var. *dendron*. The phenomenon is universal. In Figure 34 we show bursts of sectors during growth of chiral branching and branching of the swarming bacteria *P. vortex*.

We use the term "mutant" to describe the population in the emerging sectors. We have not verified the existence of a genetic difference between the bacteria in the sector and those in the rest of the colony. We have, however, verified that the phenotypic difference between the two populations is inheritable, using inoculation.

In [97] we have shown that it is sometimes possible to relate the shape of the sector, and the way it "bursts" out of the colony, with the specific kind of advantage that the mutants possess over the original bacteria.

In order to generalize the NLD model to study mutants, we must introduce two fields, for the densities of the wild-type bacteria ("type 1") and the mutants ("type 2"), and allow some probability of transition from wild-type to mutants. In the absence of chemotaxis, the equations for the bacterial density of the two

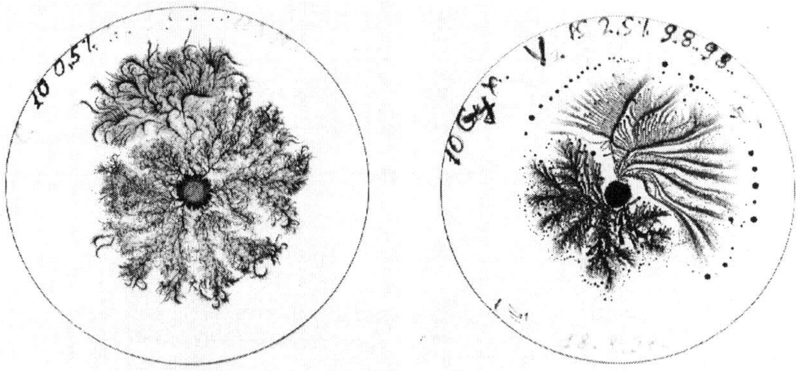

Fig. 34. — Emerging sectors in branching colonies. Left: chiral branching of *P. dendritiformis* var. *chiralis*. Right: vortex branching of *P. vortex*.

strains will be written (with subscript denoting bacteria type):

$$\frac{\partial b_1}{\partial t} = \nabla(D_1(b)\nabla b_1) + \varepsilon_1 n b_1 \Theta(b-\beta) - \mu_1 b_1 - F_{12} \tag{30}$$

$$\frac{\partial b_2}{\partial t} = \nabla(D_2(b)\nabla b_2) + \varepsilon_2 n b_2 \Theta(b-\beta) - \mu_2 b_2 + F_{12} \tag{31}$$

where $D_{0_{1,2}} = D_{0_{1,2}} b^k$ $(b = b_1 + b_2)$.

Note that the mutant strain b_2 includes a "source" term F_{12}, which is the rate of transition $b_1 \to b_2$, and is given by the growth rate of b_1 multiplied by a constant mutation rate (for simplicity, we do not include the process of reverse mutations F_{21}).

In Figures 35 and 36 we show typical results obtained from simulations of the model.

A fan-like sector appears, for example, when the mutant has a higher sensitivity to the chemotactic signals (denoted χ_0 in the model, see [37]), Figure 36. In this case, however, the sector is composed of a mixture of the "wild-type" and the mutants. Figure 37 displays the result of Communicating Walker simulation for this case (the discrete model shows a higher tendency for segregation here, a feature we are currently unable to explain). Note the similarly of both models' results with experimental observations.

7. THE EFFECT OF ANTIBIOTICS ON BACTERIAL BRANCHING GROWTH

Recently we have studied bacterial cooperative organization inder antibiotic stress [99]. In medicine, resistance to antibiotics is usually considered a qualitative property: either a specific strain of bacteria is resistant to a specific

Fig. 35. — Mutant with a higher growth rate in a branching colony: numerical simulation of the continuous model. The mutant (dark) erupts in a fan-like sector from the wildtype (light) colony. Model parameters for the wild-type: $D_{0_1} = 0.1$, $k = 1$, $\epsilon_1 = 1.0$, $\beta = 0.0001$, $\mu_1 = 0.15$, mutant has $\epsilon_2 = 5.0$. Initial nutrient level $n_0 = 1.0$.

antibiotic or it is not. The bacteria are generally regarded as being "resistant" if they can tolerate the maximal concentration of the antibiotic which is non-toxic to the treated humans or animals. However, it is known that bacterial colonies are often more resistant than the individual cells, thereby blurring the borderline between resistance and sensitivity. Moreover, antibiotics affect cells differently depending on the growth and the physiological state of the cells. Our results show that susceptibility to antibiotic should be regarded as quantitative property: the bacteria react to antibiotic even in concentrations far below the critical effective concentrations that stop their growth.

Although the available arsenal of Antibiotics contains a variety of functionally and structurally different compounds, resistance mechanisms have been developed for almost all of them. Resistance may involve decreased entry of the drug, changes in the receptor (target), metabolic inactivation or modification of the drug or production of "dummy" targets to sequester the compound and lower it's bioavailability [100]. Resistance can be a cooperative process.

Fig. 36. — Mutant with a higher sensitivity to repulsive chemotactic signaling, in a branching colony: numerical simulation of the continuous model. The mutant erupts in a fan-like sector from the colony of wildtype bacteria. However, the sector is not a segregated area, and contains wildtype bacteria as well. Model parameters for the wild-type: $D_{0_1} = 0.1$, $k = 1$, $\epsilon_1 = 1.0$, $\beta = 0.0001$, $\mu_1 = 0.2$, $\chi_{0_1} = 0.5$, mutant has $\chi_{0_2} = 1.0$. Initial nutrient level $n_0 = 1.0$.

The most compelling example is metabolic inactivation of an antibiotic occurring outside the cells: a significant reduction in the drug's concentration can be done only with high concentration of the inactivating agent which requires a dense population of bacteria.

Pathogenic bacteria form 2D films on many systems such as the cornea and sclera of the eye, in the urinary system, in the respiratory system (especially the lungs), on medical devices, etc. In [99] we study the response of the colonial growth dynamics to the presence of antibiotics under various stressful conditions, when the colony is grown on a surface. From comparison between the simulations of the models and the experimental observations we can gain insights about the response of the bacteria to the antibiotic stress. From our existing model of chiral growth, for example, we conjecture that the changes in the colonial patterns caused by the antibiotic co-trimoxazole (Septrin, trademark of Wellcome) are due to elongation of the bacterial cells. This is consistent with the method of action of Septrin, which interferes with bacterial metabolism and might mimic starvation conditions – conditions under which the bacteria are usually longer.

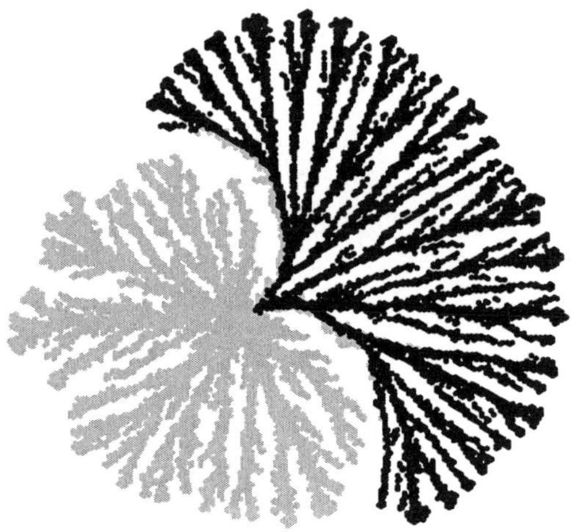

Fig. 37. — Mutant with a higher sensitivity to repulsive chemotactic signaling, in a branching colony: results of the Communicating Walker model. As in the continuous model, the mutant erupts in a fan-like sector from the colony of wildtype bacteria.

In most of our experiments we have studied the effects of Septrin, usually at a level of 0.0003% (3 μg/ml). Three phenomena are observed:

1. the rate of colonial expansion is lower;
2. the growth becomes much more ramified, *i.e.* there is a decrease in the fractal dimension;
3. for growth on harder surfaces in the presence of Septrin, there is more pronounced weak chirality.

In Figure 38 we demonstrate the second phenomenon. For these growth conditions, the fractal dimension D_f changes from $D_\mathrm{f} = 2$ to $D_\mathrm{f} \approx 1.8$ in the presence of Septrin, while the growth velocity changes only slightly. Two phenomena (weak chirality and decrease in fractal dimension) are observed in Figure 39. Note that in the presence of Septrin the velocity (colonial expansion rate) is reduced to less than half of that of the growth without antibiotics.

There is also a pronounced effect of Septrin on the growth patterns of the $\mathcal{V}$ morphotype. In Figure 40 we show a typical example. Septrin causes the pattern to become much more radially oriented and eliminates most of the side branches.

Theoretical studies of the effects of antibiotics are presented in [99].

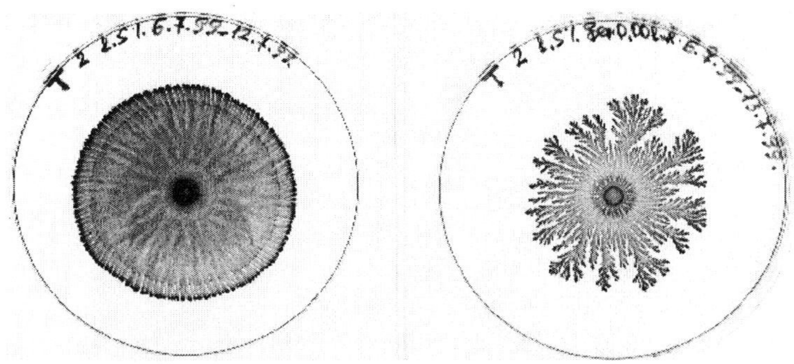

Fig. 38. — The effect of Septrin on the fractal dimension of the $\mathcal{T}$ morphotype (left: without antibiotic, right: with 3 μg/ml Septrin). $\mathcal{T}$ morphotype grown at 2 g/l peptone and 1.5% agar. Growth times are 6 days (left) and 7 days (right).

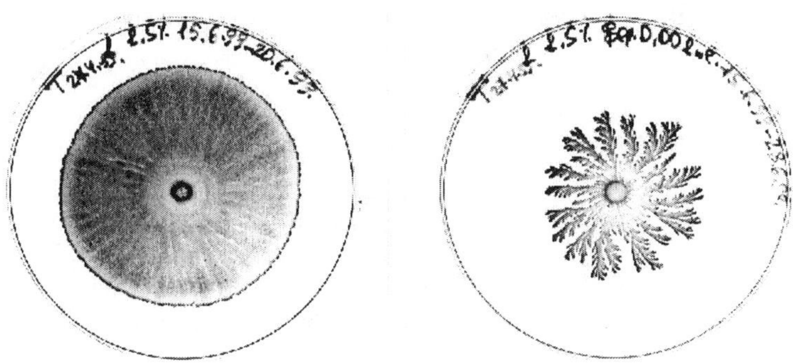

Fig. 39. — Similar to Figure 38, for growth at 1 g/l peptone. Growth times are 5 days (left) and 10 days (right).

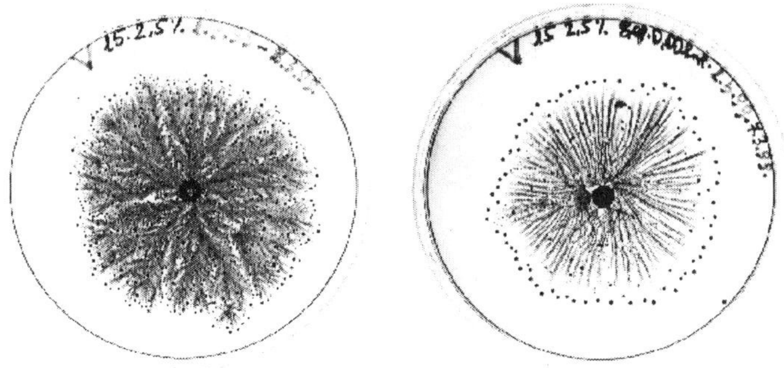

Fig. 40. — The effect of Septrin on the growth pattern of the $\mathcal{V}$ morphotype for Growth conditions of 15 g/l peptone and 2.5% agar. Growth times are 7 days (left, no antibiotics) and 6 days (right, 3 μg/ml Septrin). Note that the growth is faster under the pressure of the antibiotic.

8. CONCLUSIONS

We have presented experimental observations of branching and chiral branching growth of the swimming and lubricating bacteria *P. dendritiformis* var. *dendron* and *P. dendritiformis* var. *chiralis.* The swarming bacteria *P. vortex* also exhibit branching patterns, but of a very different nature. We have employed discrete and continuous models to capture the colonial development.

Traditionally, there has been a tendency to focus on the special features of each bacterial strain. We have learned that one can also identify universality in bacterial behavior. The universality differs from the one we are used to in physics, and it has to do with the adaptation strategies of the different strains. They all employ chemotaxis as a basic adaptive strategy. Moreover, they use both short range attractive chemotactic signaling and long range repulsive chemotactic signaling.

To conclude, there is significant progress being made towards working out the communications processes which govern the interactions between bacteria during branching growth.

Acknowledgements

I.G. and E.B. are greatful to the organizers of the meeting "Dynamics and Morphogenesis of Branching Structures" for the hospitality at Les Houches École de Physique. We have enjoyed fruitful discussions with many participants during the meeting. Identifications of the *Paenibacillus dendritiformis* var. *dendron* and genetic studies are carried in collaboration with the group of D. Gutnick. Presented studies are supported in part by a grant from the Israeli Academy of Sciences grant No. 593/95, the Israeli-US Binational Science Foundation BSF grant No. 00410-95 and the Milo and Shoshana Shefler Project.

REFERENCES

[1] E. Ben-Jacob, *Contemp. Phys.* **34** (1993) 247-273.
[2] E. Ben-Jacob, *Contemp. Phys.* **38** (1997) 205-241.
[3] E. Ben-Jacob and P. Garik, *Nat* **343** (1990) 523-530.
[4] D.A. Kessler, J. Koplik and H. Levine, *Adv. Phys.* **37** (1988) 255.
[5] J.S. Langer, *Sci* **243** (1989) 1150-1154.
[6] J. Feder, *Fractals* (Plenum, New York, 1988).
[7] B.B. Mandelbrot, *The Fractal Geometry of Nature* (Freeman, San Francisco, 1977).
[8] B.B. Mandelbrot, *Fractals: Form, Chance and Dimension* (Freeman, San Francisco, 1977).
[9] L.M. Sander, *Nat* **322** (1986) 789-793.

[10] T. Vicsek, *Fractal Growth Phenomena* (World Scientific, New York, 1989).

[11] E. Ben-Jacob, H. Shmueli, O. Shochet and A. Tenenbaum, *Physica A* **187** (1992) 378-424.

[12] E. Ben-Jacob, O. Shochet and A. Tenenbaum, Bakterien schließen sich zu bizarren formationen zusammen, edited by A. Deutsch, *Muster des Ledendigen: Faszination inher Entstehung und Simulation* (Verlag Vieweg, 1994).

[13] E. Ben-Jacob, O. Shochet, A. Tenenbaum and O. Avidan, Evolution of complexity during growth of bacterial colonies, edited by P.E. Cladis and P. Palffy-Muhoray, *Spatio-Temporal Patterns in Nonequilibrium Complex Systems*, Santa-Fe Institute studies in the sciences of complexity (Addison-Weseley Publishing Company, 1995) pp. 619-634.

[14] E. Ben-Jacob, A. Tenenbaum, O. Shochet and O. Avidan, *Physica A* **202** (1994) 1-47.

[15] H. Fujikawa and M. Matsushita, *J. Phys. Soc. Jpn.* **58** (1989) 3875-3878.

[16] H. Fujikawa and M. Matsushita, *J. Phys. Soc. Jpn.* **60** (1991) 88-94.

[17] M. Matsushita and H. Fujikawa, *Physica A* **168** (1990) 498-506.

[18] M. Matsushita, J.-I. Wakita and T. Matsuyama, Growth and morphological changes of bacteria colonies, edited by P.E. Cladis and P. Palffy-Muhoray, *Spatio-Temporal Patterns in Nonequilibrium Complex Systems*, Santa-Fe Institute studies in the sciences of complexity (Addison-Weseley Publishing Company, 1995) pp. 609-918.

[19] T. Matsuyama, R.M. Harshey and M. Matsushita, *Fractals* **1** (1993) 302-311.

[20] T. Matsuyama and M. Matsushita, *Crit. Rev. Microbiol.* **19** (1993) 117-135.

[21] E. Ben-Jacob, I. Cohen, A. Czirók, T. Vicsek and D.L. Gutnick, *Physica A* **238** (1997) 181-197.

[22] E. Ben-Jacob, I. Cohen, O. Shochet, I. Aronson, H. Levine and L. Tsimering, *Nat* **373** (1995) 566-567.

[23] E. Ben-Jacob, I. Cohen, O. Shochet, A. Czirók and T. Vicsek, *Phys. Rev. Lett.* **75** (1995) 2899-2902.

[24] E. Ben-Jacob, O. Shochet, A. Tenenbaum, I. Cohen, A. Czirók and T. Vicsek, *Fractals* **2** (1994) 15-44.

[25] E. Ben-Jacob, O. Shochet, A. Tenenbaum, I. Cohen, A. Czirók and T. Vicsek, *Nat* **368** (1994) 46-49.

[26] A. Czirók, E. Ben-Jacob, I. Cohen and T. Vicsek, *Phys. Rev. E* **54** (1996) 1791-1801.

[27] D.A. Kessler and H. Levine, *Phys. Rev. E* **48** (1993) 4801-4804.

[28] D.A. Kessler, H. Levine and L. Tsimring, *Physica D* **106** (1997) 375-388.

[29] S.A. Mackay, *J. Cell. Sci.* **33** (1978) 1-16.

[30] H. Parnas and L. Segel, *J. Theor. Biol.* **71** (1978) 185-207.

[31] M. Matsushita, J. Wakita, H. Itoh, I. Rafols, T. Matsuyama, H. Sakaguchi and M. Mimura, *Physica A* **249** (1998) 517-524.

[32] M. Mimura, H. Sakaguchi and M. Matsushita, A reaction-diffusion approach to bacterial colony formation, *preprint* (1997).
[33] K. Kawasaki, A. Mochizuki, M. Matsushita, T. Umeda and N. Shigesada, *J. Theor. Biol.* **188** (1997) 177-185.
[34] S. Kitsunezaki, *J. Phys. Soc. Jpn.* **66** (1997) 1544-1550.
[35] Y. Kozlovsky, I. Cohen, I. Golding and E. Ben-Jacob, *Phys. Rev. E* **59** (1999) 7025-7035.
[36] I. Rafols, *Formation of concentric rings in bacterial colonies*, MSc Thesis (Chuo University, Japan, 1998).
[37] I. Golding, Y. Kozlovsky, I. Cohen and E. Ben-Jacob, *Physica A* **260** (1998) 510-554.
[38] M. Tcherpakov, E. Ben-Jacob and D.L. Gutnick, *Int. J. Syst. Bacteriol.* **49** (1999) 239-246.
[39] M. Tcherpakov, E. Ben-Jacob, I. Cohen and D.L. Gutnick, *Int. J. Syst. Bacteriol.* (submitted).
[40] E. Ben-Jacob, I. Cohen and D. Gutnick, *Annu. Rev. Microbiol.* (1998) (in press).
[41] E. Ben-Jacob, I. Cohen, I. Golding and Y. Kozlovsky, Modeling branching and chiral colonial patterning of lubricating bacteria, in *Proceedings of 1998 IMA workshop: Pattern Formation and Morphogenesis* (Springer-Verlag, Berlin, 1999) (in press).
[42] A. Dukler, *Isolation and characterization of bacteria growing in patterns on the surface of solid agar*, Senior project under the supervision of D. Gutnick (Tel-Aviv University, 1993).
[43] E. Ben-Jacob, H. Shmueli, O. Shochet and D. Weiss, *Pattern Formation in Physical Sciences and Biology* (Plenum, 1993).
[44] D. van Sinderen, R. Kiewiet and G. Venema, *Mol. Microbiol.* **15** (1995) 213-223.
[45] N.J. Shih and R.G. Labbe, *Can. J. Microbiol.* **42** (1996) 628-33.
[46] E. Deak, I. Szabo, A. Kalmaczhelyi, Z. Gal, G. Barabas and A. Penyige, *Microbiol.* **144** (1998) 2169-2177.
[47] I. Cohen, A. Czirók and E. Ben-Jacob, *Physica A* **233** (1996) 678-698.
[48] J. Henrichsen, *Bac. Rev.* **36** (1972) 478-503.
[49] L. Alberti and R.M. Harshey, *J. Bact.* **172** (1990) 4322-4328.
[50] R.M. Harshey and T. Matsuyama, *Proc. Natl. Acad. Sci. USA* **91** (1994) 8631-8635.
[51] E. Hoiczyk and W. Baumeister, *Curr. Biol.* **8** (1998) 1161-1168.
[52] O. Rauprich, M. Matsushita, C.J. Weijer, F. Siegert, S.E. Esipov and J.A. Shapiro, *J. Bact.* **178** (1998) 6525-6538.
[53] A.M. Spormann, *Microbiol. Molec. Biol. Rev.* **63** (1999) 621.
[54] G.M. Young, M.J. Smith, S.A. Minnich and V.L. Miller, *J. Bact.* **181** (1999) 2823-2833.
[55] N.H. Mendelson, A. Bourque, K. Wilkening, K.R. Anderson and J.C. Watkins, *J. Bact.* **181** (1999) 600-609.
[56] D.J. DeRosier, *Cell* **93** (1998) 17-20.

[57] M. Eisenbach, *Molec. Microbiol.* **4** (1990) 161-167.
[58] H.C. Berg, *Random Walks in Biology* (Princeton University Press, Princeton, N.J., 1993).
[59] T. Matsuyama, K. Kaneda, Y. Nakagawa, K. Isa, H. Hara-Hotta and I. Yano, *J. Bacteriol.* **174** (1992) 1769-1776.
[60] F. Peypoux, J.M. Bonmatin and J. Wallach, *Appl. Microbiol. Biotech.* **51** (1999) 553-563.
[61] J.D. Desai and I.M. Banat, *Microbiol. Mol. Biol. Rev.* **61** (1997) 47-64.
[62] T. Matsuyama and Y. Nakagawa, *Colloids Surf. B: Biointerfaces* **7** (1996) 207-214.
[63] S.J. Stahl, K.R. Stewart and F.D. Williams, *J. Bacteriol.* **154** (1983) 930-937.
[64] O. Shochet, K. Kassner, E. Ben-Jacob, S.G. Lipson and H. Müller-Krumbhaar, *Physica A* **181** (1992) 136-155.
[65] O. Shochet, K. Kassner, E. Ben-Jacob, S.G. Lipson and H. Müller-Krumbhaar, *Physica A* **187** (1992) 87-111.
[66] O. Shochet, *Study of late-stage growth and morphology selection during diffusive patterning*, Ph.D. Thesis (Tel-Aviv University, 1995).
[67] E. Ben-Jacob, O. Shochet, I. Cohen, A. Tenenbaum, A. Czirók and T. Vicsek, *Fractals* **3** (1995) 849-868.
[68] E. Ben-Jacob, I. Cohen and A. Czirók, *Smart bacterial colonies*, in Physics of Biological Systems: From Molecules to Species, *Lecture Notes in Phys.* (Springer-Verlag, Berlin, 1997) pp. 307-324.
[69] E. Ben-Jacob and I. Cohen, Cooperative formation of bacterial patterns, edited by J.A. Shapiro and M. Dworkin, *Bacteria as Multicellular Organisms* (Oxford University Press, New-York, 1997).
[70] D.A. Kessler and H. Levine, *Nat* **394** (1998) 556-558.
[71] I. Cohen, *Mathematical Modeling and Analysis of Pattern Formation and Colonial Organization in Bacterial Colonies*, MSc Thesis (Tel-Aviv University, Israel, 1997).
[72] E. Ben-Jacob, I. Cohen and H. Levine, *Adv. Phys.* **49** (2000) 395-554.
[73] R.A. Hegstrom and D.K. Kondepudi, *Sci. Am.* **262** (1990) 108-115.
[74] V.A. Avetisov, V.I. Goldanskii and V.V. Kuzmin, *Phys. Today* **44** (1991) 33-41.
[75] N.H. Mendelson, *Proc. Natl. Acad. Sci. USA* **75** (1978) 2478-2482.
[76] N.H. Mendelson and S.L. Keener, *J. Bacteriol.* **151** (1982) 455-457.
[77] N.H. Mendelson and J.J. Thwaites, *J. Bacteriol.* **171** (1989) 1055-1062.
[78] N.H. Mendelson, *Sci. Progr.* **74** (1990) 425-441.
[79] J.B. Stock, A.M. Stock and M. Mottonen, *Nat* **344** (1990) 395-400.
[80] C.H. Shaw, *BioEssays* **13** (1991) 25-29.
[81] I. Cohen and E. Ben-Jacob, *Phys. Rev. Lett.* (2000) (submitted).
[82] J. Adler, *Sci* **166** (1969) 1588-1597.
[83] H.C. Berg and E.M. Purcell, *Biophys. J.* **20** (1977) 193-219.
[84] J.M. Lackiie, *Biology of the chemotactic response* (Cambridge Univ. Press, 1986).

[85] M. Eisenbach, *Mole. Microbiol.* **20** (1996) 903-910.
[86] P.A. Spiro, J.S. Parkinson and H.G. Othmer, *Proc. Natl. Acad. Sci. USA* **94** (1997) 7263-7268.
[87] J.D. Murray, *Mathematical Biology* (Springer-Verlag, Berlin, 1989).
[88] E.O. Budrene and H.C. Berg, *Nat* **349** (1991) 630-633.
[89] L. Tsimring, H. Levine, I. Aranson, E. Ben-Jacob, I. Cohen, O. Shochet and W.N. Reynolds, *Phys. Rev. Lett.* **75** (1995) 1859-1862.
[90] E.O. Budrene and H.C. Berg, *Nat* **376** (1995) 49-53.
[91] J.E. Segall, S.M. Block and H.C. Berg, *Proc. Natl. Acad. Sci. USA* **83** (1986) 8987-8991.
[92] J. Hemmingsson, *J. Phys. A* **28** (1995) 4245-4250.
[93] Y.L. Duparcmeur, H. Herrmann and J.P. Troadec, *J. Phys. I France* **5** (1995) 1119-1128.
[94] J.A. Shapiro and D. Trubatch, *Physica D* **49** (1991) 214-223.
[95] J.A. Shapiro, *BioEssays* **17** (1995) 597-607.
[96] A. Grondin, H.C. Jarell and L.R. Berube, *Arch. Microbiol.* **172** (1999) 59-62.
[97] I. Golding, I. Cohen and E. Ben-Jacob, *Europhys. Lett.* **48** (1999) 587-593.
[98] E. Boschke and Th. Bley, *Acta Biotechnol.* **18** (1998) 17-27.
[99] E. Ben-Jacob, I. Cohen, I. Golding, D.L. Gutnick, M. Tcherpakov, D. Helbing and I.G. Ron, *Physica A* **282** (2000) 510-554.
[100] S. Baron, *Medical Microbiology*, The University of Texas Medical Branch at Galveston, fourth edition (1996).

CONTRIBUTED PAPER 8

Pattern Formation Modeling of Bacterial Colonies

I.R. Cantalapiedra, A.M. Lacasta, C.E. Auguet,
A. Peñaranda and L. Ramírez-Piscina

1. INTRODUCTION

The rich variety of forms exhibited by bacterial colonies during their growth has drawn a great deal of attention [1–6]. From a theoretical point of view, the mathematical description of such diversity constitutes a stimulating non-equilibrium dynamical problem [2, 7–13]. In general, the complexity of the growth pattern increases as the environmental conditions become less favorable. Bacteria respond to adverse growth conditions by developing sophisticated strategies and higher micro-level organization in order to cooperate more efficiently. Examples of these strategies are: the differentiation into longer-motile bacteria, the production of extracellular wetting fluid, the secretion of surfactants which change the surface tension or the chemotactic response to chemical agents produced by bacteria.

In particular, colonies of the bacterium *Bacillus subtilis* OG-01 develop different kinds of patterns depending on the culture conditions [4–6]. The experiments are usually made in a petri dish, which contains a solution of nutrient and agar. A drop of bacterial solution is then inoculated in the center of the dish. The growth conditions are controlled by the initial concentration of the medium components. The agar concentration determines the consistency of the medium, which becomes harder as the amount of agar increases, and the nutrient concentration controls the bacterial reproduction. Depending on these two factors, the colony grows at a higher or lower rate, developing patterns that were classified into five types, from A to E [5]: a ramified structure (A) very similar to the patterns obtained with the diffusion-limited aggregation model (DLA) [5, 14]; a dense disk with rough interface (B), similar to the patterns obtained with an Eden model [15]; a structure of concentric rings (C) characterized by periodic dynamics [4,6]; a homogeneous disk of low bacterial density (D) and a dense branched pattern (E) similar to the dense branching morphology (DBM) found in other systems [16].

Several models have been proposed to reproduce some of the patterns exhibited by *Bacillus subtilis* [2, 7–11]. We have previously presented a model [13] that successfully reproduces the five morphologies observed in the experiments, including the ring patterns. The aim of this paper is to describe the basic ingredients of the model and the main numerical results. In addition to finding the whole morphological diagram [13], we will show that this model correctly describes the competition between colonies experimentally observed in some growth conditions.

2. DESCRIPTION OF THE MODEL

The energy and chemical balances at the cellular level involve an intricate microscopic dynamics that results in an effective macroscopic behavior here described by a continuous model. Stated in short, it consists of two coupled nonlinear reaction-diffusion equations, for bacterial density and nutrient concentration, supplemented by a third equation which models the ability of bacteria to adapt themselves to adverse conditions [13].

We consider a two-dimensional system (we neglect the thickness of the bacteria layer over the agar). The first equation of the model describes the spatio-temporal evolution of the density of bacteria. In general, it depends on birth, migration and death. The kinetics of birth by consumption of nutrient can be described by the Michaellis-Menten theory [7]. According to this theory, the uptake is linear in nutrient when the nutrient concentration is low and reaches a saturation for high values of it. When bacteria swim on a liquid agar medium, each particle performs an irregular motion called random walk, spreading out. This microscopic movement can be described by a macroscopic diffusion equation. When there is an increase in diffusion due to population pressure, the diffusion coefficient depends on the population density. Therefore, we describe the bacterial migration by means of a diffusion term with a general concentration-dependent coefficient. The bacterial death or sporulation is taken into account by means of a nutrient-dependent diffusion. In this way, bacteria become inactive (they do not reproduce nor migrate) when nutrient runs out. The second equation describes the evolution of nutrient concentration. It consists of the addition of a diffusion term with constant diffusion coefficient and a consumption term due to bacteria.

Finally, motivated by experimental observations, we have introduced the response of bacteria to adverse growth conditions in order to improve their motility. In a soft medium and high nutrient concentration, bacteria can randomly swim without difficulty, but in an adverse environment they need to adapt themselves to become longer and more motile. For intermediate conditions of semi-solid medium and sufficient nutrient, there are periods of fast growth (migration phase) and slow growth (consolidation phase) that lead to concentric ring patterns. The analysis of periodic rings suggests a dynamical scheme with hysteresis that can be outlined in the following way: during the consolidation phase, the population of longer and more motile bacteria increases in order to overcome the opposition to the movement. When this population exceeds a certain value, enhanced movement becomes possible and a migration phase begins. Then, a progressive decrease in long-bacterial population ensues, until it reaches a minimum at which the "enhanced-movement mechanism" does not work. Then a new consolidation phase begins.

At high densities of long bacteria, the collisions between them can not be neglected. In this case the diffusion of bacteria depends on the bacteria concentration [11]. All these ideas have been introduced in our model by means of two basic points: (a) the diffusion coefficient of bacteria can take two different expressions depending on the long-bacterial population. One of them describes the usual diffusion of bacteria in a liquid medium and the second one describes the cooperative enhanced-movement mechanism promoted by long bacteria [13]. (b) The net production of long bacteria is described by a third equation that depends on the environmental conditions and also on the colony phase of growth. Although the bacterial response has been expressed in terms of the population of long-bacteria, other possible kinds of responses admit identical modelization. In this sense, our model is quite general.

3. NUMERICAL RESULTS

We have numerically integrated the three equations that constitute our model in a triangular lattice. The system was initially prepared by assigning a nutrient concentration value to each point of the lattice, and a certain bacterial concentration in a small central zone. In Figure 1 we show the different patterns obtained from numerical integration of the model. By changing only two parameters, which are the diffusion coefficient (related to the inverse of the agar concentration) and the initial nutrient concentration, we have reproduced the five different morphologies shown by *Bacillus subtilis* OG-01. In the region D of Figure 1, bacteria always move by usual diffusion, whereas in regions A, B and E they always move by the enhanced-movement mechanism. In region C, the two mechanisms work alternately.

For low level of nutrient and high level of agar (Fig. 1A), the colony growth is limited by the diffusion of these nutrients. This produces first branching and then competition between branches which is known as screening effect. The branches are thicker as the initial nutrient concentration increases. The prototype model that reproduces this kind of structure is the diffusion-limited aggregation (DLA) [14], which is known to form a fractal pattern with a fractal dimension $d_{\mathrm{F}} = 1.71$ [1,17]. Experiments performed by Matsushita *et al.* [5] in this region also show a fractal growth with dimension $d_{\mathrm{F}} = 1.73$ [5]. We have

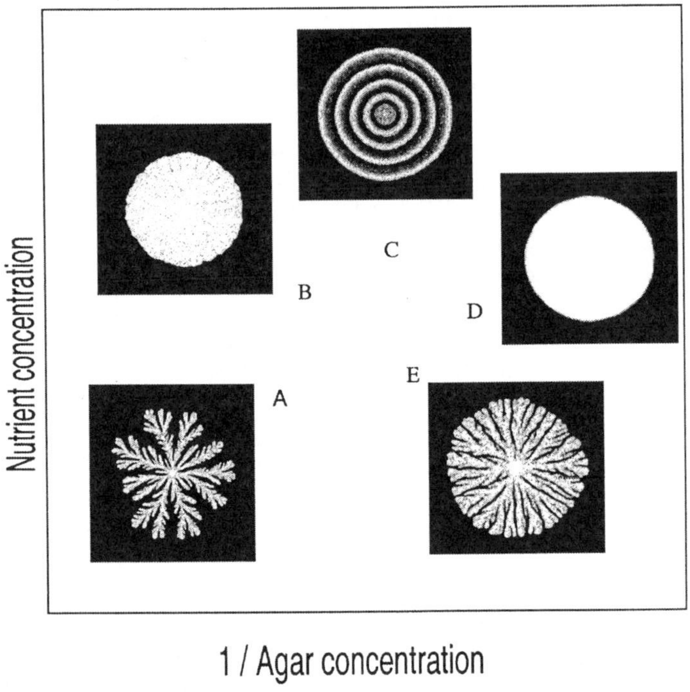

Fig. 1. — Bacterial morphological diagram obtained by numerical integration of the model as a function of nutrient concentration and surface softness. Shown patterns are: DLA-like (A), Eden-like (B), concentring rings (C), homogeneous disk (D) and DBM-like (E).

analyzed the fractal nature of the patterns obtained with our model, for the same value of agar concentration and several values of the initial nutrient, from DLA-like to Eden-like (dense disk with rough interface). We have calculated their fractal dimensions by using the box-counting method [1, 17]. We observe that the cases which correspond to low nutrient have a fractal dimension of about $d_{\mathrm{F}} = 1.73$, showing good agreement with experiments. On the other hand, there is an abrupt change between these patterns and those that are not fractal. In Figure 2 we reproduce the evolution of two colonies in the region A. The observed competition corroborates that the growth of the colony in this region is limited by the diffusion of nutrients. This repulsion can also be observed experimentally in colonies growing on the same agar plate [1].

For high levels of nutrient and agar (Fig. 1B), we obtain "Eden-like" dense rough patterns, that can be analyzed in terms of the roughness of their interfaces. It is well known that Eden structures are not themselves fractal, but their surfaces exhibit a self-affine scaling [17]. This implies that, for a long enough time, the width of the rough interface σ scales with an exponent α

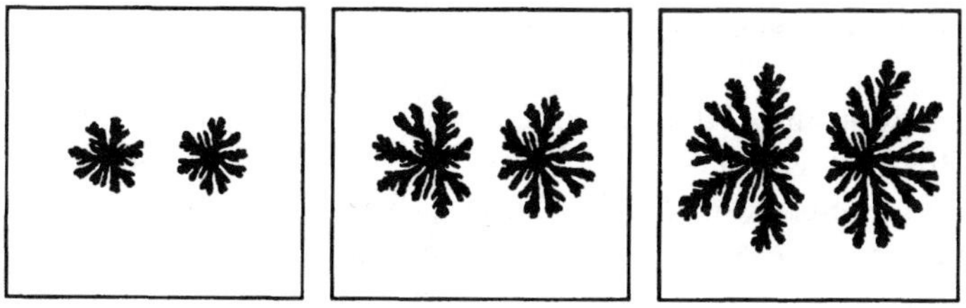

Fig. 2. — Evolution of two competing bacterial colonies for a low value of initial nutrient concentration and high level of agar concentration.

as a function of the length of the interface l ($\sigma \sim l^{\alpha}$). The roughness exponent for the Eden model is $\alpha = 0.5$. Vicsek *et al.* [15] analyzed experimental data corresponding to the region B of Figure 1. They concluded that these colony surfaces are self-affine with a roughness exponent $\alpha = 0.74$. We have checked this point for our dense rough pattern (Fig. 1B) by measuring the width σ for intervals of interface of length l. The results, as a function of l, are presented in [13]. Our results show a linear region with a slope compatible with the experimental value $\alpha = 0.74$. By increasing the diffusion coefficient, that means decreasing the agar concentration diluted in the medium, we observe a crossover from the dense rough structure (region B) to concentric rings (region C), and homogeneous disk (region D).

The ring patterns (Fig. 1C) have been obtained for intermediate values of agar and high nutrient. In this region, the bacterial response presents hysteresis and the two mechanisms of motion work alternatively, leading to cycles of migration and consolidation phases. The duration of these cycles is roughly constant for different values of nutrient and agar concentration over a wide range. This periodical dynamics generates patterns of concentric rings. Clearly the growth velocity of the colony is different between the two phases. Numerically, our model also reproduces the experimentally observed robustness of the growth-plus-consolidation period, which is barely dependent on changes in either nutrient or agar concentrations over a wide range.

From the homogeneous disk region (D), decreasing the initial nutrient concentration, the model is able to reproduce the dense branching morphology (DBM) (Fig. 1E). The ratio between branch and gap widths is constant.

In summary, we have shown that the proposed model satisfactorily reproduces the whole experimental morphological diagram. It represents a first attempt at describing the response of bacteria to adverse growth conditions and, in certain conditions, their ability to improve their motility. Furthermore, our model is adequate to describe the competition between colonies experimentally observed in some situations.

Acknowledgements

This research was supported by Dirección General de Investigación Científica y Técnica (Spain) (PB96-0241-02) and by Comissionat per Universitats i Recerca de la Generalitat de Catalunya (1999SGR00145). We also acknowledge computing support from Centre Català de Computació i Comunicacions.

REFERENCES

[1] T. Matsuyama and M. Matsushita, *Fractal Geometry in Biological Systems*, edited by P.M. Iannaccone and M. Khokha (CRC Press, 1996) p. 127.

[2] E. Ben-Jacob, *Comtemp. Phys.* **38** (1997) 205.

[3] O. Rauprich, M. Matsushita, C.J. Weijer, F. Siegert, S.E. Esipov and J.A. Shapiro, *J. Bacteriol.* **178** (1996) 6525.

[4] I. Ràfols I., *MSc Thesis* (Chuo University, Tokyo, 1998) p. 1.

[5] M. Matsushita and H. Fujikawa, *Physica A* **168** (1990) 498; M. Ohgiwara, M. Matsushita and T. Matsuyama, *J. Phys. Soc. Jpn.* **61** (1992) 816.

[6] H. Fujikawa, *Physica A* **189** (1992) 15.

[7] J.D. Murray, *Mathematical Biology* (Springer-Verlag, Berlin, 1989) p. 1.

[8] J. Wakita, K. Komatsu, A. Nakahara, T. Matsuyama and M. Matsushita, *J. Phys. Soc. Jpn.* **63** (1994) 1205.

[9] K. Kawasaki, A. Mochizuki, M. Matsushita, T. Umeda and N. Shigesada, *J. Theor. Biol.* **188** (1997) 177.

[10] M. Matsushita, J. Wakita, H. Itoh, I. Ràfols, T. Matsuyama, H. Sakaguchi and M. Mimura, *Physica A* **249** (1998) 517.

[11] I. Golding, Y. Kozlovsky, I. Cohen and E. Ben-Jacob, *Physica A* **260** (1998) 510.

[12] E. Esipov and J.A. Shapiro, *J. Math. Biol.* **36** (1998) 249.

[13] A.M. Lacasta, I.R. Cantalapiedra, C.E. Auguet, A. Peñaranda and L. Ramírez-Piscina, *Phys. Rev. E* **59** (1999) 7036.

[14] T.A. Witten and L.M. Sander, *Phys. Rev. Lett.* **47** (1981) 1400.

[15] T. Vicsek, M. Cserzö and V.K. Horváth, *Physica A* **167** (1990) 315.

[16] E. Ben-Jacob and P. Garik, *Nat* **343** (1990) 523.

[17] A.L. Barabasi and H.E. Stanley, *Fractal concepts in surface growth* (Cambridge University Press, Cambridge, 1995) p. 1.

CHAPTER XI

Dendritic Growth

A. Karma

1. INTRODUCTION AND OUTLINE

The term dendrite originates from the Greek word dendron for tree. Like trees, dendrites (Figs. 1 and 2) are highly branched structure with primary, secondary, and even tertiary branches, which are ubiquitous structures in crystal growth [1]. They can be formed by a variety of processes such as solidification, growth from a vapor, electrochemical deposition, and aggregation [2]. Patterns formed by the interface between two fluids [3] also exhibit similar branched structures if the growth is artificially made anisotropic [4], and *vice versa* dendritic crystals can be made to branch more irregularly like the interface between fluids if crystalline anisotropy is judiciously eliminated [5], as illustrated in Figure 3.

Motivations to study dendrites have been diverse. Kepler's study of snowflakes was largely driven by curiosity. In contrast, the study of solidification dendrites by material scientists has been driven by technology. The reason is that when a metallic alloy freezes dendritically during a number of process (such as casting, welding, or soldering) the freezing is accompanied by strong spatial inhomogeneities of alloy concentration in the solidified material (microsegregation) or to the formation of a new phase in the interdendritic region [6,7]. In both cases, the mechanical integrity of the final solidified product depends critically on the history of the solid-liquid interface evolution during the entire solidification process, including both the development of primary and higher order branches, and the subsequent coarsening of these branches as the interdendritic liquid solidifies.

From a fundamental viewpoint, dendritic crystallization has the strong advantage of being a natural example of branched growth whose basic physics and governing equations are completely known, at least in the simplest systems under not too far from equilibrium conditions (small supersaturation and/or supercooling) [8]. The cooperation of diffusion limited transport of mass and/or

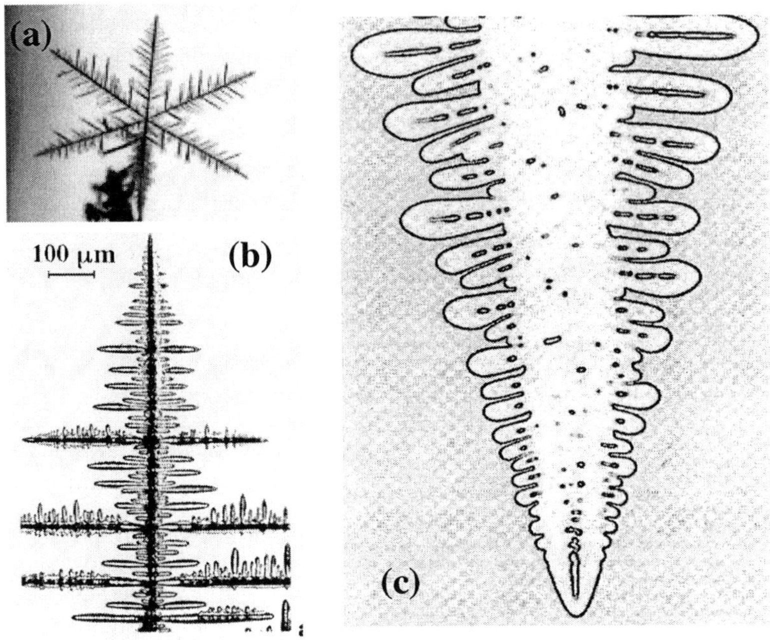

Fig. 1. — Examples of "wild-type" dendritic crystals: (a) snow crystal grown from a vapor (Tanusheva and Libbrecht, Caltech, 1997), (b) ammonium chloride crystal with an underlying cubic symmetry grown from a supersaturated solution (Chan *et al.* [9]), and (c) pure succinonitrile crystal with a cubic symmetry grown from an undercooled melt in a micrograviy experiment (Glicksman *et al.* [14]). The parabolic tip shape and the initial formation of sidebranches are clearly visible in (c).

heat on a macroscopic scale and anisotropic surface tension on an atomic scale suffices, remarkably, to produce a rich variety of branched patterns that have been extensively studied experimentally [5, 9–12, 14–18].

Early theoretical studies of dendritic growth date back to the work of Papapetrou [19] who made the crucial observation that the growth of a parabolic tip at a constant speed is compatible with diffusion limited growth, and of Ivantsov [20] who demonstrated the existence of a family of parabolic tip solutions with an arbitrary speed when surface tension is neglected. In the early nineteen eighties, the study of simplified dendrite growth models that were computationally tractable led to a fundamental understanding of how the addition of anisotropic surface tension fixes this speed [33, 34]. Building on these insights, an elaborate and quantitative theoretical understanding of dendritic growth was subsequently developed through numerous analytical [21–32] and numerical studies [33–49].

Fig. 2. — Exposed cavity revealing the side-by-side growth of a forest of dendrites after solidification of a Co-Cr alloy (Laboratoire de Micro-Analyse des Surfaces, Besançon, France). The three-dimensional structure reflects directly the underlying cubic symmetry of the crystal.

This chapter has the dual purpose to introduce the physics and governing equations of dendritic growth at an elementary level and to review our current theoretical understanding of this problem. A number of new theoretical and experimental insights obtained since earlier reviews of this topic [50–52] are emphasized.

The governing equations of dendritic growth are introduced in Section 2. The basic mechanism of morphogenesis (Mullins-Sekerka instability) that is responsible for the amplification of small bumps on an interface is discussed in Section 3. The growth of tips of needle shaped crystals that develop from such bumps is then treated in Section 4. Growth in two dimensions (2-d) is presently better understood than in 3-d because the application of the conformal mapping technique has yielded a detailed analytical understanding of Laplacian needle growth [31], which applies directly to dendritic growth at low supersaturation. Growth in 2-d is examined first, followed by a discussion of numerical and

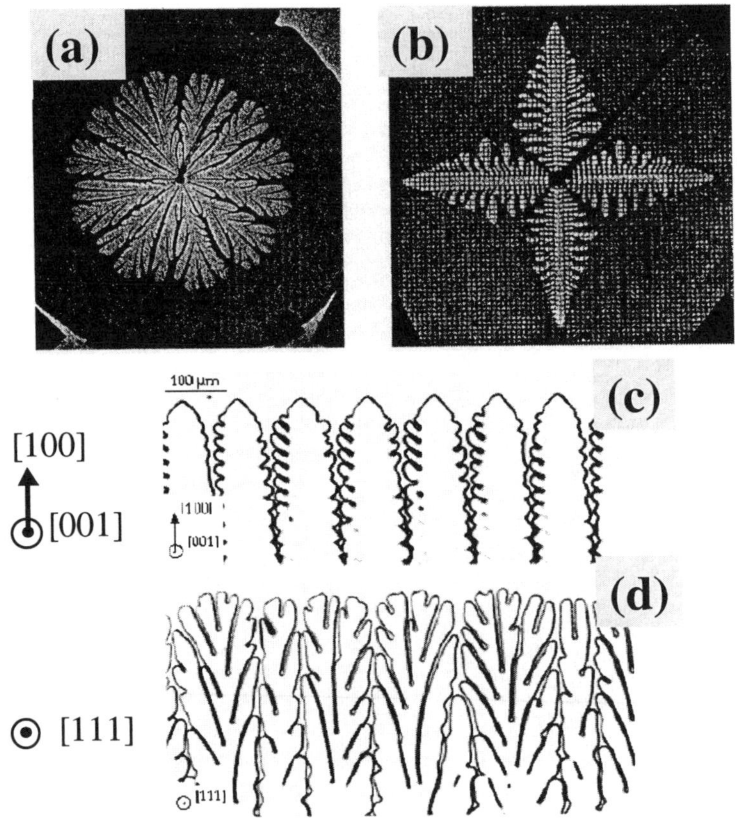

Fig. 3. — Experimental demonstrations of the crucial role of anisotropy. Patterns (a) and (b) are formed during viscous fingering in a Hele-Shaw cell consisting of two closely spaced plexiglas plates. A less viscous fluid injected at the center of the plates pushes out a more viscous fluid towards the edges (Ben-Jacob *et al.* [4]). In (a) there is no anisotropy, whereas in (b) a four fold anisotropy has been imposed by a fine ruling on one plate along the horizontal and vertical directions. Patterns (c) and (d) are formed by pulling a binary liquid mixture sandwiched between two glass plates at a constant speed in an externally imposed temperature gradient (Akamatsu *et al.* [5]), thereby forcing its growth along the axis of the gradient (directional solidification). In (c), there is a four fold anisotropy in the plane of growth, whereas in (d) the anisotropy has been effectively removed by orienting the [111] direction of the crystal perpendicular to the plates.

analytical results in 3-d. Section 5 contains a simple analysis of sidebranching (Fig. 4), which describes how bumps form behind the growing tips of mother needles, followed by a brief discussion of the survival of daughter needles that grow out of these bumps. The classification of various morphologies based

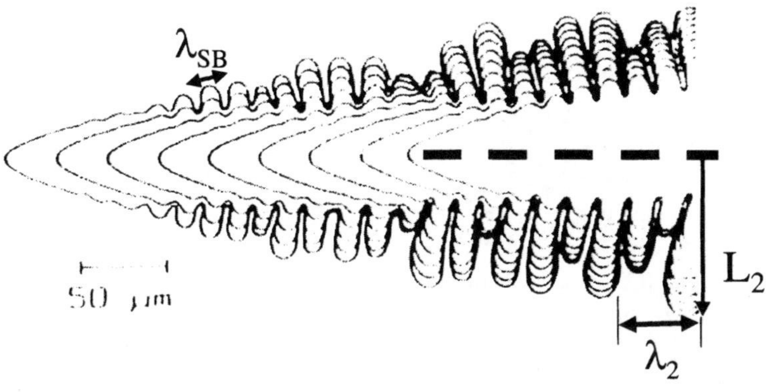

Fig. 4. — Experimental snapshots at equal time intervals of the dendritic growth of an organic crystal (Dougherty *et al.* [12]). Sidebranches appear behind the tip with a small amplitude and an initial spacing λ_{SB}. They then compete with each other for nourishment, such that the length L_2 and the spacing λ_2 of the active ones that survive this competition increases with distance behind the tip, which is also seen in Figures 1a-c.

on different growth solutions and tip splitting is then discussed in Section 6. Finally, Section 7 gives a summary of the main results and a list (far from exhaustive) of some open questions.

This article has been written to be accessible to a broad interdisciplinary readership. Thus technical aspects of theoretical analyses have been omitted for the most part, except in a few parts of Section 4 that some readers may wish to skip.

2. BASIC PHYSICS

The basic physics of dendritic crystal growth is best illustrated by considering the isothermal crystallization of a supersaturated binary liquid mixture of A and B molecules. In this case, one can think of the growing A-rich crystal as being "nourished" by A molecules, which must diffuse from the surrounding liquid, where they are in excess, to the solid-liquid interface where they are gobbled up by the crystal. A direct biological analog, more complex in its details but nonetheless similar, is the growth of a bacterial colony limited by a nutrient, which has been shown to remarkably produce dendrites if anisotropy is imposed on the agar [53].

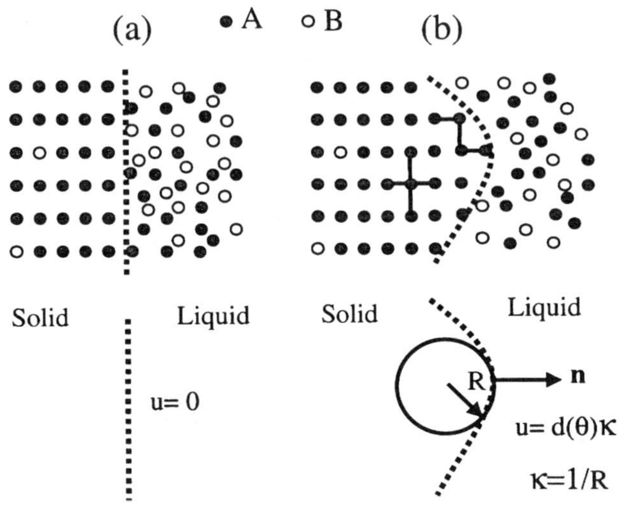

Fig. 5. — Atomic scale representation of the solid-liquid interfacial region in a binary mixture of A and B molecules. u is a dimensionless measure of the excess of A molecules in the liquid phase. When the interface is curved (b), there is a slight excess of molecules proportional to curvature as compared to when it is planar (a) since molecules are in this case less strongly bonded on average to other molecules in the solid.

2.1. Surface Tension

Let us define by $C(\vec{x}, t)$ the number of solute (B) molecules per unit volume at position $\vec{x}$ and time t, and denote by C_1^0 (C_s^0) the equilibrium concentration on the liquid (solid) side of a flat solid-liquid interface, as illustrated in Figure 5a. Typically, $C_s^0 \ll C_1^0$, because the solid is energetically less favorable to solute molecules than the liquid. To make the notation concise, it is convenient to define the dimensionless "nutrient" field in the liquid

$$u(\vec{x}, t) \equiv \frac{C_1^0 - C(\vec{x}, t)}{C_1^0 - C_s^0}, \tag{1}$$

which is positive in any liquid region where there is an excess of A molecules, and is simply zero in equilibrium. Now consider the situation where the crystal makes a protrusion into the melt and the interface is convex (looking from the liquid), as depicted in Figure 5b. In this idealized two-dimensional representation of the crystal with a cubic lattice, a molecule in the solid is bonded to 4 nearest neighbors, each bond being represented by a thick solid line. In contrast, the solid molecule that is at the tip of the protrusion is only bounded to one molecule in the solid. Similarly, the one diagonally above it to the left has only two solid bonds, and so on. Thus solid molecules in a convex region are on average less bonded to other solid molecules. It is thus energetically more favorable for them to escape into the liquid and, consequently, the liquid close to a convex region of interface will be slightly richer in A-molecules ($u > 0$) than close to a flat interface. Moreover, since the average number of broken solid bonds increases with the local interface curvature, κ, defined positively for a convex interface, u must increase with increasing κ. For a gently curved interface ($\kappa a \ll 1$ where a is the lattice spacing), we can assume that this increase is linear, or

$$u = d(\theta)\kappa. \tag{2}$$

Equation (2) is known as the Gibbs-Thomson condition where $d(\theta)$ is a microscopic capillary length that is comparable in magnitude to the range of molecular interactions and proportional to the interfacial stiffness defined below. This length generally depends on the angle, θ, between the direction normal to the interface ($\hat{n}$) and some fixed crystalline axis, for example the axis parallel to the horizontal row of atoms in Figure 5 labeled as the [100] direction. By symmetry, $d(\theta)$ must have the form

$$d(\theta) = d_0(1 - 15\epsilon\cos 4\theta). \tag{3}$$

From thermodynamic considerations, which we skip here [54], one can derive the relation $d(\theta) \approx (\gamma + \gamma'')/(k_B T C_l^0)$ for a dilute alloy, where $\gamma(\theta) = \gamma_0(1 + \epsilon\cos 4\theta)$ is the interfacial energy (*i.e.* excess free-energy of the crystal-melt interface), γ'' is the second derivative of γ with respect to θ with $\gamma + \gamma''$ being commonly referred to as the interfacial stiffness, k_B is the Boltzmann constant, and T is the isothermal growth temperature. From this we deduce the expression for the mean capillary length $d_0 \approx \gamma_0/(k_B T C_l^0)$. The primary branches of dendrites grow typically, but not always as we shall see later, along directions of maximum γ, equivalent to minimum d in two dimensions. We will see later why. For typical organic crystals, measured values of ϵ range from about 0.005 to 0.05 [10, 55, 56], and the only value computed so far for a metallic system falls within this range ($\epsilon \approx 0.02$ for Nickel [57]).

The above description of surface tension is only valid for materials where the solid-liquid interface is molecularly rough, meaning that the transition from an ordered solid to a disordered liquid occurs over several atomic layers as one moves from left to right in Figure 5a. The smooth dendrites shown in Figures 1 and 2 have molecularly rough interfaces. Other materials form facets at orientations where $\gamma(\theta)$ is no longer differentiable and equation (3) becomes invalid; for example, the form $\gamma(\theta) \sim |\theta|$ for $\theta \ll 1$ corresponds to a facet parallel to the vertical axis in Figure 5. The snow-flake is an interesting example where growth along the six hexagonal directions of ice is rough, but faceted perpendicular to the plane of growth of the six primary branches. Since facets tend to grow much slower than rough parts, the snow-flake is a flat quasi two-dimensional branching pattern growing in three-dimensional space.

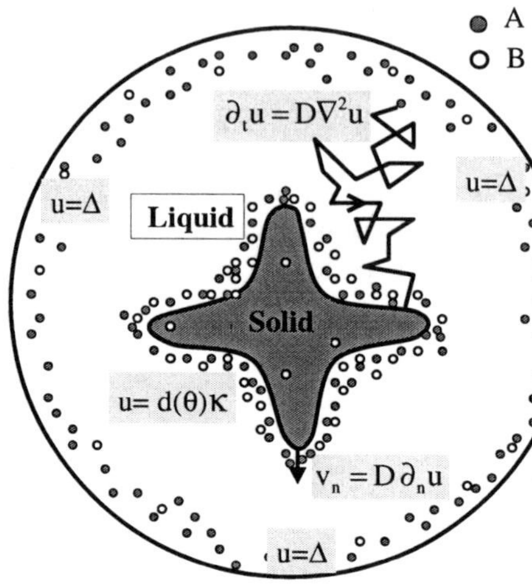

Fig. 6. — Summary of physical processes and equations that govern the crystallization of a supersaturated liquid mixture. The growth of the crystal is nourished by its surrounding liquid that contains an excess Δ of A molecules that diffuse from high to low concentration, and thus from the surrounding liquid to the solid. At the interface, u is proportional to the interface curvature κ. The proportionality constant $d(\theta) = d_0(1 - 15\epsilon \cos 4\theta)$ is a microscopic length of the order of nanometer that varies with the angle θ between the direction normal to the interface and some fixed crystalline axis (for example the horizontal row of atoms in Fig. 5). Although weak ($\epsilon \ll 1$), this variation has a dramatic effect on the growth morphology as illustrated in Figure 3. The interface velocity v_n is proportional to the nourishing flux of A molecules that feeds locally the growth of the crystal.

2.2. Diffusion

Let us now turn to the nonequilibrium growth situation where the liquid is supersaturated, with an initial spatially uniform supersaturation $u = \Delta$. If a circular crystal seed of radius $R > d_0/\Delta$ is introduced in the liquid, a concentration gradient will be automatically set up because the liquid far away from the seed will have a larger excess of A molecules than the liquid close to it, where $u = d_0/R < \Delta$. Since molecules diffuse from high to low concentration, there will be a net inward flux of A molecules towards the interface. This flux will then nourish the growth of the A-rich crystal, as illustrated in Figure 6. In contrast, for $R < d_0/\Delta$, the flux will have the opposite sign and the seed will shrink and vanish, such that d_0/Δ is the minimum radius necessary for a seed to grow, leaving out for simplicity fluctuations.

The spatiotemporal evolution of the concentration in the liquid is described by the standard diffusion equation. We briefly recall its derivation to shed light on the molecular origin of the so-called noise that we consider in the next subsection. Consider a one dimensional example where the concentration varies along x with more molecules at a point P than at a nearby point Q a distance Δx to the right of P. Molecules wander with equal probability to the left or right. Since there are more molecules at P to wander to the right than there are at Q to wander to the left, there will be a net flux, j_x, of molecules from P to Q. This flux is proportional to the difference of concentration between P and Q, or $C(x) - C(x + \Delta x) \approx -(\partial C/\partial x)\,\Delta x$, where $\partial C/\partial x$ denotes the gradient of C along x, or

$$j_x = -D\frac{\partial C}{\partial x}, \tag{4}$$

where D is the liquid diffusivity that has units of $(\text{length})^2/\text{time}$, and j_x is the number of molecules moving in the positive x direction per unit area per unit time. Next, let us calculate the change, ΔN, in the total number of molecules inside a cube of volume $\Delta x A$ (where A is the area of the faces of the cube perpendicular to x) in a time Δt, which is simply $\Delta N = [C(x, t + \Delta t) - C(x, t)]\Delta x A \approx \partial C/\partial t\,\Delta t \Delta x A$. Because molecules are neither created nor destroyed, ΔN must equal the total net number of molecules that enter or leave through the left and right faces of the cube, which we can choose at x and $x + \Delta x$, respectively, or

$$\Delta N = [j_x(x,t) - j_x(x + \Delta x, t)]\, A\Delta t \approx -\frac{\partial j_x}{\partial x}\Delta x A \Delta t, \tag{5}$$

which yields at once the continuity relation

$$\frac{\partial C}{\partial t} = -\frac{\partial j_x}{\partial x}. \tag{6}$$

Substituting j_x defined by equation (4) into the above relation, we obtain the standard diffusion equation $\partial C/\partial t = D\partial^2 C/\partial x^2$ that governs the evolution of C, or expressed in terms of u in two dimensions

$$\frac{\partial u}{\partial t} = D\nabla^2 u, \tag{7}$$

where $\nabla^2 = \partial^2/\partial x^2 + \partial^2/\partial y^2$ denotes the Laplacian and D is the solute diffusivity in the liquid. Here, we neglect diffusion in the solid that is typically several orders of magnitude slower than in the liquid (*i.e.* $D_{\text{solid}} \ll D$), and which thus has a negligible effect on the growth dynamics. It does, however, strongly influence the redistribution of solute molecules in the solid and needs to be considered to predict accurately microsegregation [6].

If the crystal seed is circular, all point on the interface will be nourished at the same rate. Whereas in the situation depicted in Figure 6, the tips of the primary branches will tend to receive more flux and grow faster. When the interface grows, it gobbles up A molecules at a rate proportional to v_n, where v_n is the velocity of the interface measured along the same normal direction. Mass conservation imposes that the number of A molecules gobbled up by the crystal must equal the number transported by the flux normal to the interface, or

$$v_{\mathrm{n}} = D\frac{\partial u}{\partial n}, \tag{8}$$

where n is a coordinate that runs parallel to the direction normal to the interface. Equations (7) and (8), together with the Gibbs-Thomson condition (2) with $d(\theta)$ defined by (3) completely determine the deterministic evolution of the crystal. The Gibbs-Thomson condition has been written here in two dimensions and its form in three dimensions can be found elsewhere [45]. The crystallization of a pure material from its undercooled melt is controlled by an identical set of equations, albeit with u being a dimensionless measure of temperature and with equal diffusivities of heat in the solid and liquid ($D_{\mathrm{solid}} = D$) [8]. The assumption that the Gibbs-Thomson condition still holds in a nonequilibrium situation is generally only valid for small growth rate of the crystal, and thus small supersaturation ($\Delta \ll 1$), which is believed to hold in many experiments. For larger growth rate, nonequilibrium effects at the interface need to be considered that are known to influence dendritic evolution [58].

2.3. Noise

In deriving the diffusion equation, we have assumed that there is not net flux from P to Q if the average number of molecules is the same at these two points. However, when a finite number of molecules wander to the left or right with equal probability, there will always be a net flux from P to Q that fluctuates from time to time in both sign and magnitude due to the random nature of the wandering, similarly to when a coin is flipped N times, the mean number of heads is $N/2$, with a standard deviation $\sim \sqrt{N}$. The standard procedure to account for this randomness is to add a fluctuating current, q_x, to the right-hand-side of equation (4), which becomes

$$j_x = -D\frac{\partial C}{\partial x} + q_x, \tag{9}$$

with a similar equation for y. The continuity relation (6) and its analog for y remain unchanged because mass conservation still holds in this case. Since molecules wander independently in all directions, q_x and q_y can be assumed to be random variables that are uncorrelated in space and time, and with each other, and that obey the same Gaussian distribution. The variance of this distribution must be chosen proportional to DC such that $\langle \Delta N^2 \rangle \approx N$ in any small volume ΔV where $N = C\Delta V \gg 1$. More details on the incorporation of fluctuations into crystal growth models can be found in [59].

It is interesting to note that the condition that fixes the variance of this distribution is a simple form of the fluctuation-dissipation theorem that has originated in a different context. Namely, Nyquist was the first to recognize that the faint *noise* that one hears when listening to a radio, which is produced by the random motion of electrons, is related to the impedance of a resistor across which the voltage is fluctuating. We refer to the wandering of molecules as "noise" even if it is inaudible, and we shall see later why it can be relevant.

3. MORPHOGENESIS

In this section, we discuss the morphological instability of the solid-liquid interface, first analyzed by Mullins and Sekerka [60], which underlies the formation of dendrites.

3.1. Planar Growth

We analyze first the growth of a planar interface and then examine its response to small perturbations in the next subsection. Let us choose the interface to move in the $+y$ direction, and denote by $y_i(t)$ and $v(t) \equiv \mathrm{d}y_i(t)/\mathrm{d}t$ its instantaneous position and velocity, respectively. The equations governing its motion are given by

$$\frac{\partial u}{\partial t} = D\frac{\partial^2 u}{\partial y^2} \tag{10}$$

$$\frac{\mathrm{d}y_i(t)}{\mathrm{d}t} = D\left.\frac{\partial u}{\partial y}\right|_{y=y_i(t)} \tag{11}$$

$$u(y_i(t),t) = 0 \tag{12}$$

subject to the boundary condition $u(\infty) = \Delta$. These equations can be solved exactly by guessing a self-similar solution of the form $u(y,t) = f(z)$, where $z = y/y_i(t)$, in which case equation (10) takes the form $\mathrm{d}^2u/\mathrm{d}z^2 + 2pz\mathrm{d}u/\mathrm{d}z = 0$ where $p = y_i(t)v(t)/(2D)$ is constant if $y_i(t) \sim \sqrt{t}$ and $v(t) = \mathrm{d}y_i(t)/\mathrm{d}t \sim 1/\sqrt{t}$. Solving this equation subject to the boundary conditions (11) and (12), or $\mathrm{d}u/\mathrm{d}z(z=1) = 2p$ and $u(z=1) = 0$, respectively, one obtains the relation

$$\Delta = 2\sqrt{p}\exp(p)\int_{\sqrt{p}}^{\infty} \mathrm{d}z\, \exp(-z^2), \tag{13}$$

between p and Δ which was first derived by Zener [61]. This solution has a simple physical interpretation. When $\Delta < 1$, there is not enough excess nourishment (*i.e.* A molecules) available to completely crystallize the liquid ahead of the interface, which follows simply from global mass conservation. Thus, as the A-rich crystal grows, the available amount of nourishment is depleted and the interface slows down accordingly. There is an equivalent way to obtain the above result in the limit $\Delta \ll 1$ which is useful to mention for what follows. In this limit the diffusion field extends ahead of the interface on a scale $\sim \sqrt{Dt}$, whereas the interface only moves by an amount $y_i(t) \sim \Delta\sqrt{Dt}$ in the same time t. Thus, we can compute the evolution of the diffusion field by assuming that the interface remains stationary at the origin, and then compute $y_i(t)$ by

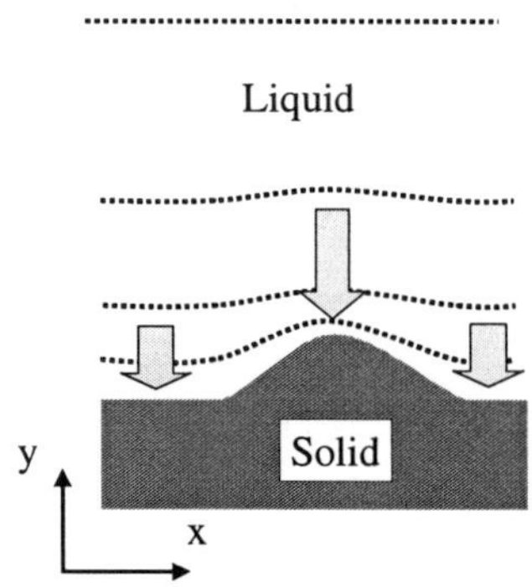

Fig. 7. — Illustration of the morphological instability mechanism. Isoconcentration lines (dashed lines) become more closely spaced ahead of a bump. Consequently, the bump receives a larger nourishing flux (arrows) than flat regions around it and becomes amplified. Surface tension has an opposing effect by increasing the concentration at the bump relative to the flatter regions nearby, and thus inducing a flux away from the tip that starves its growth.

integrating the flux at the origin over time

$$y_i(t) = D \int_0^t \mathrm{d}t' \left. \frac{\partial u(y,t')}{\partial y} \right|_{y=0}, \tag{14}$$

which is a statement of mass conservation: all the A molecules that diffused into the origin had to feed the growth of the crystal. The solution of the one dimensional diffusion equation (10) subject to the fixed boundary condition $u(0,t) = 0$ and the initial condition $u(y,0) = \Delta$ is standard, and yields that the nourishing flux at the origin decreases with time as, $\partial u(y,t)/\partial y|_{y=0} = \Delta/\sqrt{\pi D t}$. Substituting this result in (14), we obtain at once that $y_i(t) = \Delta\sqrt{4Dt/\pi}$, and thus the relation

$$\Delta = \sqrt{\pi p}, \tag{15}$$

which is identical to (13) in the small Δ limit. We make use of this result later on when we consider the steady-state growth of needle crystals.

3.2. Instability

To understand the instability mechanism, consider a bump in the planar interface illustrated in Figure 7, which could have been induced by the noise discussed earlier or some external perturbation. Prior to the formation of this bump, there is a slowly evolving concentration profile ahead of the interface. When the bump is formed, the concentration gradient just ahead of the bump becomes steeper and the rate of nourishment of the crystal increases accordingly. Consequently, the interface speeds up, which causes this gradient to become even steeper, and the interface to accelerate further. In contrast, surface tension has the opposite effect and tends to smooth out this bump. To see why, recall that the concentration is increased ahead of a convex bump of positive curvature because A-molecules are more weakly bonded to the solid on average in this region than on a flat interface. Thus there will be a net mass flux from the bump towards flat or concave regions of the interface next to it, which will tend to starve its growth. The Gibbs-Thomson condition, $u = d_0\kappa$, implies that the sharper the bump is (*i.e.* the larger κ is), the larger is this stabilizing flux. The competition between the destabilizing effect of the diffusion field and the restabilizing effect of surface tension can be analyzed more quantitatively by considering the evolution of a sinusoidal perturbation of the interface of the form $y_i(x,t) = y_i(t) + A_k(t)\sin(ky)$, where $\lambda = 2\pi/k$ is the wavelength of this perturbation and $y_i(t)$ is the unperturbed planar interface position. A standard linear stability analysis [8, 50] valid for small amplitude perturbations ($A_k(t) \ll \lambda$) yields

$$\frac{\mathrm{d}A_k(t)}{\mathrm{d}t} = \sigma_k A_k(t), \tag{16}$$

where

$$\sigma_k = vk - d_0 D k^3, \tag{17}$$

and $v(t) \equiv \mathrm{d}y_i(t)/\mathrm{d}t$ is the instantaneous velocity of the interface. If we assume that $v(t)$ varies sufficiently slowly on a time scale $1/\sigma_k$ to be assumed constant, equation (16) has for solution

$$A_k(t) = A_k(0)\,\mathrm{e}^{\sigma_k t}, \tag{18}$$

which implies that the amplitude grows or decays exponentially in time dependending on whether σ_k is positive or negative, respectively. Equation (17) implies that $\sigma_k > 0$ ($\sigma_k < 0$) for $k < k_\mathrm{s}$ ($k > k_\mathrm{s}$) where $k_\mathrm{s} = \sqrt{v/(d_0 D)} = 2\pi/\lambda_\mathrm{s}$, or equivalently that the interface is unstable (stable) for long (short) wavelength perturbations $\lambda > \lambda_\mathrm{s}$ ($\lambda < \lambda_\mathrm{s}$) where the "stability length" $\lambda_\mathrm{s} = 2\pi\sqrt{d_0 l}$ is the geometric mean of d_0 and the diffusion length $l \equiv D/v$ which measures the scale over which u varies ahead of the interface. This length is typically a few hundred microns for small Δ and consequently λ_s is of the order of micron. Although we have examined for simplicity the instability of a planar interface, a similar instability also causes a circular (spherical) crystal seed to become unstable when its radius is a few times the critical nucleation radius d_0/Δ ($2d_0/\Delta$) in 2-d (3-d) [8].

4. NEEDLE DYNAMICS

Let us now turn to the development of primary branches that follows the initial instability of a crystal seed. We shall treat in this section these branches as "needle crystals" without secondary branches, and then analyze sidebranching in the next section. This is an important simplification that seems to work in practice but has so far never been rigorously justified (nothing forbids, for example, that well developed secondary branches affect the growth of the tip of primary branches at small Δ *via* interactions mediated by the diffusion field).

Needles typically grow along directions of maximum surface energy γ and minimum interfacial stiffness $\gamma + \gamma'' \sim d(\theta)$. Since $d(\theta)$ has four minima in 2-d, the initial crystal will look like a cross with four needles growing horizontally left and right and vertically up and down. In 3-d, there are six directions of maximum surface energy that correspond to the set of $\langle 100 \rangle$ directions of the crystal (Fig. 10). It is physically logical that these directions are preferred because capillarity is less effective at smoothing out bumps along directions where $d(\theta)$ is smallest. At a more rigorous level, however, a self-consistent solution of the growth equations is needed to show that stable tips can only grow along such directions [50].

Two stages of growth should be distinguished. A first transient regime of unsteady growth during which the velocity and shape of the needles is time varying, and a second subsequent regime of steady-state growth where both are constant. Needles grow faster in the first regime because the liquid surrounding them is initially more nourishing. For small Δ, the transient regime can be extremely long and steady-state growth may not necessarily be reached in some experiments. The needle dynamics is determined by the growth and size of its tip region, which can be characterized by the tip speed $V(t)$ and the tip radius of curvature $\rho(t)$. In 2-d, a first tip condition can be derived that expresses the product ρV^2 in terms of a flux intensity factor, $F(t)$, which measures the strength of the mass flux that feeds the growth of the tip region on a scale, intermediate between ρ and the diffusion length, where surface tension effects are negligible. Such a condition was first derived for Laplacian growth [27,31], but also applies to dendritic growth for small supersaturation because the diffusion equation becomes identical to Laplace's equation on a scale smaller than D/V [28]. A second condition that fixes $\rho^2 V$ can then be derived by a self-consistent solution of the growth equations on the scale of the tip where surface tension effects become crucially important. It is expressed mathematically as a solvability condition, which predicts that $\rho^2 V$ is a function of the strength of anisotropy [25, 26, 50]. Together, these two conditions fix $V(t)$ and $\rho(t)$ uniquely in terms of $F(t)$, which can then be computed separately for the transient regime and for steady-state growth where the first condition becomes identical to the Ivantsov relation [20]. We first discuss these two conditions, and then consider their applications to different cases. We then extend the discussion to 3-d where the current state of knowledge is less advanced.

The reader should be warned that the x-y coordinates are defined differently in different subsections of this section. In 2-d, x is always defined to measure distance along the growth axis of the needle from the tip (*i.e.* with the origin $x = 0$ at the tip and the $+x$ direction pointing towards the crystal), except when discussing the early stage growth (Sect. 4.2.1) where x measures the

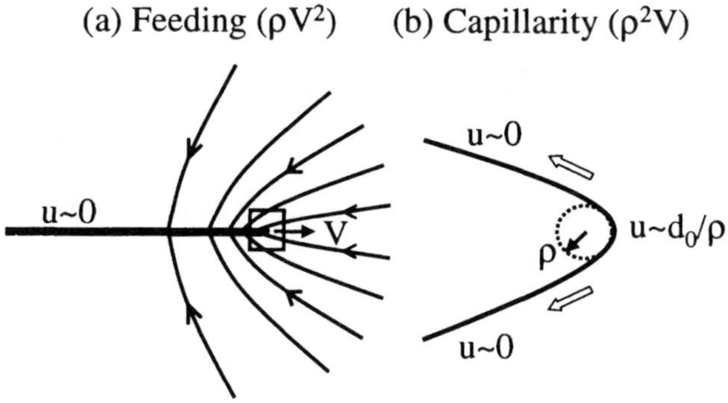

Fig. 8. — Physics underlying two separate tip conditions that determine the tip speed V and tip radius of curvature ρ of needle crystals in two dimensions and viscous fingers in anisotropic Hele-Shaw flow (Hakim [31]). On a scale large compared to ρ (a), the needle appears sharp. The rate of nourishment of the tip region determines the product ρV^2 which only depends on the needle geometry (in general the length and origin of all needles present that interact *via* the diffusion field). (b) shows a magnification of the tip region delimited by the square in (a). On the tip scale ρ, there is a competition between the nourishing flux and the surface tension induced diffusive flux from the tip to flatter regions behind it. The latter flux prevents the tip from becoming too sharp. The balance of these two fluxes fixes the product $\rho^2 V$, which depends on the strength ϵ of anisotropy. ρV^2 is only constant (independent of in time) for a mature needle growing at a constant speed, whereas $\rho^2 V$ becomes constant as soon as a needle is formed and stays constant during its entire development.

position along the growth axis of the needle but with $x = 0$ at the center of growth. In 3-d, the x-y plane is perpendicular to the growth axis, z.

4.1. Tip Growth Conditions

4.1.1. *Feeding:* ρV^2

The condition that determines ρV^2 involves the evaluation of the diffusive flux that feeds the growth of the tip on a scale large compared to ρ. On such a scale the needle appears as a thin line, as illustrated in Figure 8a. To see why, consider for simplicity the steady-state regime where the needle shape is a parabola defined by $x = y^2/(2\rho)$ where x measures the distance from the tip along the growth axis of the needle (with the origin $x = 0$ at the tip) and y measures the half width of the needle at x. For a needle of length L, the width scales as $W \sim \sqrt{\rho L}$, and the width to length ratio $W/L \sim \sqrt{\rho/L}$ becomes negligibly small when $L \gg \rho$. As we shall see shortly, the same is true in the unsteady regime, even though W/L scales differently with ρ/L. We can proceed to evaluate the tip flux by solving the diffusion equation (7), with the boundary condition that $u = 0$ on the needle because surface tension effects are negligible on this scale. In the neighborhood of the tip of this line, the diffusion equation can be approximated by Laplace's equation, $\nabla^2 u = 0$, corresponding physically to the fact that diffusion is essentially instantaneous on the scale of the tip. The solution of Laplace's equation near the tip of a sharp needle is well-known and has a square-root singularity of the form $|\vec{\nabla} u(\vec{r})| \sim 1/\sqrt{|\vec{r}|}$ where $|\vec{r}|$ is the distance from this tip. Consequently, the concentration gradient normal to this line can be written in the form

$$\left.\frac{\partial u}{\partial y}\right|_{y=0} = \frac{F}{\sqrt{d_0\, x}} \qquad \text{for} \quad x \to 0, \tag{19}$$

where d_0 has been introduced to make F a dimensionless measure of the mass flux that feeds the growth of the tip, *i.e.* $\int_0^a \mathrm{d}x\, \partial u/\partial y|_{y=0} \approx 2F\sqrt{a/d_0}$, where a is some arbitrary short distance behind the tip $\sim\rho$. Moreover, F depends solely on the behavior of u far away from the branch. As we shall see below, F depends on time in the unsteady regime and is constant in the steady-state growth regime. We are now left to show that F uniquely fixes the product ρV^2. For this, we can use again the mass conservation condition (14) of Section 3.1 since the region behind the tip on a scale large compared to ρ is essentially planar. In the present context, this condition determines how far the interface has moved normal to the growth axis of the needle in a time $t = x/V$. Changing the variable of integration from t to x, we obtain

$$y_i(x) = D \int_0^x \frac{\mathrm{d}x'}{V} \left.\frac{\partial u}{\partial y}\right|_{y=0} = 2F(D/V)\sqrt{x/d_0}, \tag{20}$$

where $y_i(x)$ is now the half width of the branch. This result implies that the tip shape is a parabola of radius $\rho = 2(FD/V)^2/d_0$, and thus that

$$\rho V^2 = 2(DF)^2/d_0. \tag{21}$$

4.1.2. *Surface Tension: $\rho^2 V$*

The physics that fixes $\rho^2 V$ is the competition of diffusion and surface tension (capillarity) on the scale ρ of the tip, where the actual interface shape now comes into play as illustrated in Figure 8b. The diffusive flux that nourishes the growth of the tip scales as $j_\mathrm{d} \sim D\partial u/\partial n \sim V$ in the tip region, whereas the surface-tension-induced flux that starves the tip, thereby preventing it from becoming too sharp, scales as $j_\mathrm{s} \sim D(u_\mathrm{tip} - u_\mathrm{side})/\rho$, where u_tip and u_side denote the value of u at the tip and a distance of order ρ away on the side, respectively. Using the fact that $u_\mathrm{tip} \approx d_0/\rho$, and $u_\mathrm{side} \approx 0$ since the curvature on the side is small, we obtain that $j_\mathrm{s} \sim Dd_0/\rho^2$. We can therefore measure the relative strength of these two fluxes by a dimensionless parameter $\sigma \sim j_\mathrm{s}/j_\mathrm{d} \sim (\lambda_\mathrm{s}/\rho)^2$, defined such that

$$\rho^2 V = 2Dd_0/\sigma. \tag{22}$$

Following fundamental insights in dendritic growth that emerged in the early nineteen eighties [33, 34], and subsequent analytical [21, 25, 26] and numerical

[35–38] studies since then, it is now clear that σ must be determined by a mathematically self-consistent solution of the growth equations describing a tip of fixed shape moving at constant velocity V. In the determination of such a solution, the computation of σ takes the form of a solvability condition, which corresponds physically to the requirement that the dendrite shape be smooth at the tip. For a completely isotropic surface tension ($\epsilon = 0$), there turns out to be no solution. In this case, dynamical simulations and experiment (Fig. 3) reveal the existence of a tip splitting behavior, or the growth of a doublon structure [5, 40] described in more details later on. For a finite anisotropy strength, however, solutions only exist for a discrete set of values of σ, and only the one corresponding to the largest σ (largest V) among those is linearly stable against infinitesimal perturbations [51], and thus physically admissible. The other solutions are generally linearly unstable against tip splitting modes. The value $\sigma(\epsilon)$ corresponding to the stable tip is a monotonously increasing function of ϵ, with the result $\sigma(\epsilon) \sim \epsilon^{7/4}$ in the limit $\epsilon \to 0$ [21, 25, 26] (also see [50] for a pedagogical exposition).

The fact that the determination of σ only involves a region of the tip of size $\sim \rho$ has an important dynamical consequence. Namely, the value of $\rho^2 V \sim 1/\sigma(\epsilon)$ is established on a short time scale $\tau \sim \rho^2/D$. Thus, even though $\rho^2 V$ is generally computed from a solvability condition that assumes a constant V, its value can be assumed constant even if V changes slowly on the scale of τ. This will hold as long as $V/|\mathrm{d}V/\mathrm{d}t| \gg \rho^2/D$, which implies that the magnitude of the acceleration or deceleration of the tip ($|\mathrm{d}V/\mathrm{d}t|$) cannot be too large. In numerical simulations, this condition turns out to be satisfied at a very early stage during branch development in both two [27, 46] and three ([48] and Fig. 10) dimensions.

4.2. Applications

Combining the two tip growth conditions (21) and (22), we obtain an independent determination of the tip velocity and radius

$$V = \frac{D}{d_0} (2\sigma)^{1/3} F^{4/3} \tag{23}$$

$$\rho = d_0 (2/\sigma^2)^{1/3} F^{-2/3}. \tag{24}$$

Next, we compute F in chronological order of branch development, with the early stage growth transient regime followed by steady-state growth.

4.2.1. Early Stage Growth

Consider the equiaxed growth of the cross depicted schematically in Figure 9a, with four needles of equal length $L(t)$. Since a diffusing molecule moves on average a distance $\sim (Dt)^{1/2}$ in a time t, the initial growth of the crystal is fed by a circular region of liquid of radius $R(t) \sim (Dt)^{1/2}$ that is *independent* of the supersaturation Δ. In contrast, $L(t)$ grows initially much slower than this

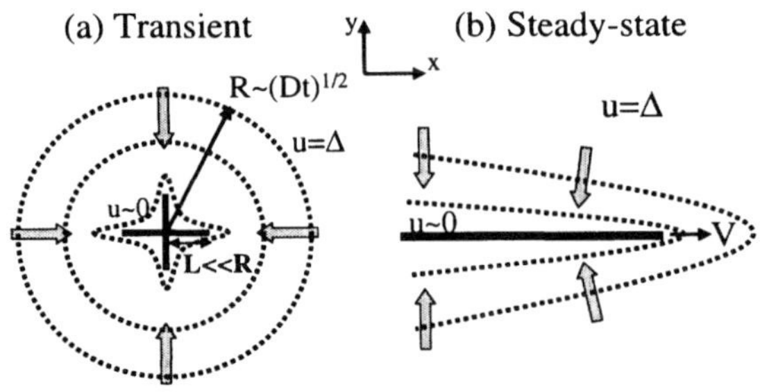

Fig. 9. — Schematic representation of the equiaxed development of four needles fed by a circular diffusion field (a), and of the steady-state growth of one of these needles (b). The dashed lines are contours of equal concentration.

radius because the concentration gradient between the tip, where $u \approx 0$, and the far field region, where $u = \Delta$, vanishes as $\Delta \to 0$. Consequently, on the scale of the cross, diffusion can be assumed to be quasi-instantaneous, and the diffusion equation (7) can be approximated by Laplace's equation, $\nabla^2 u = 0$, with the boundary conditions $u = 0$ on the cross, and $u = -A_{\rm t} \ln r/(2\pi D)$, for $L(t) \ll r \ll R(t)$, far from the cross, where r is the radial distance from the growth center and $A_{\rm t} \equiv {\rm d}A/{\rm d}t$ is the time rate of change of the area $A(t)$ of the crystal. The last condition is simply the statement of global mass conservation, which imposes that A_t be equal to the total cylindrically symmetric flux $D2\pi r \partial u/\partial r|_{r \gg L(t)}$, which feeds the crystal. The mathematical solution of Laplace's equation subject to these boundary conditions can be obtained exactly by conformal mapping technique [27], yielding

$$F(t) = \lim_{x\to L} \sqrt{d_0(L-x)}\, \frac{\partial u}{\partial y}\bigg|_{y=0} = \frac{A_t}{4\pi D}\sqrt{\frac{d_0}{L(t)}}, \tag{25}$$

where $x = 0$ at the growth center and $x = L$ at the tip of the needle. (Note that the result $F(t) \sim (A_{\rm t}/D)\sqrt{d_0/L}$ follows dimensionally from the behavior of the concentration near the needle tip (Eq. (19)), and a detailed calculation is only necessary to obtain the prefactor.) To complete the calculation, we need the time rate of change, $A_{\rm t}$, of the area of the crystal. This quantity can be obtained by considering the evolution of the circularly symmetric diffusion field on a scale $\sim(Dt)^{1/2}$. This evolution is described by Zener's self-similarity solution for a growing circle in two dimensions, which yields that the radius of this circle grows as $R = \sqrt{4p(\Delta)Dt}$ where the Péclet number $p(\Delta) \equiv R({\rm d}R/{\rm d}t)/(2D)$ is implicitly defined by

$$\Delta = p\, e^p \int_p^\infty \frac{e^{-s}}{s}\, {\rm d}s. \tag{26}$$

This implies that the area of the crystal grows linearly with time $A(t) = \pi R^2 = 4\pi p D t$, where $p(\Delta) \approx \Delta/|\ln\Delta|$ for $\Delta \ll 1$, and thus finally that

$$F(t) \approx \frac{\Delta}{|\ln\Delta|}\sqrt{\frac{d_0}{L(t)}}. \tag{27}$$

Combining this result with equation (23), and using the fact that $V(t) = \mathrm{d}L(t)/\mathrm{d}t$, we obtain $\mathrm{d}L/\mathrm{d}t \sim L^{-2/3}$ which yields after integration the final result

$$L(t) = \left[\frac{250\,\sigma}{27}\,\frac{\Delta^4}{|\ln\Delta|^4}\,\frac{D^3}{d_0}\right]^{1/5} t^{3/5}. \tag{28}$$

Thus, the branch length grows $\sim t^{3/5}$ or equivalently $V \sim t^{-2/5}$, which implies that growth slows down with time. Since the area grows linearly in time, the branch width must grow as $W \sim t^{2/5}$. Moreover, it is possible to reconstruct the interface shape $y_i(x,t)$ by utilizing the complete expression for the flux $\partial u/\partial y$ normal to the branch, rather than just the limiting form at the tip that determines $F(t)$ [27]. The result is a self-affine scaling form

$$y_i(x,t) = t^{2/5}Y(x/t^{3/5}), \tag{29}$$

where an expression for the function Y can be found in [27]. This form implies that the interface shape looks the same at any time during the transient if the x and y coordinates are scaled by the branch length and width, respectively. The existence of this scaling behavior has been recently confirmed by numerical simulations [48].

Finally, we had argued earlier that σ could be assumed to be constant during the transient as long as the magnitude of the branch deceleration is not too large, or $V/|\mathrm{d}V/\mathrm{d}t| \gg \rho^2/D$. Using equations (22) and (28), it is easy to show that this condition is equivalent to $L(t) \gg d_0/\sigma$, which always holds for small Δ since the radius of the critical nucleus $\sim d_0/\Delta$.

4.2.2. *Steady-State Growth*

The initial transient eventually ends when the needle catches up the expanding circular diffusion field. This will occur when $L(t_c) \sim (Dt_c)^{1/2}$, which defines the time t_c of cross-over to the steady-state growth regime *via* equation (28). For $t \gg t_c$, the branch becomes surrounded by non-circular iso-concentration lines that appear stationary when viewed in a frame moving at the tip velocity V (with $x = 0$ at the origin and $+x$ pointing into the crystal parallel to the growth axis), as shown schematically in Figure 9b. An expression for F can simply be obtained by noting again that the interface grows essentially as a planar front behind the tip. The result of Section 3.1 for the flux normal to a planar interface $\partial u(y,t)/\partial y|_{y=0} = \Delta/\sqrt{\pi D t}$ ($\Delta \ll 1$) can therefore be used with $t = x/V$, which yields

$$F(t) = \lim_{x\to 0}\sqrt{d_0\,x}\,\left.\frac{\partial u}{\partial y}\right|_{y=0} = \Delta\sqrt{\frac{d_0 V}{\pi D}}. \tag{30}$$

Using equations (23) and (24), we obtain the final expressions

$$\frac{Vd_0}{D} \approx \frac{2\sigma\Delta^4}{\pi^2}, \tag{31}$$

$$\frac{\rho}{d_0} \approx \frac{\pi}{\sigma\Delta^2}, \tag{32}$$

valid for $\Delta \ll 1$. Let us now recover, as announced, the 2-d Ivantsov relation between $p = \rho V/(2D)$ and Δ. Using equations (31) and (32), we obtain that $\Delta = \sqrt{\pi p}$, which indeed coincides with the small Δ limiting form of the 2-d Ivantsov relation

$$\Delta = 2\sqrt{p}\,\exp(p)\int_{\sqrt{p}}^{\infty} dz\,\exp(-z^2), \tag{33}$$

valid for arbitrary Δ. Interestingly, this expression coincides exactly with Zener's expression (13) for the Péclet number of a planar growth front (Sect. 3.1). This follows simply from the fact that the interface grows as a planar front behind the tip for arbitrary Δ.

4.3. Extension to Three Dimensions

Let us now briefly consider the extension of the above results to 3-d, which is typically the case of experimental relevance.

4.3.1. *Tip Growth Conditions*

As in 2-d, the balance of anisotropic surface tension and diffusion on the scale of the tip fixes the value of σ as a function of anisotropy strength *via* a solvability condition, and the same scaling $\sigma \sim \epsilon^{7/4}$ continues to hold in the limit $\epsilon \to 0$ [25, 26]. Moreover, as illustrated by the numerical results in Figure 10, σ becomes independent of time at a very early stage of morphological development of the primary needles, and can therefore be assumed to remain constant. Thus the second condition $\sigma =$ constant continues to hold.

In contrast, there is no direct equivalent of the first condition $\rho V^2 \sim F^2$. One important difference is that the solution of Laplace's equation with $u = 0$ on the needle is not well defined in 3-d if the needle is a line. Thus, unlike in 2-d, the flux intensity factor cannot be simply formulated in terms of the needle geometry, *i.e.* the set of lengths of all needles. Whether it is possible to derive a simplified 3-d description of interface dynamics parameterized in terms of thick needles (as opposed to lines) remains an interesting open question.

4.3.2. *Early Stage Growth*

The qualitative behavior of the early stage growth transient is similar as in 2-d. As illustrated in Figure 10, the needles decelerate as the nourishment becomes depleted and their tip radius increases concomitantly. The lack of a second condition does not make it possible to rigorously describe this transient regime, but numerical results indicate that a scaling behavior may exist with

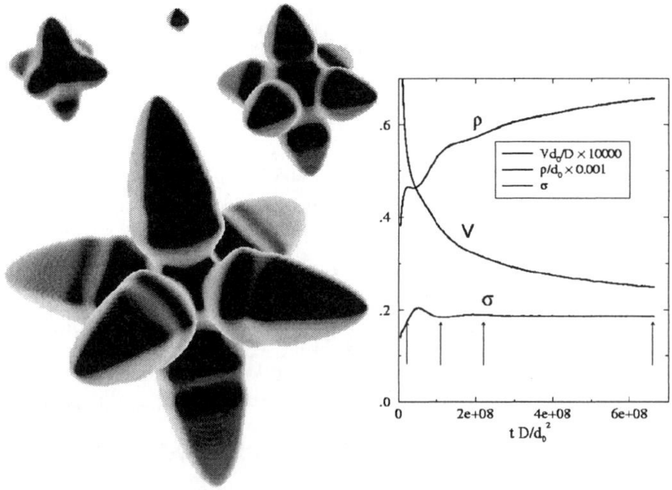

Fig. 10. — Snapshots of a three-dimensional simulation of dendritic growth in a pure undercooled melt for a dimensionless undercooling $\Delta = 0.05$ (analogous to the supersaturation) and a 2.5% anisotropy ($\epsilon = 0.025$), together with time plots of ρ, V, and $\sigma = 2Dd_0/(\rho^2 V)$ (Plapp and Karma [48]). Snapshots are shown at the times corresponding to the arrows and the diffusion field extends spatially on a scale an order of magnitude larger than the structure.

different exponents than in 2-d [48]. Moreover, a simple analytic prediction of these exponents can be made by simply using the fact that σ is constant (second condition) and arguing that the volume of the 3-d needle structure should increase with time as a growing sphere, and thus proportionally to $t^{3/2}$, as long as $L(t) \ll (Dt)^{1/2}$. This simple theory predicts that $L(t) \sim t^{2/3}$ in reasonable quantitative agreement with simulation [48], but the existence of a true self-affine scaling behavior of the needle shape remains unproven.

4.3.3. Steady-State Growth

Steady-state growth is comparatively better understood. V and ρ can simply be computed by using σ =constant together with the 3-d Ivantsov relation [20]

$$\Delta = p \exp(p) \int_p^\infty \mathrm{d}z \, \frac{\exp(-z)}{z} \tag{34}$$

where $p = \rho V/(2D)$. This relationship also describes the Péclet number $R(\mathrm{d}R/\mathrm{d}t)/(2D)$ of a growing 2-d circle of radius $R(t)$, which is the asymptotic cross-sectional shape of the needle crystal far behind the tip in the absence of surface tension anisotropy, in the same way that the 2-d needle crystal Ivantsov relation also describes the Péclet number of a planar front. Furthermore, for small Δ, $p \approx \Delta/|\ln \Delta|$, such that the 3-d analog of equations (31, 32) are

$$Vd_0/D \approx 2\sigma(\Delta/\ln \Delta)^2, \quad (35)$$

$$\rho/d_0 \approx |\ln \Delta|/(\sigma\Delta), \quad (36)$$

in the limit $\Delta \ll 1$. Experimental observations in the weakly anisotropic pure organic crystal succinonitrile (SCN), grown in an undercooled melt, yield a Péclet number and a small Δ value of $\sigma \approx 0.0175$ that agree reasonably well with, respectively, the Ivantsov relation and the prediction of solvability theory. For very small Δ, measured Péclet numbers fall slightly below the Ivantsov prediction and a perturbative analysis based on an anisotropic needle crystal shape has yielded an improved agreement with experiment [64]. Values of σ smaller then predicted by either theory or simulation have been reported for the more anisotropic crystal Pivalic acid (PVA) [14]. Numerical simulations, in contrast, agree reasonably well with solvability theory over a wide range of strength of anisotropy that encompasses ϵ values for both SCN and PVA [45,48,49]. The source of this discrepancy remains to be elucidated.

4.3.4. *Tip Shape*

The 3-d shape deviates from a paraboloid of revolution because of the anisotropy in the surface tension as illustrated in Figure 11. This figure shows a magnification of the tip region of a needle once it has reached steady-state. An analytical solution of the steady-state growth equations in the tip region [26,63] predicts that the amplitude of this deviation is independent of the strength of anisotropy, and has the form

$$z = -\frac{r^2}{2} + A_4 r^4 \cos 4\phi, \quad (37)$$

where (r, ϕ) are the polar coordinates in the plane perpendicular to the growth axis as shown in Figure 11, all lengths are measured in units of ρ, and $A_4 = 1/96$ [63]. The first term on the right-hand-side of equation (37) represents a paraboloid of revolution and the second term is the anisotropic deviation from this paraboloid. Numerical simulations produce a tip morphology that is well fitted by equation (37) and that is, to a good accuracy, independent of the strength of anisotropy (except at the very tip) [49]. The amplitude $A_4 \approx 0.004$, however, is about two and a half time smaller than predicted by solvability theory, and in good agreement with the experimentally measured value for SCN, as shown in Figure 11.

4.3.5. *Fins*

The steady-state shape of 3-d needles becomes even more anisotropic further away from the tip and develops a missile shape with four "fins" in the two planes $x = y = 0$. Brener has remarked [28] that the shape of this fin can be calculated by noting that the 2-d cross-section of the needle perpendicular to z should evolve with increasing distance $|z|$ from the tip as a two-dimensional

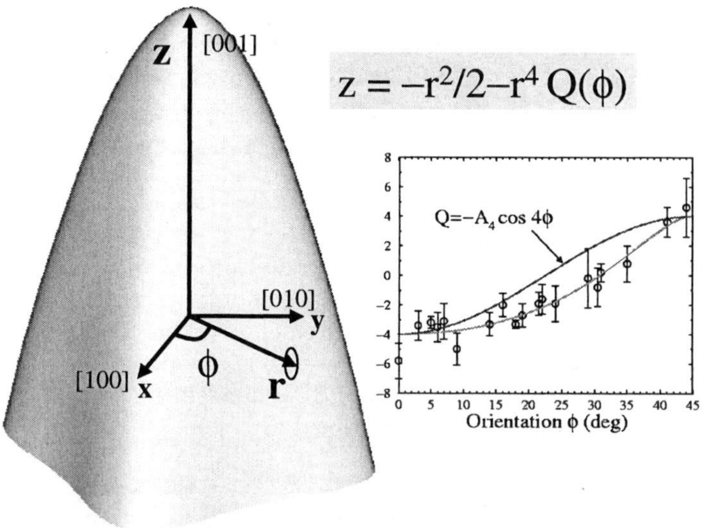

Fig. 11. — Simulated three-dimensional dendrite tip morphology (Karma *et al.* [49]). The departure from a paraboloid of revolution is well fitted by a single azimuthal mode $Q(\phi) = -A_4 \cos 4\phi$, whose amplitude $A_4 \approx 0.004$ is independent of the anisotropy strength and undercooling (supersaturation) if $\Delta \ll 1$. The "projected" shape (gray line) that one would observe by walking around the dendrite standing up parallel to the growth axis is in good quantitative agreement with experimental measurements (open circles and error bars, LaCombe *et al.* [16]).

growth shape in time $t = |z|/V_3$, where V_3 denotes here the steady-state tip speed of the 3-d needle crystal. This mapping is of course only justified far enough away from the tip where the component of the heat flux along z can be neglected. Letting $t = |z|/V_3$ in equation (28), using equation (35) to eliminate V_3, and measuring z and the needle half width x in units of ρ_3 defined by equation (36), we obtain the fin shape

$$z = -B\left(\frac{\sigma_3(\epsilon)}{\sigma_2(\epsilon)}\right)^{1/3} |x|^{5/3}, \tag{38}$$

where $B = 3(4^{1/3})/5$ [66] and where $\sigma_d \equiv 2Dd_0/(\rho_d^2 V_d)$ for $d = 2, 3$. The ratio σ_3/σ_2 approaches a constant in the limit $\epsilon \to 0$ with $(\sigma_3/\sigma_2)^{1/3}$ close to unity, such that equation (38) yields the approximate prediction $z \approx -0.95\,|x|^{5/3}$. The fin shape obtained from numerical simulations was found to be independent of ϵ and well fitted by the power law [49]

$$z \approx -0.68\,|x|^{5/3} \qquad (|z| < 20). \tag{39}$$

The difference of prefactor is not unexpected since the analytical prediction (38) is only strictly valid quantitatively in a region far behind the tip where the cross-sectional shape contains well-developed needles, whereas the numerical fit (39) was performed in a region behind the tip where these needles are just starting to be formed [49]. We note that equation (39) is very close to Brener's original prediction $z = -(3/5)|x|^{5/3}$ [28] that was used as a basis of comparison between theory and numerics in [49]. His prediction, however, becomes $z \approx -0.95\,|x|^{5/3}$ with the correct prefactor $B = 3(4^{1/3})/5$ [66]. Thus, we conclude here that equation (38) does not quantitatively describe the fin shape in the range $|z| < 20$ where this shape plays a crucial role in the onset of sidebranching, as discussed in the next section. The fit (39) should be used instead.

On the experimental side, shape measurements of Xenon dendrites by Bisang and Bilgram [15] have yielded the fit $z = -0.58 \pm 0.04|x|^{1.67\pm0.05}$. In addition, by a detailed analysis of images of pure SCN dendrites grown in microgravity [14], Li and Beckermann [65] have obtained fits of the projected area fraction

$$A_{\mathrm{p}} = 0.847 \pm 0.04\,|z|^{1.60\pm0.03} \qquad (|z| < 30), \tag{40}$$

$$A_{\mathrm{p}} = 0.578 \pm 0.04\,|z|^{1.72\pm0.03} \qquad (|z| > 30), \tag{41}$$

which is defined by $A_{\mathrm{p}} = \int_z^0 x(z)\,\mathrm{d}z$. These fits are to be compared to the analytical prediction $A_{\mathrm{p}} \approx 0.65\,|z|^{1.6}$ obtained using equation (38) with $(\sigma_3/\sigma_2)^{1/3} \approx 1$.

5. BRANCHING

5.1. Sidebranching

5.1.1. *Birth*

Consider the situation depicted in Figure 12 where the tip of a needle is perturbed periodically in time, such that the tip position oscillates at some fixed period T (angular frequency $\omega = 2\pi/T$) around its mean position. Since the tip keeps growing at velocity V as it oscillates, as viewed from a fixed frame, perturbations will travel down the sides at $-V$, as viewed from an observer that rides on the tip. Thus repeated tip oscillations will cause a wavy sinusoidal spatial perturbation of wavelength $\lambda = VT$ to travel along the needle as viewed from this observer. This is similar to waves of wavelength $\lambda = cT$ emanating from the end of a horizontal rope that is waved vertically at period T. The amplitude of such a wavy perturbation will grow exponentially in time as it travels down the sides because of the Mullins-Sekerka instability mechanism discussed in Section 3.2, *i.e.* the mass flux that impinges on the sides nourishes better and better sharper and sharper bumps. Therefore, one would expect sidebranches to form, even if the initial amplitude of the perturbation is small.

The main question becomes then what causes the tip to oscillate? One possibility is that the deterministic growth dynamics is intrinsically oscillatory. Another is that the random motion of diffusing molecules, or some other source

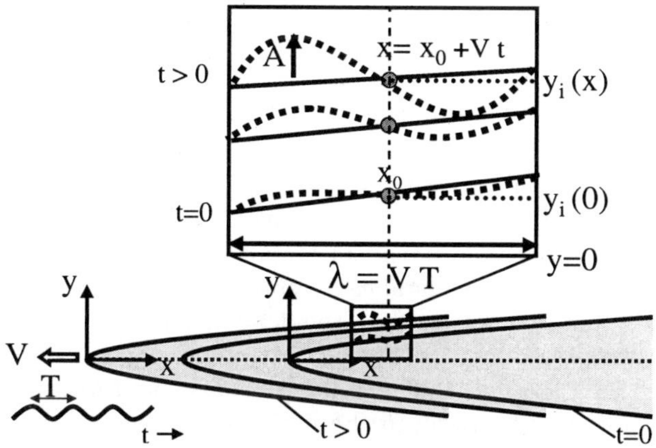

Fig. 12. — Illustration of the amplification of a time periodic perturbation of the tip of a steady-state dendrite. One period T of oscillation causes a sinusoidal deformation of wavelength $\lambda = VT$ behind the tip (dashed line) by conservation of the number of nodes. This deformation then grows in amplitude at a fixed position in a stationary frame by the same instability mechanism as in Figure 7. As viewed from a moving $x-y$ frame with the origin anchored at the tip, this perturbation appears to travel along the needle as it develops.

of noise, causes the tip to undergo aperiodic oscillations with a broad range of frequencies present. Purely deterministic tip oscillations have indeed been found in some simplified toy models of dendritic growth, such as the geometrical model where the interface velocity depends on the curvature κ and its second derivative, $\mathrm{d}^2\kappa/\mathrm{d}s^2$ (which mimics surface tension), where s measures the arclength along the interface [33]. Numerical simulations of realistic models, however, have so far consistently found [40,43,45,46,48,49] that dendrite tips do not oscillate, thereby pointing to noise as the likely source of sidebranching. Moreover, some experiments support this scenario [12].

The current theory of noise-induced sidebranching [24,29] exploits some ideas first developed by Zeldovich and co-workers to analyze the stability of flame fronts [67]. These ideas were subsequently extended by Pelcé and Clavin to treat the stability of various curved fronts (including viscous fingers and dendrites) for a fixed frequency perturbation [23]. Langer then remarked the importance of considering the amplification of an aperiodic perturbation that contains a broad range of frequencies, the key observation being that only the low frequencies are amplified far from the tip [24]. Brener and Temkin [29], in turn, recognized the importance of extending Langer's result to the fin shape described in Section 4.3.5 to obtain agreement with experiment, since the amplification rate is in general sensitive to the shape. In what follows, we derive expressions for the initial sidebranching amplitude A and wavelength λ (*i.e.* the mean initial spacing between sidebranches) that are valid for an arbitrary needle shape. The present analysis is less elaborate than the analyses of Langer [24] or Brener and Temkin [29], but yields results identical to those they obtained for specific needle shapes.

Let us first consider the amplification of a perturbation of fixed frequency ω, and thus fixed spatial wavelength $\lambda = 2\pi V/\omega$. An important simplification, is that the perturbation develops in a region behind the tip where the interface grows essentially as a planar front. We can therefore use the results of Section 3.2 to calculate how much its amplitude has grown in a time t if it was originally centered at a distance x_0 from the tip at time $t = 0$. Viewed from an $x - y$ frame that moves with the dendrite tip, this is equivalent to the perturbation traveling along the needle a distance $x = x_0 + Vt$, as illustrated in Figure 12. Using the fact that $k = 2\pi/\lambda = \omega/V$, and making the change of variable $x = Vt + x_0$, equation (16) becomes

$$V\frac{\mathrm{d}A_\omega}{\mathrm{d}x} = \left(\frac{v(x)}{V}\omega - \frac{d_0 D}{V^3}\omega^3\right) A_\omega, \tag{42}$$

where $v(x)$ is the normal velocity of the needle side at position x. The unperturbed needle shape $y_i(x)$ is constant in time and translates along x at velocity V, such that purely geometric considerations imply that $v = V\cos\theta$ where θ is the angle between the local normal to the interface and the growth axis. In the region of interest where the sides are nearly flat, $v(x)/V = \cos\theta \approx \mathrm{d}y_i/\mathrm{d}x$. Substituting this form into equation (42), and integrating once this equation by separating A and x, we obtain

$$A_\omega(x) = A_\omega(0)\,\mathrm{e}^{S(x,\omega)} \tag{43}$$

where we have defined,

$$S(x,\omega) = y_i(x)\omega/V - x d_0 D\omega^3/V^4, \tag{44}$$

and assumed that $x \gg x_0$ and $y_i(x) \gg y_i(0)$ sufficiently far from the tip. An important feature of the above results is that the amplitude of a fixed frequency perturbation always vanishes sufficiently far from the tip because $y_i(x)$ grows slower than x (*i.e.* $y_i(x) \sim x^{1/2}$ for a parabola and $y_i(x) \sim x^{3/5}$ for the fins). Thus S always becomes negative at large enough x for a fixed ω, and the amplitude of the perturbation vanishes. For a broad band noise, however, arbitrarily low frequencies are excited, and thus $A_\omega(x)$ is amplified for arbitrarily large x. To obtain the observable amplitude, $A_{\mathrm{SB}}(x)$, we must therefore sum the contribution of all possible frequencies, which is justified as long as the perturbation is small enough in amplitude. This sum is dominated by the frequency ω_c which maximizes $S(x,\omega)$. Letting $S'(x,\omega_\mathrm{c}) = 0$ (where prime denotes differentiation with respect to ω) yields at once the expression

$$\omega_\mathrm{c} \approx \left(\frac{y_i(x)V^3}{3d_0 D\,x}\right)^{1/2}, \tag{45}$$

which determines the mean sidebranch spacing at a distance x from the tip, $\lambda_{\text{SB}} = 2\pi V/\omega_{\text{c}}$. The sidebranching amplitude, in turn, is proportional to $e^{S(x,\omega_{\text{c}})}$. However, to obtain the prefactor of this exponential, we need to sum the contribution of frequencies close to ω_{c} by expanding $S(x,\omega)$ in the form $S(x,\omega) = S(x,\omega_{\text{c}}) + S''(x,\omega_{\text{c}})(\omega - \omega_{\text{c}})^2/2$ and integrating over ω. This prefactor depends generally on the type of broad band noise perturbing the tip and is proportional to $\sqrt{\sigma/(C_{\text{l}}^0 d_0)}$ if this noise is due to the random motion of molecules. Using the definition $d_0 \approx \gamma_0/(k_{\text{B}}TC_{\text{l}}^0)$, we obtain finally

$$A_{\text{SB}}/\rho \approx \bar{A}\exp\left[\frac{2}{3}\left(\frac{2[Y_i(X)]^3}{3\sigma X}\right)^{1/2}\right], \tag{46}$$

$$\lambda_{\text{SB}}/\rho \approx \pi\left(\frac{6\sigma X}{Y_i(X)}\right)^{1/2}, \tag{47}$$

where we have defined $X = x/\rho$ and $Y_i(X) = y_i(x/\rho)/\rho$, and the dimensionless noise strength

$$\bar{A} = \sqrt{\sigma k_{\text{B}}T/(\gamma_0\rho^2)}. \tag{48}$$

The same result applies to thermal dendrites growing in a pure undercooled melt except that we must replace σ by 2σ in all the above expressions because equation (17) becomes $\sigma_k = vk - 2d_0Dk^3$ with diffusion (of heat) in the solid. Let us assume that sidebranches become visible at a distance from the tip, X_{SB}, where $A_{\text{SB}}/\rho \approx 0.1$ [24]. For an undercooling $\Delta = 0.003$, the noise amplitude is $\bar{A} = 3.6 \times 10^{-7}$ for SCN. Using the tip shape $X = Y_i^2/2 - 0.004\,Y_i^4$ (defined by Eq. (37) with $A = 0.004$) and $\sigma \approx 0.0175$, we obtain $X_{\text{SB}} = 13$, in reasonable agreement with $X_{\text{SB}}^{\text{expt}} \approx 10$ in the SCN dendritic growth experiments [10]. The numerically determined fin shape $X = 0.68\,Y_i^{5/3}$ [49], which overlaps well with the former over this range of X yields $X_{\text{SB}} = 16$. In contrast, a simple paraboloid $X = Y_i^2/2$ yields $X_{\text{SB}} = 43$.

5.1.2. *Survival*

After birth, sidebranches compete for nourishment and only some of them survive to form fully developed needles. At any given time, one can therefore distinguish between branches with "active tips" that keep growing and inactive ones that stop growing and even shrink. The mean spacing, λ_2, between active branches, and the length of these branches, L_2, increases monotonously with distance x from the tip until x becomes comparable to the diffusion length D/V, where V is the growth rate of the primary branch. For $x > D/V$, active branches no longer compete for nourishment and grow at the same velocity V as primary branches, and λ_2 remains constant. For $x < D/V$, simple scaling arguments lead to the prediction that both λ_2 and L_2 should increase linearly with x [29]. A detailed analysis [65] of experimental measurements on SCN crystals, however, shows a slower growth of active tips, and a scaling behavior $L_2/\rho \approx 0.688(x/\rho)^\nu$ with an exponent $\nu = 0.859$ different from unity.

Li and Beckermann have attributed this discrepancy to the diffusive flux parallel to the growth axis [65].

5.2. Tip Splitting

Let us recall that σ is nearly independent of Δ and is determined by the strength of anisotropy ϵ, whereas the tip radius $\rho \sim d_0|\ln\Delta|/(\sigma\Delta)$ decreases as Δ increases. Since the effective noise amplitude $\bar{A} \sim 1/\rho$, sidebranches will tend to form increasingly closer to the tip as Δ increases. If sidebranches form too close to the tip, the tip will split. A rough criterion for tip splitting to occur is therefore that the sidebranching amplitude becomes comparable to ρ at a distance of order ρ from the tip. Using equations (36, 46), and (48), and the fact that $\sqrt{k_B T/(\gamma_0 d_0^2)} \sim 1$, we obtain the condition that tip splitting will occur when $\Delta > \Delta_c$, where

$$\frac{\Delta_c(\epsilon)}{|\ln\Delta_c(\epsilon)|} \sim [\sigma(\epsilon)]^{-3/2}\exp\left(-c_1/\sqrt{\sigma(\epsilon)}\right), \tag{49}$$

which is valid in the limit $\epsilon \to 0$ where $\sigma(\epsilon) \sim \epsilon^{7/4}$ and c_1 is a constant of order unity. The above expression predicts that Δ_c vanishes exponentially fast in this limit, and thus that dendrites growing in more weakly anisotropic materials should tip split at a smaller driving force. However, $\sigma \approx 0.0175$ even for a weakly anisotropic material like SCN with $\epsilon \approx 0.005$. Consequently, Δ_c is of order unity and tip splitting is not observed under typical experimental conditions where $\Delta \ll 1$. Actually, even for such a small anisotropy, equation (49) is outside its strict asymptotic range of validity for small ϵ. Moreover, when Δ is of order unity, dendrite tips grow four orders of magnitude faster than when $\Delta \sim 0.01$, and the assumption of local thermodynamic equilibrium at the interface breaks down. In this case, there are velocity dependent corrections to the Gibbs-Thomson and mass conservation conditions that need to be taken into account. However, a theory of noise-induced sidebranching and tip splitting that includes such non-equilibrium effects remains lacking.

6. MORPHOLOGICAL CLASSIFICATION

In the previous sections, we have focused our attention on the commonly observed wild-type dendrites with a hierarchical structure of primary, secondary, and higher order branches that form by the emission of sidebranches behind stable tips. However, excluding nonequilibrium effects at the interface and facets, other growth "mutants" are possible for two reasons. The first, just discussed, is that noise can induce tip splitting which typically yields disordered structures. The second, equally important, is that there exists additional purely deterministic solutions to the steady-state growth equations, even in the absence of noise [30, 40, 44].

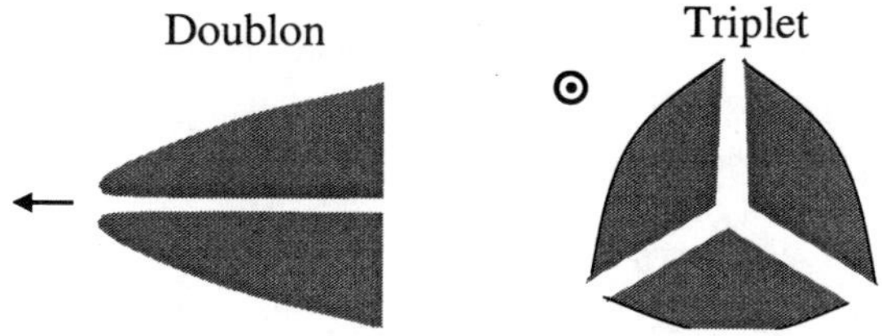

Fig. 13. — Schematic representation of a two-dimensional doublon growing leftward and a three-dimensional triplet growing out of the page. These diffusion-limited "growth mutants" can exist without anisotropy, or for sufficiently large supersaturation if anisotropy is present.

6.1. Doublons and Multiplets

In 2-d, a distinct steady-state growth solution is the so-called doublon [30, 40] that consists of two mirror symmetric fingers with a narrow liquid channel that runs along the axis of symmetry between the two fingers, as illustrated in Figure 13. The two key aspects of the doublons are that (i) they only exist for $\Delta > \Delta_{\text{doublon}}$ where [30]

$$\Delta_{\text{doublon}} \sim \epsilon^{1/4}, \tag{50}$$

and (ii) they grow faster than wild-type dendrites in their regime of existence. Moreover, in an isotropic system ($\epsilon = 0$), where wild-type dendrites cease to exist, a self-consistent solution of the growth equations yields $\sigma \sim p^{-5/2}$ for doublons [30]. The sides of doublons, away from the tip, grow essentially as a planar interface and the steady-state interface shape is therefore parabolic with an effective tip radius ρ determined by the same Ivantsov relation as wild-type dendrites, $p = \rho V/(2D) \approx \Delta^2/\pi$, from which one can deduce the scaling laws $\rho \sim d_0/\Delta^7$, and $V \sim D\Delta^9/d_0$ for the steady-state tip velocity. Since σ depends on the flux that impinges on the tip, the growth conditions used to characterize the unsteady dynamics of wild-type dendrites do not appear applicable to doublons in the isotropic case.

In 3-d, numerical simulations indicate [44] that the analog of the doublon is likely to be the triplet that is shown schematically in Figure 13, although higher order multiplets with more than three "petals" have also been found [44]. Nothing is known analytically as to the regime of existence of these structures but it is plausible, by analogy to 2-d, that they only exist for Δ larger than some anisotropy-dependent threshold.

Experimentally, anisotropy can be eliminated by forcing a crystal with a cubic symmetry to grow between two glass plates with the [111] orientation perpendicular to the plates [5], as shown in Figures 3d and 14c. For such a geometry, there is no second order anisotropy ($\epsilon = 0$) in the two-dimensional plane of growth parallel to the glass plates, and the expected doublon mutant is indeed produced, as illustrated in Figure 14c. Another way to knock out

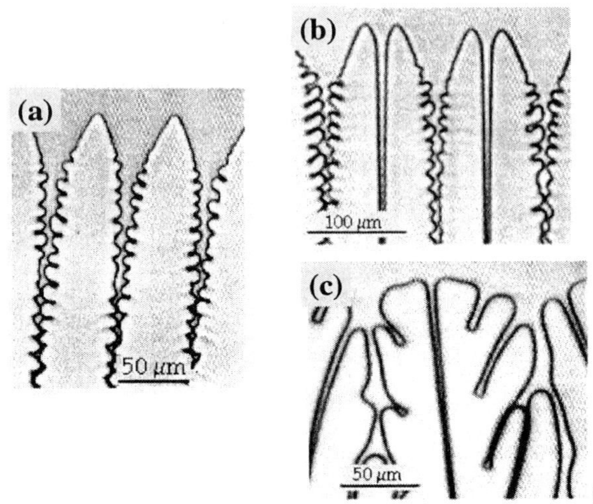

Fig. 14. — Dendrites (a) and doublons (b-c) in the same directional growth experiment as in Figures 3c-d (Akamatsu *et al.* [5]). In (a) and (b) anisotropy is present and the growth rate is larger in (b) than (a). In (c), anisotropy has been eliminated by the same choice of crystal orientation as in Figure 3d.

anisotropy is to work with fluid systems. The classic prototype is viscous fingering that can produce fractal DLA-like patterns, as shown in Figure 3a, but no doublon because the analog of the concentration field is the pressure that obeys Laplace's equation. A fluid analog that produces doublons, however, is dewetting which maps onto isotropic diffusion-limited growth [68]. Experiments that knock out anisotropy in 3-d appear to be difficult.

6.2. Compact and Fractal Structures

Let us now add noise into the picture. Whether the underlying structure is the wild-type dendrite or the doublon (or a triplet in 3-d), two different growth regimes can generally be distinguished, one where the tips are stable and sidebranches form on the sides, and another where the tips are destroyed by noise. In the first regime, the structure can only be assumed to be fractal on scales intermediate between the tip radius ρ and the diffusion length D/V, where V is the steady-state tip velocity. As we have discussed, sidebranches only compete in this range of scales, and become independent of each other when their spacing exceeds D/V. Thus, on scales larger than D/V, the structure appears as a compact object with a well-defined envelope growing at a constant velocity. The solid fraction inside this compact structure is simply determined by overall mass conservation and is equal to Δ. In the second regime, tips are destroyed

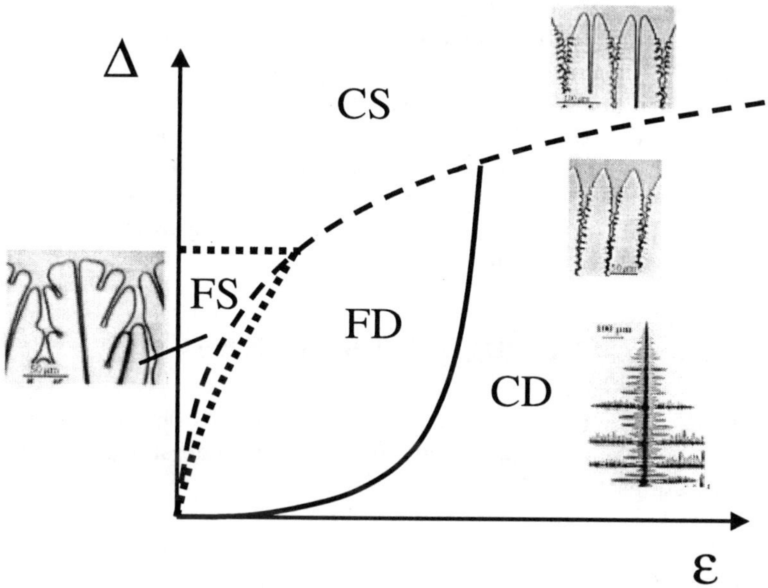

Fig. 15. — Morphology diagram distinguishing different growth morphologies as a function of supersaturation and anisotropy (Brener *et al.* [32]). The compact dendrite (CD) is the commonly observed wild-type morphology that is fractal on scales smaller than the diffusion length D/V, where V is the steady-state tip speed, and compact on a larger scale. The fractal dendrite (FD) has a tip destroyed by noise and thus no upper cutoff scale D/V. The building block of the compact seaweed (CS) is the doublon and the fractal seaweed (FS) has the same building block with a destroyed tip region. The thick solid line and the dashed line correspond to equations (49) and (50), respectively, and analytical estimates of the dotted lines that limit the region of existence of FS structures are given in [32].

by noise and there is no longer a well-defined tip velocity. Consequently, there is no obvious large scale cutoff above which the structure appears compact.

Based on these observations, Brener *et al.* [32] have proposed a morphology diagram, shown schematically in Figure 15, which distinguishes four basic growth morphologies: (i) the compact dendrite, which is the wild-type dendritic structure familiar to metallurgists with nearly paraboidal growth tips, (ii) the fractal dendrite with continuous tip splitting that resembles more DLA clusters, (iii) the compact seaweed made up of doublons with stable tips, and (iv) the fractal seaweed made up of doublons with tips destroyed by noise, where the "seaweed" terminology originates from the resemblance of these more isotropic structures with fingers separated by narrow channels to actual seaweed. The seaweed has also been termed dense branching morphology in earlier references cited in [32].

The line separating compact dendrites and doublons in this plane is defined by equation (50) and the one separating compact and fractal dendrites is simply equation (49) for the onset of tip splitting of wild-type dendrites. An expression for the line between compact and fractal seaweed can similarly be derived by assuming that the transition between these two structures occurs when sidebranches form within one tip radius of the doublon tip. Finally, the line separating fractal dendrites and fractal seaweed has been estimated by using scaling arguments to derive expressions for the growth velocities of the envelopes of these two structures [32].

7. SUMMARY AND FUTURE PROSPECTS

We conclude by summarizing fundamental aspects of dendritic growth (discussed in the previous sections) that are presently understood, and then list a few open questions.

- Dendrites form because an initially small bump on an interface gathers more mass flux from its surrounding medium than flat regions nearby, in complete analogy to the gathering of electrical flux at the tip of a lightening rod. More flux feeds faster growth, and faster growth causes a sharper bump that gathers even more flux, and so on;
- This feedback amplifies exponentially the amplitude of the bump until its growth is cut off by surface tension that creates an opposite flux from the tip of the bump to flatter regions behind it. Thus bumps naturally develop into needle shape objects that can only grow at a constant speed V along predetermined symmetry axes if surface tension is anisotropic. Although this last statement appears intuitively obvious, understanding quantitatively the role of anisotropy in dendritic growth has necessitated a self-consistent mathematical solution of the growth equations (solvability theory) which is itself not intuitive;
- Small fluctuations, which are unavoidably present because of random molecular motion or small inhomogeneities, cause small sidebumps to become amplified on the side of mother needles. These bumps then develop into daughter needles (secondary branches) that can themselves generate other offsprings (tertiary branches) by the same mechanism, thereby giving rise to a highly hierarchical branched structure;
- Shortly after birth, daughter needles compete for nourishment and only a finite fraction survive to grow into large mother needles. The structure is fractal, *i.e.* appears the same at different magnification, only up to the scale where daughter needles reach full maturity. This scale is set by the diffusion length D/V, which is the size of a region of space that can feed the steady-state growth of one tip, or equivalently the distance between needles that no longer compete for nourishment. On larger scales, the

structure looks like a compact porous object with a constant average solid fraction;

- The tip growth of needles, characterized by the tip speed V and tip radius of curvature ρ, is controlled by two key factors, with a separate growth condition rigorously associated with each one in two dimensions. The first is the intensity of the flux from the surrounding that nourishes the growth of the entire tip region, and fixes the product ρV^2. The second, is the balance of the surface-tension-induced flux and the nourishing flux at the curved tip of this region, which fixes $\rho^2 V$ as a function of the strength of anisotropy *via* a solvability condition. This product is constant in time in both two and three dimensions as soon as tips are formed. The flux intensity, in contrast, generally varies in time after birth of a needle and is determined by its growth environment. It only becomes constant in time when a needle reaches full maturity and has established its own territory;
- Fluctuation-induced sidebumps form closer to the tip when the nourishing flux is larger, and the tip splits when two bumps essentially form too close to its apex on either side. Tip splitting requires this flux, and thus the supersaturation to exceed a threshold that depends on the surface tension anisotropy. The smaller the anisotropy, the smaller is the flux needed to split the tip, and this threshold vanishes in isotropic materials;
- It is possible to classify growth morphologies based on the existence of additional steady-state growth solutions (doublons and triplets in two and three dimensions, respectively), which exist even for zero anisotropy, and on whether or not the tips of these solutions and standard needles are destroyed by fluctuations.

Despite these advances, dendritic growth still has its mysteries. For example, it is known experimentally that dendrites can suddenly change growth directions when the supersaturation exceeds some threshold as illustrated in Figure 16 [9]. This transition has been modeled by incorporating the attachment kinetics of atoms at the interface with an anisotropy that favors a different growth direction than the one favored by surface tension [4]. Recent atomistic simulations [69], however, indicate that both anisotropies should favor the [100] direction in cubic crystals. Another possibility, more recently conjectured [70], is that it is the orientation dependence of solute trapping (*i.e.* the fact that a growing crystal incorporates more solute in the solid than one in equilibrium [71]) that is responsible for this change of growth orientation. However, the way in which all three anisotropies (in surface tension, attachment kinetics, and solute trapping) compete to produce different structures in different growth regimes, and under what condition tip splitting occurs when all three are present, remains largely unknown.

The range of possible growth morphologies becomes even richer when misorientations are present, as during the growth of feathery dendrites with an underlying twin structure in aluminum based alloys [72], or when the growing

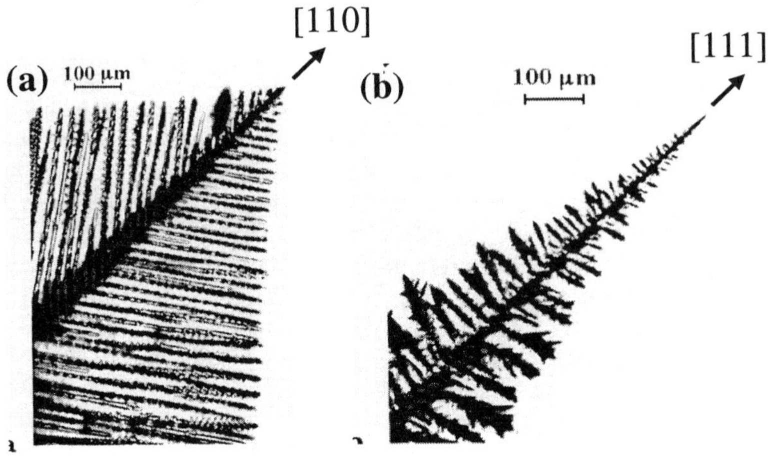

Fig. 16. — Morphologies resulting from the growth of the same ammonium chloride crystal as in Figure 1b, but for a large (a) and a very large (b) supersaturation (Chan *et al.* [9]).

interface is partially faceted [73,74] as in a wide range of other materials. Moreover, fluid flow, which speeds up mass or/and heat transport on large scales has been shown to influence morphological evolution [75,76] and its effect is only starting to be realistically modeled numerically [77,78]. Finally, the transition from cellular to dendritic array structures and the interaction between dendrites during directional growth in an externally imposed temperature gradient still remains poorly understood in comparison to the unconstrained growth of an isolated dendrite in an open geometry [7].

In conclusion, although many basic fundamental aspects of dendritic growth are understood qualitatively, and even quantitatively in many cases, there exists an extremely wide variety of other structures still largely unexplored. Such a diversity is a consequence of the fact that small changes at the microscopic scale of the interface have dramatic consequences at the macroscopic scale of the pattern, not so unlike mutations in biological systems.

Acknowledgements

The author acknowledges the support of the U.S. Department of Energy and NASA in the preparation of this manuscript.

REFERENCES

[1] R. Trivedi and A. Karma, *Encyclopedia of Applied Physics* **23** (1998) 441.
[2] See the article by P. Meakin in this book.

[3] *Dynamics of Curved Fronts*, edited by P. Pelcé (Academic Press, 1988).
[4] E. Ben-Jacob and P. Garik, *Nat* **343** (1990) 523.
[5] S. Akamatsu, G. Faivre and T. Ihle, *Phys. Rev. E* **51** (1995) 4751.
[6] W. Kurz and D.J. Fisher, *Fundamentals of Solidification* (Trans Tech, Switzerland, 1989).
[7] W.J. Boettinger, S.R. Coriell, A.L. Greer, A. Karma, W. Kurz, M. Rappaz and R. Trivedi, *Acta Mater.* **48** (2000) 43.
[8] J.S. Langer, *Rev. Mod. Phys.* **52** (1980) 1.
[9] S.K. Chan, H.H. Reimer and M. Kahlweit, *J. Cryst. Growth* **32** (1976) 303.
[10] C. Huang amd M.E. Glicksman, *Acta Metall.* **29** (1981) 701; **29** (1981) 717.
[11] R. Trivedi and K. Somboonsuk, *Acta Metall.* **33** (1985) 1061.
[12] A. Dougherty, P.D. Kaplan and J.P. Gollub, *Phys. Rev. Lett.* **58** (1987) 1652.
[13] Y. Couder, F. Argoul, A. Arneodo, J. Maurer and M. Rabaud, *Phys. Rev. A* **42** (1990) 3499.
[14] M.E. Glicksman, M.B. Koss and E.A. Winsa, *Phys Rev. Lett.* **73** (1993) 573.
[15] U. Bisang and J.H. Bilgram, *Phys. Rev. Lett.* **21** (1995) 3898.
[16] J.C. LaCombe, M.B. Koss, V.E. Fradkov and M.E. Glicksman, *Phys. Rev. E* **52** (1995) 2778.
[17] W. Losert, B.Q. Shi and H.Z. Cummins, *Proc. Nat. Acad. Sci. USA* **95** (1998) 431; *Proc. Nat. Acad. Sci. USA* **95** (1998) 439.
[18] M. Georgelin and A. Pocheau, *Phys. Rev. E* **57** (1998) 3189.
[19] A. Papapetrou, *Zeitschrift für Kristallographie* **92** (1935) 89.
[20] G.P. Ivantsov, *Dokl. Akad. Nauk SSSR* **58** (1947) 567.
[21] J.S. Langer, *Phys. Rev. A* **33** (1986) 435.
[22] P. Pelcé and Y. Pomeau, *Stud. Appl. Math.* **74** (1986) 245.
[23] P. Pelcé and P. Clavin, *Europhys. Lett.* **3** (1987) 907.
[24] J.S. Langer, *Phys. Rev. A* **36** (1987) 3350.
[25] A. Barbieri and J.S. Langer, *Phys. Rev. A* **39** (1989) 5314.
[26] M. Ben Amar and E. Brener, *Phys. Rev. Lett.* **71** (1993) 589.
[27] R. Almgren, W.S. Dai and V. Hakim, *Phys. Rev. Lett.* **71** (1993) 3461.
[28] E. Brener, *Phys. Rev. Lett.* **71** (1993) 3653.
[29] E. Brener and D. Temkin, *Phys. Rev. A* **51** (1995) 351.
[30] M. Ben Amar and E. Brener, *Phys. Rev. Lett.* **75** (1995) 561.
[31] V. Hakim, *Scale Ivariance, Interfaces, and Non-Equilibrium Dynamics*, edited by A. McKane *et al.*, *NATO ASI Series B: Physics* **344** (Plenum Press, New York, 1995) pp. 63-83, and references therein.
[32] E. Brener, H. Müller-Krumbhaar and D. Temkin, *Phys. Rev. E* **54** (1996) 2714.
[33] D.A. Kessler, J. Koplik and H. Levine, *Phys. Rev. A* **31** (1985) 1712.
[34] E. Ben-Jacob, N.D. Goldenfeld, B.G. Kotliar and J.S. Langer, *Phys. Rev. Lett.* **53** (1984) 2110.

[35] D.A. Kessler and H. Levine, *Phys. Rev. B* **33** (1986) 7687.
[36] D.I. Meiron, *Phys. Rev. A* **33** (1986) 2704.
[37] M. Ben Amar and B. Moussallam, *Physica D* **25** (1987) 155.
[38] D.A. Kessler and H. Levine, *Acta Metall.* **36** (1988) 2693.
[39] R. Kobayashi, *Physica D* **63** (1993) 410.
[40] T. Ihle and H. Muller-Krumbhaar, *Phys. Rev. E* **49** (1994) 2972.
[41] J.A. Warren and W.J. Boettinger, *Acta Metall. Mater. A* **43** (1995) 689-703.
[42] S.-L. Wang and R.F. Sekerka, *Phys. Rev. E* **53** (1996) 3760.
[43] A. Karma and W.J. Rappel, *Phys. Rev. E* **53** (1996) R3017.
[44] T. Abel, E. Brener and H. Müller-Krumbhaar, *Phys. Rev. E* **55** (1997) 7789.
[45] A. Karma and W.J. Rappel, *Phys. Rev. E* **57** (1998) 4323.
[46] N. Provatas, N. Goldenfeld and J. Dantzig, *Phys. Rev. Lett.* **80** (1998) 3308.
[47] A. Karma and W.-J. Rappel, *Phys. Rev. E* **60** (1999) 3614-3625.
[48] M. Plapp and A. Karma, *Phys. Rev. Lett.* **84** (2000) 1740-1743.
[49] A. Karma, Y.H. Lee and M. Plapp, *Phys. Rev. E* **61** (2000) 3996.
[50] J.S. Langer, *Chance and Matter*, Lectures on the Theory of Pattern Formation, Les Houches, Session XLVI, edited by J. Souletie, J. Vannimenus and R. Stora (North Holland, Amsterdam, 1987) pp. 629-711.
[51] D. Kessler, J. Koplik and H. Levine, *Adv. Phys.* **37** (1988) 255.
[52] J.S. Langer, *Sci* **243** (1989) 1150-1153.
[53] E. Ben-Jacob, H. Shmueli, O. Shochet and A. Tenenbaum, *Physica A* **187** (1992) 378.
[54] *e.g.* see D.P. Woodruff, *The Solid Liquid Interface* (Cambridge University Press, London, 1973).
[55] M.E. Glicksman and N.B. Singh, *J. Cryst. Growth* **98** (1989) 277.
[56] M. Muschol, D. Liu and H.Z. Cummins, *Phys. Rev. A* **46** (1992) 1038.
[57] J.J. Hoyt, M. Asta and A. Karma, *A method for computing the anisotropy of the solid-liquid interfacial free energy* (2000) preprint.
[58] E.A. Brener and V.I. Mel'nikov, *Adv. Phys.* **40** (1991) 53.
[59] A. Karma, *Phys. Rev. E* **48** (1993) 3441.
[60] W.W. Mullins and R.F. Sekerka, *Appl. Phys.* **35** (1964) 444.
[61] C. Zener, *J. Appl. Phys.* **20** (1949) 950.
[62] E.R. Rubinstein and M.E. Glicksman, *J. Cryst. Growth* **112** (1991) 84.
[63] E.A. Brener and V.I. Mel'nikov, *JETP* **80** (1995) 341.
[64] G.B. McFadden, S.R. Corriel and R.F. Sekerka, *J. Cryst. Growth* **208** (2000) 726.
[65] Q. Li and C. Beckermann, *Phys. Rev. E* **57** (1998) 3176; *Acta Mater.* **47** (1999) 2345.
[66] The present value of $B = 3(4^{1/3})/5$ corrects the erroneous value $B = 3/5$ derived in reference [28] and used subsequently to analyze sidebranching in reference [29].
[67] Ya.B. Zel'dovich, A.G. Istratov, N.I. Kidin and V.B. Librovich, *Combust. Sci. Technol.* **24** (1980) 1.

[68] See the article by C. Misbah in this book.
[69] J.J. Hoyt, B. Sadigh, M. Asta and S.M. Foiles, *Acta Mater.* **47** (1999) 3181.
[70] K.A. Jackson (private communications).
[71] M.J. Aziz and T. Kaplan, *Acta Metall.* **36** (1988) 2335.
[72] S. Henry, T. Minghetti and M. Rappaz, *Acta Mater.* **46** (1998) 6431.
[73] M. Adda-Bedia and V. Hakim, *J. Phys. I France* **4** (1994) 383.
[74] M. Adda-Bedia and M. Ben Amar, *Phys. Rev. E* **51** (1995) 1268.
[75] M.E. Glicksman and S.C. Huang, *Convective Transport and Instability Phenomena*, edited by J. Zierip and H. Oertel Jr. (Karlsruhe, West Germany: G. Braun, 1982) pp. 557-573.
[76] P. Bouissou, B. Perrin and P. Tabeling, *Phys. Rev. A* **40** (1989) 509-512.
[77] R. Tonhardt and G. Amberg, *J. Cryst. Growth* **194** (1998) 406.
[78] X. Tong, C. Beckermann and A. Karma, *Phys. Rev. E* **61** (2000) R49.

CONTRIBUTED PAPER 9

Sidebranching in Solutal Dendritic Growth

R. González-Cinca, L. Ramírez-Piscina,
J. Casademunt and A. Hernández-Machado

1. INTRODUCTION

Dendritic growth in solidification is one of the most studied problems in branching morphogenesis [1, 2]. In this contribution we study sidebranching induced by fluctuations in dendritic growth. We compute the amplitude of sidebranching induced by internal (equilibrium) concentration fluctuations in the case of solidification with solutal diffusion. This amplitude could be too small to explain previous experiments. We investigate the effects of other possible sources of fluctuations (of external origin) by introducing a non-conserved noise in a phase-field model. Results reproduce the characteristics of sidebranching found in both experiments and theory. The way noise is introduced in the phase-field model appears to be qualitatively fair in order to investigate sidebranching phenomena in experiments where this is due to the selective amplification of noise, regardless of its origin.

The question of how frequency and amplitude of sidebranching are determined is not fully solved yet. Two scenarios have been proposed to explain the origin of sidebranching. One of them states that periodic deterministic oscillations at the tip generate correlated branches at both flanks of the dendrite. The other scenario proposes that sidebranching is due to selective amplification of fluctuations (of either internal or external origin) near the tip of the dendrite. In this case, branches appear mostly uncorrelated. In this contribution we will study this second scenario by means of a phase-field model [3].

Dougherty *et al.* [4] studied the sidebranching in NH_4Br dendrites, where apparently random variations in phase and amplitude of the branches were observed. They determined the amplitude of the sidebranching and its mean wavelength by looking at the power spectrum of the data obtained by measuring the half-width of the dendrite at a fixed distance z from the tip at different times.

In this decade there has been an increasing use of phase-field models to tackle the dendritic growth problems. They are a very useful and practical tool to simulate such kind of processes and a good alternative to the integrodifferential equation which can be derived from the classical sharp interface model. In the phase-field models an order parameter or phase field ϕ is defined, which permits to avoid boundary conditions at the interface. An example of a dendrite obtained by means of a phase-field model is shown in Figure 1. The deepest insights into the study of sidebranching with a phase-field model have been carried out very recently in [5]. They included thermal noise in a two-dimensional phase-field model of solidification controlled by heat diffusion, in a way consistent with both bulk and interfacial equilibrium fluctuations, like it had been done previously with the sharp interface model equations.

Due to the lack so far of explicit predictions of sidebranching amplitudes for solutal dendrites, experiments of these dendrites have only shown qualitative agreement with theoretical results. In this contribution we address the question of sidebranching characteristics in presence of non-internal *vs.* internal fluctuations. First of all, we obtain a prediction of the effects of internal noise on sidebranching amplitudes for solutal dendrites. A comparison of the theory with available quantitative experimental results shows that there are serious indications that in some experiments internal thermodynamical fluctuations could not account for the observed sidebranching activity. In that case some other source of fluctuations, of external origin, should be invoked. We then study some of the consequences derived from adding a non-conserved noise source in a two-dimensional phase-field model. This noise is of very different nature than what one should employ to give account of internal fluctuations [5]. However our simulations qualitatively reproduce the properties of the noise-induced sidebranching derived analytically and observed experimentally [4]. We conclude that despite thermal noise could not always be the main origin of dendritic sidebranching, its qualitative characteristics are common for noise-induced sidebranching independently of its origin.

2. SIDEBRANCHING IN SOLUTAL SOLIDIFICATION

Available theoretical predictions on sidebranching amplitudes have been formulated for dendrites grown from a pure substance and controlled by heat diffusion. We will consider here dendrites appearing in isothermal growth of mixtures controlled by diffusion of solute.

Sidebranching amplitude is predicted to depend on the distance z along the dendrite axis as [5,6]

$$A(z) = \rho \overline{S} \exp\left(\frac{2}{3}\left[\frac{x_0(z)}{3\sigma^* z \rho^2}\right]^{1/2}\right) \tag{1}$$

Fig. 1. — Dendrite obtained with the phase-field model.

where ρ is the tip radius, $x_0(z)$ is the shape of the dendrite, and the operating mode of the dendrite is given by the parameter σ^* defined by

$$\sigma^* = 2D\, d_0/\rho^2 V \tag{2}$$

where V is the selected velocity, d_0 is the capillary length and D is the diffusion coefficient. In analogy with the case of thermal dendrites, we find that the dimensionless noise amplitude for a diluted solutal dendrite reads

$$\overline{S}^2 = \frac{2C_L^{\rm eq} D}{\left(\Delta C^{\rm eq}\right)^2 \rho^4 V}, \tag{3}$$

where $C_L^{\rm eq}$ is the equilibrium concentration in the liquid phase and $\Delta C^{\rm eq}$ is the equilibrium miscibility gap.

There are no much quantitative experimental results available on sidebranching in solutal dendrites. We focus on the experiments performed in [4] with ammonium bromide dendrites growing from supersaturated aqueous solution in isothermal conditions. When we compare the prediction of equations (1, 3) with the experimental results on the amplitude of the sidebranching shown in Figure 4 of reference [4], one can observe that the predicted amplitude when it is due to statistical noise is at least one order of magnitude smaller than the amplitude experimentally observed. Thus we are lead to conclude that thermodynamical fluctuations could be insufficient to explain the sidebranching amplitude in some experiments.

3. SIDEBRANCHING CHARACTERISTICS

We have performed simulations of dendritic growth by employing a phase-field model for solidification. Noise has been introduced in a non-conserved way, as opposed to the usual case of internal noise, to account for more general external noise sources. Branches at both sides of the dendrite have been found to be weakly correlated among them. The amplitude of the sidebranching have been computed in the same way as in [4] and in Figure 2 it is shown its dependence on the distance to the tip. The qualitative behavior coincides with the theoretical predictions, that is, it increases as $\exp(z^{2/5})$ in the linear regime and as z in the nonlinear regime [5, 6]. The sidebranching wavelength has also been computed and it also shows an overall qualitative agreement with theoretical predictions [5,6] (see Fig. 3). It increases as $z^{1/5}$ in the linear regime and as z in the nonlinear regime. The distance to the tip where nonlinear effects start to play an important role coincides in Figures 2 and 3.

4. CONCLUSIONS

From direct comparison with experiments of solutal dendritic growth, we have seen that internal fluctuations could be insufficient to explain actual sidebranching activity. In order to study the effect of other noise sources of external origin, we have used a phase-field model with non-conserved noise.

As a first step we have obtained an estimation of sidebranching amplitudes originated by internal fluctuations in solutal diluted dendrites. The resulting values appear as much smaller (at least one order of magnitude) than observed

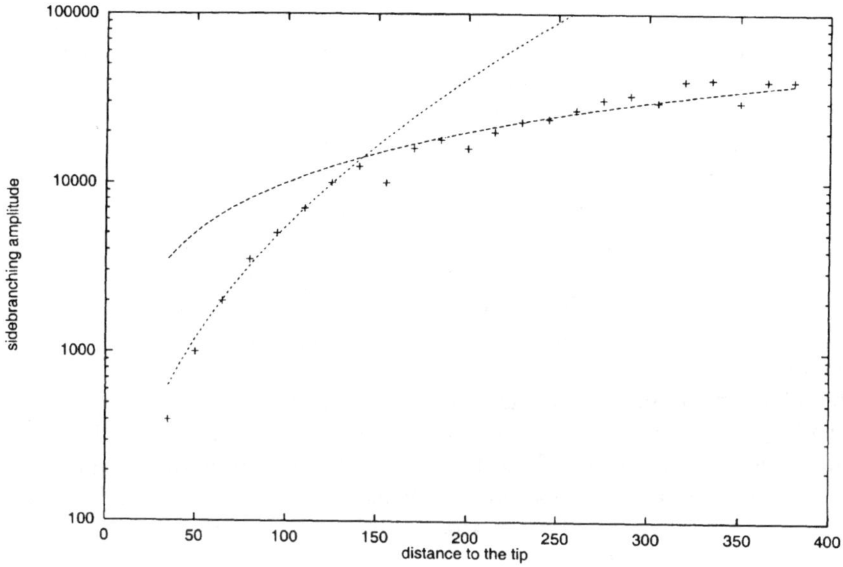

Fig. 2. — Sidebranching amplitude as a function of z (dimensionless units). Points correspond to simulation results and lines are theoretical predictions.

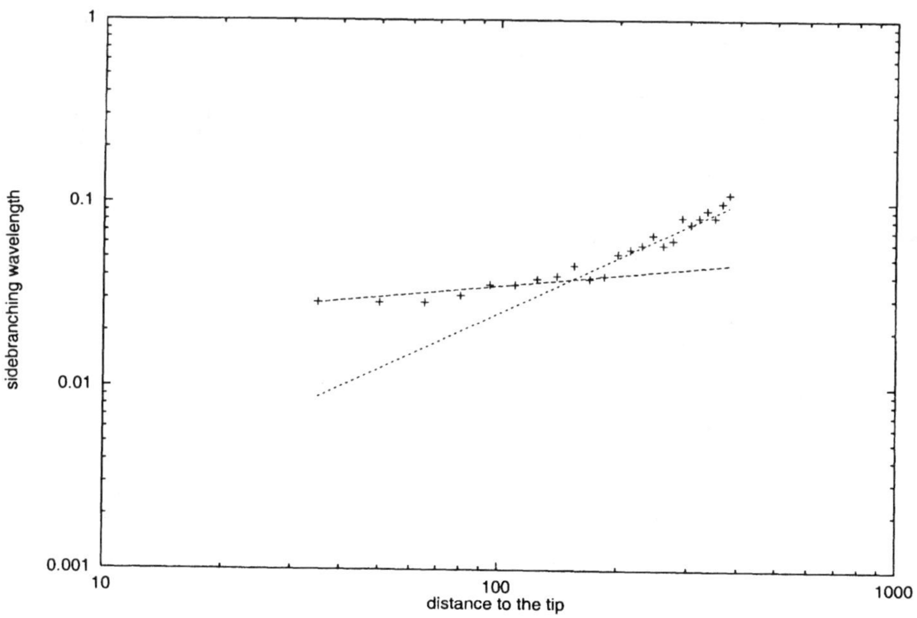

Fig. 3. — Sidebranching wavelength as a function of z (dimensionless units). Points correspond to simulation results and lines are theoretical predictions.

in some experiments (performed in concentrated solutions). That can be attributed to the effect of other noise sources, of external origin. To confirm this conclusion, it should be necessary to make use of quantitative experimental results obtained in more diluted conditions.

Simulations of the model with external noise reproduce qualitatively well previous experimental an analytical results. In particular, we have reproduced the dependence of the amplitude and the wavelength of the sidebranching on the distance from the tip, confirming the validity of what had been previously obtained with a linear theory where internal thermal noise had been considered. Thus the qualitative behavior of the amplitude of the sidebranching, when this is due to the selective amplification of fluctuations, is independent of the origin of the noise.

The phase-field model appears as a versatile method to study dendritic growth in general and sidebranching characteristics in particular. It has been shown to be adaptable for taking into account thermodynamical fluctuations [5], but more simple modifications (such as the one employed in this work) appears to be appropriated to qualitatively reproduce the behavior of sidebranching, regardless of the noise origin.

In summary, our results indicate that further experimental information, particularly for diluted conditions, would be useful for a more quantitative conclusion on the origin of noise in sidebranching experiments.

Acknowledgements

Authors thank the Dirección General de Investigación Científica y Técnica (Spain) (Projects PB96-0241-C02-02, PB96-1001-C02-02 and PB96-0378-C02-01), Comissionat per a Universitats i Recerca (Spain) (Projects 1999SGR00145, 1999SGR00041) and European Commission (Project TMR-ERB FMRX-CT 96-0085) for financial support.

REFERENCES

[1] E.A. Brener and V.I. Melnikov, *Adv. Phys.* **40** (1991) 53.
[2] J.P. Gollub and J.S. Langer, *Rev. Mod. Phys.* **71** (1999) 6396.
[3] R. González-Cinca, L. Ramírez-Piscina, J. Casademunt and A. Hernández-Machado, *Sidebranching induced by external noise in solutal dendritic growth*, preprint (2000).
[4] A. Dougherty, P.D. Kaplan and J.P. Gollub, *Phys. Rev. Lett.* **58** (1987) 1652.
[5] A. Karma and W.-J. Rappel, *Phys. Rev. E* **60** (1999) 3614.
[6] E. Brener and D. Temkin, *Phys. Rev. E* **51** (1995) 351.

CONTRIBUTED PAPER 10

Experimental Study of Sidebranching in Directional Solidification

M. Georgelin and A. Pocheau

1. INTRODUCTION

Directional solidification of an alloy stands as a well-defined physical problem involving both fundamental interest and metallurgical applications [1–5]. In addition, it provides liquid-solid interfaces showing many branching structures reminiscent of those occurring in much more complex systems as living organisms. However, although the equations describing the evolution of the liquid-solid interface have been well-known for a while, the origin of a number of solidification phenomena still remains object of active study. Among these stands the phenomenon of sidebranching in which we shall be especially interested here.

2. SECONDARY CELL INSTABILITIES

Directional solidification consists in pulling at controlled velocity V a thin film of alloy from a hot region (liquid phase) towards a cold region (solid phase). Here, the sample thickness is about 50 μm and the melt alloy is a plastic crystal: impure succinonitrile. For the present transparent material, the liquid-solid interface can be observed with a microscope without perturbing the system. At low velocity, the solidification front is planar. Above a critical velocity V_c of about one μm/s, a well-known primary instability – the Mullins-Sekerka instability – destabilizes the planar interface and restabilizes it onto an array of cellular structures (Fig. 1). At still higher velocities, secondary instabilities modify the shape and dynamics of the cellular array. Our purpose here is to discuss the origin and the nature of one of them: the sidebranching instability.

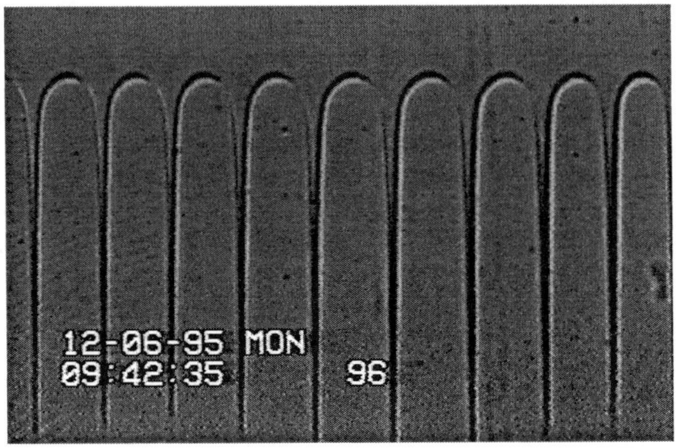

Fig. 1. — Array of solidification cells. Image width is 1 mm and pulling velocity is 10 μm/s, *i.e.* $6.3V_c$.

At first, it is worth emphasizing that the basic state of our study, *i.e.* the array of solidification cells, stays beyond present theory. Indeed, as often in physics, only limiting cases allow analytical determination. Here, at high velocity, an analogy with free growth suggests that cell shape is nearly parabolic [6–8]. Conversely, at low velocity, analogy with viscous fingering yields a Saffman-Taylor shape to be proposed for cells [9]. However, both these limits are irrelevant to our intermediate regime of velocity, so that experiment and simulation presently stand as the two most efficient ways for addressing our system.

Before focusing attention on sidebranching, let us mention that this phenomenon belongs to one of the various secondary instabilities occurring on cellular arrays. These instabilities actually lead to different events as oscillation, parity breaking, wavelength doubling, or combination of these modes. Each of them occurs in a particular range of control parameters represented by one domain in the instability diagram of Figure 2. Here, each instability domain actually corresponds to the existence of a specific phenomenon, *e.g.* creation or disappearance of a cell, parity breaking (doublets), oscillation with period doubling (2λ–O) [10] or dendrites. Note that dendrites appear at high velocity and large cell spacing, and that sidebranch amplitude is more or less developed (Fig. 3) depending on the proximity to the border line of the dendritic domain.

3. SIDEBRANCHING

3.1. Critical Curve

Following Figure 3, some definite states regarding sidebranching may be experimentally determined: the states lying in between cells (no visible sidebranch)

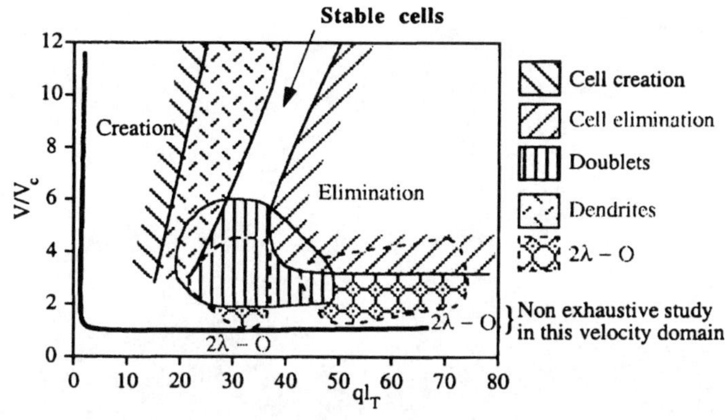

Fig. 2. — Diagram of secondary instabilities in directional solidification at $G =$ 110 K/cm. V velocity, V_c critical velocity of Mullins-Sekerka instability, l_T thermal length, $q = 2\pi/\Lambda$, Λ cell width. Elimination (resp. creation) occurs for thin (resp. large) cells. Many of these instabilities display subcritical features, so that instability domains can overlap.

and dendrites (some visible sidebranches), *i.e.* the states involving marginal sidebranching. Our goal has been to locate them with respect to control parameters so as to be able to address sidebranching from the features revealed by its onset [11].

First note that large cells show more developed dendrites than thin cells. Accordingly, onset of sidebranching actually depends not only on pulling velocity V and thermal gradient G, but also on cell width Λ. To locate marginal sidebranching with respect to control parameters, we used the fact that, depending on history, cell spacing Λ can noticeably vary. For each value of G, we then scanned a range of values of V and Λ so as to discriminate between cells showing no sidebranch ("stable" domain) from dendrites showing some ("unstable" domain). The separation line between the two domains corresponds to the critical curve relevant to the onset of sidebranching. It is reported in Figure 4 for two different values of G. Curves show that increasing any of the control parameters Λ, V, G promotes sidebranching: the three control parameters stand as destabilizing factors for sidebranching [11, 12].

3.2. Noise-Amplification Theory

A great amount of theoretical studies [13, 14] has been devoted to physical understanding of sidebranching in the case of *free* growth, *i.e.* growth from an *isothermal* undercooled melt, at $G = 0$. It is proposed that perturbation at cell tip gets amplified by a linear mechanism analogous to the Mullins-Sekerka primary instability. However, as the unperturbed front is no longer planar but parabolic, kinematic of growth at constant velocity V and constant mean shape

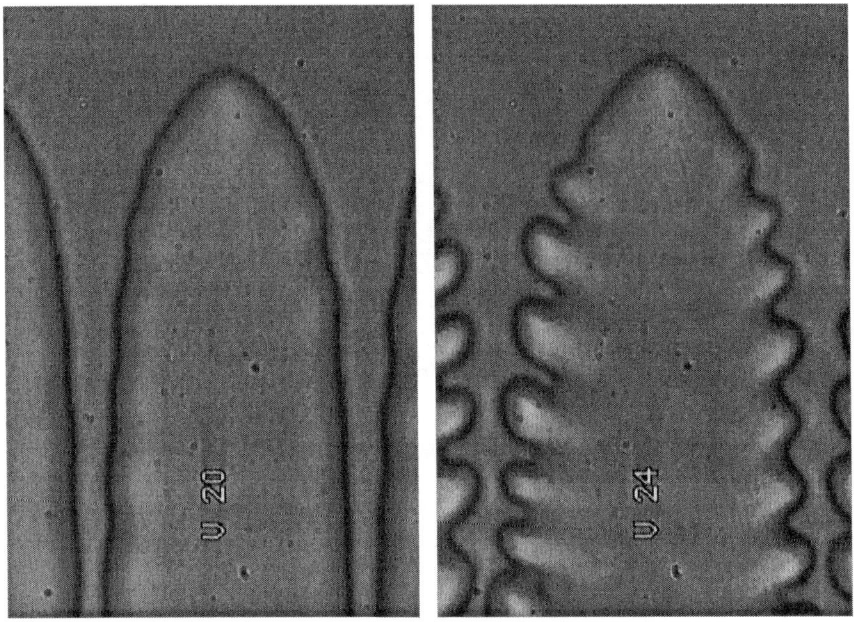

Fig. 3. — Depending on velocity, cell width or thermal gradient, dendrites show sidebranches either displaying a very small amplitude or a well developed amplitude. Both: $G = 78$ K/cm. Left: $V = 20$ μm/s, $\Lambda = 120$ μm. Right: $V = 24$ μm/s, $\Lambda = 145$ μm.

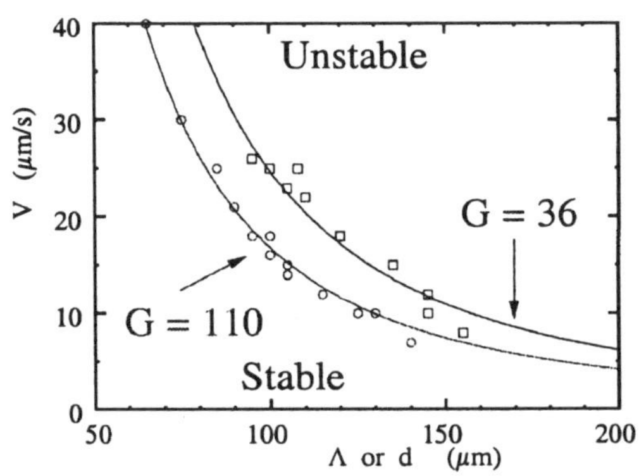

Fig. 4. — Critical curves for $G = 36$ K/cm (squares) and 110 K/cm (circles).

implies advection of perturbations along the interface towards cooler regions. The theoretical conclusion is that, provided anisotropy of surface tension is accurately taken into account, the amplification rate on isolated freely growing dendrites is actually large enough to produce the observed sidebranches, starting from thermodynamical fluctuations.

This noise-driven mechanism was transposed to directional solidification about 15 years ago [15]. This direct transposition may however be questioned owing to the important differences between the two kinds of growth, in particular regarding characteristic lengths: three of them in directional growth (capillary length d_0, thermal length l_T and diffusion length $l_D = D/V$) but only two of them in free growth (d_0 and l_D).

In directional growth, competition between amplification and advection results in a growth factor of perturbation Γ which integrates all the amplification rate σ undergone during advection at velocity v_τ from cell tip to cell sides:

$$A_{\text{side}} = A_{\text{tip}} e^{\Gamma} \tag{1}$$

where A_{tip} (resp. A_{side}) is the sidebranch amplitude at the tip (resp. at the side of the dendrite), and where amplification factor Γ is:

$$\Gamma = \int_{\text{cell shape}} \frac{\sigma}{v_\tau} \mathrm{d}s \tag{2}$$

s denoting the curvilinear coordinate along the interface.

Analysis of this integral shows that a low cut-off scale must be introduced to exclude the infinite growth which formally arises for perturbations just located at cell tip ($v_\tau = 0$). Taking for cut-off the lowest unstable wavelength λ_s of primary instability yields $\Gamma \approx \rho/\lambda_s$ where ρ denotes tip curvature radius.

For large enough amplification factor Γ, sidebranches get visible; for weak enough Γ, they don't. Marginal sidebranching stands in between and corresponds to *constant* value of Γ. This property provides a way for testing the relevance of the theoretical model, by comparing the experimental critical curve to the shape of iso-Γ surface.

To determine Γ, variable λ_s may be computed from Mullins-Sekerka dispersion relation but the radius of curvature ρ must be measured [16]. Doing so, we obtain power law variation of ρ with respect to control parameters Λ, V, G: $\rho \propto V^{-0.33}\, G^{0.5} \Lambda^{0.75}$, in agreement with previous independent measurements [2].

3.3. Questioning Theory

Predicted variation of the amplification factor Γ determined above may be put in the form $\Gamma \propto V^{0.17} G^{0.5} \Lambda^{0.75}$. Here, iso-$\Gamma$ surface should fit with the experimental determination of the marginal surface of sidebranching: $VG^{0.25}\Lambda^2 =$ constant. Comparison for each of the three control parameters gives:

- increasing Λ or G enhances the amplification factor Γ, in qualitative agreement with observations;
- theory predicts that influence of G and Λ on Γ have the same order of magnitude. However, experiment reveals a much weaker effect of G compared to that induced by Λ. Hence, the theory *overestimates* the destabilizing influence of G compared to Λ;

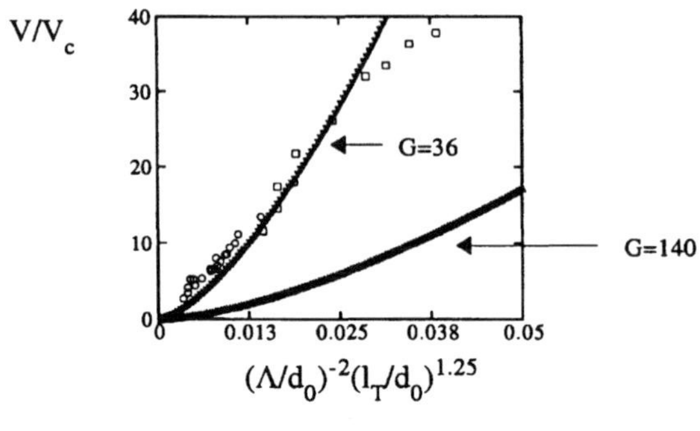

Fig. 5. — Incompatibility between theoretical values at two different thermal gradient, and experimental points. For simplicity, coordinates are taken so that all experimental points appear on the same line whatever the value of G. Theory gives us the possibility of freely choosing a prefactor. We choose it so as to obtain a good agreement between all experimental points and the theoretical prediction at $G = 36$ K/cm. However, the second curve corresponding to another gradient then clearly disagrees: theory does not recover experimental onset of sidebranching.

- more crucial is the confrontation regarding velocity effects: prediction and observation agree qualitatively on the destabilizing role of V, but strongly differ quantitatively. In particular, as the V-exponent in the predicted expression for Γ is very small, the velocity should have the weakest effect on sidebranching. This is undoubtly *contradicted* by all available experiments;
- the disagreement between theory and experiment is quantified in Figure 5. Here, for two gradients, iso-Γ surfaces are compared with the experimental critical curves. For simplicity, coordinates are taken so that all experimental curves appear on the same line. Then, choosing the unknown prefactors so that agreement is obtained for one gradient $G = 36$ K cm^{-1} reveals a large *disagreement* for the other one $G = 140$ K cm^{-1}, *beyond* experimental errors.

4. CONCLUSION

We have tested a noise-induced theory proposed for directional sidebranching by transposition of a scheme initially built in free growth. This theory is insufficient to satisfactorily recover the marginal sidebranching states that have been characterized in our experiment. This leaves the appealing problem of sidebranching unsolved, even in the present physical context where the basic ingredients of growth are all known.

Improvements of theory might go beyond the linear growth assumption to include non-linear effects. Then, sidebranch amplitude would no longer be proportional to tip perturbation, possibly even in the marginal regime analyzed in our experiment. Non-linearity might also yield other kinds of interaction between different parts of the interface. In particular, coupling between sides and tip of a cell might end in a feed-back non-linear amplifier displaying a limit cycle. Then, the dendrite would no longer behave as a noise-amplifier, but as an oscillator.

REFERENCES

[1] J.S. Langer, *Rev. Mod. Phys.* **52** (1980) 1.
[2] B. Billia and R. Trivedi, *Handbook of Crystal Growth*, Vol. 1, Chapter 14 (Elsevier Science Publishers, Ville, 1993).
[3] W. Kurz and R. Trivedi, *Acta. Metall. Mater.* **38** (1990) 1.
[4] X.W. Qian and H.Z. Cummins, *Phys. Rev. Lett.* **64** (1990) 3038; L.M. Williams, M. Muschol, X. Qian, W. Losert and H.Z. Cummins, *Phys. Rev. E* **48** (1993) 489.
[5] M.A. Eshelman, V. Seetharaman and R. Trivedi, *Acta. Metall.* **36** (1988) 1165; K. Somboonsuk, J.T. Mason and R. Trivedi, *Metall. Trans. A* **15** (1984) 967.
[6] A. Papapetrou, *Z. Krist.* **92** (1935) 89; G.P. Ivantsov, *Dokl. Akad. Nauk. SSSR* **58** (1947) 56.
[7] C. Huang and M.E. Glicksman, *Acta. Metall.* **29** (1981) 701.
[8] C. Huang and M.E. Glicksman, *Acta. Metall.* **29** (1981) 707.
[9] P. Pelcé and A. Pumir, *J. Cryst. Growth* **73** (1985) 337.
[10] P. Kopczynski, W.-J. Rappel and A. Karma, *Phys. Rev. Lett.* **77** (1996) 3387; M. Georgelin and A. Pocheau, *Phys. Rev. Lett.* **79** (1998) 2698.
[11] M. Georgelin and A. Pocheau, *Phys. Rev. E* **57** (1998) 3189.
[12] M. Georgelin and A. Pocheau, *Eur. Phys. J. B* **4** (1998) 169.
[13] E.A. Brener, M.B. Geilikman and D.E. Temkin, *Zh. Eksp. Teor. Fiz.* **94** (1988) 241 [*Sov. Phys.-JETP* **67**(1988) 1002].
[14] P. Pelcé and P. Clavin, *Europhys. Lett.* **3** (1987) 907; J.S. Langer, *Phys. Rev. A* **36** (1987) 3350.
[15] S.K. Sarkar, *Phys. Lett. A* **117** (1986) 137.
[16] Precise determination of Γ must take into account tip undercooling. Analysis however shows that this does not modify Γ, *J. Cryst. Growth* **206** (1999) 215.

CHAPTER XII

Branching During Dewetting and Wetting

C. Misbah

1. INTRODUCTION

Wetting and spreading of a liquid film is an important problem in technical science (properties of lubricants, paints and so forth). At the same time it raises a fundamental problem of nonequilibrium statistical and hydrodynamical nature. Extensive studies have been devoted to various properties of wetting [1], dewetting and dynamics of spreading. Langevin (stochastic) equations, as well as Monte-Carlo simulations [2] and hydrodynamical [3] models have been used to study spreading of a droplet on a substrate. It has been shown that the film front advances in such a way that the radius grows with time like[1] $\sqrt{t}$. Moreover it has been admitted since a long time (and often supported by several observations) that the film contour remains rather circular (except in the presence of pinning centres).

We show here that the contour of a dewetting (or wetting) front is unstable. We show that the problem can be mapped onto that of a growth limited by diffusion, which is known to lead to branched structures (of dendritic type). The experimental observation by Samid-Merzel [4] reveals that dewetting results in highly complex branched structures. Some of the patterns are reminiscent of those encountered in crystal growth when crystalline anisotropy is small, or absent (a situation which can be introduced within the model).

We shall first recall the notion of wetting at equilibrium, and discuss the general conditions under which two films of different thicknesses coexist, a situation met in the above cited experiment [4]. We shall then develop a mapping of the dewetting (or wetting) problem onto that of crystal growth. Though the work of [5] and [6] can be taken over, given the broad audience, we have

[1] The precursor film, a question not addressed here, is found in the Langevin model as well as in MC simulations to move as t instead.

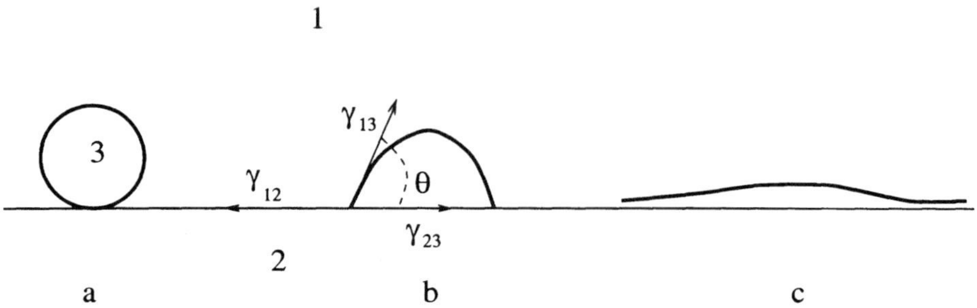

Fig. 1. — A schematic view of the three situations a): non wetting, b): partial wetting, c) complete wetting.

felt it worthwhile to present the various steps that lead to the equivalence between wetting and crystal growth. This confers to the similarity of branching in wetting and in crystal growth a profound basis. Having established the mapping, we shall give then a general argument on why a dewetting front in the situation encountered in [4] can suffer a morphological instability. The mapping will allow immediately to state that a circular contour should move with a radius increasing as $\sqrt{t}$, while the envelope of a morphologically unstable contour should move as t. We discuss some interesting questions for future investigations. It must be noted that since this lecture is intended for a broad audience, some physicists readers might find the discussion of some questions rather elementary.

2. DEFINITION OF THE CONTACT ANGLE AND SPREADING PRESSURE

Suppose that a small droplet (small enough so that gravity can be ignored; in general by small we mean less than few mm) is deposited on some substrate as shown in Figure 1. Three situations may occur:

(i) the droplet does not wet at all the substrate (case a);

(ii) the droplet partially wets the substrate (case b); we say that there is a *partial wetting*;

(iii) the droplet flattens and wets completely the substrate (case c); this corresponds to *complete wetting*.

Let us denote by "1" the ambient atmosphere, "2" the substrate, and "3" the droplet. Let γ_{ij} denote surface tensions between i and j (for example γ_{12} is surface tension between ambient atmosphere and the substrate). Surface tension is a quantity which is tangential to the surface, as shown in Figure 1. At equilibrium the total force along the substrate must be equal to zero. A projection of the forces on the substrate leads to $\gamma_{12} - \gamma_{23} - \gamma_{13}\cos(\theta) = 0$, or

$$\cos(\theta) = \frac{\gamma_{12} - \gamma_{23}}{\gamma_{13}} = 1 + \frac{S}{\gamma_{13}} \tag{1}$$

which is the classical Young condition. Here S is defined as

$$S = \gamma_{12} - (\gamma_{13} + \gamma_{23}) \tag{2}$$

θ is the contact angle, and S is known as the *spreading pressure* for the following reason. Suppose that $S < 0$, this means that $\gamma_{12} < \gamma_{13} + \gamma_{23}$. In that case it costs less energy to have an interface between 1 and 2 (atmosphere and substrate), than having the droplet in between (creating thus two interfaces 1–3 and 2–3). This entails that the droplet should not wet too much the substrate. In the extreme limit when $S = -\gamma_{13}$, from (1) we obtain a contact angle $\theta = \pi$, which corresponds to situation *a* in Figure 1: no wetting occurs. By increasing S a contact angle smaller than π is possible, and we say that there is a *partial wetting*. Conversely if $S > 0$ ($\gamma_{12} > \gamma_{13} + \gamma_{23}$), it is more favorable to create as much contact between 1 and 3, and thus between 2 and 3 as possible, than having the atmosphere in contact with the substrate. In that case the droplet would wet the substrate. In agreement with this argument, one can see from (1) that the Young condition can not be satisfied for $S > 0$ and we have complete wetting.

3. WHAT DETERMINES WETTING PROPERTIES?

Whether a droplet wets or not a substrate strongly depends on the precise interaction energies. Suppose that we focus on a part of the fluid which is in contact with the substrate as shown in Figure 2. There are two types of interactions. The first one is the classical van der Waals interaction which is referred too also as an apolar interaction (sometimes also known under the Lifshitz-van der Waals energy). The excess free energy which is necessary to build the two interfaces with the three phases is denoted as $G^{\mathrm{W}}_{123} = -A'/12\pi h^2$, where h is the film thickness, and A' is known as the effective Hamaker constant which is defined in terms of the Hamaker constants, A_{ij} for various binary interactions [5]. Another kind of interaction must be taken into account, which is the polar one. This interaction has been variously described as the hydration pressure, hydrophobic interaction, hydrogen bonding, acid-base interaction, etc., and is important for films of polar liquids (*e.g.* water) bounded by polar or apolar media. The polar interaction has the following form $G^{\mathrm{P}}_{123} = G^{\mathrm{P}}_{0}\exp\left[(d_0 - h)/l_0\right]$ where G^{P}_{0} is a constant describing the polar part of the adhesion energy at contact ($h = d_0$), and l_0 is a correlation decay length. The total excess energy per unit area reads (by combining the two interactions) as

$$G^{\mathrm{W}}_{123} = -A'/12\pi h^2 + G^{\mathrm{P}}_{0}\exp\left(d_0 - h\right)/l_0 \tag{3}$$

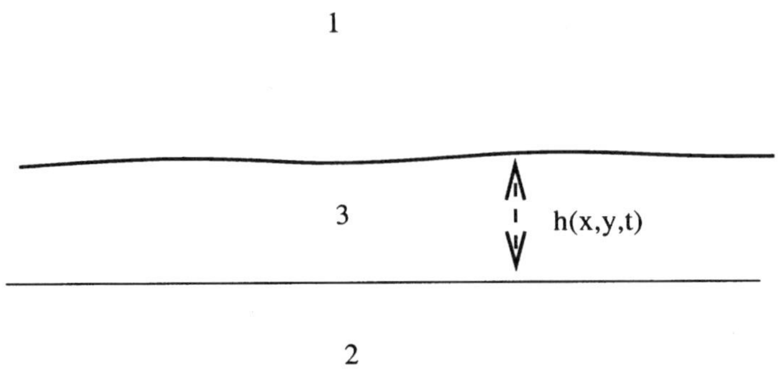

Fig. 2. — A schematic view a film of thickness h wetting the substrate.

or equivalently

$$G_{123}^{\mathrm{W}} = S^{\mathrm{W}}(d_0^2/h^2) + S^{\mathrm{p}} \exp{(d_0 - h)/l_0} \tag{4}$$

where S^{W} and S^{p} are the apolar and polar components of the spreading coefficient, and we have $S = S^{\mathrm{W}} + S^{\mathrm{p}}$. Thus a negative value of S^{W} $(A' > 0)$ implies attractive van der Waals forces between phases 1 and 2, whereas positive S^{W} corresponds to repulsion. The same reasoning holds for the polar part. We have seen above that $S < 0$ does not favor wetting, while $S > 0$ implies complete wetting. Reference [5] (and references therein) describe signs of S^{W} and S^{p} and corresponding physical systems.

The interesting situation is the one where the two interactions S^{W} and S^{p} have opposite signs. The case of interest here is the one where $S^{\mathrm{W}} > 0$ (the van der Waals forces encourage spreading) and $S^{\mathrm{p}} < 0$ (polar forces defavor wetting). Water films on most apolar and polar substrates fall in this category. Examples of such substrates include mica, cellulose acetate, human serum albumin... It is interesting to note that neglecting polar interactions would lead to the conclusion that wetting of water on all substrates is always favorable, which is clearly not the case as observed in everyday life. Thus interactions that impede spreading must be operating, these are nothing but the polar interactions.

3.1. Coexistence Film/Film

The question is now to determine whether or not the total interaction energy is concave or convex as a function of the film thickness h. A convex shape is a signature of *instability*[2] . Taking the second derivative of (4) with respect to h

[2] In a mechanical analogy if we place a solid sphere on a convex shape – say the shape has the following equation $Y = -x^2$ the sphere in the position corresponding to the top of the mound is unstable.

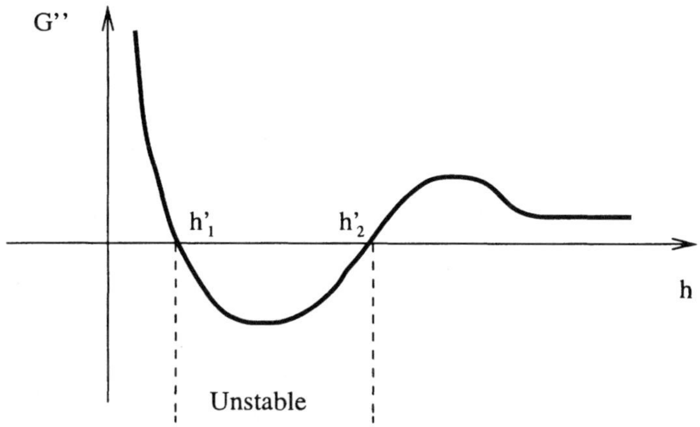

Fig. 3. — A schematic view of the second derivative of G, denoted here G'', as a function of the film thickness.

and requiring it to be negative for instability yields (from now on we omit the subscript "123")

$$\frac{\partial^2 G}{\partial h^2} = 6S^{\mathrm{W}}(d_0^2/h^2) + S^{\mathrm{P}}(h^2/\ell_0^2)\exp\left[(d_0 - h)/\ell_0\right] < 0. \tag{5}$$

It is a simple matter to check that if $S^{\mathrm{W}} > 0$ and $S^{\mathrm{P}} < 0$ then this inequality is satisfied for a certain range of h, as shown in Figure 3. The domain of film thickness between h_1' and h_2' is unstable and a film prepared initially in that range would break up into two domains, one with thickness close to (but smaller; see below) h_1', and another thicker film with thickness close to (but larger) h_2'. The result of Figure 3 can be shown differently by plotting directly G as a function of h. We show qualitatively in Figure 4 that two films with thicknesses h_1 and h_2 may coexist; actually the domains between h_1 and h_1' and between h_2 and h_2' are metastable. The coexistence values h_1 and h_2 are fixed by the classical common tangent rule. The figure shows typically a situation where two "phases" coexist with each other, a situation which is reminiscent of problems of first order phase transitions. The value of the slope of the common tangent in Figure 4 is nothing but the film pressure [1] for coexistence between the two films, and is also the pressure of the ambient atmosphere. Changing the ambient vapor pressure may favor one "phase" over the other.

4. EXPERIMENTAL FACTS

The experiment performed by Samid-Merzel *et al.* [4] has consisted in evaporating water from clean mica substrates. By acting on temperature, and thus

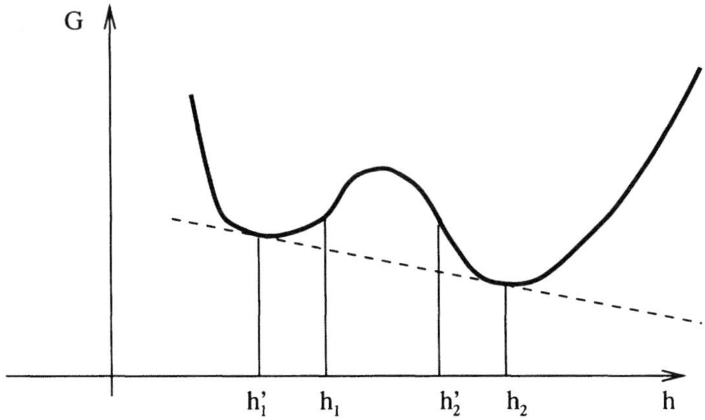

Fig. 4. — A schematic view of G as a function of the film thickness.

on pressure, it has been observed that under unsaturated vapor pressure, the reaming layer has become very thin (a few tens of nm), and suddenly dry patches (actually the authors believe that the "dry" patches are covered by a few monolayers of water) nucleate and expand to create regular patterns (see later), some of them very similar to those observed during dendritic growth without (or with very weak) crystalline anisotropy. The present situation is thus the invasion of a film of few tens of monolayers by a thiner film (of few monolayers), very much like the problem of invasion of a solid nucleus in an undercooled melt, or a snowflakes in a saturated vapor in the atmosphere. The question still remains to show whether the similarity between patterns in crystal growth and those exhibited by the invading dewetting front is accidental or is there any hidden universality that leads to common behaviors. This is the aim of the next section.

5. MAPPING OF FRONT DEWETTING DYNAMICS ONTO CRYSTAL GROWTH

During wetting or dewetting hydrodynamical flow occurs in the two films. Figure 5 shows schematically a thin film invading a thicker one. The aim of this section is to present a mapping of this problem onto that of a crystal growth problem limited by heat or mass diffusion (the contribution of Karma addresses the crystal growth problem).

5.1. The Stokes Equation

Dynamics of the dewetting front is limited by hydrodynamics flow. Let $h(x, y, t)$ denote the film thickness (which is h_1 far away behind the invading front, and

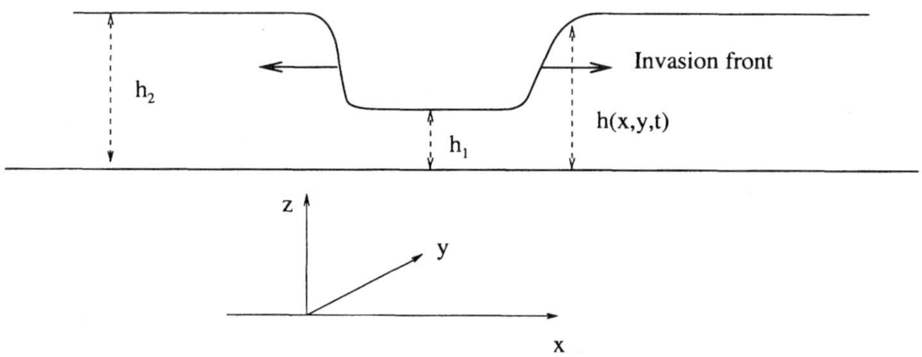

Fig. 5. — A schematic view of the dewetting front invading the substrate.

h_2 far away in front of the invading front; see last section). In the most general case the fluid velocity field $\mathbf{v}$ obeys the Navier-Stokes equation

$$\rho\left[\frac{\partial \mathbf{v}}{\partial t}+\mathbf{v}.\vec{\nabla}\mathbf{v}\right]=-\vec{\nabla}p+\eta\nabla^2\mathbf{v} \tag{6}$$

where ρ is the fluid velocity, η the viscosity and p the pressure field ($\vec{\nabla}$ is the operator gradient (**grad**) and ∇^2 the Laplacian).

Let v_0 be a typical value of the fluid velocity, h_0 is the mean value of the thick film thickness, and let $\tau = h_0/v_0$ a typical time scale for a stationary motion. The acceleration in the NS equation (6) is of order a velocity divided by time, $v_0/\tau = v_0^2/h_0$, so that the left hand side is of order $\rho v_0^2/h_0$. The viscous term has a dimension of η multiplied by a velocity (v_0) divided by a distance square, so that its typical value is of order $\eta v_0/h_0^2$. The acceleration term (the inertial term) divided by the viscous term is of order of $\rho v_0/(\eta h_0) \equiv Re$, which is nothing but the Reynolds number. In the experimental situation the invasion velocity is of order 10 μm/s $= 10^{-3}$ cm/s, while the thickness is at most $h_0 \sim$ 100 nm $= 10^{-5}$ cm, and the water kinematic viscosity $\nu \equiv \eta/\rho = 10^{-2}$ cm^2/s. Theses values lead to an estimate for the Reynolds number of $R_e \sim 10^{-6} \ll 1$. That is to say inertial effects are small in comparison to viscous ones, so that the l.h.s. of equation (6) can legitimately be set to zero. The full Navier-Stokes equation reduces to a linear equation known as the Stokes equation

$$-\vec{\nabla}p+\eta\nabla^2\mathbf{v}=0. \tag{7}$$

5.2. The Lubrication Approximation

The second step is the recognition that the flow occurs in a thin film, so that the fluid velocity undergoes variations in the z-direction which are much stronger than those occurring in the plane (x, y). Indeed the fluid velocity which is of order 10 μm/s must vanish on the substrate within a distance of order 100 nm, while in the plane the mean velocity, though inhomogeneous, remains of the same order everywhere. Let L denote the linear dimension of the film in the (x, y) plane, the velocity gradient in the plane $\partial \mathbf{v}/\partial x \sim \partial \mathbf{v}/\partial y \sim v_0/L$, while in the z-direction the gradient is of order $\partial \mathbf{v}/\partial z \sim v_0/h_0$. Since $L \gg h_0$, the gradient in the plane can be neglected. That is to say the Laplacian can be approximated by

$$\nabla^2 \mathbf{v} = \left[\frac{\partial^2}{\partial x^2} + \frac{\partial^2}{\partial y^2} + \frac{\partial^2}{\partial z^2}\right] \mathbf{v} \simeq \frac{\partial^2}{\partial z^2} \mathbf{v}. \tag{8}$$

This is known as the *lubrication approximation* that is used whenever the velocity variation is large in a film thickness in comparison to that in the plane. Using the lubrication approximation, we can write the Stokes equation (7) for the three components of the velocity v_x, v_y, v_z:

$$\begin{aligned} \eta \frac{\partial^2}{\partial z^2} v_x &= -\frac{\partial p}{\partial x} \\ \eta \frac{\partial^2}{\partial z^2} v_y &= -\frac{\partial p}{\partial y} \\ 0 &= -\frac{\partial p}{\partial z} \, . \end{aligned} \tag{9}$$

$$\tag{10}$$

The z-component of the velocity is zero, since the flow occurs in the plane (x, y) only. The last equation implies that $p(x, y, z) = p(x, y)$, *i.e.* the pressure depends on the variables in the plane only. This automatically implies that we can integrate the first two equations with respect to z and obtain

$$v_x = \frac{1}{2\eta}[z^2 - 2zh]\frac{\partial p}{\partial x} \, , \qquad v_y = \frac{1}{2\eta}[z^2 - 2zh]\frac{\partial p}{\partial y} \tag{11}$$

where we have used the following boundary conditions $v_x(x, y, z = 0) = v_y(x,y, z = 0) = 0$, expressing the no-slip condition on the substrate, and $\partial v_x(x, y, z = h)/\partial z = \partial v_y(x, y, z = h)/\partial z = 0$ stating that the free surface is shear-free; we assume the free surface to be nearly planar.

The mean velocity (over the thickness) in the x-direction is defined as

$$\langle v_x \rangle = \frac{1}{h} \int v_x \mathrm{d}z = -\frac{h^2}{3\eta}\frac{\partial p}{\partial x} \tag{12}$$

and a similar expression for $\langle v_y \rangle$. Omitting the symbol "$\langle ... \rangle$" the vector field (the mean velocity) can be written as

$$\mathbf{v} = -\frac{h^2}{3\eta}\vec{\nabla} p \tag{13}$$

which is also known as Darcy's law. This type of law is used for a Hele-Shaw cell, porous media (rocks etc.). Here and in what follows ∇ refers to

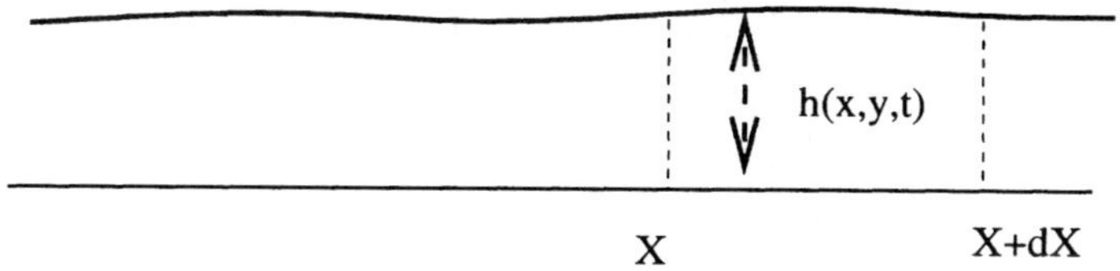

Fig. 6. —

the two dimensional gradient in the plane (x, y). It must be emphasized, and seen below, that the instability under question has *nothing to do* with the front instability in a Hele-Shaw geometry (see Couder's Chapter)! which is due to a viscosity contrast, and that the invading phase must have a lower viscosity. Here the situation is just the opposite since the effective viscosity is $3\eta/h_2^2$ in the thick phase and $3\eta/h_1^2$ in the thin one. Since $h_2 > h_1$, the effective viscosity of the invading phase, $3\eta/h_1^2$ is *larger* than that of the invaded one, $3\eta/h_2^2$. A Hele-Shaw problem would predict *stability*! The situation here is more subtle, and we shall consider it in the next section.

5.3. Mass Conservation

The present problem is not steady and though the fluid is everywhere incompressible, it would be a mistake to use the incompressibility condition in the usual form of a divergence $\nabla.\mathbf{v} = 0$, which, combined to the Darcy law would lead to $\nabla^2 p = 0$, a Laplacian dynamics. Here the mass contained within a small element does not change within time, except close to the front where one goes abruptly from one height h_1 to another h_2. Thus mass changes only because the height changes. Let us write the evolution equation for the height. We consider a one dimensional geometry, and a generalization will follow immediately.

Let us consider a fluid element between x and $x + \Delta x$ of height h having a mass ΔM as shown in Figure 6. We clearly have $\Delta M = \rho h \Delta x$ (per unit length in the orthogonal, y-direction). The mass change during a time interval dt is related to the mass current across the vertical lines (of height h) at x and $x + \Delta x$:

$$\frac{\partial \Delta M}{\partial t} = [j_x(x) - j_x(x + \Delta x)]h = -\frac{\partial j_x}{\partial x} \tag{14}$$

where $j_x(x)$ is the mass current and $j_x h$ is the flux across the vertical line. The mass current is related to the velocity field by $j_x = \rho v_x$. A similar reasoning can be made along the y direction and leads to a contribution to the mass variation given by $-\frac{\partial j_y}{\partial y}$, so that the full mass conservation equation has the form

$$\frac{\partial h}{\partial t} = -h\vec{\nabla}.\mathbf{v} = \vec{\nabla}.\left[\left(\frac{h_0^3}{3\eta}\right)\nabla p\right] \tag{15}$$

where we have made use of (13) together with the fact that we consider leading order contribution in h, so that we set h_0^3 instead of h^3, since as we will see below the pressure gradient is related to the height gradient. Here h_0 designates the mean thickness in the phase under consideration (actually we use below the fact that the thin film is almost dry and we disregard hydrodynamic flow within it accordingly).

5.4. A Constitutive Law

In order to close the system equations we need to specify the constitutive equation that relates the height to the pressure. The distribution of the pressure inside the liquid film is given by [1]

$$p = p_{\mathrm{g}} - \Gamma \nabla^2 h(x,y,t) + \frac{\partial G(h)}{\partial h} - \frac{\partial G(z)}{\partial z} \tag{16}$$

p_{g} is the gas pressure, the term proportional to Γ is the Laplace term which is the capillary pressure if the top of the film is not planar, the one given by $\partial G(z)/\partial z$ is the force associated with the van der Waals and polar energies at some point inside the film, while the last term ensures that $p(h)$ is given by Laplace's law. Note that inside the film at some given point the vertical force acting on any volume element vanishes, $-\partial p/\partial z - \partial G/\partial z = 0$. Neglecting the capillary term (the top surface is nearly planar), we can write

$$\vec{\nabla} p = \vec{\nabla} \frac{\partial G}{\partial h} = \left[\frac{\partial^2 G}{\partial h^2}\right] \vec{\nabla} h. \tag{17}$$

To leading order (this means if linear irreversible thermodynamics applies), we must evaluate the prefactor of the driving force (the gradient of the height) by its value at equilibrium if the phase under consideration (*e.g.* the wet part) were covering alone the substrate. For the wet part we must evaluate $\partial^2 G/\partial h^2$ at $h = h_2$ while for the "dry" one at $h = h_1$. In other words the quantity $\partial^2 G/\partial h^2$ represents the concavity at the point of common tangent (see Fig. 4); that concavity is positive. Plugging (17) into (15) we finally obtain

$$\frac{\partial h}{\partial t} = D(h_0) \nabla^2 h \tag{18}$$

where $D(h_0) = (h_0^3/3\eta)\partial^2 G(h_0)/\partial h_0^2$ is an effective diffusion constant, and where h_0 is the width of the wet area (noted h_2 before), and since the dry area contributes negligibly to mass transport (its width is at least 100 times smaller than that of the dry area), we shall ignore it in what follows. Here we have evaluated all prefactor at the coexistence line. We have shown for the moment that the height (which is the quantity that allows a distinction between the two phases) obeys *a diffusion equation*, precisely as in in crystal growth where the mass (or temperature) obeys a diffusion equation. The fact that we neglect transport in the thin phase amounts to considering a mapping of the present problem onto the *one-sided* crystal growth. We shall now write the boundary conditions at the invading front. At the contour the fluid velocity in the wet part must coincide with that of the front itself. The knowledge of the normal velocity of the front is sufficient to describe dynamics. Making use of Darcy's law (13) we can write that the normal velocity of the contour is given by v_n

$$\mathbf{v_n} = -\frac{h^2}{3\eta}\vec{\nabla}p.\mathbf{n} = -D(h_0)\mathbf{n}.\vec{\nabla}h. \tag{19}$$

Finally, along the contour, bonds are not all fulfilled, so that it can be viewed as a line of separation between the two phases. Thus it implies a line tension effect. Due to Laplace's law the pressure difference on both sides of the contour $p_2 - p_1$ is proportional to $\gamma\kappa$ where γ is the line tension, and κ the contour curvature. More precisely we have $p_2 - p_1 = a\gamma\kappa$ where a is a length which measure the height of the "step" between the two phases. Using the fact that the pressure inside each phase is proportional to $h - h_0$ according to (16). We simply need to develop that equation for small deviations about the equilibrium height and arrive straightforwardly to

$$h = h_0 - d_0\kappa \tag{20}$$

where $d_0 \equiv a\kappa/(\partial G(h_0)/\partial h_0^2)$, is a length scale, usually known under the name of a capillary length. Equations (18–20) constitute the set of equations that describe dynamics of the dewetting front. This set of equation is formally equivalent to the one-sided model of crystal growth [7]. This set must be supplemented with the boundary condition at long distances ahead of the front: $h(x = \infty, y = \infty) = h_\infty < h_0$. The quantity $\Delta = (h_0 - h_\infty)/h_0$ is the analogue of the dimensionless supersaturation which drives crystallization from a vapor (snow flake, for example).

6. OBSERVED PATTERNS

Once the mapping is made onto that of a crystal growth problem, we can just use the results accumulated in that field. Usually in crystal growth limited by diffusion the typical morphology is a snowflake-like pattern. The snowflake pattern requires anisotropy, which is due to the crystal structure beneath. The anisotropy is necessary to stabilize each dendritic branch (which is nearly parabolic). That anisotropy can enter the surface tension γ. In the present problem anisotropy is absent since there is no underlying order, and thus one has to compare the pattern to that obtained in the problem with vanishing anisotropy. If anisotropy is absent a variety of patterns in crystal growth can be obtained depending on the degree of supersaturation [8, 9]. For a large enough supersaturation [9] ($\Delta \sim 0.8$) the late stage dynamics is presented on Figure 7, and is compared to the finding of the experiment performed in [4] (Fig. 8). For lower supersaturation a rather fractal-like pattern is obtained.

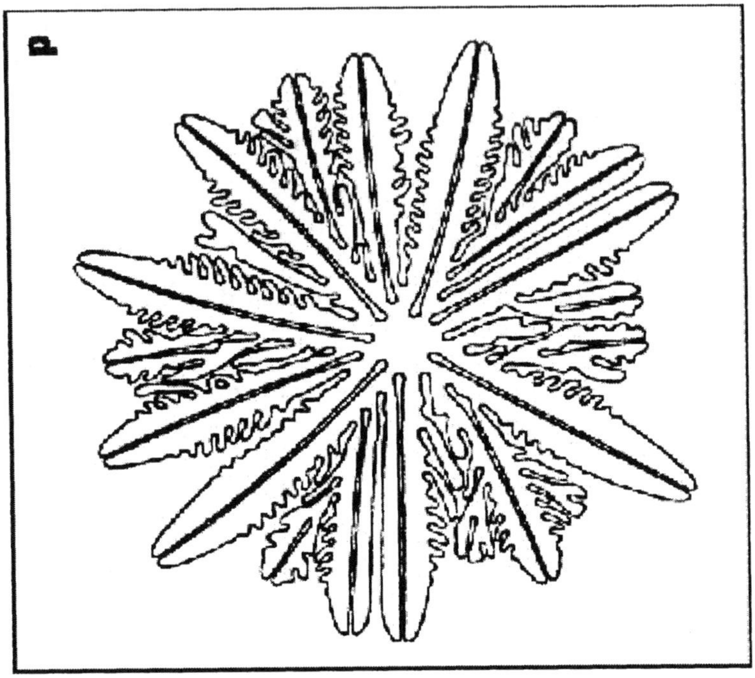

Fig. 7. — The pattern exhibited by a front in crystal growth with no anisotropy and with $\Delta = 0.8$, reference [9].

It is likely that the effective supersaturation in the experiments is large enough. If it were possible to change the supersaturation, anisotropy... we expect all possible patterns encountered in crystal growth to occur during dewetting as well. It will be a task for future investigations to find experimental systems and/or protocols in the context of dewetting in order to explore a much larger region of parameter space.

7. THE STABLE ENVELOPPE OF A WETTING OR DEWETTING FRONT MOVE AS $\sqrt{t}$

In this section we discuss the conclusion that follow from the mapping. Firstly it is instructive to note that if a front remains straight, or circular, then the front position, after transient have decayed, move asymptotically as $\sqrt{t}$. Let us suppose that the front is straight and is moving along the $x - direction$ (see Fig. 5 for example). Equations (18–20) (which is the mapped version for dewetting or wetting) admits the similarity solution as an asymptotic solution. That is to say the front position $x_{\mathrm{f}} \sim \sqrt{t}$, with $h(x,t) = h(x/\sqrt{t})$ is a solution.

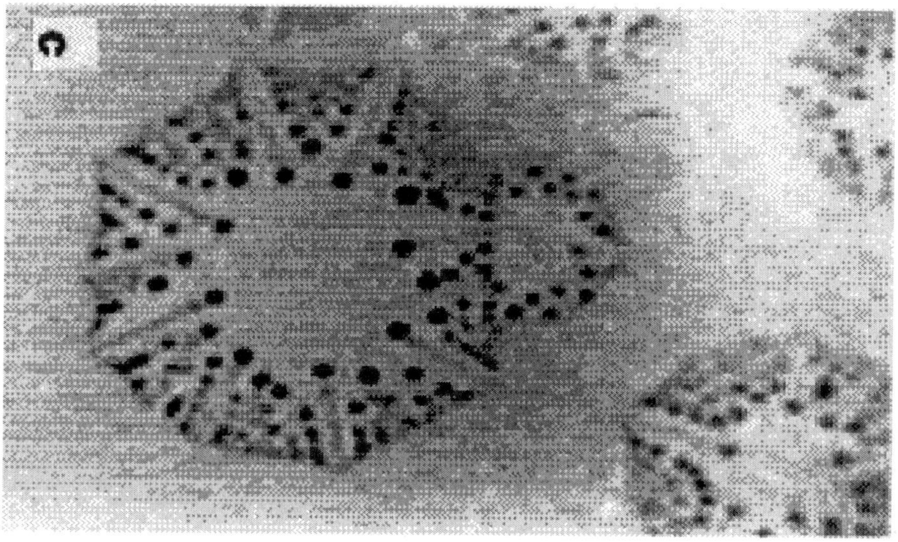

Fig. 8. — The pattern exhibited by a dewetting front in the experiment by [4].

In other words, the dynamical variable which is the height does not depend on x and t separately but on the combination $x/\sqrt{t}$ (the height at a given time is similar to the height at one later time but at a different position). For that purpose we set $X = x/\sqrt{t}$, and simple differentiation shows that the diffusion equation (18) becomes autonomous with respect to time and obeys

$$X\frac{\partial h}{\partial X} = -D\frac{\partial^2 h}{\partial X^2} \tag{21}$$

while the boundary condition (19) becomes (by setting $x_\mathrm{f} = A\sqrt{t}$))

$$A = D\frac{\partial h}{\partial X} \tag{22}$$

and the height is fixed at the wetting front by relation (20)

$$h(x = x_\mathrm{f} = A\sqrt{t}) = h(X = A) = h_0. \tag{23}$$

Equation (21) can be solved in principle (in terms of the Erf function) with two integration constants. The two boundary conditions fix the two constants. The condition at $x = \infty$ ($h = h_\infty$) fixes a relation between A and the two other constants. It is not the purpose of this paper to give the details of the solution but just to lay stress on the fact that a similarity solution solves the set of equations.

The same reasoning can be made with a circular geometry by looking to solutions of the form $h(r/\sqrt{t})$ where r is the polar coordinate. A small difference

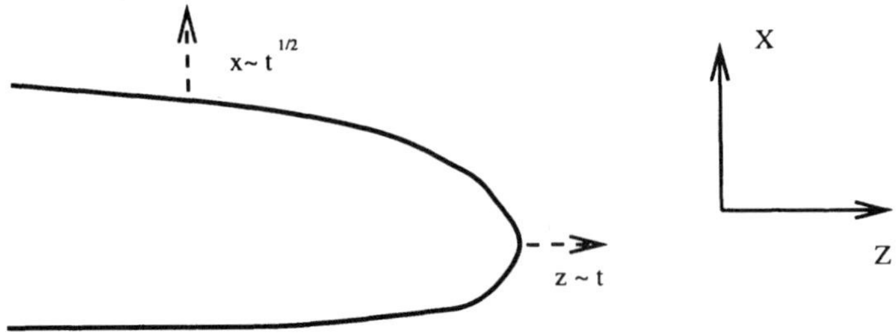

Fig. 9. — A schematic view of the a curved front, the tip of which must move proportionally to time.

occurs between the straight front and the circular one, however. Indeed, an additional contribution due to curvature in (20) leads to a term of the form d_0/r_f, and since $r_f \sim \sqrt{t}$ this term vanishes asymptotically. Thus the radius of a circular front must move after some transients as $\sqrt{t}$.

8. BRANCHING OF A FRONT

The question now arises of why the front becomes unstable and in that case what would the branching have as a consequence? Suppose that we still consider the situation where the invading "phase" is almost dry, as done above. Then the problem is that of crystal growth in the one sided limit, and it is known to be inherently unstable; the Mullins-Sekerka instability (see lecture by Karma). We give here a heuristic argument on why in general the envelope position will move linearly in time. For that purpose we refer to Figure 9. Suppose that locally the initially planar front becomes unstable and is curved. On top of the deformation the front shape is approximately parabolic with a certain radius of curvature r_c. In the tail the curvature is small, and locally everything is as if the front were planar. As we have seen above a planar front has a position which goes as $\sqrt{t}$. Since at the tip of the front the shape is parabolic ($z \sim x^2$), and if one requires that the front must move in a shape preserving manner, then this means that the tip position must change proportionally with time, since $x \sim \sqrt{t}$.

9. DISCUSSION

We have shown how the dewetting problem can be mapped onto that of the crystal growth one. The invading front moves in such a way that transport occurs mostly in front of it. During shrinkage, however, the contour must be stable given the analogy with the problem of crystal growth. Since the thin (invading) film has a larger effective viscosity, the Hele-Shaw effect is stabilizing during invasion, but destabilizing during shrinkage. But the Mullins Sekerka effect is just opposite! Since there is to our knowledge no known instability during shrinkage, we speculate that the Mullins-Sekerka effect is dominant. Also during wetting with a thin film in front (say a precursor-like film) no instability is known on simple homogeneous fluids although the Hele-Shaw effect is destabilizing. On the other hand, the type of morphology found experimentally is very reminiscent of that found for a simulation of a growing crystal, a fact which reinforces the belief that the instability is of a diffusive type.

In summary, Branching of the DLA-type, or of snowflake nature... seems to embrace a wide range of systems ranging from physics to biology. Whether all visually similar branchings have a profound basis, or are just accidentally similar, is a major question which must be addressed. At least for the system considered here, a positive statement is beyond any doubt.

Acknowledgements

I would like to acknowledge several exchanges with S. Lipson. I acknowledge stimulating discussions with E. Brener and H. Müller-Krumbhaar.

REFERENCES

[1] P.G. de Gennes, *Rev. Mod. Phys.* **57** (1985) 825.
[2] D.B. Abraham, P. Collet, J.D. Connoick and F. Dunlop, *Phys. Rev. Lett.* **65** (1990) 195.
[3] J.F. Joanny and P.G. de Gennes, *J. Phys. France* **47** (1986) 121.
[4] N. Samid-Merzel, S.G. Lipson and D.S. Tannhauser, *Physica A* **257** (1998) 413.
[5] A. Sharma and A.T. Jameel, *J. Coll. Interf. Sci.* **161** (1993) 190.
[6] H. Müller-Krumbhaar, H. Emmerich and E. Brener (preprint).
[7] P. Pelcé, *Dynamics of curved fronts* (Academic Press, Inc., New York, 1987).
[8] T. Ihle and H. Muller-Krumbhaar, *Phys. Rev. E* **49** (1994) 2972.
[9] R. Kupferman, O. Shochet and E.B. Jacob, *Phys. Rev. E* **50** (1994) 1005.

CONTRIBUTED PAPER 11

Viscous Fingering in a Gel

A. Lindner, P. Coussot and D. Bonn

1. INTRODUCTION

When a less viscous fluid pushes a more viscous fluid in a Hele-Shaw cell, the interface between the two fluids develops an instability leading to the formation of fingerlike patterns, called viscous fingers. This is the so-called Saffman-Taylor or viscous fingering instability [1]. The width of these viscous fingers is, for Newtonian fluids, determined by the capillary number $Ca = \Delta\mu U/\gamma$ which represents the ratio of viscous forces over capillary forces; $\Delta\mu$ is the viscosity difference between the two fluids, U the finger velocity and γ the surface tension. The viscous forces tend to narrow the finger, whereas the capillary forces tend to widen it: the width of the finger decreases with increasing finger velocity. Due to its relative simplicity the viscous fingering instability has received much attention as an archetype of pattern forming systems, both theoretically and experimentally [1, 2] and is by now well understood.

A whole different class of problems was uncovered somewhat later, when the instability was studied for non-Newtonian fluids, where a wide variety of strikingly different patterns are found [3]. Most natural and industrial materials as for example glues, paints, mud, etc. are non-Newtonian fluids. It is thus also from a practical point of view important to understand the instability in such "complex fluids". Fingering experiments using foams, clay pastes, slurries and polymer gels reveal branched, fractal or fracture-like patterns [4]. The physical origin of the very different structures is so far ill understood, mainly because most of these fluids exhibit multiple non-Newtonian properties, which were not determined simultaneously.

The specific examples mentioned above are all believed also to exhibit a yield stress: if they are sollicitated below a certain critical (yield) stress, they do not flow, but respond elastically to the deformation. This might thus be at the origin of some of the observed patterns. Theory on yield stress fluids has revealed that the Saffman-Taylor instability is modified drastically [5].

We investigate the instability for a typical yield stress fluid: a polymer gel [7]. The results reveal the existence of a yield stress dominated fingering regime for low finger velocities, and a viscous regime for high velocities [6]. The width of the fingers in the yield stress regime is shown to be well described by the most unstable wavelength that follows from a linear stability analysis for yield stress fluids [5, 6].

2. EXPERIMENT

The rheological behavior of the fluids is characterized with a Reologica Stress-Tech rheometer using a parallel plate geometry (gap 2 mm, diameter 20 mm). To do so we measure the shear rate ($\dot{\gamma}$) as a function of the applied stress [10] (Fig. 2a). Below a critical (yield) stress one observes an essentially elastic response and within the accuracy of the apparatus a zero shear rate is obtained. For stresses beyond the yield stress, the stress (and consequently also the viscosity) varies with the shear rate as for a typical shear thinning fluid. Fluids exhibiting the combination of these two behaviors are typical yield stress fluids, usually described by the Herschel-Bulkley [5] model for the stress: $\sigma = \sigma_{\mathrm{y}} + \alpha\dot{\gamma}^n$, where σ_{y} is the yield stress.

The viscous fingering experiments are performed in a rectangular Hele-Shaw cell consisting of two glass plates separated by a thin Mylar spacer, fixing the plate spacing b, which can be varied from $b = 0.125$ to 1 mm. The channel width W is either 2 or 4 cm. The cell is filled with the fluid, which is subsequently driven from the cell using compressed air as the less viscous fluid. The fingers are captured by a CCD camera coupled to a VCR, allowing for measurement of the finger width w as a function of the finger velocity.

In the fingering experiments for low velocities one finds ramified structures consisting of different fingers. One repeatedly observes more than one finger propagating in parallel through the cell for a significant period of time. Eventually one of the fingers will screen the others, which stop moving. The finger that still moves subsequently destabilizes again, and the whole process starts over again. The observed finger width increases with increasing plate spacing (Fig. 1a).

For high velocities, a very different regime is observed. Only a single, stable finger propagates along the centerline through the cell. In addition the finger shape is very similar to that for classical viscous fingers (Fig. 1b).

To quantify the transition between the low and high velocity regime, we measure the average finger width as a function of the finger velocity (Fig. 2b). It follows that for low velocities ($U < 0.05$ cm/s) the finger width is indeed independent of the velocity, corresponding to the ramified fingers described above. Following the motion of air bubbles it seems that flow was almost negligible in some parts of the fluid. The yield point is not exceeded everywhere

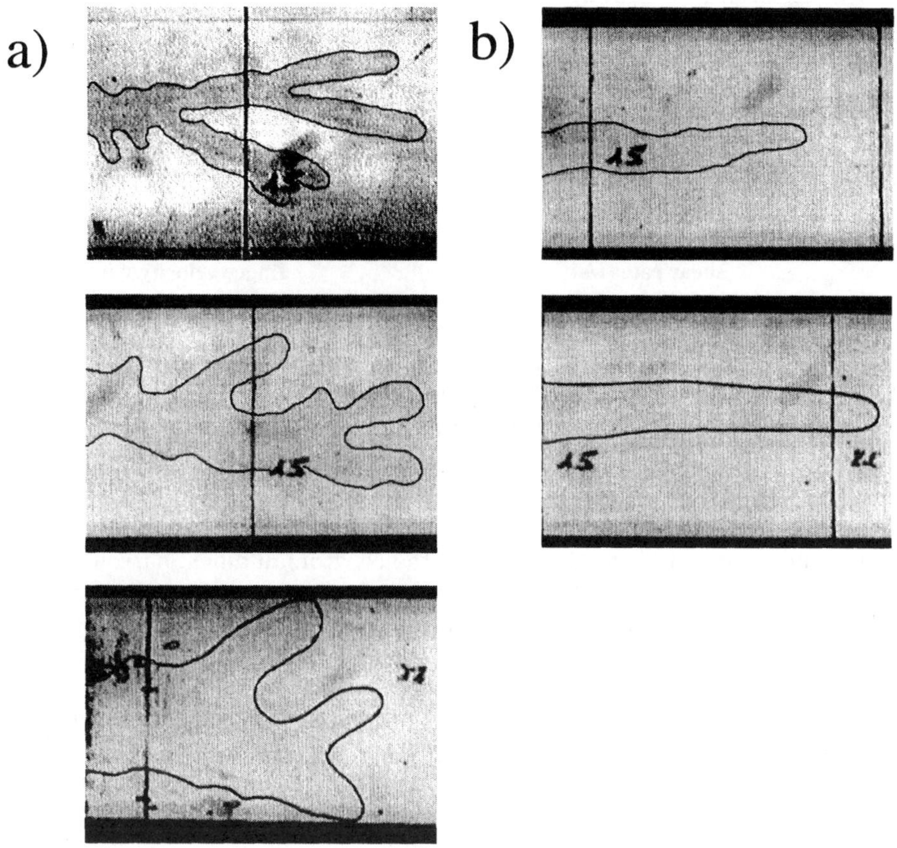

Fig. 1. — Snapshots of the viscous fingers at a) low velocities ($U < 0.01$ cm/s) for $W = 4$ cm; plate spacings of $b = 0.125$, 0.25 and 0.75 mm from top to bottom and b) at high velocities ($U > 1$ cm/s) for $W = 4$ cm, plate spacings of 0.25 and 0.75 mm from top to bottom.

and the regime is thus referred to as the yield stress regime. This also implies that the finger does not "feel" the presence of the walls delineating the canal in the Hele-Shaw cell, neither necessarily the presence of the other fingers that propagate in parallel. This provides an intuitive reason for why multiple fingers can occur. For higher velocities ($U > 0.05$ cm/s), where one finds one single finger in the centerline of the cell, the finger width decreases with increasing finger velocity. In this regime the viscous forces overcome the yield stress and the fluid should behave like an ordinary viscous fluid exhibiting shear thinning. The observed decrease of the finger width indeed indicates a viscous behavior

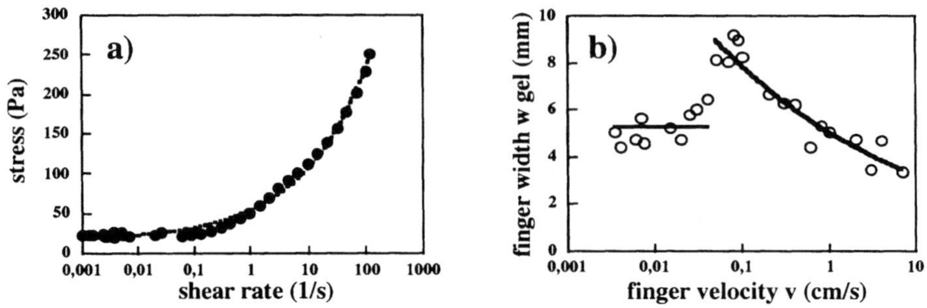

Fig. 2. — a) Rheological measurements of the stress as a function of the shear rate, the curve represents a fit to the Herschel-Bulkley model ($\sigma_y = 16$ Pa, $\alpha = 38$ Pas^n, $n = 0.38$; b) finger width *versus* finger velocity (on a logarithmic scale) for $W = 4$ cm and $b = 0.5$ mm.

of the fluid. Compared to the results for the Newtonian fluids, narrower fingers are found. Similar results have already been found for other shear thinning fluids, such as for example polymer solutions [8]. In conclusion we are able to identify two different regimes. In the following we will concentrate on the yield stress regime observed for low velocities.

A linear stability analysis of the flat interface has been performed for yield stress fluids. It can be demonstrated that the dispersion equation is in fact very similar to the Newtonian case [5, 6]. For the latter the wavelength of maximum growth is given by $\lambda_m \sim \sqrt{\gamma b^2/\mu U}$ [9], where $\mu U/b \simeq \mu\dot{\gamma}$ represents a characteristic viscous stress. For $U \to 0$ the viscous stress vanishes and λ_m diverges, leading to very large fingers at low velocity. For yield stress fluids, the viscous stress should be replaced by $\sigma = \sigma_y + \alpha\dot{\gamma}^n$ from the Herschel-Bulkley model. The wavelength of maximum growth is then again simply found from the ratio of the capillary forces to the total viscous forces, which now include the yield stress, and follows as: $\lambda_m \sim \sqrt{\gamma b/\sigma}$. An exact analysis leads to $\lambda_m = 2\pi\sqrt{3\gamma b/(2\sigma)}$ [5].

This has a number of interesting consequences. First, the result shows that the wavelength remains finite even at vanishing velocities simply because σ contains the finite yield stress. This also explains why for low velocities when the yield stress dominates over the viscous stresses ($\alpha\dot{\gamma}^n/\sigma_y \ll 1$), the finger width, which at least shortly after destabilization should correspond to the wavelength of maximum growth, is independent of the velocity. The other consequences are that the finger width should scale as the square root of the plate spacing and be independent of the channel width.

In order to test these predictions, experiments were performed varying the plate spacing b from 0.125 mm to 1 mm for channel widths of $W = 2$ cm and $W = 4$ cm. For each geometry we measured the finger width as a function

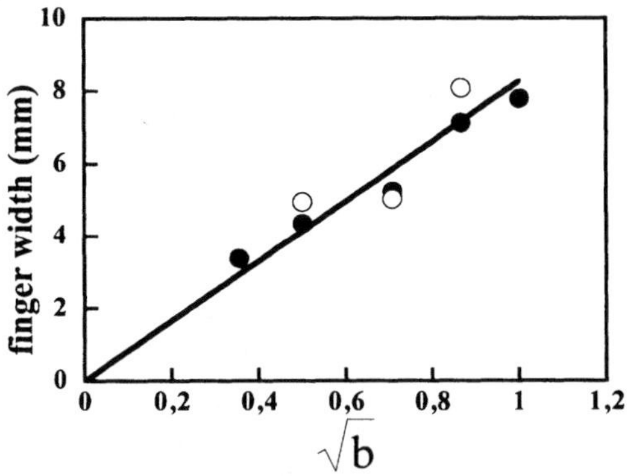

Fig. 3. — Finger width *versus* $\sqrt{b}$ for channel widths of 4 cm (•) and 2 cm (○).

of the finger velocity leading to graphs similar to Figure 2b. The results are depicted in Figure 3 where the averaged finger width w is given as a function of $\sqrt{b}$. Indeed, the finger width shows the dependence on the channel geometry predicted by the theory. More quantitatively, one can obtain the yield stress from the slope of the straight line, when equating σ to the yield stress and identifying the finger width with $\lambda_m/2$. If these identifications are made, one obtains $\sigma_y = 15.6$ Pa, in good agreement with the value $\sigma_y = 16$ Pa obtained independently from the rheological measurements. We thus conclude that in the yield stress regime, the simple theory does well in describing the experimental data.

3. DISCUSSION AND CONCLUSION

In conclusion, we have demonstrated that the Saffman-Taylor instability is drastically modified in yield stress fluids, leading to very branched patterns at low velocity, where the yield stress plays an important role. The results in the yield stress regime can be understood quantitatively from a linear stability analysis. For higher velocities only a single stable finger is observed. Fractal patterns also observed in yield stress fluids [4] occur as secondary instabilities at high speeds, for which the viscous stresses dominate the yield stress and are thus probably not related to the yield stress of the fluid.

Acknowledgements

We thank Jacques Meunier, Hamid Kellay and Martine Ben Amar for helpful discussions. LPS de l'ENS is UMR 8550 of the CNRS, associated with the universities Paris 6 and Paris 7. LMSGC is UMR113 LCPC-CNRS. A. L. has benefited from a scholarship from the DAAD under the HSP III.

REFERENCES

[1] P.G. Saffman and G.I. Taylor, *Proc. R. Soc. A* **245** (1958) 312; P.G. Saffman, *J. Fluid Mech.* **173** (1986) 73; J.W. McLean and P.G. Saffman, *J. Fluid Mech.* **102** (1881) 455.

[2] D. Bensimon, L.P. Kadanoff, S. Liang, B.I. Shraiman and C. Tang, *Rev. Mod. Phys.* **58** (1986) 977; G.M. Homsy, *A. Rev. Fluid Mech.* **19** (1987) 271; Y. Couder, *Chaos, Order and Patterns*, edited by R. Artuso *et al.* (Plenum press, New York, 1991) p. 203.

[3] K.V. McCloud and J.V. Maher, *Phys. Rep.* **260** (1995) 139.

[4] H. Van Damme, E. Lemaire, O.M. Abdelhaye, A. Mourchid and P. Levitz, *Non-linearity and breakdown in Soft Condensed Matter*, edited by K.K. Bardan *et al.*, *Lecture Notes in Phys.* **437** (Springer, 1994); E. Lemaire, P. Levitz, G. Daccord and H. Van Damme, *Phys. Rev. Lett.* **67** (1991) 2009; J. Nittmann, G. Daccord and H.E. Stanley, *Nat* **314** (1985) 141; S.S. Park and D.J. Durian, *Phys. Rev. Lett.* **72** (1994) 3347; H. Zhao and J.V. Maher, *Phys. Rev. E* **47** (1993) 4278.

[5] P. Coussot, *J. Fluid Mech.* **380** (1999) 63.

[6] A. Lindner, P. Coussot and D. Bonn, *Phys. Rev. Lett.* **85** (2000) 314.

[7] *A commercial hair gel* (Miss Helen Blue).

[8] A. Lindner, D. Bonn and J. Meunier, *Phys. Fluids* **12** (2000) 256.

[9] R.L. Chuoke, P. van Meurs and V. Van der Pol, *Pet. Trans. AIME* **216** (1959) 188.

[10] P. Coussot and C. Ancey, *Rheophysics of pastes and suspensions* (EDP Sciences, Les Ulis, 1999) (in French).

CONTRIBUTED PAPER 12

Branching Transition in Viscous Fingering with a Liquid Crystal

R. Folch, J. Casademunt and A. Hernández-Machado

1. INTRODUCTION

The phenomenon of branching has already proven to have some universality throughout this book. Misbah has even shown in the present chapter how the whole dynamics of a problem – including this branching phenomenon – can be mapped onto a different one in some approximation. In this spirit, the use of a unifying mathematical formulation in different physical models of branching can serve as a basis to understand why so different problems share the same branching phenomenon.

One such example of common mathematical tool have been the phase-field models. Inspired by models of dynamic phase transitions and originally developed for the solidification problem (treated in Karma's chapter), they not only provide a bridge between diffuse- and sharp-interface growth phenomena in phase transitions, but have recently also been extended [1, 2] to a sharp-interface problem where growth is not linked to any phase transition, namely that of viscous fingering.

Eventually, the answer to the origin of branching in physical models has turned out to be the amplification of initially small perturbations on the interface, and, anisotropy, the key common feature which, if strong enough, switches the branching from the tip-splitting to the side-branching mode [3–6].

In this contribution, we map (Sect. 2) a minimal model for the anisotropy in viscous fingering with a liquid crystal to the more studied anisotropy in surface tension, normally linked to solidification or viscous fingering in etched cells. This explains why a similar transition from tip-splitting to side-branching with increasing anisotropy is observed in viscous fingering experiments with a liquid crystal [3] and theory and simulations of solidification [4] or viscous fingering experiments in etched cells [5]. To confirm that, we finally use the mentioned extension of phase-field models to viscous fingering to numerically integrate the resulting problem (see Sect. 3), and we do find a transition which is consistent with the mentioned experiments. More details can be found in [6].

2. MODEL

In the nematic phase of a liquid crystal, its elongated molecules are locally oriented, giving rise to anisotropy in the viscosity and surface tension. The degree of orientation depends on the proximity of the other phase(s), namely the isotropic (and for some liquid crystals the smectic), *i.e.*, it still depends on temperature, and so does the anisotropy, mainly in the viscosity [3]. Therefore one should be able to explain the tip-splitting/side-branching transition as a function of temperature in the nematics by means of the anisotropy in the viscosity alone.

In a viscous fingering experiment using a liquid crystal as the more viscous fluid, the longest dimension of the molecules forms a small angle with the velocity of the liquid crystal, except maybe for the neighborhood of the interface, where it might follow its normal direction. So, as a first approximation, one can consider a viscosity which depends only on the direction of the velocity. Velocity dependent viscosities make the flow non-Laplacian, which would require to develop a separate phase-field model for it. However, in the vicinity of a finger tip we can approximate the direction of the flow for that of the finger, so that we can make a minimal model with only two different viscosities: one in the direction parallel to the finger and one in the perpendicular one. Even this simplified model is non-Laplacian, but this can be circumvented if the x and y-axis are rescaled by convenient factors, namely

$$a_x^2 \eta_x = a_y^2 \eta_y \equiv W^2 \eta, \tag{1}$$

where a_x, a_y are the rescaling factors and η_x, η_y the viscosities in the x and y directions, respectively; W is the channel width, and, η, a new, isotropic viscosity, so that the problem is now Laplacian. The anisotropy has been translated to a dimensionless surface tension of the form

$$B(\phi) = B_0 \times \left[\frac{1}{1 + (m-1)\cos^2 \phi} \right]^{3/2}, \tag{2}$$

where $m \equiv \eta_x/\eta_y$, ϕ is the angle from $\hat{x}$ to the normal to the interface in the rescaled variables, and B_0 is the dimensionless surface tension of isotropic viscous fingering except for constant m factors. This means that, even with an originally isotropic surface tension, that in the rescaled problem has a two-fold anisotropy with a very specific form given by the second factor at the r.h.s. of equation (2) (see Ref. [6] for more details).

A similar result was found in a different context, namely for the nematic-smectic B transition, where two different heat diffusivities in two perpendicular directions could be mapped to the same type of anisotropy in the capillary length [7]. However, note that here the assumption is that the growth is in the direction of lowest viscosity (because of flow alignment of the director), which results in growth in the direction of largest surface tension ($\phi = \pi/2$), whereas in [7] the situation is just the opposite: growth was found to be in the direction of lowest diffusivity because that is the direction of lowest capillary length. Also for isotropic diffusivities it is known that steady needle crystals can only grow in the direction of minimal capillary length [8], although there the anisotropy is assumed to be four-fold.

3. RESULTS

We now integrate the rescaled, Laplacian problem, by means of the following phase-field model,

$$\tilde{\epsilon}\frac{\partial\psi}{\partial t} = \nabla^2\psi + c\vec{\nabla}\cdot(\theta\vec{\nabla}\psi) + \frac{1}{\epsilon}\frac{1}{2\sqrt{2}}\gamma(\theta)(1-\theta^2) \tag{3}$$

$$\epsilon^2\frac{\partial\theta}{\partial t} = f(\theta) + \epsilon^2\nabla^2\theta + \epsilon^2\kappa(\theta)|\vec{\nabla}\theta| + \epsilon^2\hat{z}\cdot(\vec{\nabla}\psi\times\vec{\nabla}\theta), \tag{4}$$

where ψ is the stream function, whose $\psi = \text{const}$ curves are the trajectories of the fluid particles, c is the so-called viscosity contrast, θ is the phase field, and ϵ, $\tilde{\epsilon}$ are model parameters which must be small to recover the sharp-interface equations of the rescaled problem. We have defined $f(\theta) \equiv \theta(1-\theta^2)$, and $\gamma(\theta)/2 \equiv \hat{s}(\theta)\cdot[\vec{\nabla}B(\theta)\kappa(\theta) + \hat{y}]$, where $\kappa(\theta) \equiv -\vec{\nabla}\cdot\hat{r}(\theta)$, $\hat{r}(\theta) \equiv \vec{\nabla}\theta/|\vec{\nabla}\theta|$ and $\hat{s}(\theta) \equiv \hat{r}(\theta)\times\hat{z}$ are the generalization of the interface curvature and unit vectors in the direction perpendicular and parallel to the interface to any $\theta = \text{const}$ curve. In particular, the interface is identified with the $\theta = 0$ curve. This model was introduced for isotropic viscous fingering in [1] and extensively tested in [2]. From this work we know how to choose ϵ and $\tilde{\epsilon}$ so that the model yields converged results.

With this appropriate choice of ϵ and $\tilde{\epsilon}$, we investigate the transition between the tip-splitting and side-branching behaviors as both the dimensionless surface tension B_0 and its anisotropy $m-1$ are varied. We first run a steady finger with a large, isotropic dimensionless surface tension $B(\phi) = B_0 = 10^{-2}$, large enough for the finger not to destabilize for the amount of (numerical) noise we have and thus reach its steady width and velocity. Once this is achieved – see inner interface in Figures 1a and 1b –, we perform a "quench" in surface tension, *i.e.*, we instantly reduce it to some lower value. Simultaneously, we also introduce some amount of anisotropy $m-1$. The subsequent interface evolution for $B_0 = 10^{-3}$ (and $c = 0.8$, $\tilde{\epsilon} = 0.08$) is also shown within the reference frame moving with the mean interface in Figures 1a ($m = 2$) and 1b ($m = 2.25$) in the form of snapshots at time intervals 0.11.

For this value of B_0 the finger clearly destabilizes: first its tip widens and flattens and therefore slows down for any value of the anisotropy. Then, for

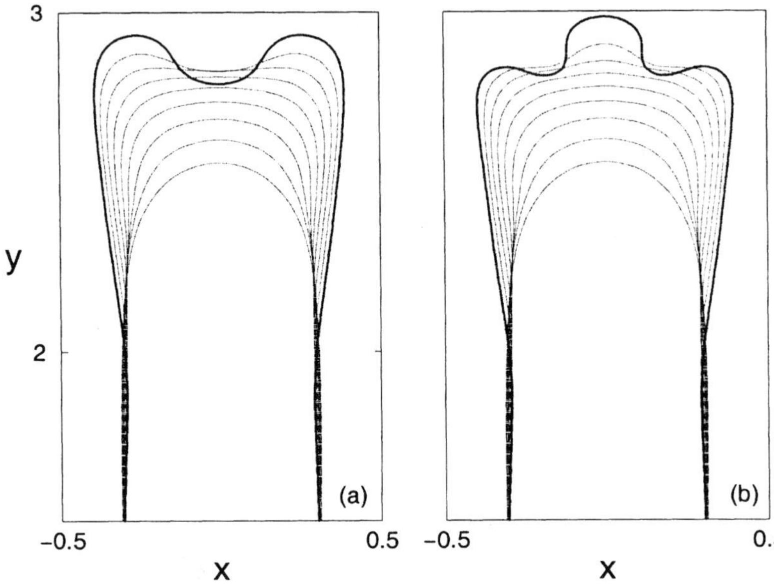

Fig. 1. — Destabilization of the tip of a finger in the reference frame moving with the mean interface, just after instantly decreasing B_0 to $B_0 = 10^{-3}$. a) Tip-splitting for low anisotropy, $m - 1 = 1$. b) Side-branching for high anisotropy, $m - 1 = 1.25$.

$m = 2$ the tip continues to flatten and slow down until its curvature and eventually its velocity in the frame of the mean interface reverse their signs (Fig. 1a), which we identify with the tip-splitting morphology.

In contrast, for $m = 2.25$ the reversal of the curvature sign takes place at some distance of the center of the channel, while at the center the curvature increases again and makes it possible for the tip to speed up again as well (Fig. 1b). We identify this reversal of the curvature sign at a distance from the center of the channel and this recovery of the tip velocity with the side-branching morphology.

In this way we systematically explore values of the dimensionless surface tension ranging from $B_0 = 10^{-2}$ down to $B_0 = 4 \times 10^{-4}$. For each value of B_0 we simulate with several values of the anisotropy $m - 1$, and we find that there is a relatively sharp transition between the tip-splitting and side-branching morphologies. In Figure 2 we show for each value of B_0 (x-axis) the two closest values of $m - 1$ (y-axis) for which the two different morphologies are observed, namely tip-splitting (circles) and side-branching (triangles). Thus we know that the transition line must lie somewhere between the circles and the triangles, and that above (larger values of $m - 1$) and left (lower values of B_0)

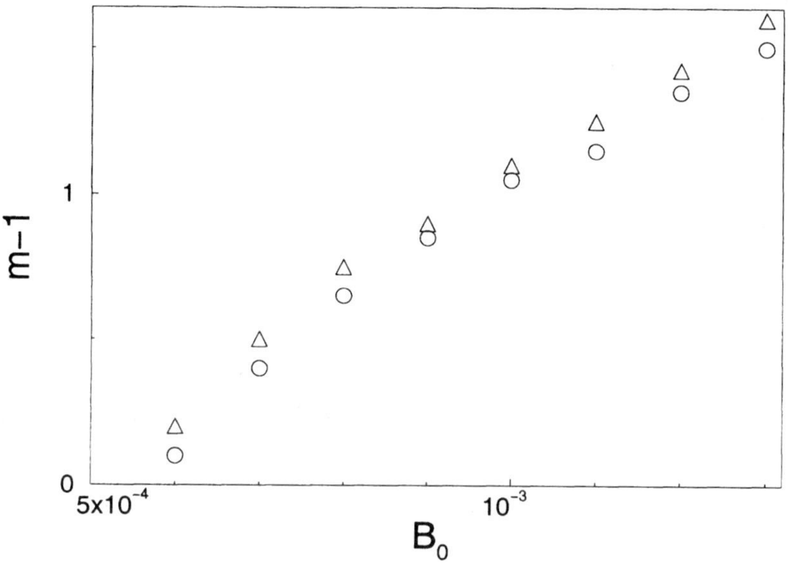

Fig. 2. — Transition between tip-splitting (circles) and side-branching (triangles) as a function of the surface tension anisotropy $m-1$ and the dimensionless surface tension B_0.

of that transition line the morphology is side-branching, and below and right of it, tip-splitting. This means that the critical anisotropy $m-1$ above which side-branching replaces tip-splitting decreases with decreasing dimensionless surface tension B_0. At least the fact that above a critical anisotropy side-branching replaces tip-splitting is consistent with experiments [3–5] (see Ref. [6] for more details).

In fact, this critical anisotropy vanishes at $B_0 \sim 5 \times 10^{-4}$, and below this value only side-branching is observed, even if one uses negative anisotropies down to $m-1=-0.9$, which correspond to a viscosity larger in the direction of growth of the finger than in the perpendicular one, and which is not the case of the liquid crystal experiments that motivated this study. ($m-1>-1$ to keep the two viscosities and therefore $B(\phi=0)$ finite and positive.)

On the other hand, for $B_0 \geq 1.4 \times 10^{-3}$ and for the time elapsed in our runs no clear side-branching is actually observed above the transition line extrapolated from lower values of B_0, whereas tip-splitting still occurs below that line. For even larger values of B_0, namely $B_0 \sim 2 \times 10^{-3}$, not even tip-splitting is observed again within the time elapsed, although the steady finger still destabilizes through the widening and flattening of its tip. Finally, for $B \geq 10^{-2}$ the finger is completely stable for the amount of noise we have, as was pointed out before.

Acknowledgements

We are grateful to Á. Buka and T. Tóth-Katona for helpful discussions. We acknowledge financial support from the Dirección General de Enseñanza Superior (Spain), Projects No. PB96-1001-C02-02 and PB96-0378-C02-01, and the European Commission Project No. ERB FMRX-CT96-0085. Simulations have been carried out using the resources at CESCA and CEPBA, coordinated by C^4. R.F. also acknowledges a grant from the Comissionat per a Universitats i Recerca (Generalitat de Catalunya).

REFERENCES

[1] R. Folch, J. Casademunt, A. Hernández-Machado and L. Ramírez-Piscina, *Phys. Rev. E* **60** (1999) 1724.
[2] R. Folch, J. Casademunt, A. Hernández-Machado and L. Ramírez-Piscina, *Phys. Rev. E* **60** (1999) 1734.
[3] Á. Buka, in *Pattern formation in liquid crystals* (Springer-Verlag, New York, 1996) p. 291.
[4] T. Ihle and H. Müller-Krumbhaar, *Phys. Rev. E* **49** (1994) 2972.
[5] E. Ben-Jacob and P. Garik, *Physica D* **38** (1989) 16.
[6] R. Folch, J. Casademunt and A. Hernández-Machado, *Phys. Rev. E* **61** (2000) 6632.
[7] T. Börzsönyi , Á. Buka and L. Kramer, *Phys. Rev. E* **58** (1998) 6236.
[8] E.A. Brener and V.I. Mel'nikov, *Adv. Phys.* **40** (1991) 53.

CONTRIBUTED PAPER 13

Mathematical Meristems: The Singularities of Laplacian Growth

M.A. Peterson

1. INTRODUCTION

Laplacian growth and the branched structures it generates have been a subject of intense investigation for many years. By Laplacian growth one means some variant of the following process: the boundary ∂D of a region D moves with a velocity $\vec{v} = -\vec{\nabla}\phi$, where ϕ solves the Laplace equation (or the diffusion equation) in the exterior of D. One must still specify boundary conditions for ϕ, for example ϕ = constant on ∂D, and $\phi \to 0$ at infinity. These conditions would describe, among other things, a solid conductor growing by electrodeposition from an infinite conducting solution along the electric field lines. The exterior of D might also include other boundary surfaces, etc.

This mathematical problem has peculiarities of nonuniqueness and instability which are still not entirely understood. An example of nonuniqueness that has been known for a long time is Saffman and Taylor's Laplacian growth analysis of the viscous finger in a Hele-Shaw cell [1]. They found no unique steady propagating solution, but rather a continuous family of them, parameterized by finger width. Experiment, however, somehow selects just one of these solutions, the one having half the channel width. In the 1980's this problem was, in a sense, resolved by the suggestion that surface tension is a singular perturbation to Laplacian growth, and hence not to be neglected. In the limit as surface tension goes to zero, the observed finger is selected from among the Saffman-Taylor solutions [2, 3]. Another suggestion that interfacial tension cannot be neglected grows out of the analysis by Mullins and Sekerka of the infinitesimal instability of a plane interface undergoing Laplacian growth [4]. They showed that the interface is unstable to every Fourier perturbation of its shape, and the higher the spatial frequency, the more unstable it is. Surface tension thus seems to be an essential part of the theory, preventing curvature singularities from developing at the interface. Just this result, by a very different method, was demonstrated by Shraiman and Bensimon in the 2-dimensional Laplacian growth of a bounded region D in the plane (so-called radial geometry) [5]. They parameterized D by a conformal map from the exterior of the unit circle onto the exterior of D. The dynamics of Laplacian growth in this formulation turns out to be just the motion of the singularities of this map, assumed to be a finite number of poles and zeros inside the unit circle. A singularity moves to the unit circle, the boundary of the physical region, in finite time, corresponding to the development of a cusp in the boundary of D. However a similar conformal mapping treatment by Mineev-Weinstein of Laplacian growth in channel geometry (*i.e.*, a Hele-Shaw cell) revealed, unexpectedly, that in this geometry the singularities might stay safely away from the physical region, so that the solution stays regular, despite the Mullins-Sekerka instability [6]. Mineev-Weinstein has also pointed out that the correct Saffman-Taylor finger is selected, without introducing surface tension [7, 8]. Peterson used the conformal mapping formulation to derive a singular limit of 2-dimensional Laplacian growth in radial geometry [9, 10]. In this limit the singularities remain on the boundary of the physical region (the unit circle), but they can *split* at any time, producing new singularities, so that the dynamics is highly nonunique. This singular limit is described in more detail below.

The above synopsis of exact results in Laplacian growth has said nothing of the voluminous literature on numerical and approximate solutions, beginning with the DLA simulations of Witten and Sanders [11], showing among other things the peculiar sensitivity of Laplacian growth to noise, anisotropy, and other perturbations. Recent progress in efficient, reliable numerical methods for this problem (phase field formalism) is described in this volume by Karma [12].

2. SINGULAR MODEL

In this model, the growing region D is a tree graph in the complex w-plane parameterized by a singular Schwarz-Christoffel map $w = H(z)$ from the exterior of the unit circle in the complex z-plane onto the exterior of D. The singularities of the conformal map H are *on* the unit circle, and consist of M branch points at $\{e^{i\beta_j}\}$ and N simple zeros of the derivative H' at $\{e^{i\gamma_j}\}$. These singularities will be referred to henceforward as β_j, γ_j, etc., as shown in Figure 1. The image of the unit circle is a degenerate polygon, enclosing zero area, and traversing every edge of the tree graph D twice, once in each direction. The image of a branch point β_j is an interior vertex of the graph, where two edges come together. Associated with each branch point β_j is an angle $\alpha_j\pi$, which is the angle through which one turns at the interior vertex as one traverses D. The image of each γ_j (simple zero of H') is an extremal vertex. The conformal

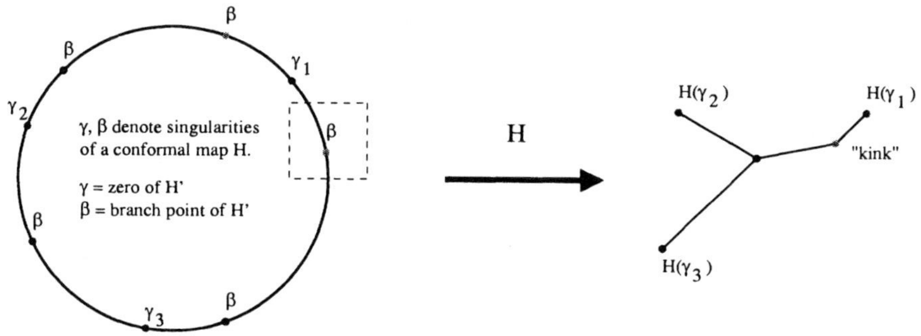

Fig. 1. — The domain of the parameterizing conformal map H in the z-plane is the exterior of the unit circle. The image of H in the w-plane is the exterior of a tree graph. The singularities of H are on the unit circle, and their images are the vertices of the tree. The branch point enclosed in a box goes to the vertex labeled "kink", for example.

map $w = H(z)$ is most simply described by its derivative H'. It is

$$\frac{\partial H}{\partial z} = \prod_{j=1}^{M}(1 - \mathrm{e}^{i\beta_j}/z)^{\alpha_j} \prod_{k=1}^{N}(1 - \mathrm{e}^{i\gamma_k}/z) \tag{1}$$

if there are M branch points and N simple zeros. There are in addition many constraints that must be satisfied if the map is to take the unit circle onto a tree graph D, among them a condition on the α's, describing the angles in the tree,

$$N + \sum_{j=1}^{M} \alpha_j = 2. \tag{2}$$

The dynamics of Laplacian growth in this limit is just the dynamics of the singularities on the unit circle. This is entirely analogous to what happens in the corresponding non-singular problem, in which the singularities of H are *inside* the unit circle [5]. The singular model follows, in fact, from the non-singular model by appropriate scaling in space and time, so that the image of the unit circle under the map H has a size always of order one, and grows with a velocity of order one. If one then moves the singularities onto the unit circle, the system continues to make sense [10]. The result is the following autonomous system of ordinary differential equations for the positions of the singularities as a function of the scaled time s:

$$\frac{\mathrm{d}\beta_j}{\mathrm{d}s} = -\frac{1}{\alpha_j}\frac{\partial \ell n\, W}{\partial \beta_j} \tag{3}$$

$$\frac{\mathrm{d}\gamma_k}{\mathrm{d}s} = -\frac{\partial \ell n\, W}{\partial \gamma_k} \tag{4}$$

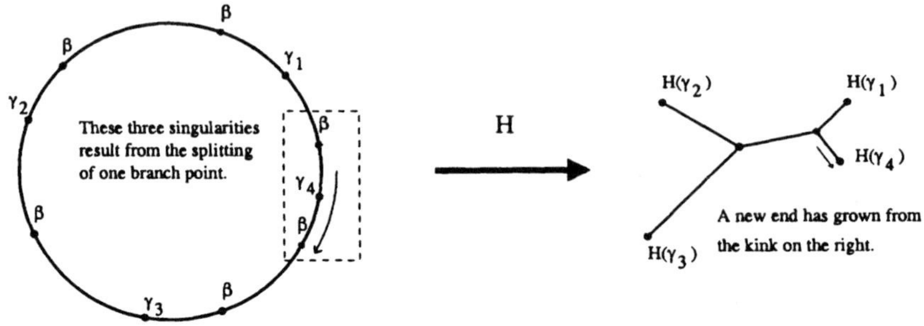

Fig. 2. — According to the dynamics of singular Laplacian growth, branch points of the conformal map H can split, giving rise to new singularities. The effect on the image is to create new branches.

where

$$W = \sum_{k=1}^{N} w_k \tag{5}$$

and

$$w_k = \prod_{j=1}^{M} \sin^{-2\alpha_j}\left(\frac{\gamma_k - \beta_j}{2}\right) \prod_{j \neq k} \sin^{-2}\left(\frac{\gamma_k - \gamma_j}{2}\right). \tag{6}$$

It is a surprising feature of the differential system equations (3–6) that the singularities of H can split to create new singularities. When this happens, the parameters M and N increase. This phenomenon can be seen indirectly by integrating the system backward in time: singularities are seen to coalesce, in finite time [10]. That is, singularities disappear by merging with other singularities. Viewed forward in time, new singularities are created by the splitting of existing singularities. A crucial difference is that the backward process, coalescence of singularities, is deterministic; the forward process, creation of singularities, is not. Creation of singularities can occur at any time. It is allowed by the (singular) differential system, but not required.

There are two basic singularity splitting processes. The first corresponds to the introduction of a kink into a growing branch, in which a γ splits off a β on either side. The kink on the right in Figure 1 might have been formed in this way if γ_1 split off the two β's on either side of it in Figure 1. The second singularity splitting process corresponds to the "budding" of a new branch at a kink, in which a β (with corresponding $\alpha > 0$) splits off a β and a γ. This process is illustrated in Figure 2. If both these processes occur, kinking and budding together, a single growing tip splits into two growing tips.

3. MATHEMATICAL MERISTEMS

The behavior of the singularities of Laplacian growth is so suggestive of botanical meristems that it is hard to resist calling them "mathematical meristems" (see [13]). By far the easiest way to describe the singular Laplacian growth process to a botanist, in fact, is to point out this analogy. Like real, botanical meristems, these singularities can create new singularities by a process of splitting, and thus control the shape of a developing branched structure. In the course of splitting, botanical meristems, like the singularities of the model, must in some concrete way lay down information at the base of a branch which determines geometrical angles, analogous to the branchpoints β with their corresponding angles α. It is natural to wonder if this analogy is more than coincidence.

To be clear, this is a question about what happens *within* meristems, not about the large scale growth of plants, which, of course, has nothing to do with Laplacian growth once the positions and angles of branches are determined. In the singular theory the morphogenesis of the growing region is controlled by what happens at just a few special singular points. If something like Laplacian growth is occurring within the meristem, the actual singularities would have to be smaller still, perhaps just a few special cells, whose presence and behavior is only revealed later by subsequent development, when it is ascribed to the meristem as a whole.

The singular Laplacian growth model emphasizes that the key processes of Laplacian growth can happen in very small regions, discrete points in the model. Thus it is suggested that the generation of geometrical pattern, characterized by definite angles, at a length scale even smaller than a botanical meristem, might still be a manifestation of Laplacian growth. The meristem would be the "universe" in which the process takes place, containing both the inside and the outside of the relevant growing region, which is a subset of the meristem. The more familiar nonsingular picture of Laplacian growth, as computed in the phase field formalism, could suggest a different picture, as pointed out recently by Fleury [14]. Here the growth of the meristem as a whole is a process like Laplacian growth, but the Laplacian field is inside, not outside, and acts as an inhibitor of growth, not a stimulant.

It is to be hoped that such ideas will have some relevance to real meristems. Laplacian growth is a very basic process, nearly universal, including, in all probability, living systems. It is well known that various cells communicate with each other by releasing diffusible substances, both in their normal functioning and during development, when structures are still forming. This is the prerequisite for something like Laplacian growth to occur: growth in response to something diffusible. Meristems in particular are small enough that pattern formation utilizing Laplacian growth could be taking place within them.

Laplacian growth, because of its sensitivity to perturbation, can take many different forms. It is not a blueprint for a structure, but rather a kind of "metastructure", or "virtual structure", which needs only the details to be filled in to become one or other real structure. It seems unlikely that biological evolution would not have discovered this mechanism and exploited it, ideally suited as it is for experimentation with form.

To stabilize Laplacian growth, to control it, and to use it to produce one particular form reliably and reproducibly, one must, in the picture suggested by the singular theory, get control of its singularities. This is what could have happened in the case of meristems: the singularities could have been realized as small, specialized tissues within the meristem, with properties characteristic of this or that species, and therefore, somehow, under control. The physical mechanisms underlying this control could include diffusion of more substances than just one, both promoters and inhibitors of growth (as suggested by Fleury [14]), interacting with each other and stabilizing each other through feedback loops of cellular response.

It has always been one of the motivations for studying Laplacian growth to learn to control it, for purposes of metallurgy, oil extraction, etc. Might biological organisms already have developed ways to control it? Just the possibility seems reason enough to pursue these questions.

REFERENCES

[1] P.G. Saffman and G.I. Taylor, *Proc. R. Soc. A* **245** (1958) 312-329.
[2] *Dynamics of Curved Fronts*, edited by P. Pelce (Academic Press, Inc., Boston, 1988).
[3] D.A. Kessler, J. Koplik and H. Levine, *Adv. Phys.* **37** (1988) 255.
[4] W.W. Mullins and R.F. Sekerka, *J. Appl. Phys.* **35** (1964) 444-451.
[5] B.I. Shraiman and D. Bensimon, *Phys. Rev. A* **30** (1984) 2840-2844.
[6] M.B. Mineev-Weinstein and S.P. Dawson, *Phys. Rev. E* **50** (1994) R24-R27.
[7] M.B. Mineev-Weinstein, *Phys. Rev. Lett.* **80** (1998) 2113-2116.
[8] M.B. Mineev-Weinstein, http://xxx.lanl.gov/abs/nlin.PS/9902007.
[9] M.A. Peterson, *Phys. Rev. Lett.* **62** (1989) 284-287.
[10] M.A. Peterson, *Phys. Rev. E* **57** (1998) 3221-3226.
[11] T.W. Witten and L.M. Sander, *Phys. Rev. Lett.* **47** (1981) 1400-1403.
[12] A. Karma, this volume.
[13] P. Laufs, this volume.
[14] V. Fleury, *C. R. Acad. Sci. Paris, Sciences de la vie* **322** (1999) 725-734.

Glossary

Actin: *Biochem.* A globulin that is present in muscle plasma and that in connection with myosin plays an important role in muscle contraction.

Actin filaments: These microfilaments have a variety of roles–they form part of the cell's cytoskeleton, they interact with myosin to permit movement of the cell, and they pinch the cell into two during cell division. In muscle contraction, filaments of actin and myosin alternately unlink and chemically link in a sliding action. The energy for this reaction is supplied by adenosine triphosphate.

Action potential: *Physiol.* The change in electrical potential that occurs between the inside and outside of a nerve or muscle fiber when it is stimulated, serving to transmit nerve signals.

Agar: 1. A gelatinlike product of certain seaweeds, used for solidifying certain culture media, as a thickening agent for ice cream and other foods, as a substitute for gelatin, in adhesives, as an emulsifier, etc. 2. Biol. A culture medium having an agar base.

Allograft: *Surg.* A tissue or organ obtained from one member of a species and grafted to a genetically dissimilar member of the same species. Also called allotransplant, homograft, homotransplant.

Angiosperm: *Bot.* A plant having its seeds enclosed in an ovary; a flowering plant.

Anisotropic: 1. *Physics* Of unequal physical properties along different axes. 2. *Bot.* Of different dimensions along different axes.

Axil: *Bot.* The angle between the upper side of a leaf or stem and the supporting stem or branch.

Axillary bud: *Bot.* A bud that is borne at the axil of a leaf and is capable of developing into a branch shoot or flower cluster. Also called lateral bud.

Axon: *Cell. Biol.* The appendage of the neuron that transmits impulses away from the cell body.

Branch: 1. A division or subdivision of the stem or axis of a tree, shrub, or other plant. 2. A limb, offshoot, or ramification of any main stem. 3. A tributary stream or any stream that is not a large river or a bayou; **dense branching**, system in which the branch formation leads to a structure filling homogeneously space, with a fractal dimension equal to the dimension of the space in which it grows; **side branching**, formation of branches on the sides of a main growing dendrite or stem.

Bud: In lower plants and animals, a protuberance from which a new organism or limb develops; in seed plants, a miniaturized twig bearing compressed rudimentary lateral stems (branches), leaves, or flowers, or all three, and protected in cold climates by overlapping bud scales. In warm climates buds may grow all year; in temperate climates they grow in summer and remain dormant in the winter. The form of winter buds (particularly the larger terminal buds on twigs) of trees and shrubs may be used to identify the species. The "eyes" of a potato are undeveloped buds.

Budding: Type of grafting in which a plant bud is inserted under the bark of the stock (usually not more than a year old). It is best done when the bark will peel easily and the buds are mature, as in spring, late summer, or early autumn. Budding is a standard means of propagating roses and most fruit trees in nurseries. It is also a vegetative propagation of plants or of (some) animals. In this book it has been sometimes used as "formation of buds".

Capillary: Microscopic blood vessel, smallest unit of the circulatory system. Capillaries form a network of tiny tubes throughout the body, connecting arterioles (smallest arteries) and venules (smallest veins). Through the thin capillary walls, which are composed of a single layer of cells, the nutritive material and oxygen in the blood pass into the body tissues, and waste matter and carbon dioxide in turn are absorbed from the tissues into the bloodstream.

Chemotropism: *Biol.* Oriented growth or movement in response to a chemical stimulus.

Capillary constant a, has units of length and gives the characteristic scale of capillary waves on a liquid-gas interface

$$a = \sqrt{\frac{2\sigma}{g\rho}}$$

σ surface tension (J/m^2)
g gravity acceleration (m/s^2)
ρ density of the medium (kg/m^3).

Capillary length: Usually denoted by d_0, is a microscopic length scale that is related to the reaction of the interface to curvature. It is usually of the order of the interface thickness, *i.e.* ~ 1 nm.

$$d_0 = \frac{\sigma \, T_{\mathrm{m}}}{\mathcal{L} \, T_{\mathrm{sc}}}$$

σ surface tension ($\mathrm{J/m^2}$)

$\mathcal{L}$ latent heat per unit volume ($\mathrm{J/m^3}$)

T_{m} melting temperature (K)

T_{sc} characteristic temperature scale for a given material.

(for a pure material : $T_{\mathrm{sc}} = \mathcal{L} \, / c_{\mathrm{p}}$; for alloys: freezing range) (K).

Chemotaxis: Orientation or movement of an organism or cell in relation to chemical agents.

Chiral: *Chem.* (for a molecule) not superimposable on its mirror image.

Collagen: Any of a group of proteins found in skin, ligaments, tendons, bone and cartilage, and other connective tissue. Cells called fibroblasts form the various fibers in connective tissue in the body. Collagen consists of groups of white inelastic fibers with great tensile strength. These fibers include fine fibrils, which are composed of even finer filaments, visible only through the electron microscope. Collagen protein contains an unusually high percentage of the amino acids proline and hydroxyproline. X-ray diffraction studies provide evidence that the protein forms a wavy band, a coiled chain with periodic, *i.e.*, repeating, arrangement of its amino acids. Cartilage is composed of fibrous collagen in an amorphous gel. The organic (nonmineral) content of bone is made up largely of collagen fibers with calcium salt crystals lying adjacent to each segment of the fiber; the fibers and salt crystals combined form a structure with compressional and tensile strength comparable to that of reinforced concrete. A group of diseases, often termed collagen, or connective tissue, diseases, involve a variety of alterations in the connective tissue fibers; rheumatoid arthritis, rheumatic fever, lupus, and scleroderma are included in this group. Some of these diseases may involve an autoimmune response, in which the immune mechanism injures or destroys the individual's own tissues. Collagen dissolved in boiling water becomes denatured to form gelatin.

Compliance: *Physics* 1. The strain of an elastic body expressed as a function of the force producing the strain. 2. A coefficient expressing the responsiveness of a mechanical system to a periodic force.

Cotyledon: *Bot.* The primary or rudimentary leaf of the embryo of seed plants; **polycotyledon**, a plant having more than two cotyledons, as certain gymnosperms.

Cytology: The study of the microscopic appearance of cells, esp. for the diagnosis of abnormalities and malignancies.

Cytoskeleton: *Cell Biol.* The network of protein filaments and microtubules in the cytoplasm that controls cell shape, maintains intracellular organization, and is involved in cell movement.

Dendrite: 1. *Petrol., Mineral.* a) A branching figure or marking, resembling moss or a shrub or tree in form, found on or in certain stones or minerals due to the presence of a foreign material. b) Any arborescent crystalline growth. 2. *Anat.* The branched part of a neuron that conducts impulses toward the cell. See diag. under neuron.

Diffusion length: For an interface whose growth is limited by the diffusion of heat or matter with diffusion coefficient $\mathcal{D}$ (m^2/s) and which advances with velocity v, the diffusion length $l_D = \mathcal{D}/v$ gives the length scale of appreciable variations of the diffusion field.

Dimensionless numbers:

L characteristic length (m)

$\mathcal{D}_\mathrm{T}$ thermal diffusivity of the fluid $(\mathrm{m}^2/\mathrm{s})$

d_T thermal diffusion length (m)

$\mathcal{D}$ molecular diffusivity of the fluid $(\mathrm{m}^2/\mathrm{s})$

k thermal conductivity of the fluid $(\mathrm{W/m^2K/m})$

c_p heat capacity of the fluid (J/kgK)

ρ density of the fluid $(\mathrm{kg/m^3})$

μ viscosity of the fluid $(\mathrm{N.s/m^2})$

V velocity of the fluid (m/s)

σ surface tension $(\mathrm{J/m^2})$.

Reynolds number Re, is proportional to {(inertial force)/(viscous force)} in a hydrodynamic flow and is used in momentum, heat, and mass transfer equations to account for dynamic similarity.

$$\mathrm{Re} = \frac{L\ V\ \rho}{\mu}.$$

Prandtl number Pr, is proportional to {(momentum diffusivity)/(thermal diffusivity)} and is used in heat transfer in general and free and forced convection calculations in particular.

$$\mathrm{Pr} = \frac{c_\mathrm{p}\ \mu}{k}.$$

PECLET NUMBER Pe, is proportional to {(bulk heat transfer)/(conductive heat transfer)} and is used in heat transfer in general and forced convection calculations in particular. It is equivalent to (Re · Pr).

$$\mathrm{Pe} = \frac{L\,V}{\mathcal{D}_{\mathrm{T}}} = \frac{L}{d_{\mathrm{T}}}.$$

CAPILLARY NUMBER Ca, is proportional to {(viscous force)/(surface tension force)} and is used in momentum transfer in general and calculations of atomization and 2-phase flow in beds of solids in particular. It is equivalent to (We/Re). Remark: do not confuse with capillary constant or capillary length.

$$\mathrm{Ca} = \frac{\mu\,V}{\sigma}.$$

In 2-phase flow, the viscosity difference between the two fluids $\Delta\mu$ is the convenient parameter.

WEBER NUMBER We, is proportional to {(inertial force)/(surface tension force)} and is used in calculations of momentum transfer in general and bubble/droplet formation and breakage of liquid jets in particular.

$$\mathrm{We} = \frac{L\,V^2\,\rho}{\sigma}.$$

SCHMIDT NUMBER Sc is proportional to {(kinetic viscosity)/(molecular diffusivity)} and is used in mass transfer in general and diffusion in flowing systems calculations in particular

$$\mathrm{Sc} = \frac{\mu}{\rho\,\mathcal{D}}.$$

Distal: Situated away from the point of origin or attachment, as of a limb or bone; terminal.

Drainage basin: Also called CATCHMENT AREA, or WATERSHED, area from which all precipitation flows to a single stream or set of streams. For example, the total area drained by the Mississippi River constitutes its drainage basin, whereas that part of the Mississippi River drained by the Ohio River is the Ohio's drainage basin. The boundary between drainage basins is a drainage divide: all the precipitation on opposite sides of a drainage divide will flow into different drainage basins.

A drainage basin provides a limited surface area within which physical processes pertinent to the general hydrology may be considered.

The climatic variables and the water and sediment discharge, water storage, and evapotranspiration may be measured; from these measurements, denudation rates and moisture and energy balances may be derived, each of which is useful in the consideration and understanding of landscape formation.

Ectopic: *Pathol.* Occurring in an abnormal position or place; displaced.

Endoderm: 1. Also called **endoblast**. *Embryol.* The innermost cell layer of the embryo in its gastrula stage. 2. *Anat.* The innermost body tissue that derives from this layer, as the gut lining. Also, **entoderm**.

Epigenesis: *Biol.* 1. The theory that an embryo develops from the successive differentiation of an originally undifferentiated structure (opposed to preformation). 2. The approximately stepwise process by which genetic information, as modified by environmental influences, is translated into the substance and behavior of an organism.

Epithelium: *Biol.* 1. A membranous cellular tissue that covers a free surface or lines a tube or cavity of an animal body and serves esp. to enclose and protect the other parts of the body, to produce secretions and excretions, and to function in assimilation. 2. A usu. thin layer of parenchyma that lines a cavity or tube of a plant.

Fibroblast: *Cell Biol.* A cell that contributes to the formation of connective tissue fibers. The fibroblasts produce three types of fibers to form the ground substance: collagen, elatin, and the reticulum.

Field: *Physics* The influence of some agent, as electricity or gravitation, considered as existing at all points in space and defined by the force it would exert on an object placed at any point in space. *Cf.* electric field, gravitational field, magnetic field. More generally a **scalar** (*e.g.* a pressure), **vector** (*e.g.* a force), or **tensor** (*e.g.* a strain), value defined at all points in space.

Follicle: *Anat.* A small cavity, sac, or gland.

Foregut: 1. *Zool.* a. The first portion of the vertebrate alimentary canal, extending from the pharynx and esophagus to the end of the stomach or gizzard and, in some animals, the anterior duodenum, functioning in the ingestion, temporary storage, and partial digestion of food. b. The first portion of the alimentary canal in arthropods and annelids, composed of ectodermal, chitin-lined tissue and usually comprising the pharynx, esophagus, crop, and gizzard. 2. *Embryol.* (in mammals) the upper part of the embryonic alimentary canal from which the pharynx, esophagus, lung, stomach, liver, pancreas, and part of the duodenum develop.

Fractal geometry: Branch of mathematics concerned with irregular patterns made of parts that are in some way similar to the whole, *e.g.*, twigs and tree branches, a property called self-similarity or self-symmetry. Unlike conventional geometry, which is concerned with regular shapes and whole-number dimensions, such as lines (one-dimensional) and cones (three-dimensional), fractal geometry deals with shapes found in nature that have non-integer, or fractal, dimensions D–linelike rivers with a fractal dimension of about 1.2 and cone-like mountains with a fractal dimension between 2 and 3. Roughly speaking, the volume, surface or length of the fractal object varies (in average) inside a sphere of radius R like R^D.

Gastrula: *Embryol.* A metazoan embryo in an early state of germ layer formation following the blastula stage, consisting of a cuplike body of two layers of cells, the ectoderm and endoderm, enclosing a central cavity, or archenteron, that opens to the outside by the blastopore: in most animals progressing to the formation of a third cell layer, the mesoderm.

Gastrulation: *Embryol.* 1. The formation of a gastrula. 2. Any process, as invagination, by which a blastula or other form of embryo is converted into a gastrula.

Gibbs-Thomson law: Links surface tension, principal curvatures and latent heat of phase transformation to the difference of temperature δT between the sides of an interface separating two phases (*e.g.* liquid-gas).

$$\frac{\delta T}{T_{\mathrm{m}}}\mathcal{L} = -\frac{(\sigma + \sigma_1'')}{R_1} - \frac{(\sigma + \sigma_2'')}{R_2}$$

$\mathcal{L}$ latent heat per unit volume (J/m^3)

T_{m} melting temperature (K)

σ surface tension (J/m^2)

R_{i} principal curvatures of the interface (m)

σ_i'' second derivative $\partial^2\sigma\ /\partial\theta_i^2$ of σ with respect to the local surface principal orientation angles.

Glia or **neuroglia:** *Anat., Cell Biol.* A class of cells in the brain and spinal cord that form a supporting structure for the neurons and provide them with insulation.

Gymnosperm: *Bot.* A vascular plant having seeds that are not enclosed in an ovary; a conifer or cycad. *Cf.* angiosperm.

Heterozygous: *Biol.* 1. Having dissimilar pairs of genes for any hereditary characteristic. 2. Of or pertaining to a heterozygote.

Homeobox: A short usu. highly conserved DNA sequence in various eukaryotic genes (as many homeotic genes) that codes for a peptide which may be a DNA-binding protein.

Homologue: *Biol.* Equivalent genes in different species. Perhaps the most obvious benefit of biomedical relevance to emerge from the availability of the *saccharomyces genome database* (SGD, the familiar baker's, brewer's, and vintner's yeasts) is the ability to quickly find yeast counterparts, or *homologues*, of genes in humans that are associated with specific diseases.

Homozygous: *Biol.* 1. Having identical pairs of genes for any given pair of hereditary characteristics. 2. Of or pertaining to a homozygote.

Hysteresis: *Physics.* 1. The lag in response exhibited by a body in reacting to changes in the forces, esp. magnetic forces, affecting it. *Cf. magnetic hysteresis.* 2. The phenomenon exhibited by a system, often a ferromagnetic or imperfectly elastic material, in which the reaction of the system to changes is dependent upon its past reactions to change.

Knock out: When we create knock-out mice in the laboratory, we destroy the gene of interest by tricking the cell. Essentially, we fool it into exchanging its normal gene for an irrelevant disrupting DNA sequence we made in the laboratory, and we do so using the cell's own recombinant machinery.

Lumen: The cavity of a tubular organ. *Ex.*: the ~ of a blood vessel.

Meristem: A formative plant tissue usu. made up of small cells capable of dividing indefinitely and giving rise to similar cells or to cells that differentiate to produce the definitive tissues and organs. **primary meristem**, meristem (as procambium) derived from the apical meristem; **ground meristem**, the part of a primary apical meristem remaining after differentiation of protoderm and procambium; **apical meristem**, a meristem at the apex of a root or shoot that is responsible for increase in length.

Mesenchyme: *Embryol.* Cells of mesodermal origin that are capable of developing into connective tissues, blood, and lymphatic and blood vessels.

Mesentery: *Anat.* The membrane, consisting of a double layer of peritoneum, that invests the intestines, attaching them to the posterior wall of the abdomen, maintaining them in position in the abdominal cavity, and supplying them with blood vessels, nerves, and lymphatics, esp. the part of this membrane investing the jejunum and ileum.

Mesoderm: *Embryol.* The middle germ layer of a metazoan (zoological group comprising the multicellular animals) embryo.

Mesomorphic: 1. Pertaining to or having a muscular or sturdy body build characterized by the relative prominence of structures developed from the embryonic mesoderm (contrasted with ectomorphic, endomorphic). 2. *Physical Chem.* Pertaining to or existing in an intermediate state, as a liquid crystal in the nematic or smectic state.

Metanephros: *Embryol.* One of the three embryonic excretory organs of higher vertebrates, becoming the permanent and functional kidney.

Microtubule: *Cell Biol.* A hollow cylindrical structure in the cytoplasm of most cells, involved in intracellular shape and transport.

Microfilament: *Cell Biol.* A minute, narrow tubelike cell structure composed of a protein similar to actin, occurring singly and in bundles, involved in cytoplasmic movement and changes in cell shape.

Mitogen: *Biol.* Any substance or agent that stimulates mitotic cell division.

Morphogen: A diffusible chemical substance that exerts control over morphogenesis esp. by forming a gradient in concentration.

Motile: *Biol.* Moving or capable of moving spontaneously: *motile cells; motile spores.*

Nematic: *Physical Chem.* (of liquid crystals) noting a mesomorphic state in which the arrangement of the molecules is linear.

Nephron: *Anat., Zool.* The filtering and excretory unit of the kidney, consisting of the glomerulus and tubules.

Newtonian fluid: *Hydrodynam.* Any fluid exhibiting a linear relation between the applied shear stress and the rate of deformation.

Ontogenesis: *Biol.* The origin and development of individual living being.

Order parameter, *Chem., Physics* Parameter used to characterize the phases of a system. An *order parameter* changes during a phase transition. *Ex.*: the *magnetization* in a ferromagnetic substance, as iron, that below a certain temperature, the Curie point, can possess magnetization in the absence of an external magnetic field (the magnetic moments of the atoms are then aligned).

Pelvis: (kidney) *Anat., Zool.* The cavity of the kidney that receives the urine before it is passed into the ureter.

Peptidase: *Biochem.* Any of the class of enzymes that catalyze the hydrolysis of peptides or peptones to amino acids.

Phase diagram: *Chem.* A graph, usually using temperature, pressure, and composition as coordinates, indicating the regions of stability of the various phases (solids in equilibrium composition, liquids, or vapour) of a system.

Phase field: A mathematical tool introduced to simplify numerical simulations of moving boundaries between two media, or phases. The phase field indicates for each point in space to which medium it belongs. Often, the phase field is proportional to a thermodynamical order parameter.

Phase transition: *Chem.*, *Physics* Change from one phase to another when some physical parameter (temperature, pressure, composition, ...) is modified. It delimits the regions of stability of the various phases.

Phenotype: *Genetics.* 1. The observable constitution of an organism. 2. The appearance of an organism resulting from the interaction of the genotype and the environment.

Phenylephrine: *Pharm.* An alpha-adrenergic stimulant, $C_{19}H_{13}NO_2$, used chiefly as a nasal decongestant.

Phylogenesis: *Biol.* The evolution of the tribe or race, or of any organ or feature in the race.

Placode: *Embryol.* A local thickening of the endoderm in the embryo, that usually constitutes the primordium of a specific structure or organ.

Plexus: A network of anastomosing or interlacing blood vessels or nerves. **solar plexus**, a nerve plexus in the abdomen that is situated behind the stomach and in front of the aorta and the crura of the diaphragm and contains several ganglia distributing nerve fibres to the viscera. **brachial plexus**, a network of nerves lying mostly in the armpit and supplying nerves to the chest, shoulder, and arm

Prune: *pruning*, To cut off or cut back parts of for better shape or more fruitful growth.

Pseudopodia: Used in locomotion and feeding of amebia, may be rounded at the tip (**lobopodia**), pointed (**filopodia**), branched and fused together (**rhizopodia**), or somewhat rigid and pointed (**axopodia**).

Rheology: The study of the deformation and flow of matter.

r.h.s.: Right hand side of an equation.

Scaling: *Phys. Math.* The manner in which a geometrical object, or a graph changes by a change of scale. *Ex.:* A fractal inside a disk of radius R, of fractal dimension D, "scales" with R like R^D; **scale invariance**, invariance of the geometrical structure by a change in scale (like after a zoom); **scaling relations**, simple power law relations between two parameters (length, temperature... on one side, surface, mass, flux... on the other side), as it is the case between R and R^D, in fractal structures.

Self-affinity: Invariance of an object by a change of scale different along different axis. *Ex.*: Rough surfaces, mountains, graph of the gain as a function of the number of tosses in games of chance...

Self-similarity: Invariance of an object by dilation. *Ex.:* The triadic von Koch curve is self similar, as it superpose to itself by a zoom of magnifications $3, 9, 27...$ The self-similarity can be statistical, when only the average on many samples remains invariant (*Ex.:* percolation cluster).

Septum: *Biol.* A dividing wall, membrane, or the like, in a plant or animal structure; dissepiment.

Smectic: *Physical Chem.* (of liquid crystals) noting a mesomorphic state in which the arrangement of the molecules is in layers or planes.

Solute: The substance dissolved in a given solution.

Soma: *Biol.* The body of an organism as contrasted with its germ cells.

Somato: A combining form meaning "body," used in the formation of compound words: somatotonia, somatopetal (which moves near to), somatofugal (which moves away from).

Strain: *Mech., Physics* Deformation of a body or structure as a result of an applied force.

Strength of materials: Measurement in engineering of the capacity of metal, wood, concrete, and other materials to withstand stress and strain. Stress is the internal force exerted by one part of an elastic body upon the adjoining part, and strain is the deformation or change in dimension occasioned by stress. When a body is subjected to pull, it is said to be under tension, or tensional stress, and when it is being pushed, *i.e.*, is supporting a weight, it is under compression, or compressive stress. Shear, or shearing stress, results when a force tends to make part of the body or one side of a plane slide past the other. Torsion, or torsional stress, occurs when external forces tend to twist a body around an axis. Materials are considered to be elastic in relation to an applied stress if the strain disappears after the force is removed. The elastic limit is the maximum stress a material can sustain and still return to its original form. According to Hooke's law, the stress created in an elastic material is proportional to strain, within the elastic limit

Stress: *Mech.* 1. The action on a body of any system of balanced forces whereby strain or deformation results. 2. The amount of stress, usually measured in *pascals*. 3. A load, force, or system of forces producing a strain. 4. The internal resistance or reaction of an elastic body to the external forces applied to it. 5. The ratio of force to area. *Physiol.* A specific response by the body to a stimulus, as fear or pain, that disturbs or interferes with the normal physiological equilibrium of an organism.

Sympodium: *Bot.* An axis or stem that simulates a simple stem but is made up of the bases of a number of axes that arise successively as branches, one from another, as in the grapevine.

Synapse: *Physiol.* A region where nerve impulses are transmitted and received, encompassing the axon terminal of a neuron that releases neurotransmitters in response to an impulse, an extremely small gap across which the neurotransmitters travel, and the adjacent membrane of an axon, dendrite, or muscle or gland cell with the appropriate receptor molecules for picking up the neurotransmitters.

Thermal diffusion length: d_T

$$d_\mathrm{T} = \frac{\mathcal{D}_\mathrm{T}}{V}$$

$\mathcal{D}_\mathrm{T}$ thermal diffusivity $(\mathrm{m}^2/\mathrm{s})$
V interface velocity (m/s).

Thermal noise, *Thermodynam., 1. Elect.* A wide spectrum of electromagnetic noise appearing in electronic circuits and devices as a result of the temperature-dependent random motions of electrons and other charge carriers. *2.* More generally the fluctuations due to the temperature-dependent random motions of the atoms in matter leading to side branching in dendritic growth, nucleation, spinodal decomposition...

Transduction: The action or process of transducing; esp.: the transfer of genetic determinants from one microorganism to another by a viral agent (as a bacteriophage).

Transgenic: Of, pertaining to, or containing a gene or genes transferred from another species: transgenic mice.

Translocation: 1. *Genetics,* A chromosomal rearrangement in which a segment of genetic material from one chromosome becomes heritably linked to another chromosome. 2. *Bot.,* The conduction of soluble food material from one part of a plant to another.

Credit: Large parts of this glossary are taken from the Web site: http://www.infoplease.com, of Information Please, 20 Park Plaza, Suite 1420, Boston, MA 02116, U.S.A.

List of Participants

S. Akamatsu, Groupe de Physique des Solides, Université Paris VII – Denis Diderot, 2 place Jussieu, Tour 23, 2e étage, 75251 Paris Cedex 05, France

R. Ballou, Laboratoire Louis Néel, CNRS, 25 avenue des Martyrs, BP. 166, 38042 Grenoble Cedex 9, France

M.-O. Bernard, Laboratoire de Physique de la Matière Condensée, École Polytechnique, 91128 Palaiseau Cedex, France

S. Bodea, Max-Planck-Institut für Mikrostrukturphysik, Weinberg 2, 06120 Halle, Germany

S. Bottin, IRPHE Service 252, Faculté de Saint-Jérôme, 13397 Marseille Cedex 20, France

G. Canright, Department of Physics, 401 Nielsen Physics Building, University of Tennessee, Knoxville, TN 37996-1200, U.S.A.

I. Cantalapiedra, Departamento de Fisica Aplicada, c/gregorio marañon 44, Universidad Politecnica de Catalunya, 08028 Barcelona, Spain

F. Cassot, INSERM, U455, Service de Neurologie, CHU Purpan, 31059 Toulouse Cedex, France

A. Cornelissen, Department of Medical Physics, University of Amsterdam, P.O. Box 22660, 1100 DD Amsterdam, The Netherlands

J.-M. Debierre, MATOP, Case 151, Faculté de Saint-Jérôme, 13397 Marseille Cedex 20, France

M. Dejmek, Laboratoire de Physique de la Matière Condensée, École Polytechnique, 91128 Palaiseau Cedex, France

S. Douady, LPS/ENS, 24 rue Lhomond, 75005 Paris, France

E. Dubois-Violette, LPS, bâtiment 510, Université de Paris-Sud, 91405 Orsay, France

V. Fafeur, CNRS EP560, Institut de Biologie de Lille/Institut Pasteur, 59019 Lille, France

G. Faivre, Groupe de Physique des Solides, Université Paris VII – Denis Diderot, 2 place Jussieu, Tour 23, 2e étage, 75251 Paris Cedex 05, France

M. Filoche, Laboratoire de Physique de la Matière Condensée, École Polytechnique, 91128 Palaiseau Cedex, France

R. Folch, Groupe de Physique des Solides, Université Paris 7, Tour 23, 2 place Jussieu, 75251 Paris Cedex 05, France

C. Frelin, Institut de Pharmacologie Moléculaire et Cellulaire, UPR 411, 660 route des Lucioles, 06560 Valbonne, France

A. Gabrielli, Laboratoire de Physique de la Matière Condensée, École Polytechnique, 91128 Palaiseau, France

V. Gafiychuk, National Academy of Sciences of Ukraine, Institute of Applied Problems of Mechanics and Mathematics, Naukova Street 3B, 290020 Lviv, Ukraine

M. Georgelin, IRPHE, Faculté de Saint-Jérôme, Service 252, 13997 Marseille Cedex 20, France

I. Golding, School of Physics and Astronomy, Tel Aviv University, 69978 Tel Aviv, Israel

R. Gonzalez-Cinca, Department of Applied Physics, Universitat Politecnica de Catalunya, Campus Nord B5, J. Girona Salgado s/n, 08034 Barcelona, Spain

I. Harper, Department of Anatomy, Monash University, Clayton, Victoria 3168, Australia

J.C. Brings Jacobsen, Department of Medical Physiology, The Panum Institute, Bldg. 10.5, Blegdamsvej 3, N 2200 Copenhagen, Denmark

G. Hentschel, Department of Physics, Rollins Research Center, Emosy University, 1510 Clifton Road, Atlanta, GA 30322, U.S.A.

K. Karamanos, Centre for Non Linear Phenomena & Complex Systems, CP. 231 Campus de la Plaine, ULB, Boulevard du Triomphe, 1050 Bruxelles, Belgium

T. Kim, University of Illinois, Department of Physics, 110 West Green Street, Urbana, Illinois 61801-0478, U.S.A.

O. Kirillova, Theoretical Department, Ulyanovskaya Str. 1, Stary Peterhof, 198904 St. Petersburg, Russia

O. Kozhin, A.B. Kogan Research Institute for Neurocybernetics, Rostov State University, 194/1 Stachka Ave., 344090 Rostov on Don, Russia

A. Lindner, Laboratoire de Physique Statistique, École Normale Supérieure, 24 rue Lhomond, 75231 Paris Cedex 05, France

P. Molho, Laboratoire Louis Néel, CNRS, 25 avenue des Martyrs, BP. 166, 38042 Grenoble Cedex 9, France

J. Müller, University of Leiden, Lorentz Institut, Department of Physics, Niels Bohrweg 2, 2333 CA Leiden, The Netherlands

A. Pereira, Université de Bordeaux, Centre de Recherches Paul Pascal, CNRS, Avenue A. Schweitzer, 33600 Pessac, France

M. Peterson, Physics Department, Mount Holyoke College, South Hadley, Massachussets 01075, U.S.A.

M. Plapp, Laboratoire de Physique de la Matière Condensée, École Polytechnique, 91128 Palaiseau Cedex, France

A. Pocheau, IRPHE, Faculté de Saint-Jérôme, Service 252, 13997 Marseille Cedex 20, France

M. Rackaitis, Research Center for Microsystems and Nanotechnology, Kaunas University of Technology, Studentu 65, LT 3031 Kaunas, Lithuania

W.-J. Rappel, Department of Physics, UC San Diego, 9500 Gilman Drive, La Jolla, CA 92093, U.S.A.

L. Schwartz, Service de Radiothérapie, Hôpital Lariboisière, 2 rue Ambroise Paré, 75010 Paris, France

A. Shahverdian, Yerevan Physics Institute, Alikhanian Brother's Street 2, 375036 Erevan, Armenia

D. Van Effenterre, Laboratoire de Physique de la Matière Condensée, Collège de France, Place Marcellin Berthelot, 75005 Paris, France

D. Warburton, Department of Surgery, Mail Stop 35 Children's Hospital, 4650 Sunset Boulevard, P.O. Box 54700, Los Angeles, California 91011, U.S.A.

T. Watanabe, HHSC1416, Columbia University, 701 West 168th Street, New York 10032, U.S.A.

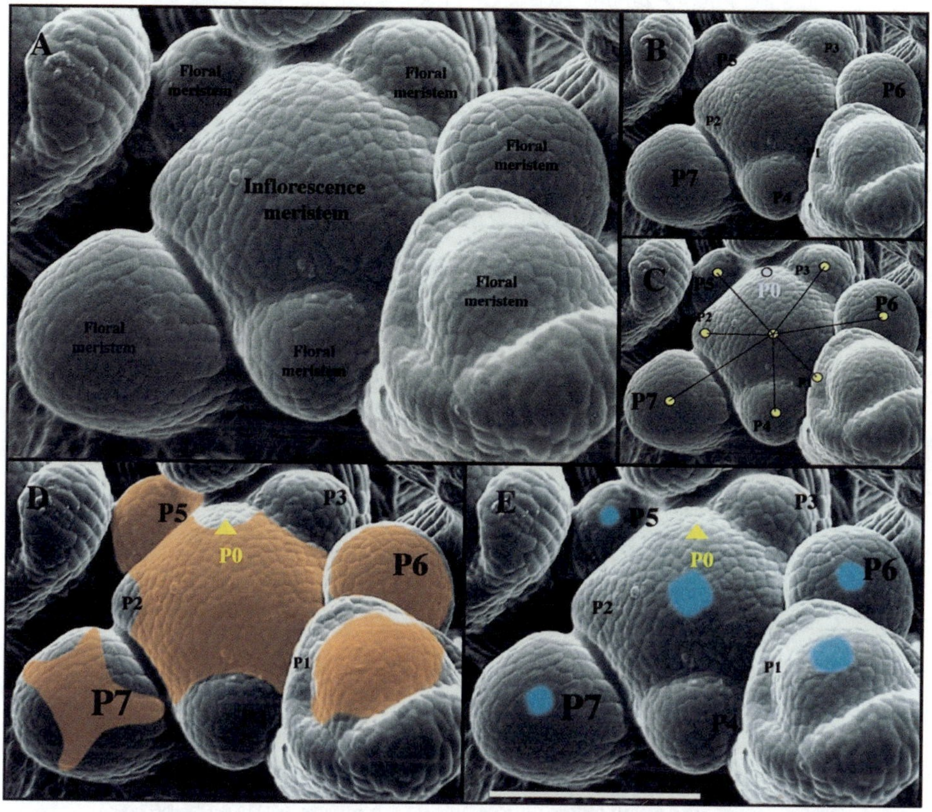

Planche 1. Formation of floral meristems in *A. thaliana*. A) Floral meristems arise on the flank of the inflorescence meristem. The outer surface of the L1 cells is visible. B) The floral meristems can be numbered from the youngest meristem visible (P1, here hidden by an older floral meristem) to the oldest (here P7). C) Floral meristems are produced in a regular pattern. This allows prediction of the position of the next floral meristem to arise called P0. D) The expression pattern of *STM* is represented in red. *STM* is expressed throughout the inflorescence meristem except where the primordium P0 is expected. No expression is visible in young floral meristems (P1 to P4) but *STM* expression is turned on again in older floral meristems (P5 onwards). E) The expression pattern of *CLV3* is represented in blue. *CLV3* is expressed in the centre of the inflorescence meristem and older floral meristem. It is not expressed in young floral meristems (P1 to P4). Bar = 100 μm. Adapted with permission from Laufs, Grandjean, Jonak, Kiêu and Traas, Cellular parameters of the shoot apical meristem in *Arabidopsis*, *Plant Cell* **10** (1998) 1375-1390.

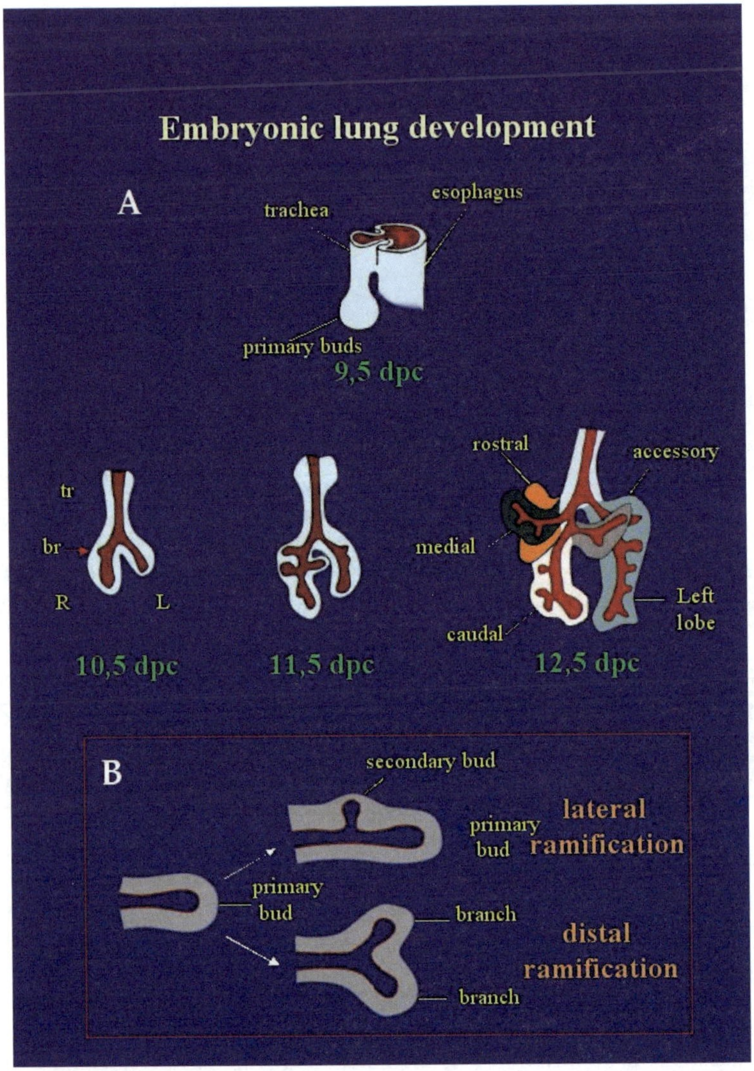

Planche 2. Lung branching morphogenesis (see Bellusci's contribution). Embryonic lung development in mouse. Lung development starts at 9.5 dpc with the emergence of two primary buds from the ventral foregut (only one is showed in the picture). At 10.5 dpc, the lung is composed of a trachea (tr) and two bronchi (br). Note the presence of a secondary bud on the right bronchus (red arrow). From 11.5 to 12.5 dpc, the lung undergoes an important process of ramification to give rise to four lobes on the right (called rostral, medial, accessory and caudal lobes) and one lobe on the left (see text).

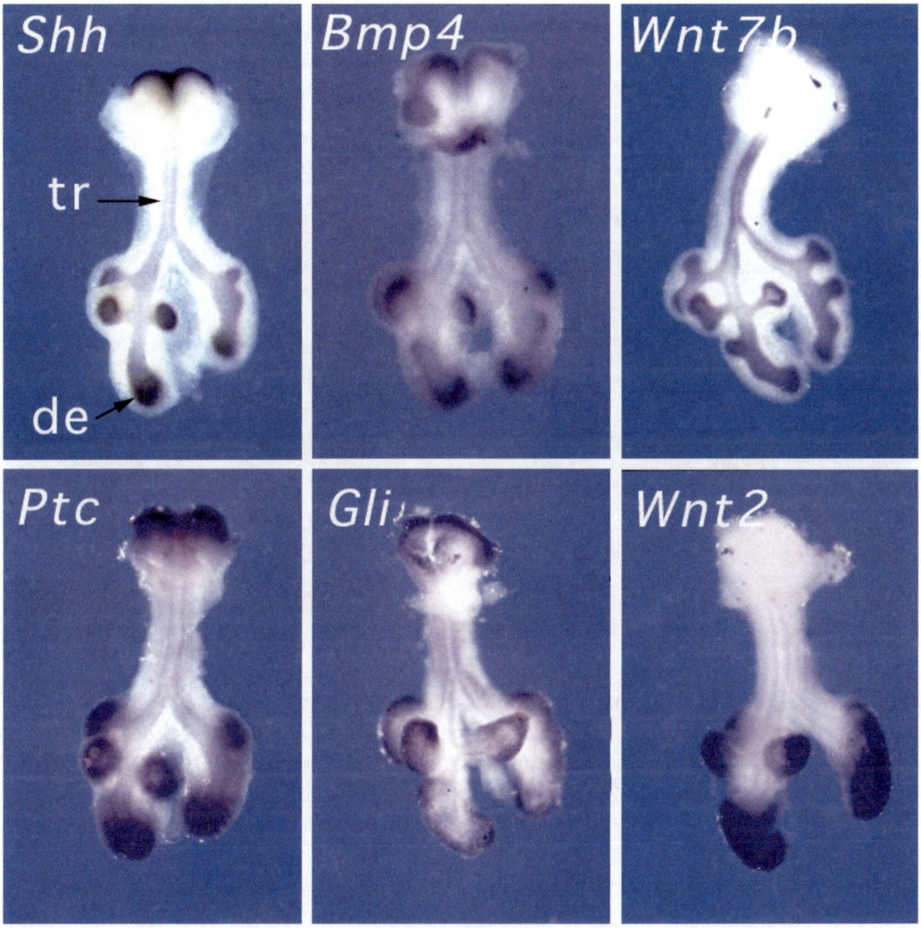

Planche 3. Embryonic lung development. Examples of gene expression in the 11.5 dpc lung. Class I genes such as Sonic hedgehog, Bone Morphogenetic Protein 4 and Wnt7b are expressed in the epithelium (the sheet of cells forming the branched structure). Note the high level (darker spots) of gene expression in the distal epithelium (de) compared to that seen in the trachea (tr). Class II genes such as Patched, Gli and Wnt2 are expressed in the distal mesenchyme (outside the epithelium, in the surrounding tissue). This staining is obtained by whole mount *in situ* hybridization using digoxigenin-labeled antisense RNA probes.

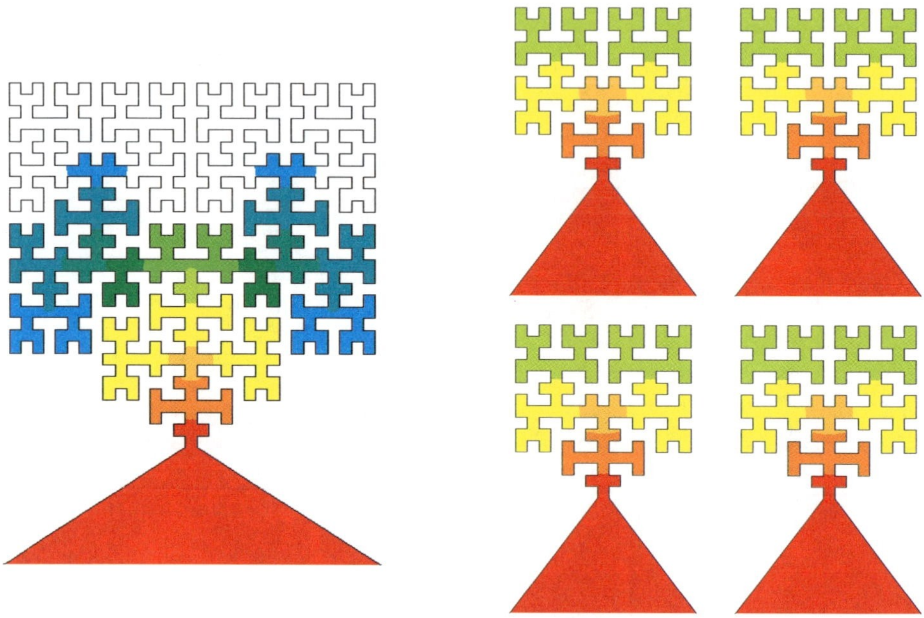

Planche 4. Illustration of the principle "smaller is better" for the acinus of a lung (see Sapoval's contribution). The representation shows a change of color at each equiconcentration value smaller than the previous one by a factor 2 (stripe representation of isobars in log scale). In the same global volume, here surface, 4 acini of generation 3 present a total conductance larger than one acini of generation 4 for $\Lambda/\ell = 200$. Most of the generation 4 acinus perimeter does not participate tin the exchange while the concentration is more uniform in generation 3 acini. For much larger values of the unscreened perimeter, the two systems would behave equivalently. This means that a lung with more smaller features, but occupying the same volume, is more efficient than a lung with larger features in the acinus.

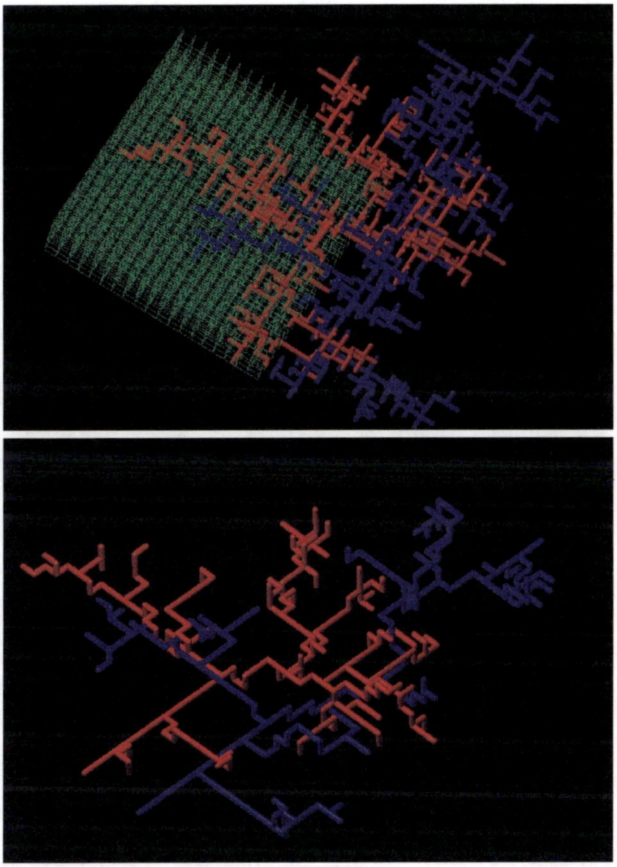

Planche 5. A (top) and B (bottom). These two plates show a process of formation and self-organization of two connected animal vasculatures (arteries in red, veins in blue). These vascular trees form by selection and replacement of small capillaires (in green). For the sake of clarity, only one fourth of the capillary mesh is shown in 5A. In 5B, only the final vasculature is shown. The replacement of small capillaries by larger vessels ("pruning") obeys a dielectric-breakdown scheme, in the direction of high shear stress (see text by Fleury and Schwartz). In addition, capillary regression is taken into account. The flow is injected in the capillary bed through the tips of the arteries, which serve as source term of pressure (and hence of shear) for vein growth. In a 3D capillary mesh, the final network is spontaneously interlaced. While this model is rather simple, it is interesting in that it provides a natural way of simultaneous formation of both arterial and venal vasculatures. It also gives a good starting point for further complications (or, in other words, for improvement through evolution by natural selection).

Planche 6

1. Francis Hallé
2. Patrick Laufs
3. Chaouqi Misbah
4. Roger Folch
5. Laurent Schwartz
6. Marcus Dejmek
7. Vincent Fleury
8. Vasyl Gafiychuk
9. Bernard Sapoval
10. Tomoko Watanabe
11. Véronique Fafeur
12. Alexander Melker
13. Paul Meakin
14. Marc-Olivier Bernard
15. Annemiek Cornelissen
16. Pierre Molho
17. Mark Peterson
18. Jean-Pierre Ternaud
19. Kostantinos Karamanos
20. Andrea Gabrielli
21. Marcel Filoche
22. Sabine Bottin
23. Jean-Marc Debierre
24. Jean-François Gouyet
25. Inmaculada Cantalapiedra
26. Anke Lindner
27. Elisabeth Dubois-Violette
28. Simona Bodea
29. Gabriel Faivre
30. Geoffrey Canright
31. Ashot Shahverdian
32. Silvère Akamatsu
33. Alain Pocheau
34. Jüdith Müller
35. Marc Georgelin
36. Olga Kirillova
37. Savério Bellusci
38. Alain Karma
39. Mindaugas Rackaitis
40. Rafik Ballou
41. Ian Harper
42. Wouter- Jan Rappel
43. Antonio Pereira
44. George Hentschel
45. Ido Golding
46. Eshel Ben Jacob
47. David Warburton
48. Jens Christian Brings Jacobsen
49. Yves Couder
50. Stéphane Douady
51. Marc Leonetti
52. Damien Van Effenterre
53. Darryl Kropf
54. Oleg Kozhin

Planche 7

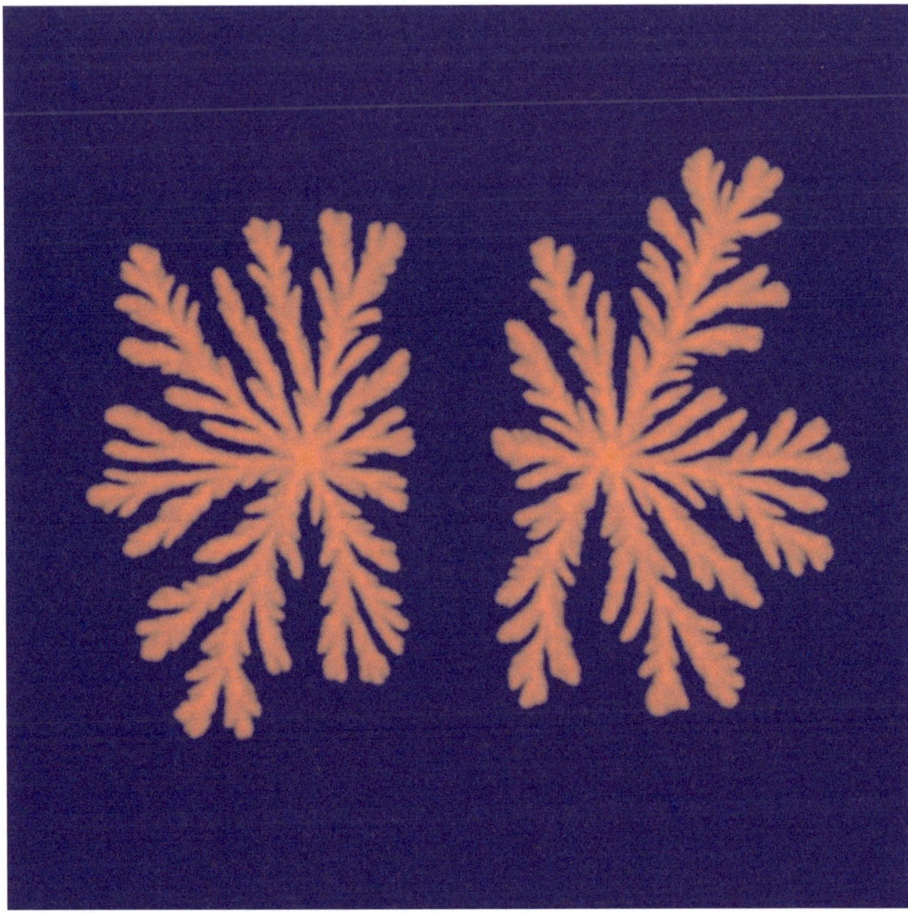

Planche 8A. Growth of bacteria. The image shows the numerical integration of a mathematical model consisting of reaction-diffusion equations for bacterial colony growth. It reproduces the competition between two colonies, inoculated at the same time, for an initially low value of the nutrient concentration and a hard medium. The shape of the colonies is akin to the pattern known as a DLA aggregate, and it is quite similar to real patterns (see Contributed Paper 8).

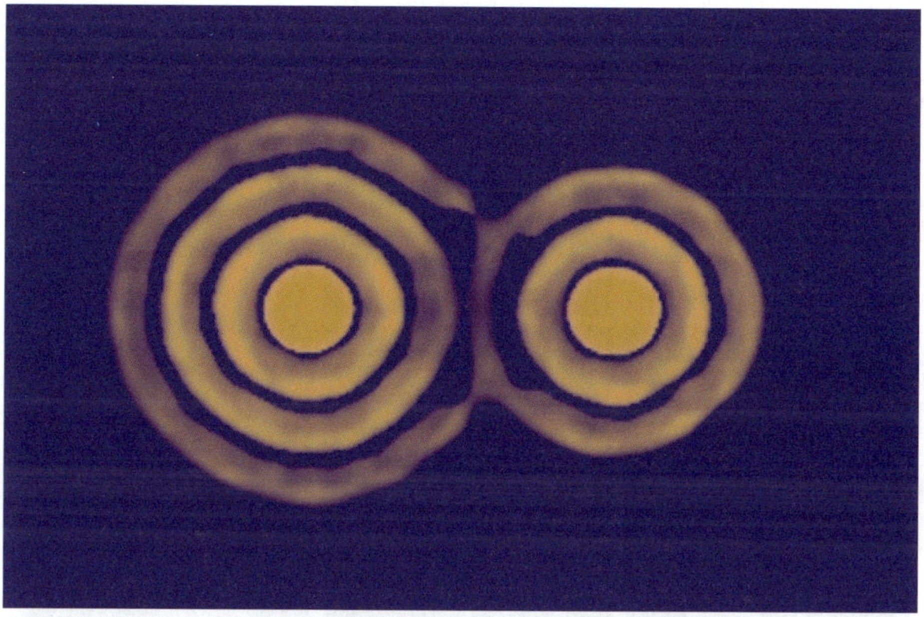

Planche 8B. In this simulation, two colonies are inoculated at different times, in an initially hight level of the nutrient concentration, on a semi-solid medium. The colonies show spatially periodic structures with the same periodicity but they remain out of phase with each other throughout development (see Contributed Paper 8).

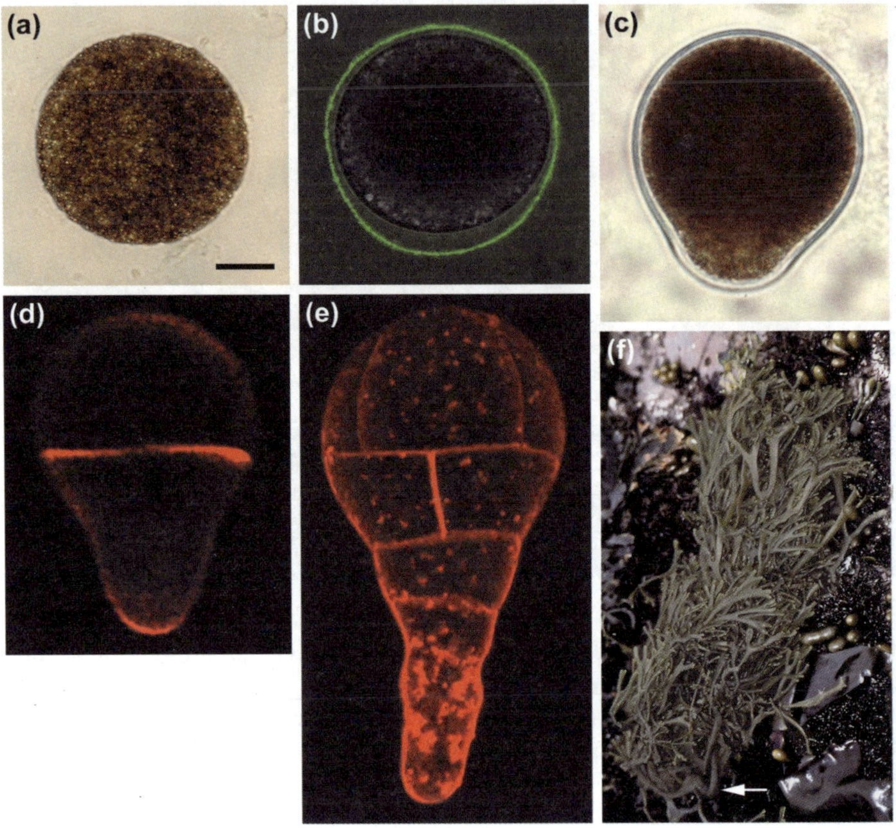

Planche 9. Early development in *P. compressa*. (a) Bright field image of an unfertilized egg. (b) Confocal image of a young zygote bearing polar adhesive (outlined with fluorescent microspheres). The region of thickest adhesive (bottom of the zygote) indicates the rhizoid pole. (c) Bright field image of a germinated zygote with emerging rhizoid. Zygote has been plasmolyzed to visualize the cell wall (d, e) Confocal images of embryos in which plasma membranes have been labeled with FM 4-64. (d) The first division is transverse to the growth axis and partitions the zygote asymmetrically. (e) Division pattern in an older embryo is nearly identical to that in higher plants. (f) The mature alga is approximately 0.5 m in length. Arrow denotes the holdfast, derived from the rhizoid, that attaches the mature plant to the rock; the leafy fronds derive largely from the thallus. Scale bar = 25 μm for (a-e).

Imprimé en France. - JOUVE, 18, rue Saint-Denis, 75001 PARIS
N° 288548U. Dépôt légal : Avril 2001